T0296916

The Seismic Analysis Code

A Primer and User's Guide

The Seismic Analysis Code (SAC) is one of the most widely used analysis packages for regional and teleseismic seismic data. Now, for the first time, this book provides users at both introductory and advanced levels with a complete guide to SAC, enabling all users to make best use of this powerful tool.

The book leads new users of SAC through the steps of learning basic commands, describes the SAC processing philosophy and presents its macro language in full. The text is supported throughout with example inputs and outputs from SAC. For the more experienced and ambitious practitioners of the code, it also describes SAC's many hidden features, including advanced aspects of its graphics, its file structure, how to write independent programs to access and create files, and use of the methods SAC provides to integrate external processing steps into production-type data analysis schemes. Tutorial exercises in the book engage users with their newly acquired skills, providing data and code to implement the standard methods of teleseismic shear-wave splitting and receiver function analysis.

Methodical and authoritative, this combined introduction and advanced tutorial guide is a key resource for researchers and graduate students in global seismology, earthquake seismology and geophysics.

GEORGE HELFFRICH is a professor of seismology in the School of Earth Sciences at the University of Bristol. His research interests include using observational seismology to study features of the crust, mantle and core. Recently, he has based his analysis techniques on large-scale seismic array data, using SAC as the primary seismological data analysis tool. Before embarking on his research career, Professor Helffrich was a programmer who developed and supported mainframe operating systems. Bringing this experience to the seismological realm, he has contributed to the development of SAC for over 20 years.

JAMES WOOKEY is a research fellow and lecturer in the School of Earth Sciences at the University of Bristol. His research focuses on observational seismology, particularly seismic anisotropy, applied to problems from the inner core to oil reservoirs, with a recent focus on Earth's core–mantle boundary region. Dr. Wookey has spent much of his research career developing and applying novel methods for analyzing seismic data, and comparing them with predictions from mineral physics and geodynamics to better understand Earth processes. His experience with SAC spans 15 years, including as contributor to its development.

IAN BASTOW is a lecturer in seismology in the Department of Earth Science and Engineering at Imperial College, London. His research focuses primarily on the analysis of broadband seismological data from networks of temporary seismograph stations to better understand the Earth's crust and mantle. Dr. Bastow has worked extensively on tectonic problems concerning the seismically and volcanically active East African rift system, as well as on the development of Laurentia, the Precambrian core of North America. He has been a user of SAC for over a decade.

THE SEISMIC ANALYSIS CODE

A PRIMER AND USER'S GUIDE

GEORGE HELFFRICH
School of Earth Sciences, University of Bristol, United Kingdom

JAMES WOOKEY
School of Earth Sciences, University of Bristol, United Kingdom

IAN BASTOW
Department of Earth Science and Engineering, Imperial College London, United Kingdom

Shaftesbury Road, Cambridge CB2 8EA, United Kingdom
One Liberty Plaza, 20th Floor, New York, NY 10006, USA
477 Williamstown Road, Port Melbourne, VIC 3207, Australia
314–321, 3rd Floor, Plot 3, Splendor Forum, Jasola District Centre, New Delhi – 110025, India
103 Penang Road, #05–06/07, Visioncrest Commercial, Singapore 238467

Cambridge University Press is part of Cambridge University Press & Assessment, a department of the University of Cambridge.

We share the University's mission to contribute to society through the pursuit of education, learning and research at the highest international levels of excellence.

www.cambridge.org
Information on this title: www.cambridge.org/9781107613195

First published 2013

Internal book layout follows a design by G. K. Vallis

A catalogue record for this publication is available from the British Library

ISBN 978-1-107-04545-3 Hardback
ISBN 978-1-107-61319-5 Paperback

Additional resources for this publication at www.cambridge.org/helffrich

Contents

Preface		***page*** **ix**
Acknowledgements		**xi**
1	**Introduction**	**1**
1.1	What is SAC?	1
1.2	History and development	2
1.3	Alternatives to SAC	2
1.4	SAC variants	3
1.5	Requirements and installation	4
1.6	Scope of this book	4
2	**The SAC data format**	**5**
2.1	Philosophy and structure	5
	SAC file format	5
	Alphanumeric and binary forms	5
	Interconversion of formats	6
2.2	Conversion from other data formats	6
	GSE files	6
	SEG Y, MSEED, GCF and CSS formats	7
2.3	Byte-order issues	8
2.4	SAC file layout	10
3	**The SAC processing philosophy**	**11**
3.1	Phases of a typical analysis task	11
	Organize	11
	Interact	13
	Process	14
	Display	14
3.2	Command summary for each phase	14
3.3	Further information about SAC commands	15

4 **Basic SAC commands** **17**
- 4.1 Command style 17
- 4.2 Command history 17
- 4.3 Reading and writing data 18
  - Reading examples 18
  - Writing data 19
- 4.4 Plotting and cutting 19
  - Devices 19
  - Windows and window placement 20
  - Plotting data 21
  - Cutting data 24
  - Permanent plots 24
- 4.5 Picking 25
- 4.6 The file header 25
  - Time representation 27
  - Listing 27
  - Changing 28
  - Writing 29
- 4.7 Trace preparation and resampling 29
  - De-glitching 29
  - Mean and trend removal 30
  - Resampling 30
- 4.8 Rotation 31
- 4.9 Frequency-domain operations and filtering 32
  - Filtering 34
  - Designing filtering strategies 35
- 4.10 SAC startup files 35
- 4.11 SAC utility programs 37

5 **SAC macros** **38**
- 5.1 Macros and invoking them 38
- 5.2 Writing a simple macro 39
- 5.3 Tracing macro operations 39
- 5.4 Searching for macros 40
- 5.5 Decision making in macros 40
- 5.6 Variables in macros 41
  - Types and scope 41
  - Setting 41
- 5.7 Expressions 42
  - Syntax 43
  - Built-in functions 43
  - Escape character 48
  - Evaluation order 48
  - Conditions 49
- 5.8 Suspension, resumption and escape from macros 50
- 5.9 Operating system interaction 51
- 5.10 Looping commands 52

	WHILE	53
	WHILE READ	54
	Escaping from loops	54
	DO	55
5.11	Macro parameters	57
	Positional	57
	Keyword	59
	Recursion	60
5.12	Advanced operating system interaction	61
6	**Accessing SAC functionality and data from external programs**	**64**
6.1	Automating SAC execution	64
	Running SAC from the shell	64
	Automation of SAC execution in the shell using scripting	65
6.2	Accessing SAC data in external programs	68
	Accessing SAC data from Fortran using the `sacio` library	68
	`sacio90`: object-oriented SAC data interaction in Fortran	70
	Other languages	71
6.3	Accessing SAC functionality in Fortran programs	72
7	**Graphical data annotation**	**74**
7.1	Plot annotation	74
	Seismograms	74
	Composite plots	77
7.2	Annotating plots with graphical elements	79
	Graphical elements	80
	Assembling graphical elements	80
	Parameters controlling graphical elements	83
7.3	Using `PLOTC`	83
8	**Array data handling**	**86**
8.1	SAC subprocesses	86
8.2	The signal stacking subprocess	87
	Trace collections	87
	Adding, deleting and changing traces	88
	Plotting record sections	89
	Stacking	90
	Saving stack data and uncertainties	95
	Picking data in stacks	95
8.3	Array maps	101
8.4	Beamforming	102
8.5	Travel-time analysis	106
9	**Spectral estimation in SAC**	**109**
9.1	Spectral estimation	109
9.2	The spectral estimation subprocess	110
	Correlation	110

Spectrum 112
Saving the correlation and the spectrum 113

10 Three-dimensional data in SAC 115
10.1 The concept of 3D data 115
10.2 Spectrograms 115
10.3 Contour plots 117
10.4 Composite 3D data plots 119
10.5 Properties of 3D data 120
10.6 Writing 3D data files 121

11 Implementation of common processing methodologies using SAC 123
11.1 Seismic anisotropy and shear wave splitting 123
Overview 123
11.2 Shear wave splitting analysis 123
Parameter estimation methodologies 124
Macro and auxiliary program design 125
SAC implementation 127
11.3 Receiver function analysis 127
Overview 127
Estimation methodologies 129
Macro and auxiliary program design 130
SAC implementation 130

Appendix A **Alphabetical list of SAC commands 135**
Appendix B **Keyword in context for SAC command descriptions 142**

References 167
Index 170

Color plate section is found between pp. 108 and 109.

Preface

One of the most widely used analysis packages for regional and teleseismic seismic data is SAC (the Seismic Analysis Code). It was developed in the 1980s by nuclear test monitoring agencies in the United States, who freely made the source code and paper documentation available to academic users. From this initial distribution, the analysis package became popular in academic circles due to its ease of use and suitability for research data analysis in seismology and geophysics.

SAC's documentation was on ring-bound paper shipped along with the nine-track half-inch source code tapes that you received in the post. The academics in receipt of them generally made copies from the master document and distributed them to colleagues and students. They also tutored new users on the use of SAC, guiding them through their first session and then left them to the documentation. Consequently, much of the knowledge of SAC was passed tutorially from an experienced user, not unlike trade apprenticeship. Those intrepid enough to find and read the documentation usually exceeded their tutors' ability. The far fewer who delved into the source code learned of undocumented features of great utility. The usual reasons for this puzzling knowledge gap apply: in software development, documentation always lags feature development and, for SAC, new releases were sporadic and focused on new capability.

The original SAC documentation consisted of: (1) tutorial guide; (2) command table; (3) detailed command descriptions; (4) SAC file structure internals; (5) auxiliary program guide for programs to turn graphics to hard-copy form. Only the first of these was of any help to the new SAC user. Moreover, it is only in the tutorial guide that SAC's very powerful macro capabilities were described, but then only superficially. Consequently, the typical new user response after reading it, trying a few things out with SAC and getting confused, was to lean over to the nearest grad student and ask for help. SAC's most powerful capabilities could only be learned that way, if at all.

This book aims to remedy this continuing state of affairs when learning to use SAC. Despite the emergence of the web and web-based documentation for SAC, those available reproduce the paper ones as of about 1990. Despite some SAC development since then, no new documentation at the novice level, and little at the command description level, has been written. The main documentation effort today by the IRIS Consortium focuses on enhancing detail in the individual command descriptions. This book is for new SAC users to lead them through the steps of learning basic commands, to describe the SAC processing philosophy and to describe the SAC macro language in full. All ideas are introduced with example input and output from SAC. For the more experienced user, the book describes the advanced features of SAC graphics: graphical interaction with traces and annotation of displays of traces with auxiliary data. We also describe the powerful, but under-appreciated feature of SAC's array data handling facilities and spectral analysis methods. Also for the experienced user, we show how to write independent programs to access and create files, and how to use the methods that SAC provides to integrate external processing steps into production-type data analysis schemes. We show this with descriptions of code and SAC macros that implement the standard methods of teleseismic shear-wave splitting and receiver function analysis. Example commands and macros for tryout and text for programs is flagged using the following scheme:

Ex-0.1

```
example text here
```

This material is linked from the publisher's web site (www.cambridge.org/helffrich) for download, keyed to the identifier in the box.

We originally developed this material to teach a three-day course in SAC data processing to new PhD students in geology and geophysics. Following the successful reception of the course, we decided to turn the lecture notes and exercises into a more permanent version that redresses the shortcomings of existing SAC documentation. We hope that this book will be of use for incoming and existing PhD students, or in undergraduate courses on seismic data processing. If it is to hand at every seismologist's desk, it will have achieved its aim.

I salute the original authors of SAC, Bill Tapley and Joe Tull, for its design simplicity and implementation clarity. There can be no better testimony to the quality of a piece of software than its being in continuous use for over 30 years.

Acknowledgements

All of us came to use SAC through some variant of my own introduction to it. My thesis advisor, Seth Stein, came by me in the computer room one day, dropped the SAC manual into my lap and said, "You might be interested in using this." At least I think it was him, but some of my fellow PhD students might have different recollections that involve their agency. I apologize for any faulty memory. In the case of IB and JW, Graham Stuart and Mike Kendall were the responsible supervisory parties.

More recently, Frank Scherbaum helped to fill in some blanks in the history of SAC's distribution. I thank him for fingering back through some dusty filing cabinets finding documents he thought he'd never need to look at again – and frankly, didn't want to.

It was Mike Kendall's suggestion that led to us teaching a course on SAC's use. His administrative duties prevented him from contributing to the book, but he deserves the blame for our increased workload and the credit for the book's inspiration when we try to trace its inception. The course went on the road, and Dave Eaton bravely funded its import and field-testing at the University of Calgary.

Finally, I thank Cheril Cheverton, the better grammarian, for reading through the first draft of the book and making countless useful suggestions despite not knowing the subject at all. That kind of criticism is the best.

George Helffrich, for the authors

Citations for development of SAC should include one or more of these articles:

Goldstein, P., and Snoke, A. (2005). SAC availability for the IRIS community. *Electronic Newsletter*, Incorporated Institutions for Seismology, Data Management Center.

Goldstein, P., Dodge, D., Firpo, M., and Minner, L. (2003). SAC2000: signal processing and analysis tools for seismologists and engineers. Invited contribution to *The IASPEI International Handbook of Earthquake and Engineering Seismology*, ed. W.H.K. Lee, H. Kanamori, P.C. Jennings, and C. Kisslinger. London: Academic Press.

CHAPTER ONE

Introduction

1.1 WHAT IS SAC?

SAC is an acronym for the Seismic Analysis Code, a command line tool for basic operations on time series data, especially seismic data. SAC includes a graphical interface for viewing and picking waveforms. It defines a standard type of file for storing and retrieving time series and also reads files written in other data formats (SEG-Y, MSEED, GCF) used during field collection of seismic data. SAC is self-contained and does not rely on network access for any of its capabilities, including documentation, which makes it useful for field data quality control.

SAC reads data formats (CSS and GSE formats) used by nuclear test monitoring agencies. It also contains programming language constructs that provide basic methods for developing elaborate, multi-step analysis methodologies. Collectively, these features make SAC a useful interactive platform upon which customized analytical methods may be built and prototypical procedures may be developed.

SAC is widely known. The IRIS Data Management Center (DMC), one of the largest whole-Earth seismological data repositories in existence, allows data to be requested in SAC form. The instrument response information provided by the DMC's SEED reading program, *rseed*, is usable by SAC in pole-zero or *evalresp* form. Owing to SAC's longevity, a rather large and well debugged software tool ecosystem has evolved around its file format. One such tool, *jweed*, searches for and retrieves data held by the DMC. SAC data and SAC-compatible instrument response information are among its output options. This and many other programs developed by individual researchers make SAC a natural choice for time series data analysis.

1.2 HISTORY AND DEVELOPMENT

SAC initially was developed in the early 1980s by the Livermore and Los Alamos National Laboratories in the Treaty Verification Program group. The developers were led by W. C. Tapley and Joe Tull, and the package incorporated parts of Dave Harris' XAP program (Harris, 1990). SAC's Fortran source code was distributed to interested academics as it became a tried, tested and valuable system for non-commercial seismic data processing. In this early distribution epoch, there was a collegial agreement between users and maintainers to send bug fixes and improvements to the developers in exchange for use. In 1986, SAC came to GH's notice, who used it in his thesis work. By about 1990, SAC had become the *de facto* standard analysis system for academic whole-Earth seismologists worldwide.

Beginning in about 1992, development of SAC was increasingly taken over by Livermore and access to the source code became restricted through distribution agreements with the lab. This culminated in the last source code release, version 10.6f, around 2003.

During this period, SAC's Fortran code base was converted to C language using an automatic Fortran to C conversion tool called *f2c*. This was apparently done in the belief that Fortran was too restrictive a language and that it hampered further development of SAC's features. Livermore continued SAC development using the C code base with a view to future commercial sales of a seismic data analysis product called SAC2000. In this version, SAC's functionality was to be extended by adding, among other features (Vergino and Snoke, 1993), a processing log database that would record all steps in transforming a trace from a raw form into a processed form. The design allowed for the processing steps to be tentatively saved and then either committed to memory or rolled back to an earlier state. During this period, distribution of SAC's source code ceased due to ongoing development and commercial licensing restrictions.

In about 1998, the IRIS Consortium recognized that SAC's core user community, essentially IRIS' membership, had no guarantee of affordable access to SAC's source code. IRIS began negotiation with Livermore to develop two strands, one with database features for use by nuclear monitoring organizations and another without database features for academic use. The commercialization efforts focused on the database-enabled version. In 2005, IRIS took up support and development of SAC2000 without the database features, leading to a SAC version called, in this book, SAC/IRIS. There was no academic community interest in the commercial SAC release, and Livermore's support for SAC2000 was eventually withdrawn.

During this time, the Fortran 10.6d code base (later integrated with 10.6f) continued to be maintained and developed at the University of Bristol. The bugs that existed in the Fortran code base were gradually eliminated (though they continued to exist in the C source version) and SAC's functionality continued to expand, particularly in the area of array processing of interest to Bristol researchers. This is the version documented in this book.

1.3 ALTERNATIVES TO SAC

gSAC is a name inspired by the GNU Project's free versions of the C compiler (gcc) and the Fortran compiler (gfortran). gSAC was developed by Bob Herrmann at Saint Louis University in response to SAC's monolithic internal structure and its previously closed source distribution, using advances in computer platforms. Over a period of about six weeks in 2004, gSAC re-implemented SAC's basic seismic trace manipulation functionality from scratch. Now gSAC is a group effort that provides documented tools for manipulating seismic traces which happen to be stored in SAC's file format.

Seismic-Handler was developed at the Seismological Observatory Gräfenberg by Klaus Stammler and several contributors. It is a software package that defines a set of waveform modification programs on a common data format and a scripting language to string together the programs to produce an analysis stream. There is an interactive graphical user interface for observatory purposes (e.g., daily routine seismicity analysis) and a command line version for scientific research. In 2008 the Deutsche Forschungsgemeinschaft (German Research Foundation) funded a project for further development of Seismic-Handler.

SEISAN is a package similar in structure to Seismic-Handler but oriented for use by observatories involved in routine seismic analysis. The system comprises a complete set of programs and a simple database for analyzing earthquakes from analog and digital data. SEISAN includes graphical user interaction facilities on waveform data to locate events, to edit events, to determine spectral parameters, seismic moment, and azimuth of arrival from three-component stations and to plot epicenters. A database search functionality exists to extract and operate on the data for particular events. Most of the programs can operate both conventionally (using a single file with many events) or on a database. SEISAN contains integrated research-type programs like coda Q, synthetic modeling and a complete system for seismic hazard calculation. The system is freely available for all non-commercial use. SEISAN was developed at the University of Bergen by Jens Havskov and Lars Ottemöller.

SU/Seismic UNIX is an open source seismic utilities package supported by the Center for Wave Phenomena at the Colorado School of Mines. The package provides an instant seismic research and processing environment for users running UNIX or UNIX-like operating systems. The package is a set of independent programs that exchange data in a common format through pipes. It has no graphical interface. The original developers were Stockwell and Cohen, though now it is an open source project with many contributors.

AH is a UNIX-inspired set of basic seismic operations (reading, filtering, decimation, etc.) on an input stream to transform data that is then delivered onto an output stream. It is based on UNIX pipes. The system was developed in the early 1990s by Dean Witte and Tom Boyd of Lamont-Doherty/Columbia University. The C language source code is still available online, although the package is no longer actively maintained.

PITSA was written by Frank Scherbaum (then at the University of Munich, now at the University of Potsdam) in the early 1990s and runs on IBM PCs and Sun workstations. It offers utilities for simple trace manipulation, like shifting or scaling of traces, adding or concatenating traces, stacking and others. Internally, PITSA uses a data format geared toward earthquake seismology, and it also reads plain ASCII and SEED files. PITSA's graphical user interaction is based on menus, utilizing dialog boxes and pop-up menus. PITSA no longer seems to be maintained, but its source code is available.

MATLAB offers a time series toolbox.

R is a free, open source system with built-in graphics oriented toward statistical analysis. It includes a time series library, and loadable libraries exist to read and write SAC files.

DIY tools. SAC's documented file structure has a useful collection of library routines usable from C, Fortran and Python that make it easy to develop personal tools for analyzing SAC time series. See Chapter 6 for further information.

1.4 SAC VARIANTS

Fortran SAC. This is SAC as implemented in Fortran source code. It was distributed by its developers up to version 10.6f. Source code for this version was also distributed in restricted form in some versions of the IASPEI Software Library circa 2003.

SAC2000. This is SAC translated from Fortran into C source code and subsequently maintained in C. Database capabilities were the principal development addition to this version and some new commands were also implemented. This version of SAC is no longer distributed.

SAC/IRIS. This is derived from SAC2000, without the database capabilities. It is actively maintained by the SAC development team under the aegis of the IRIS Consortium and is distributed by IRIS.

MacSAC (SAC/BRIS). This variant is derived from the 10.6d Fortran source code distribution and represents a superset of the capabilities of SAC/IRIS. The principal extensions relative to SAC/IRIS are in the capabilities of the macro language and significantly expanded handling capabilities for array data.

1.5 REQUIREMENTS AND INSTALLATION

SAC/IRIS is distributed through a licensing agreement with the IRIS Consortium. Distributions are available in source and binary forms from IRIS. Binary distributions are available for 32- and 64-bit Linux systems, 32- and 64-bit Macintosh systems and Solaris systems. Windows users must build from source in the Cygwin environment.

MacSAC (SAC/BRIS) is presently available in prepackaged distributions for MacOSX systems from 10.2 onward. The system automatically builds itself under MacOS, Solaris, FreeBSD and Linux systems from source releases based on 10.6d.

1.6 SCOPE OF THIS BOOK

The SAC command repertoire is vast – over 200 in all. We will not attempt to cover them all in this book. Our goal is rather to provide an introduction to SAC's basic concepts and its basic command set. Consequently, many commands are not included. An exhaustive list of commands with a keyword index of concepts derived from the command description appears in the appendix. This list, along with the help information built into SAC, will hopefully lead you to speculate on, find and use SAC's broader functionality.

The first five chapters of this book constitute basic material. The following chapters emphasize SAC features not adequately documented anywhere else. Our goal here is to supply that documentation and to show, with examples, the occasionally surprising utility of those features.

The final chapter is a description of how SAC may be used to implement some of the standard data analysis procedures used by seismologists on teleseismic data. Though the procedures are eminently serviceable as they stand, their inclusion here is to serve as examples of how SAC's capabilities may be used more effectively. They represent the distillation of over 20 years of experience with SAC and will reward study and reflection. Looking into the future, one can imagine developing procedures for analyzing station noise levels from long sequences of MSEED field data blockettes, for aspects of ambient noise study and for normal mode seismology.

Note that some capabilities provided by SAC require significant skill to use properly, particularly correcting for instrument response. Unfortunately, the subtleties of this topic are beyond the scope of a SAC-oriented text such as this one. Other problematic areas include attenuation correction and earthquake location. SAC provides features that either perform these functions or link with standard programs that do them. Again, the scope widens significantly when these topics are included, and so, with apologies, we have opted to omit them here. Sorry.

CHAPTER

TWO

The SAC data format

2.1 PHILOSOPHY AND STRUCTURE

SAC file format

A seismic data trace is a set of data points that is continuous in time but that may not have been sampled at an even rate. SAC's simple approach to seismic data deems that there is one seismic trace per file. Each file contains a header that describes the trace (also known as metadata) and a section that contains the actual data. The header occupies a fixed-length position at the beginning of each file, followed by the variable-length data section.

Header data is of mixed type: integer and real values, logical (true/false) values, categorical values (distinct properties like explosive source, nuclear source, earthquake source), or text (station code, event identification, wave arrival type). The seismic data in a trace is a sequence of real-valued numbers representing the sampled physical property.

Alphanumeric and binary forms

There are both binary and alphanumeric (character) formats for a SAC file. The binary version is a more compact format that is efficiently read and written, while the alphanumeric format is easier for user programs to read and write.

The alphanumeric data format[1] is intended for ease of reading and writing and for transfer between different machine types. In this format, the header data is organized into

[1] Word processors should not be used to work with SAC alphanumeric data files. These are not in RTF or Word format, even though they contain text. The relevant format is TXT (in DOS terminology) and any editing software used to prepare or edit an alphanumeric file should be capable of writing this file type, which is devoid of typesetting formatting information.

sections based on the type of data, with each section subdivided into lines. While not an intuitive organization, this makes it easier to read and write the header data because all of the data on each line of the file is of the same type: integer, real, etc.

In contrast, the binary SAC format is compact and efficient to read and write. In this format, the header data is either integer or real floating point numbers (in IEEE 754 standard format). SAC binary data is always stored in single precision (32-bit) IEEE 754 floating point format.

Interconversion of formats

Converting to and from SAC files of different format is easy with SAC. From alphanumeric to binary,

```
SAC> read alpha <afile>
SAC> write binary <bfile>
```

From binary to alphanumeric,

```
SAC> read binary <bfile>
SAC> write alpha <afile>
```

2.2 CONVERSION FROM OTHER DATA FORMATS

GSE files

SAC is capable of reading files in other common data exchange formats. One common format generated by AutoDRM software (Kradolfer, 1996) is the GSE format (Group of Scientific Experts) (GSETT-3, 1995), in use by the United Nations' International Monitoring System, part of the Comprehensive Nuclear Test-Ban Treaty framework. This alphanumeric format can be directly read and written by SAC. To read GSE data into SAC and rewrite it in SAC format,

```
SAC> readgse <gfile>
SAC> write <sfile>
```

and to write GSE-format data when the original form is SAC,

```
SAC> read <sfile>
SAC> writegse <gfile>
```

Note that there is a separate SAC command to read and write GSE data; the `READ` command is not used.

Many traces can be present in one GSE file. The file usually is a string of messages produced by an AutoDRM and sometimes is the text of an e-mail. `READGSE` will skip the parts of the file that do not contain trace data and read the traces contained in the file into memory. The various traces may be selected for further processing by writing them as SAC files individually.

The conversion between GSE and SAC formats is not isomorphic. Some information is lost in the conversion from GSE into SAC format. Multiple origins for an event are lost; SAC only has a single origin associated with each trace. Arrival picks in a trace in excess of 10 are also lost. Any information loss is reported by SAC to the user, however.

SEG Y, MSEED, GCF and CSS formats

Unlike the various SAC formats and the GSE format, the other formats cannot be written by SAC, only read by it.

SEG Y

The SEG Y data format is based on a standard developed by the Society of Exploration Geophysicists for storing geophysical data. The first definition of the format was in 1973 and it was documented in 1975 (Barry *et al.*, 1975). Defined when computers were larger than automobiles and when magnetic tape was the most portable recording medium, many variants emerged as different user communities modified it to versions suited to more modern storage media. The PASSCAL project defined one of these variants to handle field data gathered during passive seismic experiments. This is the version read by SAC. Briefly, it has no 3600-byte reel header, requires that the trace code header value be seismic (type code 1) and allows only one waveform per file.

To read SEG Y data, use a variant of SAC's `READ` command,

```
SAC> read segy <sfile>
```

which reads the trace from the designated SEG Y file.

MSEED

MSEED (mini-SEED) (IRIS, 2006) is a format defined by the International Federation of Digital Seismograph Networks principally for data archiving and exchange. The data format is also output by field dataloggers. Consequently, this makes SAC a useful tool for field computers for field data quality control.

To read MSEED data, use a variant of SAC's `READ` command,

```
SAC> read mseed <mfile>
```

MSEED data is commonly collected in short packets, each with a time stamp and a sequence of samples starting at that time. Consequently, the data might not be continuous across packet boundaries. Any data overlaps or gaps will be reported by SAC when read.

GCF

Another data format encountered in the field is GCF (Güralp Compressed Format)[2]. This is a raw data format output by dataloggers manufactured by Güralp Systems. To read GCF data, use a variant of SAC's `READ` command,

```
SAC> read GCF <gfile>
```

GCF data is collected in blocks of 1024 characters. Consequently, the data might not be continuous across block boundaries. As with MSEED, any data overlaps or gaps will be reported.

CSS

CSS format is a format defined by the Center for Seismic Studies and is a waveform database format used by the International Monitoring System, with separate files for trace metadata

[2] Defined in Güralp online information at `http://www.guralp.com/gcf-format/`

and trace data (Anderson *et al.*, 1990). Consequently, at least three separate files must be grouped to describe a trace. The grouped files are associated through one master waveform metadata file. The metadata files have names with the same prefix but different suffixes. Files with suffixes `.wfdisc` and `.origin` pertain to the waveform metadata and the origin information for the event, respectively. A third file, whose location is given in the `.wfdisc` file, contains the data for one or more traces. Each trace is described by one line in the `wfdisc` file, so one CSS metadata file can describe one or more traces.

To read a CSS file in SAC, provide the name of the `wfdisc` file, either with or without the suffix:

```
SAC> readcss <cssdata>
```

This implies that given a file prefix of `cssdat`, files called `cssdat.wfdisc` and `cssdat.origin` also exist. Alternatively, the full `wfdisc` file can be specified on the `READCSS` command, which is useful when pattern matching for a group of files.

Many traces can potentially be present in a CSS file collection. Using suitable `READCSS` command options, subsets of the traces may be chosen based on the channel name, station name and frequency range of the sensor. By default, `READCSS` reads all of them.

2.3 BYTE-ORDER ISSUES

Information in SAC's file header unfortunately does not include an explicit indication of the data's type of binary format. Contemporary processors organize data in memory in either most-significant byte order (big-endian; SPARC or PowerPC order) or least-significant byte order (little-endian; DEC/Intel order).

The added complexity in reading and writing data in the proper way may lead to many errors in user-written programs that read and write SAC files. One strategy to avoid byte-order problems is to read and write them in alphanumeric format. This has the additional advantage that the data is easily verified simply by viewing the file as text. Alphanumeric format is a useful intermediate form to convert data from another format not recognized by SAC to make it SAC-readable.

For programs that analyze or change data in a processing tool chain, there is more of an emphasis on performance. In this case, working with the binary format is more efficient. SAC provides a set of user-callable subroutines for easy access to header and data in SAC binary files. These routines handle the byte-order related problems automatically and can cope with past or future changes to the data format in SAC binary files (changes in the binary file format are being considered as of November 2012 and will probably occur in 2013 or 2014). The routines are provided in library form along with SAC. See Chapter 6 for further details about accessing SAC file information from programs and SAC's built-in help entry for its library of user-callable subroutines (`HELP LIBRARY`).

Unfortunately, SAC binary file byte order is not specified in any standard. Because SAC was originally developed on big-endian machines (Sun, Prime and Ridge workstations), this is the expected byte order, and any files written by SAC will be in big-endian order no matter what the underlying machine type is. Some versions of SAC (notably SAC/IRIS) will write binary files in the machine's native format. If one is encountered, SAC will report that the format is unexpected. A utility program (called `sactosac` and distributed with SAC) is available to switch to the proper byte order. See SAC's help information for its utility programs (`HELP UTILITIES`) for further usage information.

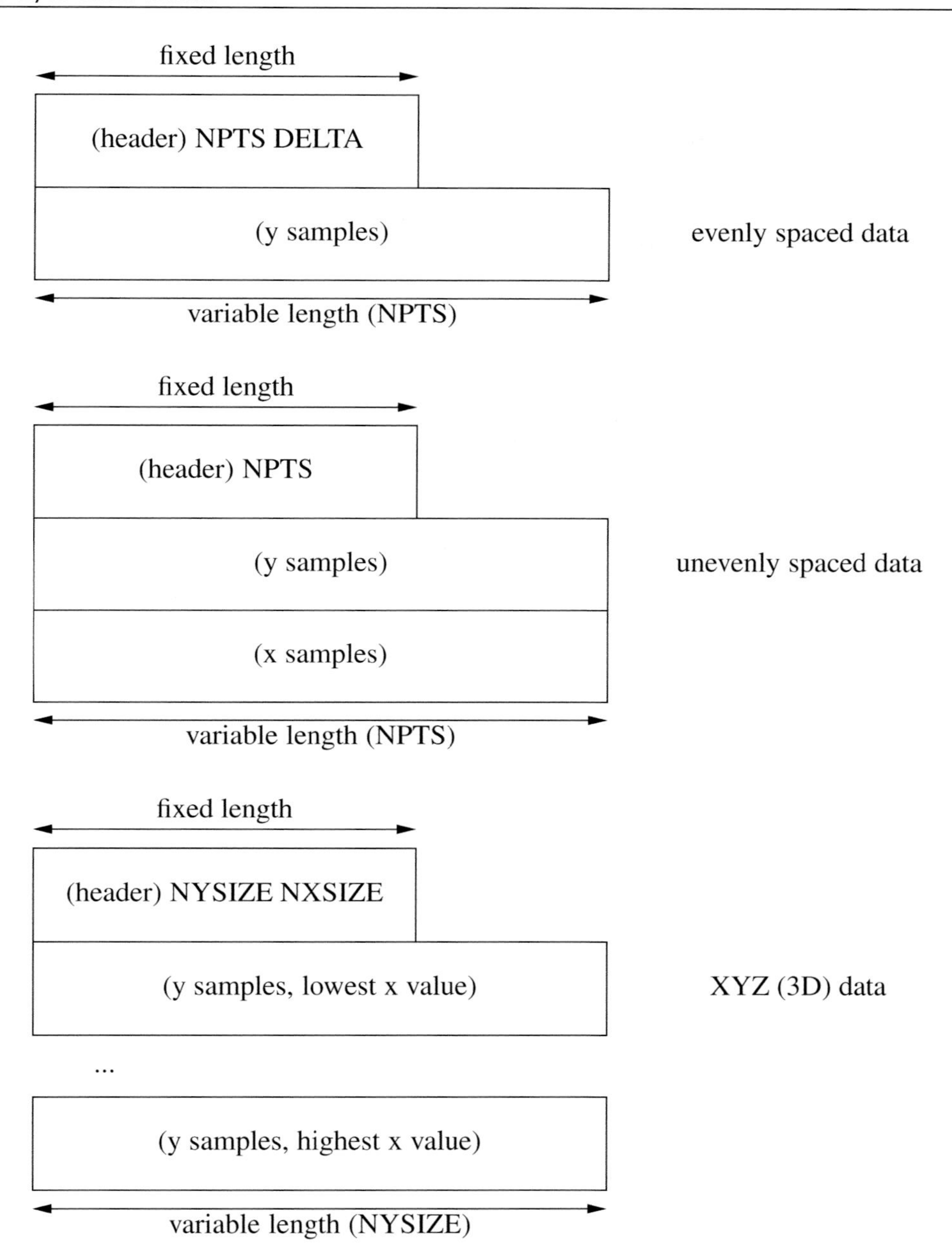

Figure 2.1 Schematic layout of SAC files of different types. All types have a fixed-length header and a variable-length data portion. For evenly spaced data, the `NPTS` entry in the file header specifies the number of data points (sampled every `DELTA` seconds). For unevenly spaced data, `NPTS` in the header specifies the number of samples, but the sampled values appear first in the data section followed by the time at which the sample was taken. For XYZ, or 3D data (see Chapter 10), the sampled values in the Y direction at the lowest X value appear first, followed by samples for the next-higher X up to `NXSIZE` X values. `NYSIZE` gives the Y dimension.

2.4 SAC FILE LAYOUT

SAC provides essentially three types of data files. The metadata-containing header is fixed length. Depending on the number of data points and the type of the file indicated in the header, a variable number of data values will follow in the file. How many data values are present depends on the type of the file. Simple time series (the most common type – IFTYPE of ITIME) have a fixed number of samples (NPTS) that are separated by a fixed sample interval (DELTA). In contrast, unevenly sampled data files (IFTYPE of IXY) have a fixed number of samples separated by random time intervals. Consequently, DELTA has no significance, and the sample values are followed by the individual times at which the samples were made. The third type of file is XYZ (or 3D data – IFTYPE of IXYZ), where samples are made on a fixed spatial grid of (X,Y) positions. In this case, the header provides the number of X and Y sample positions (NXSIZE and NYSIZE), and sample values follow for each of the NXSIZE × NYSIZE points. The ideas are sketched in Figure 2.1.

CHAPTER

THREE

The SAC processing philosophy

3.1 PHASES OF A TYPICAL ANALYSIS TASK

SAC has a complete set of reliable and well tested commands to document, view, process, transform and save data and results. The command set is confusingly large at first sight. Rather than list them all, we'll describe the steps one typically takes when processing a large seismic dataset. The commands that SAC provides relevant to each of the steps will be introduced to explain what they do and to suggest a way to remember them. The basic steps are as follows:

- Organize – standardize/prepare data for analysis.
- Interact – select a portion of the waveform to operate on.
- Process – calculate a property / estimate a parameter / transform data.
- Display – graphically show the result of processing to verify its proper functioning and to validate the data to the analyst.

Organize

Naming

The processing phase of a seismic experiment usually begins with a data delivery via hard disk, e-mail attachment, USB stick or whatnot containing data in some format. Typical formats encountered nowadays are SEED, a *tar* archive containing SAC files associated with a particular day or a particular event, or perhaps raw GCF or MSEED data straight from a field experiment. The files are rarely named appropriately. For example, a typical file group of three-component data produced by the *rseed* program (distributed by the IRIS DMC to read SEED datasets), might be:

```
2012.055.12.34.56.7777.YW.MAIO.01.BHE.Q.SAC
2012.055.12.34.50.6666.YW.MAIO.01.BHN.Q.SAC
2012.055.12.34.54.5555.YW.MAIO.01.BHZ.Q.SAC
```

representing data from a broadband field experiment recorded by the station MAIO. Rather than coping with the long file names, key them to a particular earthquake suitable for the teleseismic study in mind. For example, the event might be tersely represented by a two-digit year code, month code and day code: yymmdd. Then transform the long file names into a station name, event name and component name (BHE, BHN, BHZ) of the following form:

```
MAIO.120224.BHE
MAIO.120224.BHN
MAIO.120224.BHZ
```

This is best done using UNIX commands provided by the shell. For example, with the file names provided above, a useful set of *sh* commands to use to rename the files might be:

```
for c in BHE BHN BHZ ; do
   mv 2012.055.*MAIO*.$c.Q.SAC MAIO.120224.$c
done
```

Given SAC's command syntax, it is best to organize data with the component name at the end of the file name and the station name at front. Thus, using the *ls* command it is possible to search for file names that recorded a particular event:

```
ls *.120224.BHZ
```

or for all events recorded at a particular station:

```
ls MAIO.*.BHZ
```

Event information

After the files are properly named, standardize the data for each event. For example, it is possible to set the event information (origin time, latitude, longitude, depth and magnitude) into each file header for traces that recorded a particular earthquake.

Station and component information

Other items associated with a trace are the station location (latitude, longitude and elevation) and the orientation of the sensor. Some stations, particularly temporarily deployed ones, often do not have their N and E components pointing accurately north and east. This is important information to correct in the file header if it has not been provided automatically by the data center.

Trace length and sample rate

It might be necessary to cut the data into time windows that all have the same number of samples and have the data recorded at the same sample rate. This requires decimating or interpolating the data to the same time interval between samples and then throwing away data at the beginning or end of the trace that extend beyond the desired time window for analysis.

Discontinuous data

Sometimes the data is not continuous but is chopped up into files that contain data segments that last for a fixed time. Thus, another organizational chore might be to join the separate files into a single file representing the continuous data stream.

De-glitch data

Electronics problems with the seismometer's recording system will occasionally lead to spikes or dropouts in the seismic data stream. These so-called glitches must be edited out of the data before it is analyzed because the glitch can resemble a signal. Data glitches were more common in the era of analog data recording methods, but with digital electronics, they are becoming more rare.

Instrument response

Depending on the type of analysis, the instrument response might need to be removed in order to turn the data stream into displacement, velocity or acceleration time series. Alternatively, the instrumentation might not be homogeneous throughout a seismic network, and the analyst might want to cast the data stream as though it was recorded by a single type of instrument. This requires using the instrument response to modify the data.

Remove data mean and trend

Often a signal will have a nonzero mean or a very long-period baseline change that gives it a slope. These can thwart a signal analysis scheme and, if so, must be removed from the data before it is analyzed.

Filtering

Almost all data analysis schemes must be restricted to a range of frequencies in the data. Consequently, filtering of the time series into lowpass, highpass, or bandpass data is necessary.

Picking

Sometimes it is useful to add travel-time picks to the time series data that are predicted by some model of the process being studied. This is typically done before analysis.

Write as SAC files

Finally, the data and its metadata (event origin information, preliminary picks, etc.) is written as a SAC file. This puts the data into a homogeneous form that simplifies the following analysis.

Interact

After the data is organized, there is usually a step involving human assessment and interaction with the data. At this stage, it is useful to display the data graphically because this is less error prone than, say, typing in numbers read off a trace, and because people are visually adept and can easily evaluate a graphic. Typical interactions are as follows:

- Data quality control (QC) – discarding noisy/unsuitable traces.
- Windowing – defining a time window in the time series.
- Picking – picking a signal onset.

For these interactions, the information is presented graphically but the ultimate result is numerical or text. Data may be saved in memory for further user interaction, in the SAC file header as information associated with the trace, or in an external file. For QC steps, good file names are saved to be used as input for the next analysis step, and bad file names are removed from the analysis collection.

Process

This is the heart of an analysis task. There is usually a specialized external program, separate from SAC, to which a SAC file is passed for processing. A good program will provide text output that the user can read during the analysis for informational or verification/debugging purposes. The program should also produce output for validation of the analysis. In graphical form it is easy to confirm whether the processing was successful.

Processing might also combine a sequence of SAC analysis elements packaged to yield a useful result or to allow input to the external analysis program. For example, spectrum estimation, stacking of multiple traces, waveform picks, or trace sample arithmetic might be required.

Display

The best way to confirm that processing worked properly is through graphical output. The display should follow the processing step so that the analyst can see whether the processed signal is good. Because SAC has built-in commands to plot time series and two-dimensional data, the display step usually involves reading a file that the processing step wrote. Consequently, SAC provides a simple library of subroutines for programs to use to write files.

3.2 COMMAND SUMMARY FOR EACH PHASE

These processing steps provide a framework for introducing SAC's basic command set.

Organizing data

- Read data: `READ`, `READGSE`, `READCSS`
- Change data in file headers: `CHNHDR`
- Cut/merge data to common length: `CUT`, `CUTIM`, `MERGE`
- Put data on a common time base: `CHNHDR ALLT`, `SYNCHRONIZE`
- De-glitch: `RGLITCHES`
- Remove mean and trend: `RMEAN`, `RTREND`
- Taper: `TAPER`
- Filter operations: `LOWPASS`, `HIGHPASS`, `BANDPASS`, `BANDREJ`
- Instrument response: `TRANSFER`
- Add travel-time picks: `TRAVELTIME`
- Preview and winnow: `PLOT1`, `MESSAGE`, `SETBB` using *reply*
- Write data: `WRITE`

Interacting with the user

- Graphical interactions: `PLOTPK`, `PLOTRECORDSECTION`
- Text interactions: `MESSAGE`, `SETBB` using *reply*
- Graphical display: `PLOTC`, `XLABEL`, `YLABEL`, `TITLE`, `AXES`, `TICKS`, `COLOR`, `LINE`

Processing

- Invoke a UNIX command/external program: SYSTEMCOMMAND, $RUN ... $ENDRUN
- SAC processing facilities: APK, SPE subprocess commands, SSS subprocess commands

Display

- Graphical interactions: PLOTPK, PLOTRECORDSECTION
- Text interactions: MESSAGE, SETBB using *reply*
- Graphical display: PLOTC, XLABEL, YLABEL, TITLE, AXES, TICKS, COLOR, LINE

3.3 FURTHER INFORMATION ABOUT SAC COMMANDS

SAC itself has a built-in help facility that may be used to obtain more information about SAC commands. The most basic command is HELP COMMANDS. It provides a list of all of SAC's commands, which admittedly may be overwhelming for the new user, but can remind the experienced user of that forgotten command name.

More useful is HELP APROPOS followed by a word. SAC scans its online help for every command with the word in its description. For example, commands relating to the GSE file format may be found in this way:

```
SAC> help apropos gse

READGSE - Read data files in GSE 2.0 format from disk into memory.
WRITEGSE - Write data files in GSE 2.0 format from memory to disk.
SAC>
```

This lists two SAC commands, READGSE and WRITEGSE, that are related to GSE file handling.

Once a potentially useful command is found, more detailed help can be obtained by the HELP command. Continuing with the previous example, information about the command to read GSE files is obtained like so:

```
SAC> help readgse

 SUMMARY:
 Read data files in GSE 2.0 format from disk into memory.

 SYNTAX:
 READGSE [MORE] [VERBOSE {ON|OFF}] [SHIFT {ON|OFF}]
         [SCALE {ON|OFF}] [DIR <dir>] <file> ...

 INPUT: ...

 DESCRIPTION:
     See the READ command for general details about file reading.

 A single GSE file can contain more than one trace.  All present
 in a GSE file will be read into SAC memory.  Waveform formats of
 INT, CM6, CM7 and CM8 can be read.  The following GSE data
 messages can be handled: WAVEFORM, STATION, CHANNEL, ARRIVAL,
 ORIGIN.  If the file begins with a WIDx record, it is ...
```

The output shows SAC providing the command summary, its syntax and expected input, and then a description of what it does.

If only the command syntax is wanted, use the SYNTAX command:

```
SAC> syntax readgse

 SUMMARY:
 Read data files in GSE 2.0 format from disk into memory.

 SYNTAX:
 READGSE [MORE] [VERBOSE {ON|OFF}] [SHIFT {ON|OFF}]
         [SCALE {ON|OFF}] [DIR <dir>] <file> ...

 SAC>
```

There are also online sources of information about SAC commands. A useful one is located at the IRIS DMC, accessible through the following URL (active as of November 2012):

```
http://www.iris.edu/software/sac/commands/func_commands.html
```

CHAPTER FOUR

Basic SAC commands

4.1 COMMAND STYLE

SAC commands are typed from the command line or read from a file. After each command is processed, SAC reads another command from its input source until it is told to stop or the input is exhausted.

Commands are single verbs (e.g., READ, WRITE), or a compound phrase (e.g., FILTER-DESIGN). Abbreviations exist for the longer or commonly used command names. A series of options that control the command's actions follow the command name. Command names or options may be typed in upper or lower case. However, when file names appear in commands, case does matter, and SAC preserves it.

White space separates options and the command name, and can even precede the command name. This is useful for indenting groups of commands for documentation purposes.

Multiple commands may be placed on the same line separated by the ";" (semicolon) character and will be processed left-to-right as they appear on the command line. Any command whose first character is * is a comment and is ignored. Thus the string ;* introduces a comment in the command listings that follow.

SAC supplies default command options if they are not specified. Command options, once set, stay in force for future uses of the same command. This provides a way to tailor personal command defaults. SAC can read a file of commands setting your personal defaults before it reads the input. They will be described in detail in Section 4.10.

4.2 COMMAND HISTORY

If told to, SAC will remember typed commands and allow them to be re-entered without typing them in full. The text of previous commands may be recalled and corrected or modified and re-entered as a new command. By default, this only affects commands in the current SAC session. To recall commands across sessions, save the transcript in a file using the `TRANSCRIPT` command. To turn on transcription and save the commands in the file "`~/.saccommands`", type:

```
SAC> TRANSCRIPT HISTORY FILE ~/.saccommands
```

The file can be local to your present working directory or global. Change the file name to choose where it is placed. The previous example sets up a global transcript in a file called "`.saccommands`" in your home directory. The default history file location, however, is a file local to your working directory called, "`.sachist`".

The command history is lost if SAC stops unexpectedly (by being killed or interrupted). Only when explicitly stopped (by `QUIT` or finding the end of the input) is the transcript saved for future use. See `HELP TRANSCRIPT` for a more general listing of processing steps.

4.3 READING AND WRITING DATA

Rather unsurprisingly, the basic command to read data is `READ`. It is abbreviated `R`.

```
SAC> help read

 SUMMARY:
 Reads data from SAC, SEGY, GCF or MSEED data files on disk into
 memory.

 SYNTAX:
 READ <options> {<file>|<wild>|<url>} ...
```

`READ` can read one or multiple traces. It defaults to SAC binary format, but `READ` can read files in alphanumeric format, XML format, and other standard types (see Chapter 2). When reading new files, they replace the traces stored in memory, but this can be changed to add further traces to the collection in memory.

Files are specified by their full names or by using a wildcard to describe a pattern to apply to the file names and only reading the ones that match. The wildcards are the standard UNIX ones:

- * matches any string of characters including an empty string;
- ? matches any single (non-null) character;
- [A,B,C] matches any character (or string) in the list.

Traces are read in the order specified. File names and wildcards may contain directory prefixes.

Reading examples

Here are a few examples that show different ways to use `READ`. Names can be wildcarded

```
SAC> R SWAV.BH?
SWAV.BHE SWAV.BHN SWAV.BHZ
```

or listed

```
SAC> R SWAV.BHE SWAV.BHN SWAV.BHZ
```

The `READ MORE` option appends traces to the set in memory

```
SAC> R SWAV.BHE
SAC> R MORE SWAV.BHN
SAC> R MORE SWAV.BHZ
```

Directory references can be relative or absolute and even include wildcards

```
SAC> R /tmp/SWAV.BH?
SAC> R ../SWAV.BH?
```

Writing data

The `WRITE` command outputs the current traces stored in memory. It is abbreviated `W`.

```
SAC> help write

 SUMMARY:
 Writes data in memory to disk.

 SYNTAX:
 WRITE <options> <namingoptions>
```

The `<namingoptions>` represent different ways to name the files that are output from memory. They can be a list of explicit file names

```
SAC> R SWAV.BH?
SAC> W SWAV.BHE SWAV.BHN SWAV.BHZ
```

or output and overwritten using the file names they were read with

```
SAC> R SWAV.BH?
SAC> W OVER
```

or with the file names changed by appending (or prepending) a string

```
SAC> R SWAV.BH?
SAC> W APPEND .new
SWAV.BHE.new SWAV.BHN.new SWAV.BHZ.new
```

4.4 PLOTTING AND CUTTING

Devices

One of SAC's strengths is its ability to make plots on a variety of computer types and plotting devices. Devices are started (or switched to) using the `BEGINDEVICES` command (abbreviated `BD`):

```
SAC> help begindevices

 SUMMARY:
 Begins plotting to one or more graphics devices.
```

```
SYNTAX:
BEGINDEVICES [PREVIOUS]
             [[MORE] {SGF|XWINDOWS|MACWINDOWS|TERMINAL}]
```

Useful devices are Xwindows, SGF and Mac. Xwindows is the graphics system built on the Xwindows library, available on most UNIX systems. SGF (SAC Graphics File) is a hardcopy device that saves plot descriptions to a file that can be converted to PostScript and viewed or inserted in documents. Mac is a native viewer for Apple MacOSX systems (Quartz) and may not be available on all versions of SAC. (Terminal uses the features of an old (Tektronix) type of terminal that could mix graphical output with text input and output. The *xterm* terminal emulator preserves these features through the "Tek window" feature.)

Graphical plots will not be made unless there is an open graphics device. SAC provides a way to designate a graphics device to open if and when a need for one arises. Not only is this convenient, it means less screen real estate is taken up by a window that does not yet contain any information. The `SETDEVICE` command does this. `SETDEVICE name` sets the default device to any of those available through the `BEGINDEVICES` command.

Windows and window placement

The `XWINDOWS` and `MACWINDOWS` graphics devices allow multiple windows to be open – up to five in the present version of SAC. This can be helpful during an analysis where two graphical displays are required. For example, one window might contain an average waveform trace from a group of stations recording a particular event, and another window might contain the trace for one station. As a quality control step, visual inspection for similarity to the average waveform might lead to selection or rejection of the station for inclusion in the analysis.

Windows have a size and a position on the graphics screen. SAC allows control of the size of a window and where it is placed before it opens. Tell SAC where to put a window by using the `WINDOW` command:

```
SAC> help window

 SUMMARY:
 Sets the location and shape of graphics windows.

 SYNTAX:
 WINDOW <n> [XSIZE <xlo> <xhi>] [YSIZE <ylo> <yhi>]
```

The number `<n>` (1-5) defines that window's properties. The `XSIZE` and `YSIZE` provide the size and location of the window. They represent the low and high X and Y coordinates in terms of a relative coordinate system where the bottom left is (0,0) and the top right is (1,1).

The first window to open is numbered 1 and opens at the behest of the `BEGINDEVICES` command. To switch between windows, use the `BEGINWINDOW` command with one of the numbered windows. Plots will keep appearing in that window until it is closed (using the `ENDWINDOW` command) or until the window is switched using another `BEGINWINDOW`. `ENDWINDOW` removes the window from the screen, whereas `BEGINWINDOW` either opens a new window or switches future graphical output to it leaving the previous window visible.

Plotting data

The basic plotting commands provided by SAC are PLOT, PLOT1 and PLOT2. SAC plots time series with the time axis horizontal and the amplitude axis vertical. The command variants differ in how data from multiple traces are plotted on the device. PLOT plots one trace per window and pauses between traces. PLOT1 (abbreviated P1) organizes all the traces in a single display that shares a common time axis and plots them all in one window. PLOT2 organizes all the traces in a single display and overlays them on a single time and amplitude axis. For example, a set of three traces from built-in datasets in SAC may be displayed in different ways by PLOT1 and PLOT2 (see Fig. 4.1):

```
SAC> datagen sub regional knb.? ;* Three traces of built-in data
knb.e knb.n knb.z
SAC> plot1
SAC> plot2
```

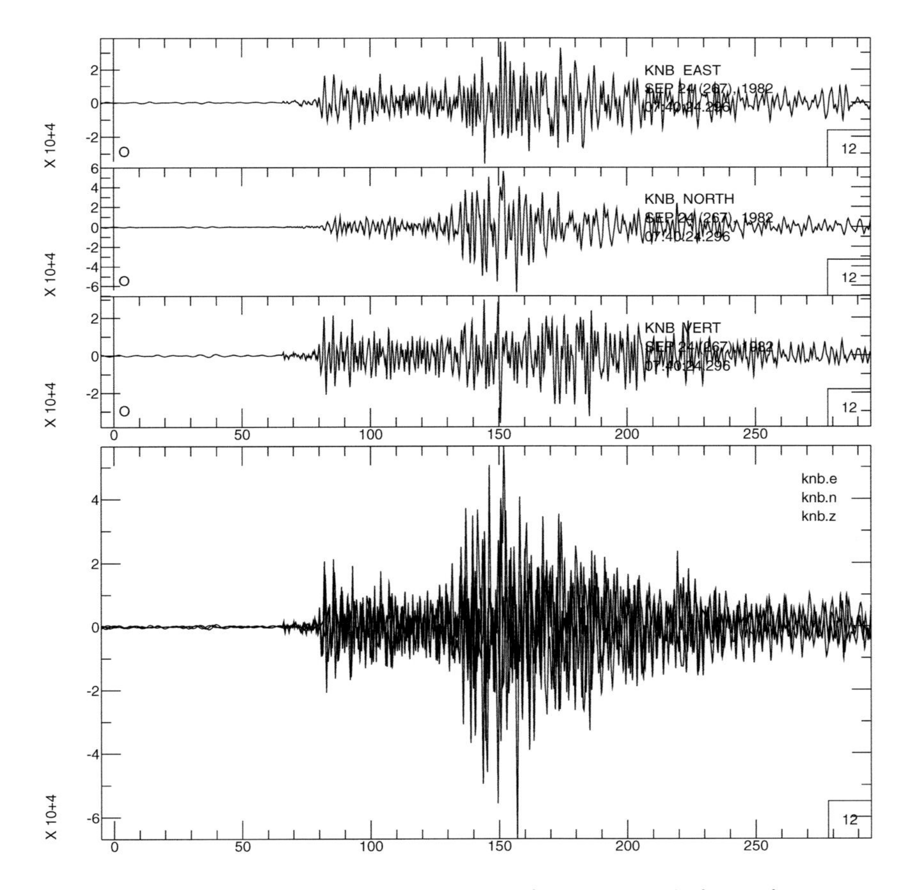

Figure 4.1 The plot made of three seismograms by PLOT1 (*top*) shows the traces in panels consisting of each trace, but sharing a common time axis. In contrast, PLOT2 (*bottom*) plots the traces overlaid in a single plot. Boxed numbers in lower right of each trace indicate a decimation factor for plotting purposes.

Many commands exist for configuring the plot size, axis labeling, plot positioning and making sub-plots in an individual display. Use `HELP APROPOS` to find further information about them.

By default, `PLOT1` plots all the traces in memory in a single display. This might yield unreadably small plots if there are many traces. Larger numbers of traces may be organized into smaller groups for display. The `perplot` option controls this. For example,

```
SAC> plot1 perplot 5
```

would limit the display to no more than five traces per page.

When plotting multiple traces in a single window, the traces might not share the same start and end times. If so, it might be useful to plot the data on a relative time scale referenced to the start time of each trace. The `absolute` and `relative` options to `PLOT1` and `PLOT2` affect this.

The `XLIM` command sets limits on the time range of the data displayed in a plot. Similarly, the `YLIM` command limits the amplitude range displayed. For fixed time limits, use

```
SAC> xlim 0 10; p1
```

The default behavior is to scale the amplitudes of individual traces to the maximum vertical space available in a plot. To make the scale the same on all traces contained in a single plot, use

```
SAC> ylim all; p1
```

To restore the default behavior, use

```
SAC> ylim off; p1
```

SAC provides controls on the style of line used to draw time series data. When plotting more than one time series on the same time axis, different line styles may be chosen to distinguish them. `LINE LIST` defines one or more line styles to be used successively in a multi-trace `PLOT2` plot. The `LINE INCREMENT` option instructs SAC to cycle through the line style list. The commands below produce the multi-trace plot in Figure 4.2.

Ex-4.1

```
SAC> datagen sub regional knb.[z,n,e]     ;* Three traces
SAC> rmean
SAC> line list 1 3 3 increment on
SAC> xlim 60 70
SAC> plot2
SAC> datagen sub regional knb.z; rmean    ;* One trace
SAC> line fill black/yellow               ;* Hymenopteral choice
SAC> plot1
```

The figure also shows another useful `LINE` option used to fill the area above and below the zero level in a trace. `LINE FILL` defines a pair of colors used to fill each trace. When

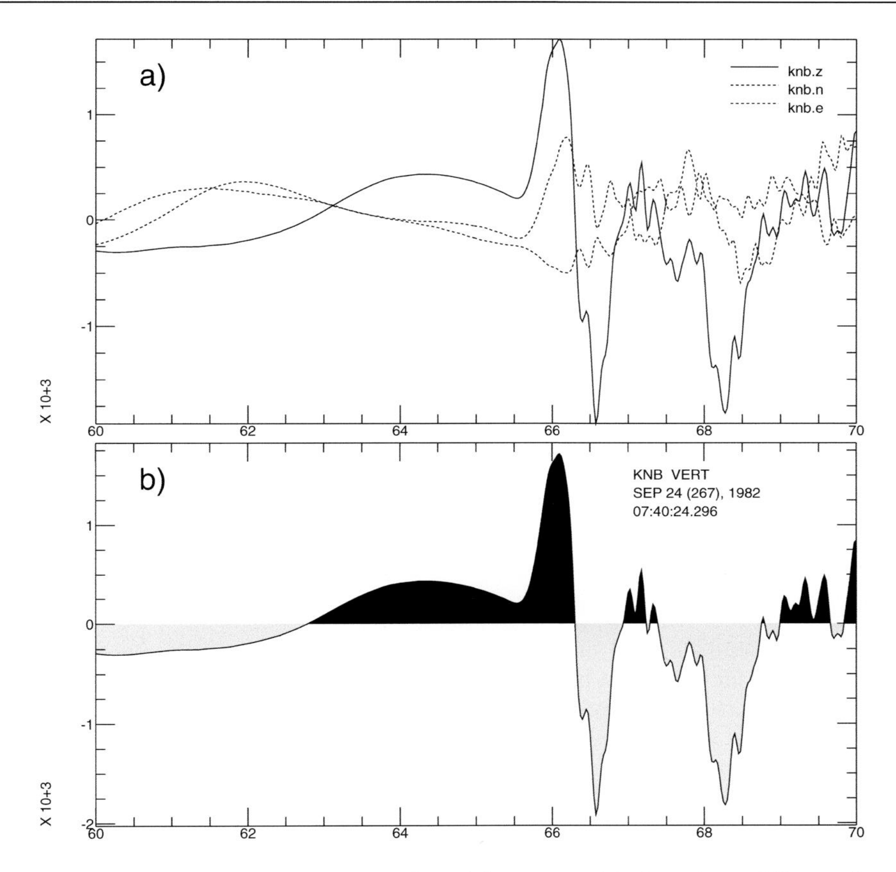

Figure 4.2 The three seismograms plotted by PLOT2 (*a*) show how different line styles enhance comprehension of the display. The line style for the horizontal component traces is different from the vertical component trace. PLOT2 shows this in the legend at top right in the display frame. Another line style feature that SAC provides to improve trace feature visibility is color-filled traces (*b*). The positive and negative areas may be filled with the same color, different colors, or limned selectively. (For color version, see Plates section.)

filling is active, it may be preferable to omit the line itself, leading to displays where trace power is the primary visual impact of the data.

There are other uses for color in plots. Trace lines may be colored. Different trace lines on a multi-trace plot may be drawn in different colors as well. The COLOR command controls these features with similar options as LINE. The basic color palette contains the colors WHITE, RED, GREEN, BLUE, YELLOW, CYAN, MAGENTA and BLACK, any of which can be selected for a line color or used in a color list. REPORT COLOR lists the color options, and Section 10.4 shows how to define your own colors.

Traces are decimated when plotted to reduce display latency (Fig. 4.1). The speed of modern computers makes a nonsense of this feature for screen displays, but it still is relevant when operating SAC remotely (as with a remote desktop on a network-connected machine) or when preserving plots (SGF files) to reduce data storage. The QDP command controls the decimation extent.

Cutting data

It is often desirable to cut information from the beginning and ending of traces to limit the data to a common time interval and to set a common time base among the traces. The CUT command enforces that only a subset of the file is read into memory. For example,

```
SAC> cut 1000 1500; r ACKN.BH?
```

would restrict the time interval from 1000 to 1500 seconds in the data. Only data in that time window is read into memory. This means that if the traces are subsequently written out, only those data will be written. So be careful – if you overwrite your original files, the data outside of the time window will be lost.

CUT and file header rewriting (the WRITEHDR command) may also interact unfavorably. SAC prevents making permanent file header changes when the header information and the data disagree. To avoid problems relating to CUT, a typical processing sequence might resemble the following:

```
SAC> r ACKN.BH?
SAC> cut 1000 1500; r          ;* Repeat previous read and cut
SAC> w prepend SHORT_
SAC> cut off; r SHORT_ACKN.BH?
```

A cut, once set, applies to all subsequent reads. Cut windows (called partial data windows by SAC) set for one file might not be appropriate for all files if, for example, a subsequent file does not contain data at the time of one of the cut limits. The CUTERR command controls what SAC should do if such errors arise. One option is to fill the missing data requested with zero. This is useful for padding data to constant length when read. It also can be handy when cutting or padding data in memory that was already read from a file when used with the CUTIM command.

SAC provides quite flexible ways to express time windows. They may be specified by times in the file or by offsets from significant markers in it. The markers can be the time of the first (B) or last (E) data point or of any pick (Section 4.5). The limit may also include an offset from the marker in seconds or number of points. Thus XLIM B N 20 represents the first 20 data points of the file, and CUT B +1 E -1 represents all but the first and last second of the file. See HELP CUT for full details. Similarly specified time windows also govern the behavior of the MTW, RGLITCHES, RMS and WIENER commands.

Permanent plots

The only use of the SGF device is for making permanent plots. When SGF is among the active devices, SAC saves a description of each plot in a file. The file, processed with one of SAC's utility programs (see HELP UTILITIES), may be converted into a form for printing or inclusion into a document for publication. The most widely usable output forms are PostScript (PS) or Encapsulated PostScript (EPS) files, though Portable Document Format (PDF) files may also be useful for document preparation.

To produce a permanent plot, activate the SGF device:

```
SAC> bd more sgf        ;* Add SGF device to active device set
SAC> p1                 ;* Make plot
SAC> bd previous        ;* Restore to previous active devices
```

Note that using the more or previous option with the BEGINDEVICES command saves or restores, respectively, the active devices.

After making the plot, SAC puts a file in the current working directory called `f001.sgf`. Subsequent plots to the SGF device will produce file names with increasing numbers in sequence. It is possible to change where SGF files are stored by choosing a different file name prefix (here it is simply `f`) and to change the sequence number used as well. See the `SGF PREFIX` and `SGF NUMBER` command for further details. The default values reset when SAC is rerun.

To translate the SGF files to PostScript, use the `sgftops` command (this is a UNIX command, note):

```
sgftops f001.sgf myplot.ps
```

Similarly, use `sgftoeps` to convert to EPS, and use `sgftopdf` to convert to PDF.

4.5 PICKING

Seismologists commonly use travel-time picks to mark events in a time series for further analysis. This might be the arrival time of a seismic wave, the peak amplitude in a waveform, or the begin and end time of a time window containing a signal of interest. SAC provides ways to make and save time picks by putting them into the SAC file header. The best way to make a pick is graphically, by pointing at the trace with a mouse and clicking to make the pick. The command provided for this is `PLOTPK`. Its function is to produce a plot of the trace and to start a graphical interaction with the user. SAC shows the trace with the current `XLIM` and `YLIM` options in force and displays a crosshair tied to the mouse. A keyboard character defines a pick at the current crosshair position or instructs SAC to magnify the trace (or collapse it back to its prior magnification). The most basic use of `PLOTPK` is to type a character on the keyboard to save the current crosshair position. The character identifies the name of the pick to associate with the time in the trace. The most commonly used pick names are `A`, `F` and `T0` ... `T9`. See Table 4.1 for a complete list of pick names and `PLOTPK` functions.

The most commonly used keyboard responses are:

- q - quit;
- x - pick first or second X (time) limit for a zoom window;
- o - unzoom to previous window;
- a - set `A` pick value;
- f - set `F` pick value.

Picks are stored in the file headers *in memory* and must be written out to be permanently associated with the files. By default, a pick is only associated with the trace in the crosshairs. To associate a time pick with all visible traces, use the `PLOTPK MARKALL` option. As with multiple plots, the `PLOTPK PERPLOT` option may be used to restrict the number of traces simultaneously visible to prevent excessive vertical shrinkage.

4.6 THE FILE HEADER

The file header stores metadata about the trace. These relate to:

- the trace itself (number of points `NPTS`, sample interval `DELTA`);
- the reference time of the trace (year `NZYEAR`, Julian day `NZJDAY`, ...) and the trace's relation to it (begin time `B`, end time `E`);

Table 4.1 PLOTPK cursor options

Char.	Description
K	Finish PPK
Q	Finish PPK
N	Go to next trace of multiple trace display
B	Go to previous trace of multiple trace display
X	Define bound of new time window (left first, then right)
O	Return to last time window (memory of the five most recent windows is kept)
L	List time and amplitude of current cursor position
A	Define A pick
Tn	Define Tn pick (n is a digit 0–9)
D	Define first motion DOWN
U	Define first motion UP
I	Define first motion quality impulsive
E	Define first motion quality emergent
+	Define first motion slightly UP
-	Define first motion slightly DOWN
	(Blank) define first motion unknown
@	Cancel pick information accumulated so far
P	Define P-wave arrival time
S	Define S-wave arrival time
F	Define coda length
J	Define noise level
Z	Define zero level
C	Characterize pick quality of first arrival: [I|E][U|+|-|D][0-3] uses C1, C6 from APK (C1 is weight for previous difference, C6 is weight for previous mean absolute value); set background level to 1.6 times mean absolute value amplitude. Motion sense determined if excursion is $> 4\times$ background level or if absolute value of derivative changes in a consistent sense for > 3 points. Pick quality determined by height of first three peaks relative to the background level (P1, P2, P3 here) and derivative relative to background (here K) at pick (P or S): quality 3 if $K \leq 1/2$; quality 2 if $K > 1/2$ and $P1>2$; quality 1 if $K > 1/2$ and $P1 > 3$ and ($P2 > 3$ or $P3 > 3$); quality 0 if $P1 > 4$ and ($P2 > 6$ or $P3 > 6$)
G	Write pick information into HYPO file
H	Write pick information into HYPO file
W	Compute waveform (sets type W)
V	Compute waveform (sets type WAWF)
M	Compute waveform (sets type MWF) uses first five peaks after pick to determine an average decay rate per unit time and extrapolates it to zero to determine end of waveform

- the event (origin offset O, event depth in km EVDP, latitude EVLA, longitude EVLO, event ID KEVNM);
- the station (latitude STLA, longitude STLO, elevation STEL, name KSTNM, sensor specifier KHOLE);
- the ray path (azimuth AZ, back-azimuth BAZ, range in degrees GCARC, distance in km DIST).

Header information is either set manually or is automatically computed based on other header information (for example, AZ, BAZ and GCARC are calculated whenever STLA, STLO, EVLA, EVLO, are available or change, under control of LCALDA). Header items can have an UNDEF value, which represents the unset state. See HELP HEADER for a full list of header items.

Time representation

SAC references all times in a file relative to the zero time in the file, which is represented to the precision of a millisecond. Every other time in the file is expressed as an offset, in seconds, from the file's zero time (Fig. 4.3). This provides a very simple way to ensure time consistency between picks in a trace. However, the times are stored as single precision IEEE floating point numbers, which have only about six decimal significant figures, though their magnitude range may vary over $10^{\pm 38}$. Consequently, time precision degrades with distance from the file zero time. This is not a noticeable problem unless picks are made in very long data files. For example, times more than 1000 s from file zero only have a precision of ± 1 s. (SAC may change its internal way of storing pick information in the near future to remedy this.)

Listing

The command LISTHDR (abbreviated LH) lists header information for the files currently in memory. By default, this lists all non-UNDEF (defined) entries for all files. The list can be

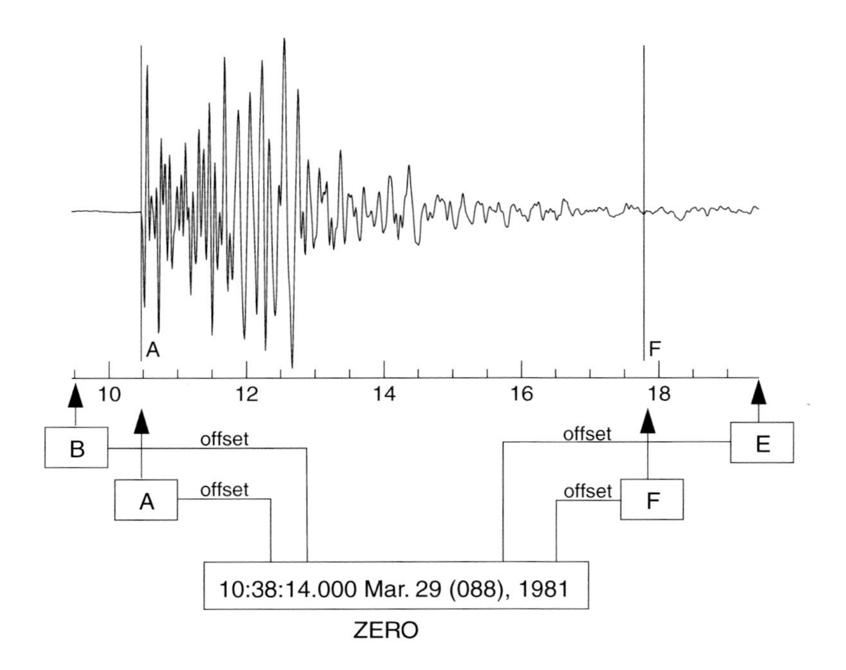

Figure 4.3 All times in a file are relative references to a time (ZERO) that has millisecond resolution. All other times (event origin time, picks (A, F), data begin B, and end E) are relative to that reference. Each time is stored in the file as an offset, in seconds, from the zero time of the file.

tailored by naming specific header variables. For example, the file zero date and time may be listed by

```
SAC> funcgen seismogram
SAC> lh kzdate kztime

  FILE: SEISMOGR
  --------------
    kzdate = MAR 29 (088), 1981
    kztime = 10:38:14.000
```

Changing

To change header information, use CHNHDR (CH).

```
SAC> help ch

 SUMMARY:
 Changes the values of selected header fields.

 SYNTAX:
 CHNHDR  [FILE <n> ...] [ALLT <v>] <field> {<value>|UNDEF} ...
```

By default, the information for all files is changed, but it may be restricted to specific files given by their order in memory. The header indicated by *<field>* is changed to *<value>*. For example, to set event information, use

```
SAC> ch evla 35.2 evlo 135.5 evdp 420.0
```

Values representing times (picks, event origin time) are values in seconds as an offset from the file ZERO time. Hence when changing a time, a real number in seconds is usually used. However, when setting an event origin time, the arithmetic is complicated by day, month and year boundaries that may appear in the trace. Time values may be specified as either GMT or as YMD times, and the arithmetic to determine the correct offset will be done automatically. For YMD times, the time is specified by seven integers: year, month, day, hour, minute, second and millisecond. For GMT times, the time is specified by six integers: year, Julian day, hour, minute, second and millisecond. (The Julian day is the day number in the year (1-365 or 1-366) and may be determined by using the -j option to the UNIX *cal* command.) Some examples might be

```
SAC> chnhdr O -45
SAC> chnhdr O YMD 2006 3 16 13 45 28 600
SAC> chnhdr O GMT 2006 075 13 45 28 600   ;* Same time as previous
```

All the times in a file header can be changed if needed for a new reference. The ALLT option does this to a specified time shift

```
SAC> chnhdr allt 415.045
```

The SAC command SYNCHRONIZE also does this, but only to the latest file start time in a group of files.

Writing

When header values change, only the copy in memory changes. The file version does not change unless the file is rewritten with WRITE or the updated header information is written with WRITEHDR (abbreviated WH). For example, the following commands change the file origin time and write the updated header information to the file without changing the trace data:

```
SAC> chnhdr O GMT 2006 075 13 45 28 600; wh
```

The following example sets station and event information into the header for a file and updates the header information in the file:

```
SAC> r ACKN.BHZ
SAC> chnhdr stla 64.9915 stlo -110.871
SAC> chnhdr evla 36.520 evlo 70.84 evep 183.0
SAC> chnhdr O GMT 2004 096 21 24 04 000; wh
SAC> lh O

   FILE: ACKN.BHZ
   --------------
     o = -415.0

SAC> chnhdr allt 415; wh ;* All times now relative to event origin
```

4.7 TRACE PREPARATION AND RESAMPLING

Seismic data is rarely recorded in a directly analyzable form. Instrument problems, long-period noise, temperature variations and electronic interference all imprint seismic data – especially data recorded during field deployments – with unwanted features. Data might also be recorded at different sample rates at different stations in a network, or at unnecessarily high rates that generate unprocessable amounts of data. Finally *all* data contains noise, which masks, to some extent, the signal of interest. This section is an overview of the trace preparation steps that remove many of these annoyances before data analysis. In essence, the operations remove information from the traces that lie outside the frequency range of the signal being sought.

De-glitching

SAC's RGLITCHES command removes spikes from data. Spikes are single samples whose values exceed the local data mean by many orders of magnitude and are usually caused by electronic noise in the recording circuitry (Fig. 4.4). Glitches are identified when data exceeds a particular absolute threshold value or differs from a running mean by more than a threshold value, or are discovered by a method that finds that a particular single sample difference is larger than the average intersample difference over a fixed time window. Glitches usually set the signal to a constant level, so a threshold method usually works best to remove the glitch. A specific time window may be set to restrict glitch removal to an interval where one is visible.

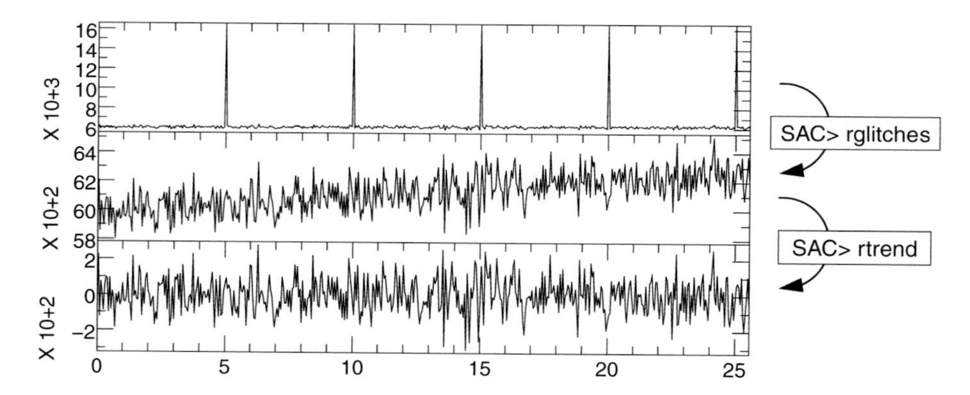

Figure 4.4 The plot made of three seismograms shows traces in successive stages of data cleanup prior to processing. (*top*) The raw data contains glitches every 5 s where the signal reaches very large values due to electronic problems. Operating on the data with the RGLITCHES command removes them. (*middle*) The repaired data contains a slope presumably due to instrument instability and has a nonzero mean. (*bottom*) After RTREND removes the trend and mean of the data, the resulting trace is centered around zero.

Mean and trend removal

Seismic data frequently displays long-period drift due to diurnal changes in ground conditions. Long-period drift can also arise when instruments are not in observatory conditions where the vault temperature is not stabilized. Expansion of the metallic parts of the instrument's casings can cause tilt, leading to a long-period signal. Thus the signal mean changes with time and can be significant during long-term drift (Fig. 4.4). Seismic signals often contain long-term trends for similar reasons.

To deal with these, SAC provides RMEAN, which simply removes the mean, and RTREND, which removes the mean and the trend and reports the slope and intercept of the best-fitting line of the data. If the TERSE option is invoked, the command sets blackboard variables (see Section 5.6) that may be used to retrieve these quantities for further processing.

Resampling

A typical data processing step is to standardize the sample rate to a value that is sensible for the intended use. There are many reasons to want to do this. File bloat reduction is one consideration, but compatibility with software-imposed sample rate restrictions in a particular analysis package is another.

Interpolation is used to attain higher sample rates and is a straightforward and easy-to-understand process. The method that SAC uses (from Wiggins, 1976) ensures continuous derivatives through the interpolated data values. The command syntax is

```
SAC> help interpolate

 SUMMARY:
 Interpolates evenly or unevenly spaced data to a new
 sampling rate.
```

```
SYNTAX:
INTERPOLATE [DELTA <d>] [EPSILON <v>] [BEGIN {<b>|OFF}]
            [NPTS {<n>|OFF}]
```

which shows that the new (higher) sample rate may be chosen through a lower INTERPOLATE DELTA value. INTERPOLATE is the command to use to convert unevenly spaced files (X-Y data pairs) into evenly spaced files, upon which spectral analysis may be done.

Obtaining lower sample rates requires more care due to aliasing of high-frequency values to lower. For example, suppose a sine wave with a frequency of 20 Hz and sampled at 40 Hz is downsampled to 1 Hz simply by using the value of every 40th sample. If the first sample chosen corresponds to a peak of the sine wave, every succeeding lower-rate sample will also be at a peak, and the trace will apparently have a nonzero mean when, on average, the sine wave mean is zero. To prevent the higher-frequency information aliasing down to lower frequencies, filtering must simultaneously be done. SAC implements such a strategy by providing finite impulse response filters to filter and simultaneously downsample through the DECIMATE command. The set of decimation factors is small (2-7), so large downsampling rates must be arranged by repeating the command. For example, 100 samples per second data could be reduced to one sample per second by decimating by 5, 5, 2 and then 2.

```
SAC> funcgen random 1 npts 1001 delta 0.01
SAC> lh delta npts b e


  FILE: RANDOM01
  --------------
    delta = 0.10E-01
     npts = 1002
        b = 0.0
        e = 10.010
SAC> decimate 5; decimate 5; decimate 2; decimate 2
SAC> lh delta npts b e


  FILE: RANDOM01
  --------------
    delta = 1.0
     npts = 11
        b = 0.0
        e = 10.0
```

4.8 ROTATION

Sometimes, seismograms are not recorded in the orientation most convenient to a particular processing method, by either accident or design. Three-component sensors – in theory, at least – record the full vector motion of the ground. Motion information recorded on any three orthogonal component axes can be rotated to form any other axis orientation.

For ease of installation, seismic instruments usually have their component axes oriented up-down (Z), north-south (N) and east-west (E). Seismologically, due to the decoupling of the SH and P-SV wavefields in spherically symmetric, layered media, the radial (R), tangential (T) and vertical (Z) are more useful component orientations. R and T depend on the source's position relative to the instrument. The radial direction is parallel to the ground and directed along the great circle joining the source and the instrument receiving the signal from the source. The positive direction is from the source toward the receiver. Z is radially upward and T is perpendicular to the two.

SAC's `ROTATE` command does component rotation only on a pair of components. The trace pairs must have the same length and have the same sample rate. If the components are horizontal (as indicated by `CMPINC` = 90 in their header information) they can be rotated to `GCARC` (or, synonymously, `GCP`) to isolate the SH component on the T component. If the rotation is `NORMAL`, the three-component set (with Z) forms a left-handed coordinate system (Z up, R away from the source, T left facing along positive R), whereas if the rotation is `REVERSED` the three-component set forms a right-handed coordinate system (Z up, R away from the source, T to the right facing positive R). Header information for the station (`STLA`, `STLO`) and the event (`EVLA`, `EVLO`) must also be present in order to define the great circle path geometry.

Alternatively, horizontal components may be rotated through a specified angle. The `ROTATE THROUGH` command does a clockwise rotation through a specified angle in degrees. This type of rotation also can be done with one horizontal component and one vertical component (as indicated by `CMPINC` = 0 in the file header).

Three-dimensional rotations may be performed by chaining together successive `ROTATE` commands.

4.9 FREQUENCY-DOMAIN OPERATIONS AND FILTERING

A fundamental representation of a time series is a reduction to component frequencies that are summed together with appropriate amplitudes and phase lags to reconstruct it. The frequency coefficients constitute the time series' spectral amplitude or, more simply, its spectrum. Conversion from the time domain to the frequency domain is exact and is done using the digital Fourier transform (DFT).

SAC's `FFT` command does the conversion to the frequency domain and `IFFT` does the conversion back to the time domain. The frequency-domain representation of the signal may be examined graphically using the `PLOTSP` command, which plots the spectral amplitude, phase, or both. Tapering might be necessary to reduce spectral artifacts by applying a window to the data. The `TAPER` command provides this facility.

`PLOTSP`, by default, plots the amplitude spectrum and then the phase spectrum as separate plots much like the `PLOT` command does. Usually, the spectral amplitude is more diagnostic for analysis, and with the `PLOTSP AM` command, only the amplitude spectrum will be shown. For example,

```
SAC> funcgen seismogram
SAC> fft
SAC> plotsp am linlin
```

produces the (single) plot shown in Figure 4.5 on a linear scale for both the amplitude and frequency axes. Log-linear or log-log plots may be requested to show particular spectral trends.

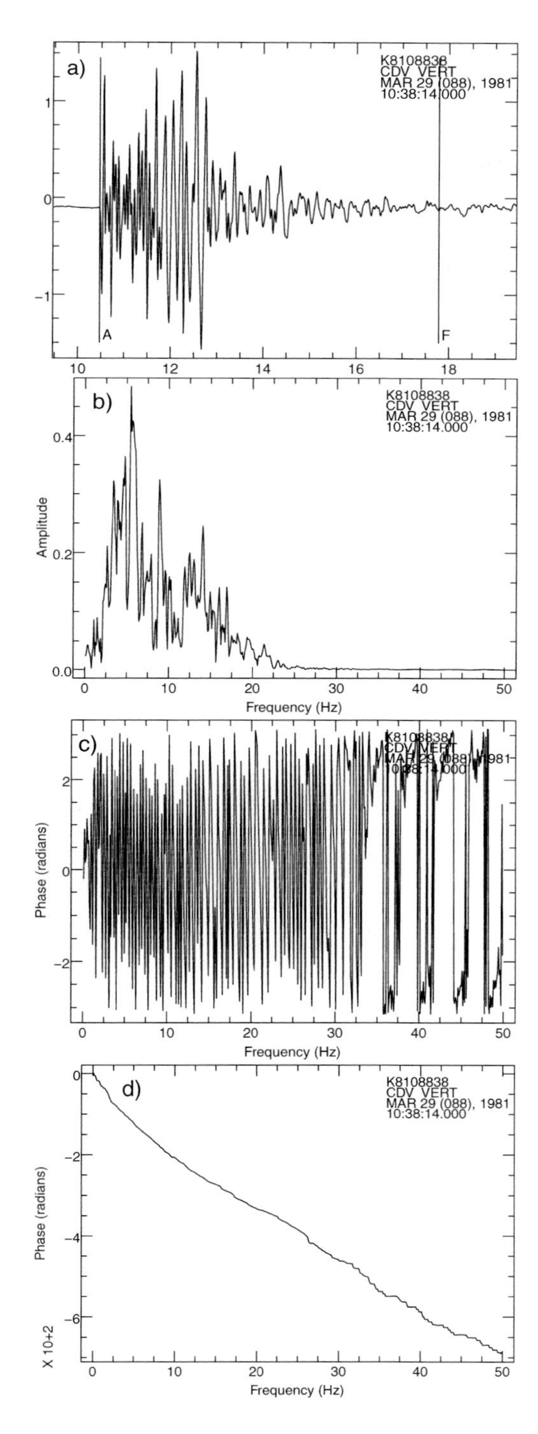

Figure 4.5 A time series (*a*), when transformed to the frequency domain by a fast Fourier transform, is represented by its amplitude spectrum (*b*) and phase spectrum (*c*,*d*). The amplitude spectrum, shown here on a linear scale, shows power peaked at 6 Hz. The raw phase spectrum (*c*) is not particularly useful. However, the unwrapped phase spectrum (*d*) shows a generally decreasing phase delay indicative of a causal signal.

The phase spectrum is usually uninterpretable viewed graphically because it varies rapidly between its limits of $\pm\pi$. If the phase is unwrapped in an unambiguous way it has more graphical significance. Thus SAC provides a combined fast Fourier transform (FFT) and phase unwrapping function through the `UNWRAP` command. This uses a technique to solve for the phase ambiguity to yield a continuous phase that is more helpful when assessing signal characteristics.

Once in its spectral representation, the trace may be saved and read using the normal SAC `READ` and `WRITE` commands. However, the spectral representation in the file is not the same as the time series representation. Often one wants to treat the spectral amplitude or unwrapped phase as a time series (in frequency) in its own right for further analysis (e.g., cepstral analysis, departure from linear phase). While this is not possible on the spectral representation, the spectra may be written as a time series and then re-read for further processing. The `WRITESP` command is useful for this purpose. The files that it writes (either the spectral amplitude, phase, or both) are formatted just as time series data, and if read as such by SAC are analyzable using SAC's command repertoire. To combine the spectra-as-time-series representations back into a true spectral representation, use the `READSP` command.

Filtering

If, using spectral analysis or some model of the expected signal, we can determine which frequencies present in our seismogram represent noise and which represent signal, we can filter out unwanted frequencies to improve our observation of the signal of interest. Filtering covers a very broad range of theory and methodology and cannot be covered even in an introductory form here. A recommended reference is Hamming (1989).

SAC provides filters to remove noise at low frequency, high frequency, in an intermediate band of frequencies or simultaneously at low and high frequencies. These are respectively called *highpass*, *lowpass*, *bandreject* and *bandpass* filters. Each of these has a particular shape in the frequency domain and may be visualized as a function that goes from one in some frequency range (the pass band) to zero in another frequency range (the stop band). The transition from the pass band to the stop band, conventionally chosen as the "3 db point," is where the signal drops to $10^{-3/20}$ of its value in the pass band (about 70% of the pass band value). This is usually referred to as the filter's "corner." Thus lowpass and highpass filters have a single corner frequency, while bandpass and bandreject filters require specification of two.

Another filter characteristic is the shape of the filter in the frequency domain. Filters that drop abruptly at the corner frequency sharply divide the signal portion from the noise. However, they introduce time-domain artifacts into the trace, distorting the signal. Thus smooth filter transitions are preferred to preserve signal fidelity. SAC provides four types of filters of increasingly aggressive shapes to separate signal from noise. They are Butterworth (`BUTTER`), `BESSEL` and two Chebyshev types, `C1` and `C2`.

There are further controls upon the steepness of the pass band to stop band transition for each shape. For mathematical reasons, this is called the number of poles that the filter has. The more poles, the steeper the transition and the more artifacts introduced in the time series after filtering.

The final filtering consideration is preserving the causality of the filtered signal. Increasingly aggressive filters introduce increasingly strong shifts in the phase spectrum that change the shape of a time-domain pulse to broaden it. This shifts peak positions to times later than the peak in the corresponding unfiltered trace. To preserve peak positions, which is important for some analyses (e.g., arrival time differences), the filter is passed twice over the trace, once forward and the second time backward. The price paid is that some high-frequency energy in a pulse appears to arrive earlier than the pulse itself, destroying causality.

Thus, the filter commands that SAC provides have many options to control the features of filters. LOWPASS and HIGHPASS filters have a single corner frequency but four choices of filter type: BUTTER, BESSEL, C1 and C2. There is a choice for the number of poles, NPOLES, and a choice for the number of passes, PASSES. Finally, for the Chebyshev filters, there are further transition bandwidths (TRANBW) and attenuation (ATTEN) values to supply. The help information for BANDPASS contains a good demonstration that explains the Chebyshev controls. Owing to its simplicity, the Butterworth filter is the default filter type.

To pass through or filter out frequencies in a particular band, SAC provides the BANDPASS and BANDREJ commands. As one would expect, these require two corner frequencies to define the frequency band. Otherwise, the commands recognize the same options as do LOWPASS and HIGHPASS.

Designing filtering strategies

The compromises involved with filtering also show that a filtering strategy must be carefully thought through before applying it to data. Experimenting is extremely important, but SAC also provides help in visualizing the consequences of a filter choice through the FILTERDESIGN command, whose abbreviation is FD. The following command produces a graphical display of the Butterworth bandpass filter

```
SAC> fd bandpass corners 5 25 npoles 2 passes 2 delta 0.01
```

The resulting display (Fig. 4.6) shows the frequency-domain visualization, the phase and group delays, and a time-domain plot of the filter acting on a delta function.

Filtering decisions are complex and the consequences of a poor filtering choice might not be apparent until late in a data analysis procedure. Consequently, you should be prepared to re-create your filtered data from the raw unfiltered seismograms if you discover problems later. To make this as painless as possible, good practice is to keep the raw data in a separate directory from the analysis directory. Then filter the data and move the result to the analysis directory, preserving the raw data. SAC's macro features, described in Chapter 5, help with this process. Write a macro that takes a trace or a single group of traces from the raw data directory, filters it, and then writes the filtered data into the analysis directory. If the filter choice proves to be problematic, go back to the macro, change the filter parameters, and then re-process the data in the raw data directory. When it comes time to document and publish the results of the analysis, the macro also serves to document the precise filtering parameters used (instead of relying on memory or a lab notebook, either of which might go tragically missing).

4.10 SAC STARTUP FILES

SAC's command set has many options that define functionality. Recall that SAC remembers the previous set of options for each command and re-uses those options when the command

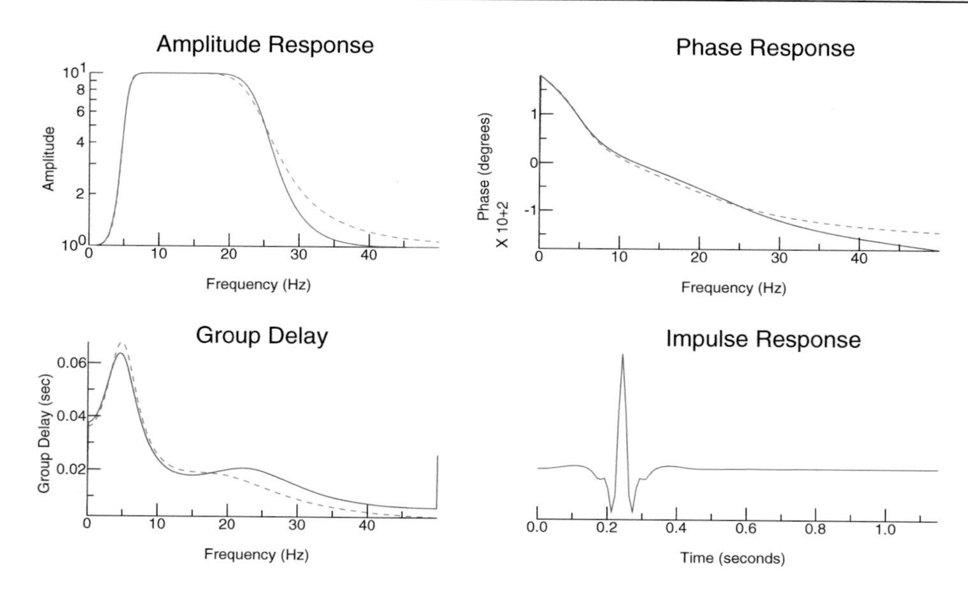

Figure 4.6 `FILTERDESIGN` output for a Butterworth bandpass filter with corners at 5 and 25 Hz acting on 100 Hz data. The filter is a relatively gentle two-pole filter and preserves peak position by being two-pass. However, the two passes spread the energy in the spike (bottom right) to both before and after the peak, whereas the original delta function signal was concentrated at the peak. Also note the filtered trace is negative, whereas the delta function is non-negative. The dashed line shows the analog filter prototype approximated by the digital filter (solid line).

is used the next time, unless changed. This suggests a mechanism by which a different, personally tailored set of default command options may be set up.

SAC reads commands from the standard input, defined in the UNIX system. This is usually the terminal keyboard, but, with input redirection, it can be a file. When invoked, any file name on the command line will be read by SAC before it reads from the standard input. In this way a file can be designated as a startup file to prepare your personal SAC environment. Typical commands to appear in the startup file are commands to define window sizes and locations (see `WINDOW`), the default graphics device (see `SETDEVICE`), the default macro search path (see `SETMACRO`), transcript location (see `TRANSCRIPT`), text size and style (see `TSIZE` and `GTEXT`), trace plotting resolution (see `QDP`), trace line style and color (see `LINE` and `COLOR`), and file header listing options (using the otherwise strange-seeming `LISTHDR FILES NONE` option).

A simple way to invoke SAC and direct it to read your personal startup file is to set up a shell alias to start SAC and have it read the file. If you use a Bourne-type shell (*sh* or *bash*), a way to achieve this is to put the command

```
alias sac='sac ~/.sacinit'
```

in your `~/.bashrc` or `~/.profile` file, which *sh* reads at startup. With this definition, when you type `sac`, SAC will read your commands after reading the ones in your initialization file `~/.sacinit`. A more flexible aliasing mechanism is to define a shell function as

```
func sac() { `which sac` ~/.sacinit $* ; }
```

which will let you add further files to the initial processing sequence (if you want) before reading commands from the standard input.

If you use a C-shell family shell (*csh* or *tcsh*), this will work when added to the shell startup file (`~/.cshrc` or `~/.tcshrc`):

```
alias sac '`which sac` ~/.sacinit \!1*'
```

4.11 SAC UTILITY PROGRAMS

SAC's utilities are auxiliary programs that do useful things with SAC time series files and SAC graphics files in the UNIX environment outside of SAC proper. They are distributed in binary form along with SAC and are installed along with it in a standard installation. See `HELP UTILITIES` for a full list of utility programs; a few of the most useful ones will be highlighted here.

When processing SAC data files during the organization phase of a data analysis project (see Chapter 3), a shell script is often used. Frequently, processing in the script depends on properties of a SAC data file. For example, the name given to a file might be taken in part from the event name in the KEVNM header variable. The script needs access to that name, so obtaining it from the file itself would be useful.

This is the purpose of the `sachdrinfo` utility: `sachdrinfo` operates on a file (or list of file names read from the standard input) and provides, on the standard output, the values for one or more file header variables provided as command line arguments. Thus it is designed as a UNIX filter for integration with other filters in a pipe. The shell provides innumerable mechanisms to retrieve and process values in a pipe, making the utility a natural tool for use by the shell.

A research project culminates with a presentation, a publication or technical report on its results. Time series graphics might be one of the displays used. Therefore, some of SAC's utilities take a SAC graphics file (SGF) and turn it into a form useful for inclusion in the presentation. The utilities `sgftops`, `sgftoeps` and `sgftopdf` take an SGF file and transform it to each of the formats implied by the utility's name. Some commercially available software may be used to edit the result to add annotation or edit the content to improve clarity. Another of SAC's utilities, `sgftoxfig`, converts the SGF file into a form editable using the free software program *xfig*. An advantage of editing the file in this way is that the organization of the graphical elements in a plot is retained. For example, a text string that represents the title of a plot is a single, text string, possibly centered, left- or right-justified. This information is retained in the figure description, facilitating editing.

CHAPTER

FIVE

SAC macros

5.1 MACROS AND INVOKING THEM

SAC has a facility to encapsulate repeated or commonly used sequences of other SAC commands into a new command with its own name. This is somewhat like UNIX shell scripts or Windows .bat files. SAC calls these procedures "macros." In addition to being a convenient shorthand for repeating a series of commands, macros extend SAC's processing capabilities in ways that are not available by typing in commands. The novel feature that macros provide is a way to define local variables that may be accessed and changed and used to make decisions about which particular SAC commands are issued by the macro. This widens the SAC command language into a type of programming language.

The basic command to invoke a macro is the MACRO command (abbreviated M):

```
SAC> M TTSAC    ;* Invoke a macro in a file named TTSAC
```

The body of a SAC macro is stored in a text file. This is not a file containing word processed text in the form of Word's .doc or Rich Text .rtf. Instead, the file must be created with a program capable of writing text files (for example, BBEditLite, TextEdit, or the UNIX commands *vi, pico, nedit*). There is no particular naming convention required for SAC macros and they are not required to end with .txt or .m.

The lines inside a macro file are either SAC commands or a few macro commands that are only recognized in macros. The character "$" introduces expressions that use features of the macro language and flags those commands usable only in macros.

5.2 WRITING A SIMPLE MACRO

To write a macro called `Hello` containing the text `Hello, World!`, first enter the following text into a file named `Hello` in the current working directory.

Ex-5.1

```
MESSAGE "The macro is starting ..."
$RUN cat -
Hello, World!
$ENDRUN
MESSAGE "... and now it is finished."
```

The commands used here are a mix of SAC commands (MESSAGE writes a message to the user) and macro commands ($RUN and $ENDRUN run an operating system command using the text bracketed between them as input).

Next, invoke the macro in SAC using the MACRO command

```
SAC> m Hello
```

which results in

```
 The macro is starting ...
Hello, World!
  ... and now it is finished.

SAC>
```

5.3 TRACING MACRO OPERATIONS

In the previous example, notice that SAC did not show the macro commands it was performing, it just did them. This is useful, particularly if a macro contains a long series of commands, but makes it difficult to debug macros during development.

To see the commands inside a macro as each is performed, use the SAC command ECHO with either ON or OFF as an argument. When ON, SAC echoes each line as it is retrieved from the macro file. Each line is also echoed after macro variables (to be discussed later) are substituted into the text of the line.

The output from the same macro with ECHO ON would be

```
SAC> echo on                ;* Turn on macro line echoing

SAC> m Hello                ;* Invoke
 m Hello
 MESSAGE "The macro is starting ..."
 The macro is starting ...
 $RUN cat -
 Hello, World!
 $ENDRUN
```

```
Hello, World!
 MESSAGE "... and now it is finished."
 ... and now it is finished.
SAC>
```

This time, SAC writes each command as it is read from the macro (the lines shown in **bold** face) and then writes the results of executing each command.

Though not shown here, SAC echoes lines that contain macro variables with a line prefixed by ==>, after variable substitution.

The ECHO command may be restricted to any or all commands, lines read from macros, and lines after any macro processing. HELP ECHO provides more details.

5.4 SEARCHING FOR MACROS

SAC searches for files containing macros in particular places and in a particular order. Because macros are simply files, the macro name is a file name. SAC searches for a macro in sequence in the following places:

- the current directory (where SAC is running);
- directories specifically indicated by SETMACRO;
- the default, built-in macro directory.

SETMACRO is therefore a key command to direct SAC to a personal repository of macros. The following commands show two uses of SETMACRO to set up and then to augment the macro search directories:

```
SAC> SETMACRO /U/kit/sacmacro   ;* Set search path

SAC> SETMACRO MORE /tmp         ;* Add another path element

SAC> * Macro search path now is current directory,
SAC> *   /U/kit/sacmacro, /tmp, and default built-in macros.
```

Note that SETMACRO MORE appends a new directory to the macro search path rather than setting a new directory for macro search. The most common use of SETMACRO is in SAC startup files (Section 4.10) where the search path is set once and then used throughout the subsequent session with SAC.

5.5 DECISION MAKING IN MACROS

Some commands make sense only when encountered inside a macro. A command that tests a condition and then goes on to perform specific other SAC commands based on the outcome of the test is silly if you type the commands yourself: you already know the outcome of the test and can surely type the correct command! Inside a macro they make more sense because you can anticipate different work flows depending on the outcome of the test. For similar reasons, commands that loop back to repeat other commands also only make sense inside macros.

The basic conditional commands are IF / ENDIF and IF / ELSE / ENDIF. These test a condition and perform the commands in the macro bracketed by IF and ELSE or ELSE and ENDIF depending on the outcome of the test.

SAC also provides a facility to select one case from many (similar to C language's *switch* command, UNIX shell's *case* command, or Fortran's *if* command), using the construct IF / ELSEIF / ELSEIF / ELSE / ENDIF. One of the bracketed groups of commands is selected depending on the condition tested on the IF or ELSEIF command.

The SAC commands DO / ENDDO and WHILE / ENDDO provide the basic looping constructs. SAC repeatedly executes the group of bracketed commands until a condition that is tested at the beginning of each loop iteration is satisfied.

Before going into specifics about these commands, we need to introduce macro variables. These are used to build up expressions that are tested when deciding which SAC commands to run.

5.6 VARIABLES IN MACROS

SAC recognizes macro variables by special characters that delimit the variable name. The variable is recognized and then replaced with its value. After replacement, the command might be echoed (see ECHO), but then it is processed as a SAC command. Therefore, SAC variables can be used to generate part or all of a SAC command, including the name of the command.

Types and scope

Suppose there is a variable called X in a SAC macro. This may be one of three types of variable. Each has a different source for its value and is flagged by a different syntax.

- **Blackboard variables** (syntax %X%). The scope of a blackboard variable is global. A user sets a blackboard variable value by using the SETBB command. The variable exists until SAC stops running or the variable is removed using the UNSETBB command.
- **Macro variables** (syntax X). The scope of a macro variable is local to a macro. Once that macro is finished running, the variable disappears and its value is lost. Users do not set values for macro variables – they are set only by other macro commands.
- **File header variables** (syntax &n,X&). File header variables are specific to each SAC trace in memory. The syntax &n,X& refers to the value X in file number n (which is a number from 1 to the number of files in memory). Values are saved in the SAC file header (see CHNHDR) and (if the file header or data is written out) continue to exist the next time SAC is run and the file is re-read. File header variables are pre-defined; you cannot create new ones, only set or retrieve values of existing ones.

Case is unimportant in SAC macro variables. $XYZZY$ and $xyzzy$ refer to the same variable and yield the same value upon substitution. The trailing variable type flag ($, & or %) may be omitted, for convenience, if white space follows it.

Setting

Blackboard variables are the most commonly used variable in macro processing. The SETBB command sets the value (and defines the variable if it does not already exist) and the syntax

%...% is used to refer to variables once set. Blackboard variables are strings, but they can be interpreted as numbers when used in a suitable context. Here is an example of a string value given to a blackboard variable and used as a string and as a number in an expression:

```
SAC> SETBB x "hello"                ;* Set x to string "hello"

SAC> MESSAGE "I said %x% to you!" ;* value of x substituted

 I said hello to you!

SAC> * Now a numerical example
SAC> SETBB sum 0                    ;* Set sum to value zero
SAC> SETBB sum (%sum% + 1)          ;* increment sum
SAC> SETBB sum (%sum% + 1)

SAC> MESSAGE "sum is %sum%"         ;* value of sum substituted

 sum is 2.000000
```

The UNSETBB command deletes the definition of a macro variable and is used to tidy up after the processing in a macro is finished.

```
SAC> SETBB x "hello"                ;* Set x to char. string "hello"

SAC> MESSAGE "I said %x% to you!" ;* value of x substituted

 I said hello to you!

SAC> UNSETBB x                      ;* unset

SAC> MESSAGE "I said %x% to you!" ;* x forgotten now

 ERROR 1201: Could not find VARS variable blackboard X

SAC>* No longer insistent, it seems
```

5.7 EXPRESSIONS

Expressions appear in many contexts in SAC macros. One use already mentioned is to evaluate conditional macro commands. However, expressions can appear anywhere in a SAC command and may be used to create parts of it. Thus macro expressions can supply numerical values for many of SAC's commands (say, filter corner frequencies), the name of a file to be read, or even the name of the SAC command to run. The EVALUATE command will evaluate numerical expressions and save them to a blackboard variable. They may be used more widely to reduce reliance on temporary blackboard variables, as described below.

Syntax

A macro expression always appears inside parentheses (...). The expressions may be nested by nesting parentheses. Numerical expressions may be evaluated using the syntax, `( x + y )`, `( x - y )`, `( x * y )`, `( x / y )`, `( x ** y )`, where `x` and `y` are strings representing numerical values (** represents exponentiation). The blackboard variable examples encountered earlier showed how expressions are used to increment a blackboard variable.

Built-in functions

In addition to arithmetic operations, expressions may also be functions. The function name is the first item following the expression's opening parenthesis, and the arguments follow the name, separated by spaces. The numerical functions include square root, exponential, logarithms, etc., and the trigonometric functions. Additionally, there are miscellaneous functions that return numerical values. Table 5.1 lists these functions. Among them, *getval* is particularly useful. It will return the value of the time series in a trace at a particular time for all files in memory. If the value of only one file is wanted, say of the third file, that may be specified by using the alternative form `(GETVAL FILE 3 19.2)`.

SAC also provides a suite of functions to manipulate strings of characters. The functions to select substrings and replace text are particularly useful for transforming file names from one form into another. For example:

Ex-5.2

```
SAC> setbb str "ABC.XYZ"
SAC> message "Letters 2 and 3 of the string are (substring 2 3 %str%)"
 Letters 2 and 3 of the string are BC
SAC> message "The Jackson Five sang (change XYZ 123 %str%)"
 The Jackson Five sang ABC.123
SAC> message "Prefix is (before . %str%); suffix is (after . %str%)"
 Prefix is ABC; suffix is XYZ
SAC> message "Blank replacing the dot yields (delete . %str%)"
 Blank replacing the dot yields ABCXYZ
SAC> message "Substitution is (change . ' ' %str%)"
 Substitution is ABC XYZ
SAC> message "All in lower case: (changecase L %str%)"
 All in lower case: abc.xyz

SAC> setbb rts "ABC 123 XYZ"
SAC> message "There are (itemcount %rts%) items in the string"
 There are 3 items in the string
SAC> message "Item 2 is (item 2 %rts%)"
 Item 2 is 123
SAC> message "The last two are (items 2 3 %rts%)"
 The last two are 123 XYZ
SAC> message "Does str exist? (existbb str)"
 Does str exist? Y
SAC> message "  how about xyzzy? (existbb xyzzy)"
   how about xyzzy? N
```

Table 5.1 Built-in numerical functions usable in expressions

Command	# args[a]	Example	Description
PI	0	(PI)	Value of π
ADD	>0	(ADD 1 2)	
SUBTRACT	>0	(SUBTRACT 2 1)	
MULTIPLY	>0	(MULTIPLY 1 2 3 4)	
DIVIDE	>0	(DIVIDE 12 3)	
MINIMUM	>0	(MINIMUM 0.78 -3.5 15.2)	
MAXIMUM	>0	(MAXIMUM 0.78 -3.5 15.2)	
SQRT	1	(SQRT 2)	Square root
EXP	1	(EXP 2)	Powers of *e*
ALOG	1	(ALOG 2.5)	Natural logarithm
POWER	1	(POWER 3)	Powers of 10
ALOG10	1	(ALOG10 1000)	Base 10 logarithm
SINE	1	(SINE 0.37)	Argument in radians
ARCSINE	1	(ARCSINE 0.37)	Result in radians
ASINE	1	(ASINE 0.37)	Synonym
COSINE	1	(COSINE 0.37)	Argument in radians
ARCCOSINE	1	(ARCCOSINE 0.37)	Result in radians
ACOSINE	1	(ACOSINE 0.37)	Synonym
TANGENT	1	(TANGENT 0.37)	Argument in radians
ARCTANGENT	1	(ARCTANGENT 1.00)	Result in radians
ATANGENT	1	(ATANGENT 1.00)	Synonym
INTEGER	1	(INTEGER 1.05)	Integer part of expression
ABSOLUTE	1	(ABSOLUTE 1.05)	Absolute value
GETTIME	1,2	(GETTIME 17.05) (GETTIME MAX 17.05) (GETTIME MIN -7.05) (GETTIME MAX) (GETTIME MIN)	Time of first occurrence of value in file, or value greater than (MAX) or less than (MIN) in file, or DEPMAX or DEPMIN if no value provided
GETVAL	1,3	(GETVAL 157.32) (GETVAL FILE N 157.32)	Data values at 157.32 s in each file in memory Data value at 157.32 s in file N; N integer > 0

[a]Number of function arguments.

The *reply* function is designed for user interaction. A macro might prompt the user for a response, read it, and deliver the response for further use. In the example shown below, the response becomes part of a message:

```
SAC> setbb q "(reply 'Type in a file name: ')"
Type in a file name:  sesame
SAC> message "The file's name is %q%"
 The file's name is sesame
SAC>
```

In this example, *reply* is used to print out a prompting question ("Type in a file name:" shown in bold) that the user responds to. The response becomes the value of the *reply* function that is printed out in the message to the user.

A list of built-in character functions is given in Table 5.2. Whether blackboard or macro variables exist may be tested using the *existbb* and *existmv* functions. The function *existbb* was used earlier in one of the examples.

The function *hdrnum* is a subtle but useful built-in function used to select particular traces for processing based on values in their headers. For example, the file header has a field called KCMPNM that contains the component name. Suppose we want to find all files in memory that have the component name LE. We could do this with the following expression:

```
SAC> r /data/hinet/2005/vel/[EIHH,EDSH].L[E,N,R,T]
SAC> lh kcmpnm

  FILE: /data/hinet/2005/vel/EIHH.LE
  ----------------------------------
    kcmpnm = LE

  FILE: /data/hinet/2005/vel/EIHH.LN
  ----------------------------------
    kcmpnm = LN

  FILE: /data/hinet/2005/vel/EIHH.LR
  ----------------------------------


  FILE: /data/hinet/2005/vel/EIHH.LT
  ----------------------------------


  FILE: /data/hinet/2005/vel/EDSH.LE
  ----------------------------------
    kcmpnm = LE

  FILE: /data/hinet/2005/vel/EDSH.LN
  ----------------------------------
    kcmpnm = LN

  FILE: /data/hinet/2005/vel/EDSH.LR
  ----------------------------------


  FILE: /data/hinet/2005/vel/EDSH.LT
  ----------------------------------
SAC> setbb files "(hdrnum kcmpnm eq LE)"
SAC> message "There are (itemcount %files%) LE files."
 There are 2 LE files.
SAC> message "They are numbers %files% in memory"
 They are numbers 1 5 in memory
```

Table 5.2 Built-in character functions usable in expressions

Command	# args[a]	Example	Description
CHANGE	3	(CHANGE X Y Z)	Changes X to Y in Z
DELETE	2	(DELETE X Y)	Deletes X in Y
BEFORE	2	(BEFORE X Y)	Returns text in Y preceding X
AFTER	2	(AFTER X Y)	Returns text in Y following X
SUBSTRING	3	(SUBSTRING X Y Z)	Returns characters X to Y in Z. Index starts with 1; X or Y may be END to indicate last character
CONCATENATE	>0	(CONCATENATE X Y Z ...)	Joins all strings together
CHANGECASE	>1	(CHANGECASE D X ...)	Change case of X to upper or lower case depending on D being UPPER or LOWER
EXISTBB	1	(EXISTBB X)	Return Y or N if blackboard variable X exists or not
EXISTMV	1	(EXISTMV X)	Return Y or N if macro variable X exists or not
HDRVAL	>2	(HDRVAL H OP X ...)	Return values of file header variable H that are in relation OP to value X. Multiple OP X pairs are allowed; the logical AND of all conditions governs selection. OP may be LT, LE, EQ, GE, GT, NE, AZ or PM for numeric header variables, EQ or NE for character and symbolic header variables. Value and X compared. For logical header values, OP may be TRUE or FALSE with no X. Operators AZ and PM must always be paired, and are an azimuthal angle comparison: AZ x PM y means that H and x must differ in angle by no more than y
HDRNUM	>2	(HDRNUM H OP X ...)	As for HDRVAL, but returns number of file in memory, not the header value
ITEM	>1	(ITEM N Y Z ...)	Select item N from blank-delimited list of items Y Z ...; Y is item 1
ITEMS	>2	(ITEMS M N X Y Z ...)	Select items M through N from blank-delimited list of items X Y Z ...; X is item 1, and N may be END to indicate last item
ITEMCOUNT	>1	(ITEMCOUNT X Y Z ...)	Count of number of blank-delimited items in list
REPLY	1	(REPLY X)	Prints X and reads user reply
QUOTEBB	1	(QUOTEBB X)	Returns blackboard variable named X, preserving line breaks, for later use with WHILE READ command
QUOTEHV	2	(QUOTEHV N X)	Returns file header variable named X for file N. No macro or blackboard variable interpretation is done on the value
QUOTEMV	1	(QUOTEMV X)	Returns macro variable X. No file header or blackboard variable interpretation is done on the value

[a]Number of function arguments.

The files named .LR and .LT apparently do not have KCMPNM set in their headers. The SETBB command sets the value of the blackboard variable to the numbers of the files that contain a KCMPNM value of LE. The function *itemcount* prints the number of files by counting how many numbers are in the list. The message expands the value of `files`, showing the full list of file numbers.

The function *hdrval* returns the actual values in the files whose header matches, not the file number. While not so useful when testing whether a component is equal or not equal to a given value, *hdrval* is more useful when making a comparison to determine whether a particular file is within a range of values. For example, suppose we want to find the back-azimuths from the station to the earthquake that are $\pm 50^\circ$ of 10°. The following commands could be used:

Ex-5.3

```
SAC> datagen sub local *.z
SAC> setbb baz "(hdrval baz az 10 pm 50)"
SAC> setbb fnum "(hdrnum baz az 10 pm 50)"
SAC> message "Back-azimuth values %baz%"
 Back-azimuth values 53.949791 324.18115
SAC> message "Back-azimuth files %fnum%"
 Back-azimuth files 1 2
```

The result shows that two files, 1–2 in the group of nine read in, have back-azimuths in the proper range. Another example showing how range comparisons may be used with *hdrval* and *hdrnum* follows, this time with earthquake depths (the EVDP file header variable):

```
SAC> lh evdp

  FILE: /tmp/ex-1
  ---------------
    evdp = 10.0

  FILE: /tmp/ex-2
  ---------------
    evdp = 50.0

  FILE: /tmp/ex-3
  ---------------
    evdp = 250.0

  FILE: /tmp/ex-4
  ---------------
    evdp = 640.0
SAC> message "(hdrval evdp ge 200)"
 250. 640.
SAC> message "(hdrnum evdp ge 200)"
 3 4
```

Table 5.3 Status function

Command	# args[a]	Example	Description
STATUS	>1	(STATUS V)	Return SAC status item V; see below for list
STATUS items			
Name	Description		
GRAPHICS D	0 or 1 depending on whether any graphics device is open (if D is absent or ANY) or whether device D is open (D can be TERM, SGF, X, SUN, MAC)		
NFILES	Number of files currently in SAC memory		

[a]Number of function arguments.

```
SAC> * files 3 and 4 in list have evdp > 200
SAC> message "(hdrnum evdp gt 35 le 200)"
 2
SAC> * only file 2 has evdp > 35 and <= 200
```

The last group of built-in functions are those that query the internal state of SAC itself. These are useful for checking whether files are in memory or whether graphics devices are active in preparation for picking a time window in a seismogram. The *status* function provides this information. Table 5.3 lists its capabilities.

Escape character

In SAC, special characters flag the beginning of macro variables and expressions. To prevent these special characters from being interpreted as an expression, SAC provides the escape character `@`. Thus, if `@` appears before `(`, `)`, `%`, `$`, `&`, or even another `@`, the following character has its normal meaning. So, to put a % into a message, use:

```
SAC>  message "This is a 7@% solution NOT a bb var"

 This is a 7% solution NOT a bb var

SAC>  message "This is a 7% solution NOT a bb var"

 ERROR 1201: Could not find VARS variable blackboard

SAC>
```

The second message causes a parsing error because SAC expects a blackboard variable to follow the `%` character.

Evaluation order

SAC evaluates expressions in a strict sequence that may be used to embed expressions inside other expressions. Expression evaluation proceeds in three stages:

- expansion of blackboard variables %...% and macro variables $...$;
- expansion of file header variables;
- evaluation of functions.

This means that blackboard or macro variables can be inside file header variables:

```
SAC> setbb x 1                                  ;* set x to 1
SAC> message "File %x% evdp is &%x%,evdp&"     ;* %x% is inside &..&
 File 1 evdp is 10.0
SAC>
```

Expressions are also nestable:

```
SAC> setbb x 1                                  ;* set x to 1
SAC> message "sqrt (%x% + %x%) is (sqrt (%x% + %x%))"
 sqrt 2.000000 is 1.414214
SAC>
```

Macro variables may be nested inside blackboard variables, making it possible to implement blackboard variable arrays:

Ex-5.4

```
SAC> sc cat showme                     ;* Show the macro text
* This is the content of a macro called "showme"
setbb x1 "x1 value" x2 "x2 value"
do i list 1 2
   message "i is $i$ and x$i$ is '%x$i$%'"
enddo
SAC> macro showme                      ;* Invoke the macro
 i is 1 and x1 is 'x1 value'
 i is 2 and x2 is 'x2 value'
SAC>
```

The macro uses the `DO` command to set the macro variable `$I$` to the values of 1 and 2. For each value, the macro prints a message that contains the value of `I` and then the value of `%x1%` or `%x2%`, depending on `I`.

Conditions

The conditional commands `IF`, `ELSEIF` and `WHILE` evaluate a condition to decide which groups of SAC commands to execute. A condition is made of two expressions and a relational operator *<e1>* `OP` *<e2>*. The relational operator `OP` may be

- EQ (equal) or NE (not equal) – numerical or character expressions;
- LT (less than), LE (less than or equal), GT (greater than), GE (greater than or equal) – numerical expressions only.

(The operator names are similar to relational operators in the Fortran programming language.)

A few examples follow to show some guidelines when writing conditions.

```
if "%x%" EQ 'debug'    ;* quotes in case string contains blanks
   setbb debug 1
endif
```

Ex-5.5

```
* Below, reply is prefixed with _ to recognize no response
*    -- a blank line typed for the reply
setbb ans "_(reply 'Enter yes or no:')" ;* prompt and get answer
if "%ans%" eq "_yes"
   message "response is yes"
elseif "%ans%" eq "_no"
   message "response is no"
elseif "%ans%" eq "_"
   message "nothing typed"
else
   message "invalid response: (after _ '%ans%')"
endif
```

5.8 SUSPENSION, RESUMPTION AND ESCAPE FROM MACROS

Macros normally run until the commands in the macro file are exhausted, but a macro also may be explicitly stopped. The `$KILL` command provides a way to conditionally end a macro. For example, a (*reply* ...) expression might prompt the user to pick the beginning and the end of a waveform as follows:

```
setbb resp "(reply 'Hit return to continue or q to abort: ')"
if '_%resp%' EQ _q
   $kill
endif
```

Macros also may be temporarily suspended and resumed. While suspended, SAC returns to reading commands from the user's terminal. The user can issue any SAC command, including running new macros – these will not interfere with the macro that is suspended. This facility provides a type of co-routine structure between the macro and the command language.

The intent of the macro suspension facility is to allow errors in the processing of a macro to be corrected. For example, suppose the macro implements a procedure to rotate the components of a three-component seismogram from the Z, N and E coordinate system into the Z, R and T coordinate system. If there is no earthquake epicenter associated with the files, this rotation is impossible. The macro might detect this condition and ask the user to add event information before proceeding further. When the user adds the event information, the macro may be resumed.

The `$TERMINAL` command suspends a macro and the `$RESUME` command resumes a suspended macro. To show that there is a suspended macro running in the background, SAC changes its command prompt to show the level of suspension. The following example shows how macros are suspended and resumed:

Ex-5.6

```
SAC> sc cat susp
* This is the contents of the "susp" macro
message "Entering the susp macro ..."
$terminal
message "...continuing the susp macro"
message "Further processing would take place here"
SAC> m susp
 Entering the susp macro ...
SAC(macro001)> report xlim

    XLIM option is OFF
SAC(macro001)> $resume
 ...continuing the susp macro
 Further processing would take place here
SAC>
```

The `susp` macro started up and was then suspended via the `$TERMINAL` command. Notice that the prompt changed from `SAC>` to `SAC(macro001)>` while the macro was suspended. After processing the `$RESUME` command, SAC picked up where the macro left off.

This facility is useful for fixing errors. However, it requires the analyst to be highly knowledgeable about SAC and its commands. Users unfamiliar with SAC's capabilities probably will not be able to use SAC's command mode usefully to fix any problems. Therefore, expert users will probably find it most helpful.

5.9 OPERATING SYSTEM INTERACTION

SAC is not a stand-alone program but instead runs as an application program hosted by a UNIX-like operating system, such as a UNIX variant, a Linux system, or in the Cygwin environment under Windows. SAC is able to issue commands to the operating system that are the same as you would type them to the system itself. SAC provides `SYSTEMCOMMAND` (abbreviated `SC`) to do this. In its simplest form (more advanced uses will be discussed later), type `SC` followed by the operating system command:

```
SAC> sc date
Thu Oct 25 12:24:28 BST 2012
SAC> sc ls
dosect.make            picks-s2ks-grh         ppmap-make
fev2.stations          picks-s2ks-na          s2ks.picks-wrong
fev3.stations          picks-s2ks-new         s3ks.picks
fev4.stations          picks-s3ks-grh         s3ks.picks-orig
flt                    picks-s3ks-new         slant
SAC>
```

This runs the UNIX *date* command to provide the date and then the *ls* command to list the files in the current directory. The result in this case is typed on the screen.

A more powerful use of SYSTEMCOMMAND is to invoke a program that is part of a larger analysis scheme embedded in SAC. For example, a program could be written to do multi-taper spectral analysis of a time series in a SAC file. The output of the program might be a SAC file containing the amplitude spectrum. After using SYSTEMCOMMAND to run the program, the resulting output file may be read into SAC and shown to the user. Packaging the analysis procedure inside the macro provides a new type of SAC command that uses external programs.

Some external programs need to have input provided to them other than through command line arguments. This is a limitation of the SYSTEMCOMMAND mechanism, which only provides for a single command line to be sent to the operating system. Of course, a SAC macro could write lines into an external file and then run the program with the file connected to standard input, but as a convenience, SAC provides a second mechanism to provide input to an external command or program. The $RUN command (only usable in a macro) names an external program to run and indicates that subsequent macro lines up to a $ENDRUN command are to be connected to the standard input of the program. For example, the *awk* program might be invoked to process a set of file names:

Ex-5.7

```
SAC> systemcommand cat tryawk
* Contents of macro "tryawk"
$run awk '{print @$1}'
&1,filename&
&2,filename&
&3,filename&
$endrun
SAC> datagen sub local cdv.[e,n,z]
SAC> m tryawk
cdv.e
cdv.n
cdv.z
SAC>
```

In this case, the macro was first listed using the *cat* command. Then SAC ran the macro, which provided the file names of the first three files in memory. The *awk* command printed the first blank delimited part of the file name (in this case, the whole name).

$RUN is useful because if the input to the program (the lines up to $ENDRUN) includes macro variables, blackboard variables, file variables, or expressions, they are evaluated before being passed to the program. This provides a simple way to get information internal to SAC passed to the program for its use. Note that because SAC uses $ to delimit macro variables, it was escaped (using @) to pass it unchanged to the *awk* command. In some cases, this might be required for command input lines as well.

5.10 LOOPING COMMANDS

SAC provides two looping constructs inside macros: DO and WHILE. (A loop is a sequence of commands that is repeated over and over by returning to the beginning after the end is reached.) The difference between DO and WHILE is that DO repeats the loop a finite number of times, whereas WHILE repeats a loop for as many times as it takes for a condition associated with the command to become false. Both looping constructs bracket the repeated commands by ending them with ENDDO.

WHILE

The WHILE command is followed by a condition that is evaluated whenever the commands making up the loop (re)start. The next example uses a WHILE loop to prompt the user for a response until a valid response is given. This illustrates the potentially unlimited number of times the loop will be re-run, depending on the response of the user.

Ex-5.8

```
message "Fee fie foe fum" "I smell the blood of an Englishman"
message "What should I do now?"

* Ask for response and validate
setbb ok no                             ;* Sets value of bb var ok
while %ok% EQ no                        ;* Tests value of bb var ok
   setbb ans "_(reply 'respond kill or pause:')"   ;* Prompt, read
   if "%ans%" EQ _kill
      setbb ok yes
   elseif "%ans%" EQ _pause
      setbb ok yes
   else
      message "Invalid response"
   endif
enddo

* Act on response
if "%ans%" EQ _kill
   message "OK, quitting ..."           ;* Macro will terminate
   $KILL
endif
if "%ans%" EQ _pause
   message "Type @$RESUME to resume"    ;* Macro will be suspended
   $TERMINAL
   message "Resuming macro ..."         ;* Resumed at this point
endif
message "If he is alive or if he is dead"
message "I'll crush his bones to make my bread"
```

If the answer is “kill,” the macro stops running. If the answer is “pause,” the macro uses the suspend/resume facility to pause for later resumption. When resumed, the macro finishes.

WHILE READ

Another use of the `WHILE` command is for reading and parsing input from another program. `WHILE` can be used to read lines of input saved in a blackboard variable, presumably from output produced by an external program. To do this, use the `READ <bb>` verb following `WHILE` to indicate a blackboard variable to be read. A list of *macro variable* names follow that will be assigned blank-delimited words on each input line. The last variable will receive all of any words remaining, separated by blanks. (This is designed like the *sh* built-in *while read* command.) Note the distinction between blackboard variables (used as input) and macro variables (the output).

How to get input from another program will be introduced later (in Section 5.12 on advanced features of operating system interaction), but to see how looping is related to command input, suppose that one line of input was read and saved in the blackboard variable `inp`. The following `WHILE` loop will process it.

Ex-5.9

```
SAC> sc cat xample
* Contents of file "xample"
setbb inp "one two three and the rest"

while read inp a b c d     ;* bb variable read is %inp%
   message "a is $a$"      ;* note $a$-$d$ macro vars, not bb vars
   message "b is $b$"
   message "c is $c$"
   message "and d is '$d$'"
enddo

SAC> m xample
 a is one
 b is two
 c is three
 and d is 'and the rest'
SAC>
```

If there were more input lines in `%inp%`, the `WHILE` loop would repeat until they were exhausted.

Escaping from loops

No matter what the type of loop, SAC's `BREAK` command will jump out of the loop and resume processing using the command that follows `ENDDO`. When used inside a conditional statement, this provides a useful way to escape a loop before the full range of iteration possibilities are exhausted. Some uses of `BREAK` will be found in later examples.

DO

In contrast to WHILE, DO loops over a finite set of values. The values might be calculated from a range designated by a begin value, an end value and an increment (if different from one). Alternatively, DO can be given a list of explicit values to iterate over as blank-separated character strings. The strings might be literals or might be the result of a pattern match on files in the present directory.

The syntax for DO loop iterating over a numerical range is

```
DO <v> FROM <x> TO <y>
```

or

```
DO <v> FROM <x> TO <y> BY <z>
```

Here <v> is the name of a macro variable and <x>, <y> and <z> are numeric expressions (<z> is implicitly 1 if absent). The macro variable takes on the sequence of values starting with <x> and ending with <y>, running the commands up to ENDDO with each value assigned to <v>. The example below shows a macro that will calculate the first ten Fibonacci numbers using a DO loop.

Ex-5.10

```
SAC> sc cat dofrom
* Contents of macro file "dofrom"
setbb fib 1.0 prefib 0
do i from 0 to 10            ;* Loop macro variable value is i
   message "Fibonacci number $i$ is (BEFORE . %fib%)"
   setbb newfib (%fib% + %prefib%)     ;* Next Fibonacci num
   setbb prefib %fib% fib %newfib%     ;* Remember previous two
enddo

SAC> m dofrom
 Fibonacci number 0 is 1
 Fibonacci number 1 is 1
 Fibonacci number 2 is 2
 Fibonacci number 3 is 3
 Fibonacci number 4 is 5
 Fibonacci number 5 is 8
 Fibonacci number 6 is 13
 Fibonacci number 7 is 21
 Fibonacci number 8 is 34
 Fibonacci number 9 is 55
 Fibonacci number 10 is 89
SAC>
```

This macro illustrates a useful tactic to use when doing arithmetic with integers. SAC does all calculations as real numbers, even if the values are integers. Consequently, (*before* . %fib%) is used to strip the trailing zeros from the floating point numbers before the result (the MESSAGE statement) is printed.

Another example of a loop that finds a power of 2 larger than a number given by an input prompt is as follows. This macro uses a loop to make increasingly large powers of 2 and then escapes from the loop when the smallest power bigger than the desired number is reached.

Ex-5.11

```
SAC> sc cat dobreak
* Contents of macro "dobreak"
setbb num "(reply 'Enter number:')"
do pow from 1 to 32       ;* Try powers of two
   setbb v (2 ** $pow$)
   if %v% GE %num%        ;* Test present one
      break               ;* Equals or exceeds number
   endif
enddo
message "Power of 2 larger than %num% is 2**$pow$ or %v%"

SAC>  m dobreak
Enter number:2096
 Power of 2 larger than 2096 is 2**12 or 4096.000
SAC>
```

The syntax for DO loop iterating over a selection of character strings is

```
DO <v> LIST <a> <b> <c> ...
```

or

```
DO <v> WILD DIR <d> <a> <b> <c> ...
```

Here <v> is a macro variable name, and <a> ... are character strings that <v> takes on successively before processing the commands up to the following ENDDO. In the case of a file pattern match, SAC scans the files in directory <d> for matches to the patterns <a> ... and collects the results into successive values for <v>.

The following example illustrates how a macro may be used to check whether the header of a SAC file is sufficiently populated to allow processing to proceed.

Ex-5.12

```
SAC> sc cat dolist
* Contents of macro "dolist"
do hdr list stla stlo evla evlo evdp scale ;* Hdr fields to check
   if UNDEFINED EQ "&1,$hdr$&"             ;* Check if set
      message "$hdr$ not set in file header"
   endif
enddo
```

```
SAC> funcgen seismogram                    ;* Generic data
SAC> m dolist
 scale not set in file header
SAC>
```

This example uses a built-in seismogram SAC has for testing, produced by the FUNCGEN command. The macro checks that the file header variables STLA, STLO, EVLA, EVLO, EVDP and SCALE are all set. If not, the macro produces an error message.

5.11 MACRO PARAMETERS

Until now, we have treated macros as files that perform a specific task when invoked by SAC. If the task requires user input, it comes from an expression that uses (*reply* ...). This section introduces more powerful capabilities of macros that make them resemble SAC commands in their own right.

The new capability is provided by arguments that may be appended to SAC macros. The arguments follow the macro name on the SAC command line. For a macro called `example`, the following commands to invoke the macro provide two arguments: a file name and a frequency.

In the first instance, the arguments are positional: the first argument is the file name and the second is the frequency.

```
SAC> m example KEVO.BHZ 2.5
```

The second example identifies the arguments by keyword that introduces a value:

```
SAC> m example file KEVO.BHZ freq 2.5
SAC> m example freq 2.5 file KEVO.BHZ  ;* order independent
```

The two examples are equivalent providing that the macro is expecting keyword parameters (note that keyword parameters are independent of position on the command line).

Arguments provided to a macro might be required or they might be optional. For example, a macro might require a file name to operate on, but it might be happy to assume that a frequency, if not otherwise given, takes on a default value.

Positional

Positional parameters are the simplest to use in a macro, so they will be introduced first. Inside the body of a macro, the macro variable `$1$` refers to the first argument that follows the macro name, `$2$` refers to the second, and so on. Thus the macro `mppos`, defined below, retrieves the values provided when the macro is invoked as follows:

Ex-5.13

```
SAC> sc cat mppos
* Contents of macro "mppos"
* Macro user provides two parameters:  SAC file and frequency
message "File name is $1$; frequency is $2$"
```

```
read $1$                                  ;* Read SAC file
rmean; rtrend terse; taper w 0.05         ;* Prepare for filtering
lowpass corner $2$ npoles 2 passes 2      ;* Filter corner frequency

SAC> m mppos /tmp/ex-1 2                  ;* Macro used here
 File name is /tmp/ex-1; frequency is 2
SAC>
```

Examine what happens if the second positional parameter (the frequency) is missing:

```
SAC> m mppos /tmp/ex-1                    ;* Second parameter missing
2?  5
 File name is /tmp/ex-1; frequency is 5

SAC>
```

The prompt "2?" requests that a value for the second parameter be provided before it is used. In this case, the number 5 is given and the macro proceeds. You can provide a default value for the second positional parameter of 5 if one is not provided by the user, as follows:

Ex-5.14

```
* Contents of macro "mpposm"
* Macro user provides two parameters:  SAC file and frequency
$default 2 5
message "File name is $1$; frequency is $2$"

read $1$                                  ;* Read SAC file
rmean; rtrend terse; taper w 0.05         ;* Prepare for filtering
lowpass corner $2$ npoles 2 passes 2      ;* Filter corner frequency
```

The `$DEFAULT` command provides default values for macro parameters if a value is not provided when the macro is invoked. This says that the parameter called `2` (the second positional parameter) is to have the default value 5. Now see what happens when the macro is invoked in different ways:

```
SAC> m mpposm /tmp/ex-1                   ;* Second par omitted
 File name is /tmp/ex-1; frequency is 5

SAC> m mpposm                             ;* Both pars omitted
1?  /tmp/ex-1
 File name is /tmp/ex-1; frequency is 5

SAC>* Prompt for first parameter which lacks a default

SAC> m mpposm /tmp/ex-1 1
 File name is /tmp/ex-1; frequency is 1
SAC>* Explicit second parameter value overrides default
```

A default was only provided for the second positional parameter. If the first one is omitted it elicits a prompt when needed. Any value provided for the second parameter overrides the default.

Keyword

Most SAC commands do not expect arguments that depend on position. The BANDPASS command, for example, has keywords that introduce the low and high corner frequency, the number of poles and the number of passes. Similarly, SAC macros provide a way to define keywords that introduce arguments to the macro. The text following a keyword becomes the value of that parameter inside the macro. To introduce keywords to a SAC macro, use the $KEYS command at the beginning of the macro. The following redefines the previous macro using keyword rather than positional parameters:

Ex-5.15

```
SAC> sc cat mpkey
* Contents of macro "mpkey"
* Macro user provides file xxxx and freq yyyy
$keys file freq
message "File name is $file$; frequency is $freq$"

read $file$
rmean; rtrend terse; taper w 0.05
lowpass corner $freq$ npoles 2 passes 2

SAC> m mpkey file /tmp/ex-1 freq 5
 File name is /tmp/ex-1; frequency is 5
SAC> m mpkey freq 2 file /tmp/ex-2
 File name is /tmp/ex-2; frequency is 2
```

This example shows that keyword arguments can appear in any order, rather than being identified by position.

As with positional parameters, prompts are made for unassigned keyword parameters when their values are needed:

```
SAC> m mpkey file /tmp/ex-1
freq? 2
 File name is /tmp/ex-1; frequency is 2
```

In this case the prompt is clearer because it prints the keyword name. To provide a default value for a keyword, use the $DEFAULT command:

```
* Macro user provides file xxxx and freq yyyy
$keys file freq
$default freq 5                    ;* default frequency is 5
message "File name is $file$; frequency is $freq$"
```

```
read $file$
rmean; rtrend terse; taper w 0.05
lowpass corner $freq$ npoles 2 passes 2
```

Recursion

Macros can be recursive, which makes some algorithms simple to express. The macro listed below solves the Towers of Hanoi problem using positional parameters to describe a series of disk moves from one stack to another. Each stack is represented by a blackboard variable that contains a sequence of numbers representing a disk. The macro calls itself to move disks from one stack to another:

Ex-5.16

```
SAC> sc cat hanoi
* Towers of Hanoi:  3 stacks called A B C in BB vars A, B, C.
*  usage: m hanoi <n> <from-stack> <to-stack>
*    or   m hanoi setup <n> <stack>
if $1$ eq setup
   setbb A "" B "" C ""              ;* Clear out stacks
   do i from 1 to $2$
      setbb $3$ "%$3$% $i$"          ;* Add digit to stack end
   enddo
   message "Initial arrangement:  %A% | %B% | %C%"
elseif $1$ eq 1
   setbb tmp "(item 1 %$2$%)"        ;* Copy from stack top item
   setbb $2$ "(items 2 END %$2$%)"   ;* Remove from stack top item
   setbb $3$ "%tmp% %$3$%"           ;* Add item to top of to stack
   message "Move %tmp% from $2$ to $3$:  %A% | %B% | %C%"
else
   m hanoi (int ($1$ - 1)) $2$ (delete $2$ (delete $3$ 'ABC'))
   m hanoi 1 $2$ $3$
   m hanoi (int ($1$ - 1)) (delete $2$ (delete $3$ 'ABC')) $3$
endif
SAC> m hanoi setup 3 A
 Initial arrangement:   1 2 3 |  |
SAC> m hanoi 3 A C
 Move 1 from A to C:  2 3 |  | 1
 Move 2 from A to B:  3 | 2 | 1
 Move 1 from C to B:  3 | 1 2 |
 Move 3 from A to C:   | 1 2 | 3
 Move 1 from B to A:  1 | 2 | 3
 Move 2 from B to C:  1 |  | 2 3
 Move 1 from A to C:   |  | 1 2 3
```

5.12 ADVANCED OPERATING SYSTEM INTERACTION

We met SYSTEMCOMMAND earlier as a way to send commands to the operating system running SAC. The command is run by the operating system while SAC waits for the result. When completed, a user or a SAC macro could then retrieve the result.

SYSTEMCOMMAND provides a further feature that, when combined with WHILE READ, provides a powerful and flexible way to retrieve output from a program and process it inside SAC. Output from any command can be retrieved and put into a blackboard variable, which is then available for processing inside SAC. The syntax of SYSTEMCOMMAND when used in this way is

```
SYSTEMCOMMAND TO <v> cmd
```

Here <v> is the name of a blackboard variable whose value will become the result of the output from cmd. This macro will parse the result of the *date* command to write out the time:

Ex-5.17

```
SAC> sc cat demosc
* Contents of the macro "demosc"
sc to out date                        ;* date output to bb var out
message "Time now is: (ITEM 4 %out%)" ;* select 4th field in %out%

SAC>  m demosc
 Time now is: 09:51:29
SAC>
```

Additionally, SAC preserves line breaks between output lines read from the command. This means that using the WHILE READ command in a loop will process the lines produced as output. The following macro shows how SAC can parse the result of the *ls* command to list the names and sizes of files:

Ex-5.18

```
SAC> sc ls -l /tmp/do*               ;* output from ls command
-rw-------  1 geo-g4  wheel  210 Jun 17 09:18 /tmp/dobreak
-rw-------  1 geo-g4  wheel  179 Jun 17 08:55 /tmp/dofrom
-rw-------  1 geo-g4  wheel  309 Jun 17 10:06 /tmp/doscls

SAC> sc cat doscls
* Contents of macro "doscls"
sc to lsout ls -l /tmp/do*           ;* output saved in bb var lsout

while read lsout fperm flink fowner fgroup fsize fmm fdd ftt fnm
   * Field 1 of output [permissions] assigned to fperm
   * Field 2 of output [links] assigned to flink
   * Field 3 of output [user name] assigned to fuser, etc.
   message "File $fnm$ size $fsize$"
enddo
```

```
SAC>  m /tmp/doscls
 File /tmp/dobreak size 210
 File /tmp/dofrom size 179
 File /tmp/doscls size 309
```

Another handy use for retrieving system command output is to generate temporary file names that will not conflict with names used by other computer users. This takes advantage of the shell's built-in variable $$, which expands into the process ID of the shell itself. A unique temporary file name can be obtained in the following way:

```
sc to scr echo /tmp/temp@$@$.sh
```

Note the use of the escape character `@` to prevent interpreting $ as a macro variable indicator. After this, references to the blackboard variable `%scr%` will provide the name of a temporary file that will not conflict with other users.

The final example is a macro that will check whether file names with particular prefixes and suffixes exist. The macro defines a shell script designed to test whether a particular file exists. It then builds a list of file names as input for that shell script. The output is a list of file names and existence verdicts: "yes" if it exists, "no" if not. The macro reads this output and the list of files and prints out whether each exists.

Note that there are many times when characters especially significant to the shell must be escaped to prevent their interpretation by SAC.

Ex-5.19

```
SAC> sc cat doexist
* Contents of macro "doexist"
$keys pfx sfx
$default sfx BHE BHN BHZ                ;* Default suffix list
sc to scr echo /tmp/temp@$@$.sh         ;* Temp script name scr

* Copy text to create a temp. shell script to check if file exists
*   Note need to use escape characters for shell script syntax.
$run cat - > %scr%
# This shell script reads a file name and checks whether it exists
#   It echos the file name and adds "yes" if it exists or "no"
while read fn ; do
   test -f \@$fn @&@& echo \@$fn yes || echo \@$fn no
done
$endrun

sc to finp echo /tmp/temp@$@$.in        ;* Temp input file finp
do s list $sfx$                         ;* Iterate on suffix list
   sc echo $pfx$*.$s$ >> %finp%         ;* Add name to input
enddo

sc to inp sh %scr% < %finp%             ;* Run script on input

while read inp fn exist                 ;* Process script output
```

```
   message "File $fn$ exists? $exist$"    ;* Report file existence
enddo

sc rm %scr% %finp%                        ;* Remove temporary files

SAC> m doexist pfx /data/hinet/2005/vel/EDSH sfx LE LN LZ LR LT
 File /data/hinet/2005/vel/EDSH.LE exists? yes
 File /data/hinet/2005/vel/EDSH.LN exists? yes
 File /data/hinet/2005/vel/EDSH*.LZ exists? no
 File /data/hinet/2005/vel/EDSH.LR exists? yes
 File /data/hinet/2005/vel/EDSH.LT exists? yes
SAC>
```

CHAPTER

SIX

Accessing SAC functionality and data from external programs

Despite the broad range of utility that SAC provides, at times it may be necessary to use external applications to augment its capabilities. In addition, it is sometimes desirable to access the functionality of SAC without interacting manually with the program, for example to include it in a longer processing workflow. In this chapter we will describe techniques and give examples of how to achieve both these ends.

Note that details of the languages and applications for which examples are shown are beyond the scope of this book. The reader should seek more specialized books for that material.

6.1 AUTOMATING SAC EXECUTION

Running SAC from the shell

Executing SAC

While a decent amount of batch processing is possible within SAC using its built-in macro language (see Chapter 5), it is often useful (for example, to access functionality built into the operating system) to run SAC using scripting languages provided by the shell (under UNIX-like environments, examples include *bash* and *tcsh*).

One way is by using startup files (see Section 4.10). After the commands in it are executed, SAC then enters interactive mode. Because (usually) there is no need for an interactive phase when scripting, it is helpful to make the last line a `QUIT` command to terminate SAC and allow control to return to the shell.

So, if the file `MyCommandsFile` contains the commands

```
r MyStation.BHN MyStation.BHE ; * read in horizontal components
rotate to gcarc               ; * rotate them to radial-transverse
w MyStation.BHR MyStation.BHT ; * output as new files
quit                          ; * terminate SAC
```

on executing

```
sac MyCommandsFile
```

SAC will read in the specified data files, perform the rotation, write out the new traces and terminate.

Errors and SAC output

If an error is encountered during the execution of the SAC macro, SAC will halt to the interactive prompt. Output will be, as in a normal SAC session, directed to the standard output (the normal UNIX output stream) by default. This can then be captured or disposed of using standard UNIX command line syntax:

```
sac MyCommandsFile > /dev/null ; # execute sac and dispose
                                   of the terminal output.
```

Graphics

Unless SAC interactive commands are needed (for example, for time picking), it is usually sensible to direct any required graphical output to SAC's file-based plotting devices (such as the `sgf` device, see Section 4.4) to prevent SAC from locking up the display as it generates plots. For example, a file containing the commands

```
r MyStation.BHN MyStation.BHE ; * read in horizontal components
rotate to gcarc               ; * rotate them to radial-transverse
begindevices sgf              ; * open the SGF graphics device
ylim all; plot1               ; * plot radial vs transverse
enddevices sgf                ; * terminate the SGF graphics device
quit                          ; * terminate SAC
```

will create the file `f001.sgf`, which is a SAC graphics format file containing a plot of the radial and transverse components.

Automation of SAC execution in the shell using scripting

The utility of being able to control SAC from the shell command line is the ability to execute it repeatedly in automated shell scripts. This enables blending shell capability (for example, file handling) with SAC's functionality. It should be noted that this is not an efficient method of processing large amounts of data because shell scripts (being interpreted languages) are slow, and there is additional overhead associated with repeatedly starting and terminating SAC. However, for many applications where smallish amounts of data are involved, or the task only needs performing once, these drawbacks are heavily outweighed by its convenience.

Example: Batch processing of data files

A simple example of this is to use shell scripts to enable batch processing of large numbers of data files. While this is also possible to do using SAC macros (see Chapter 5), it is often more convenient to use the capability provided by shell scripting.

Suppose we have a directory structure containing event selected data for a network of stations. The root directory UB contains a directory for each station (UB01 to UB05), each of which contains a folder for each event recorded (EV01 to EV10). These event folders contain three SAC files containing the data from each component for that event:

```
$ ls -Rx UB/*/*/*.SAC
UB/UB01/EV01/E.SAC  UB/UB01/EV01/N.SAC  UB/UB01/EV01/Z.SAC
UB/UB01/EV02/E.SAC  UB/UB01/EV02/N.SAC  UB/UB01/EV02/Z.SAC
...
UB/UB05/EV09/E.SAC  UB/UB05/EV09/N.SAC  UB/UB05/EV09/Z.SAC
UB/UB05/EV10/E.SAC  UB/UB05/EV10/N.SAC  UB/UB05/EV10/Z.SAC
```

To generate radial and transverse components for each of these we create a simple shell script that will traverse the directory structure and run a simple set of SAC commands in each directory:

Ex-6.1

```
#!/bin/bash
# create a SAC macro to rotate the data,
# and save in /tmp
cat << END > /tmp/RotateMacro
   r N.SAC E.SAC      ; * load the data
   rotate to gcarc ; * rotate to R-T
   w R.SAC T.SAC      ; * write out
   quit
END
# outer (station) loop
for station in UB/UB??
do
   cd $station
   # inner (event) loop
   for event in EV??
   do
      cd $event
      echo "Rotating " $event " for " $station
      # run SAC
      sac /tmp/RotateMacro > /dev/null
      cd ..
   done
   cd ../..
done
# End of script.
```

Note that in this example the SAC macro is generated at the start of the script. Not only is this neater (keeping everything in a single file), it allows use of the tools provided by the shell script to manipulate the SAC macro. An alternative script that achieves the same result as that given above looks like this:

Ex-6.2

```
#!/bin/bash
# outer (station) loop
for station in UB/UB??
do
   # inner (event) loop
   for event in $station/EV??
   do
      # create SAC macro
      echo $event
      echo "r "$event"/N.SAC "$event"/E.SAC" > /tmp/RotateMacro
      echo "rotate to gcarc"  >> /tmp/RotateMacro
      echo "w "$event"/R.SAC "$event"/T.SAC" >> /tmp/RotateMacro
      echo "quit"  >> /tmp/RotateMacro
      # run SAC
      sac /tmp/RotateMacro > /dev/null
   done
done
# End of script.
```

In this version, rather than traversing the directory structure, the script generates a new macro for each set of files and executes SAC from the root directory. This ability to create and modify macros during execution of the script significantly increases the range and capability of automation. Although beyond the scope of this book, it is possible to use shell scripting to enable parallel processing of SAC data by spawning multiple processes within a shell script.

Example: Extracting information from SAC files

This technique can also be used to extract information from a set of SAC files for external processing. Suppose, using the dataset described above, we have picked the direct S-wave arrival in each of the transverse component seismograms and stored it in the `T1` header field (see Section 4.5). We now want to extract the time pick and the epicentral distance from the event to the station from the files to a single file for plotting. This is achieved by scripting:

Ex-6.3

```
#!/bin/bash
# create a SAC macro to rotate the data,
# and save in /tmp
cat << END > /tmp/ExtractPick
   r T.SAC
   message '&1,gcarc& &1,t1&'
```

```
   quit
END
# create an empty pick file
cat /dev/null > pickfile
# outer (station) loop
for station in UB/UB??
do
   cd $station
   # inner (event) loop
   for event in EV??
   do
      cd $event
      # run SAC, and put the last line of output
      # into the pick file
      sac /tmp/ExtractPick | tail -1 >> ../../../pickfile
      cd ..
   done
   cd ../..
done
# End of script.
```

This results in a two-column text file (called `pickfile`) containing the epicentral distance and corresponding time. This could then be, for example, plotted using any of a variety of software.

6.2 ACCESSING SAC DATA IN EXTERNAL PROGRAMS

The relative simplicity of the SAC data format has made it popular for storing seismic data, especially in a working, as opposed to archival, format. This makes it simple to include, for example, independent applications into a SAC processing workflow. (Note, in the following we only discuss SAC binary files.)

Accessing SAC data from Fortran using the `sacio` library

The simplest way to interact SAC data with user applications in Fortran is to use the `sacio` library. The development of this and other libraries is integrated with the development of SAC itself, and thus any future changes in SAC format will be transparent to the user (since the interfaces will not change), making code maintenance easier.

The `sacio` library provides a level of abstraction from the SAC file. It provides the following callable subroutines.

- `RSACH`: This subroutine instructs the library to load the header (only) of a specified SAC format data. The header information is stored within the library.
- `RSAC1`: This subroutine instructs the library to load a specified (evenly spaced) SAC format data file. The trace data (length and values), the initial time and the sampling rate are returned. The other header information is retained within the library.
- `RSAC2`: This is the equivalent of `RSAC1` for unevenly spaced files.

- GETFHV: Once SAC header information has been loaded using RSACH, the header values can be queried by name from the library (see HELP HEADERS for the available variables). GETFHV fetches the floating point headers (for example, time headers: A, F and T0-T9; and event location information: EVLO, EVLA and EVDP).
- GETNHV, GETKHV and GETLHV: As above, but for querying integer, character and logical header values respectively.
- SETFHV, SETNHV, SETKHV and SETLHV: These are counterparts of the GETxHV family, which store user-specified values of the appropriate type to the header retained in the library.
- WSAC0: Output a SAC file, combining the header values currently held in the library and one or two user-supplied arrays (for evenly and unevenly spaced formats, respectively).
- WSAC1: This is a simpler write function to write an evenly spaced file with a minimum header which does not require a previously loaded trace. This is useful for quickly outputting temporary or quality-control files or files for which the header no longer applies.
- WSAC2: This is the equivalent of WSAC1 for unevenly spaced files.

The complete calling sequence for these routines is available in the SAC help system. Note that the model adopted by the sacio library means that only one SAC file at a time can be interacted with. A second RSAC1 call will overwrite the header information currently kept by the library. For this reason, concurrent interaction with multiple traces requires the user to extract (through GETxHV calls) the header values they need and store them in variables in their own code. Furthermore, for efficiency reasons, sacio library routines do not perform extensive error correction or consistency checking of the files during read/write operations.

Example: Doubling of trace amplitudes

The following Fortran code demonstrates the use of these routines to double the amplitudes in a seismogram stored in a SAC data file:

Ex-6.4

```
program thedoubler
   implicit none
   integer, parameter :: nmax = 10000
   real :: ampl(nmax),beg,dt
   integer :: npts,nerr,ip

   ! Read in the data file
   call rsac1('data.sac',ampl,npts,beg,dt,nmax,nerr)

   ! loop over the data points, doubling each
   do ip=1,npts
      ampl(ip) = ampl(ip) * 2.0
   enddo

   ! set a user character header
   call setkhv('KUSER0','DOUBLED!',nerr)
```

```
   ! write the altered file back to
   !    disk under a different name.
   call wsac0('doubled.sac',ampl,ampl,nerr)
end program thedoubler
```

It is compiled and linked with the `sacio` library with a command like:

```
gfortran -o thedoubler thedoubler.f90 /usr/local/lib/sacio.a
```

(The exact form of this command will depend on the details of the installation of SAC, the compiler used and so on.) When executed, it reads in `data.sac`, loops over the array, doubling each entry and setting a character header variable (`KUSER0`) to contain the text "DOUBLED!", and then writes the new file back to disk with the name `doubled.sac`.

`sacio90`: object-oriented SAC data interaction in Fortran

Modern versions of Fortran (Fortran90 and later) provide facilities for a more object-oriented approach to programming than was possible in earlier incarnations. This has a number of advantages for the convenience of handling SAC files. In the standard `sacio` library, seismic data are separated from associated metadata in their headers. The library `sacio90` provides data structures and routines to provide access to SAC data in a more object-oriented way. Data and associated headers are stored in a single structure, making handling complete seismograms much more convenient. It also makes concurrent handling of multiple traces simpler.

`sacio90` uses the `sacio` behind the scenes to maintain compatibility. Extensive documentation of this library is beyond the scope of this book, but an example of a program using this functionality might look like this:

Ex-6.5

```
program thedoubler
   use sacio90 ! import the SACIO90 module
   implicit none

   type (SAC1) :: trace ! SAC time-series file structure

   ! Read in the data file
   call sacio90_read('data.sac',trace)

   ! double all data points
   trace % y(1:trace % npts) = &
      trace % y(1:trace % npts) * 2.0

   ! set a user character header
   trace % kuser0 = 'DOUBLED!'
```

```
   ! write the altered file back to disk under
   ! a different name.
   call sacio90_write('doubled.sac',trace)
end program thedoubler
```

This program replicates the functionality of the previous example. The seismogram, including all of its associated headers, is returned to the user in the structure `trace`. The header variables `KUSER0` and `NPTS` and the data are stored in the fields of this structure.

Other languages

Many libraries offering access to and manipulation of SAC data files have been written over the years by the seismological community in a number of different languages. Since the SAC data format is relatively simple (compared, for example, to SEED), it is relatively easy to code one's own routines to provide simple read/write access to SAC data files. Several libraries are included in the code library accompanying this book (for example, for accessing SAC files in Python and MATLAB), but many more are distributed for other languages. This allows access to functionality provided by these languages, such as more advanced graphics capability.

Example: Variable density plot in MATLAB

The following code demonstrates using the 3D plotting facilities of MATLAB to generate a variable density plot of a set of seismograms recording a seismic phase arriving at an array. The resulting plot is shown in Figure 6.1. This uses the library MSAC.

Ex-6.6

```
% read a set of traces into an array of SAC structures.
RS = msac_mread('*.BHE') ;

% build normalised amplitude, time and distance matrices.
for itr = 1:length(RS)
   A(:,itr) = RS(itr).x1./(RS(itr).depmax-RS(itr).depmin) ;
   D(:,itr) = ones(1,RS(itr).npts).*RS(itr).gcarc ;
   T(:,itr) = (0:RS(itr).npts-1).*RS(itr).delta + RS(itr).b ;
end

% sort matrices into order of increasing epicentral distance.
[~,ind] = sort([RS(:).gcarc]) ;
AS = A(:,ind) ; TS = T(:,ind) ; DS = D(:,ind) ;

% plot as greyscale surface plot.
surf(DS,TS,AS,'LineStyle','none')
colormap(gray) ; caxis([-0.75 0.75]); axis tight
xlabel('Epicentral distance (deg)') ; ylabel('Time (s)')
shading interp ; view(0,90) ;
```

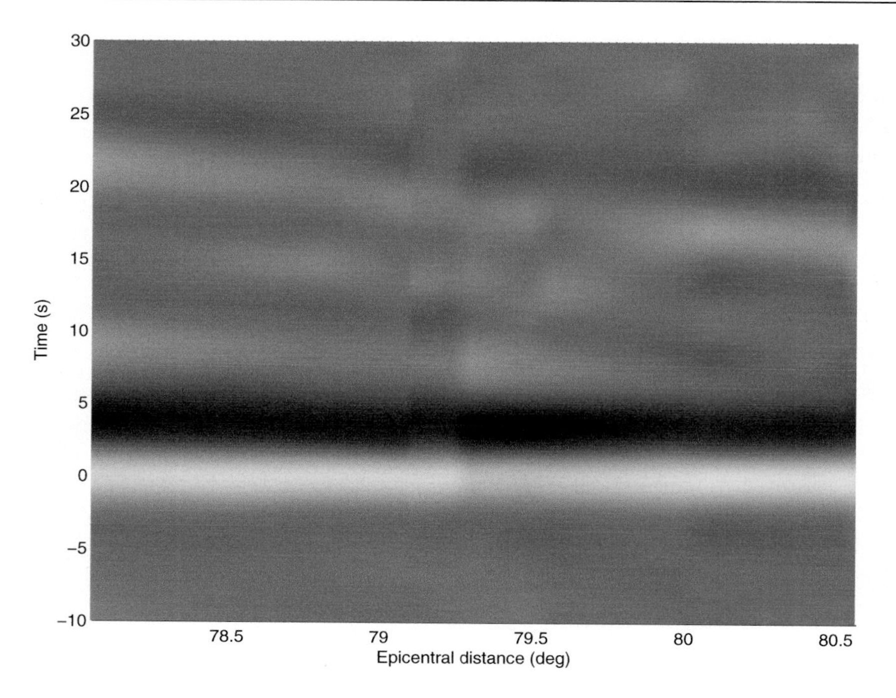

Figure 6.1 Various community-supported libraries exist for interacting with SAC in other programming languages. This example shows a variable density plot of a set of seismograms recording a teleseismic S-phase arriving across a regional array. This is generated in MATLAB, using the library MSAC to interact with SAC-formatted seismic data (see text for source code).

6.3 ACCESSING SAC FUNCTIONALITY IN FORTRAN PROGRAMS

SAC is often used as a preprocessor for seismic data destined for some other external analysis. This is often done because SAC represents a trusted (or, at least, very well tested) method for doing simple processing (filtering, for example). While access to this SAC functionality is possible using file input/output and shell scripting (see Section 6.1), for many applications the performance cost of this is prohibitive. However, SAC provides a library interface to its filtering capabilities (see Section 4.9), which are among its most used functionality. As with the `sacio` library, this is callable from the user's own Fortran programs and is implemented in the externally callable subroutine `xapiir` (see `HELP XAPIIR`).

Example: Bandpass filtering a seismogram

The following Fortran code implements a simple Butterworth bandpass filter on a seismogram:

Ex-6.7

```
! Demonstrate a simple bandpass Butterworth filter using
! the XAPIIR library distributed with SAC.
program simple_filter
```

```
   implicit none
   integer, parameter :: nmax = 10000
   integer :: nlen, npts, nerr
   real*4 :: data(nmax)
   real*4 :: beg, dt, unused

   ! filter parameters
   integer :: n_order, n_pass
   real*4 :: flo, fhi

   ! Read in the data file
   call rsac1('raw.sac',data,npts,beg,dt,nmax,nerr)

   ! Set up filter parameters
   n_order = 2   ! two pole
   n_pass  = 2   ! two pass
   flo     = 1.0 ! low cut-off
   fhi     = 5.0 ! high cut-off
   unused  = 0.0 ! does not apply for Butterworth.

   ! call the filter library, specifying a Butterworth
   ! bandpass filter.
   call xapiir(data,npts,'BUTTER', &
      unused,unused,n_order,'BP',flo,fhi,dt,n_pass)

   ! write the filtered back to disk.
   call wsac0('filt.sac',data,data,nerr)

end program simple_filter
```

This is equivalent to the following commands in SAC:

```
SAC> r raw.sac
SAC> bandpass butter corners 1.0 5.0 npoles 2 passes 2
SAC> w filt.sac
```

There is an ongoing effort to include more SAC functionality in the externally callable libraries.

CHAPTER

SEVEN

Graphical data annotation

7.1 PLOT ANNOTATION

Seismograms

For SAC, the basic display element is a seismogram plotted with time along the horizontal axis and the sampled quantity along the vertical axis. SAC's basic plotting commands, `PLOT`, `PLOT1` and `PLOT2`, supply axis tick marks numbered to show the relevant scales and connect data points with lines. SAC draws tick marks around all four sides of the plot to facilitate quantitative use of the data in the plot.

Tick marks, axes and borders

SAC uses a good default algorithm that chooses tick mark intervals and numbering based on the range of values to be plotted. If the result is unsatisfactory, tick mark intervals may be explicitly set using the `XDIV` and `YDIV` commands for the horizontal and vertical axes, respectively. The number or spacing of the tick marks may be specified using the keywords `INCREMENT` or `NUMBER`. To reset the default settings, use the `NICE` option.

The `TICKS` command selectively turns off the tick marks on each of plot's four sides using the keywords `TOP`, `BOTTOM`, `LEFT` and `RIGHT`, respectively. The omission of tick marks results in clearer plots, but they lose their quantitative clarity. `TICKS` is mostly used to make publication-quality plots for inclusion in a talk or a publication.

Grid lines across the plot (like graph paper) may be useful in some circumstances but are distracting due to the resulting clutter. To draw grid lines on plots, use the `GRID`, `XGRID` and `YGRID` commands; they are off by default. A solid or dotted grid line may be chosen in either or both of the X and Y directions. Each command recognizes the keywords `ON` or `OFF` to control their use and `SOLID` and `DOTTED` to express their form. The keywords may be used with any of the three commands. GRID controls both the X and Y axis grids, whereas XGRID and YGRID affect grid drawing on one coordinate only.

The default plot that SAC draws is of a trace surrounded by a box. The tick marks or grid lines extend from the border into the box. The X and Y coordinate axes form part of this box and may be selectively omitted by the AXES command. For example,

Ex-7.1

```
SAC> FUNCGEN SEISMOGRAM
SAC> BW 1; PLOT1
SAC> AXES ON ONLY LEFT BOTTOM; TICKS OFF ALL
SAC> BW 2; PLOT1
```

will display a seismogram with axes labeled only on the left and bottom. Figure 7.1 shows an example of the resulting plots with and without tick marks.

If all axes and tick marks are omitted, SAC produces a bare plot consisting of the data trace and any identifying labels. If a box is still desired around the plot, the BORDER command may be used to force one to be drawn. By default, there is no border because it is formed from the axes and tick choices.

Labels may be provided for the X and Y axes using the XLABEL and YLABEL commands and for the entire plot using the TITLE command. A label can be displayed or suppressed with the ON or OFF keywords and set by giving a string of text enclosed in single or double quotes. The size of the text may be given by specifying a SIZE of either TINY, SMALL, MEDIUM or LARGE. The position of the label is specified by giving a LOCATION of TOP, BOTTOM, LEFT or RIGHT. To restore label characteristics to their value prior to being changed, use the PREVIOUS option. With three labels to choose from, fairly detailed labeling is possible (and potentially confusing).

Notes can also be placed inside the data area of the plot itself, using the PLABEL command. Up to five notes can be added to a plot this way. The notes are numbered 1-5 and may be individually turned ON or OFF or set by enclosing text in single or double quotes. The text size may be specified as for the other labeling commands using SIZE. The text position is set relative to the previous note by using the keyword BELOW, or by giving a POSITION <x> <y>, where <x> and <y> are given in fractions of the window size (between 0 and 1). The note's text may be placed at a non-horizontal angle that follows <y>; the angle is in degrees clockwise from horizontal.

Publication-quality plots often require borders to be omitted. Turning off axes and tick marks will usually achieve this. It may occasionally be useful for a box to be drawn around the data in the plot nonetheless. The BORDER command may be used for this. Normally off, an explicit border may be plotted, suppressed, or restored to its previous state by using the keywords ON, OFF and PREVIOUS.

Trace information

Any seismogram generally looks like any other. Consequently, information on a seismogram plot is crucial to identify the data under scrutiny. By default, SAC adds information to each trace based on what is recorded in the file header (see Fig. 7.1). The items included (if known) are the event name, the station component names, and the zero date and time of the trace. If none of these are known, SAC uses the file name.

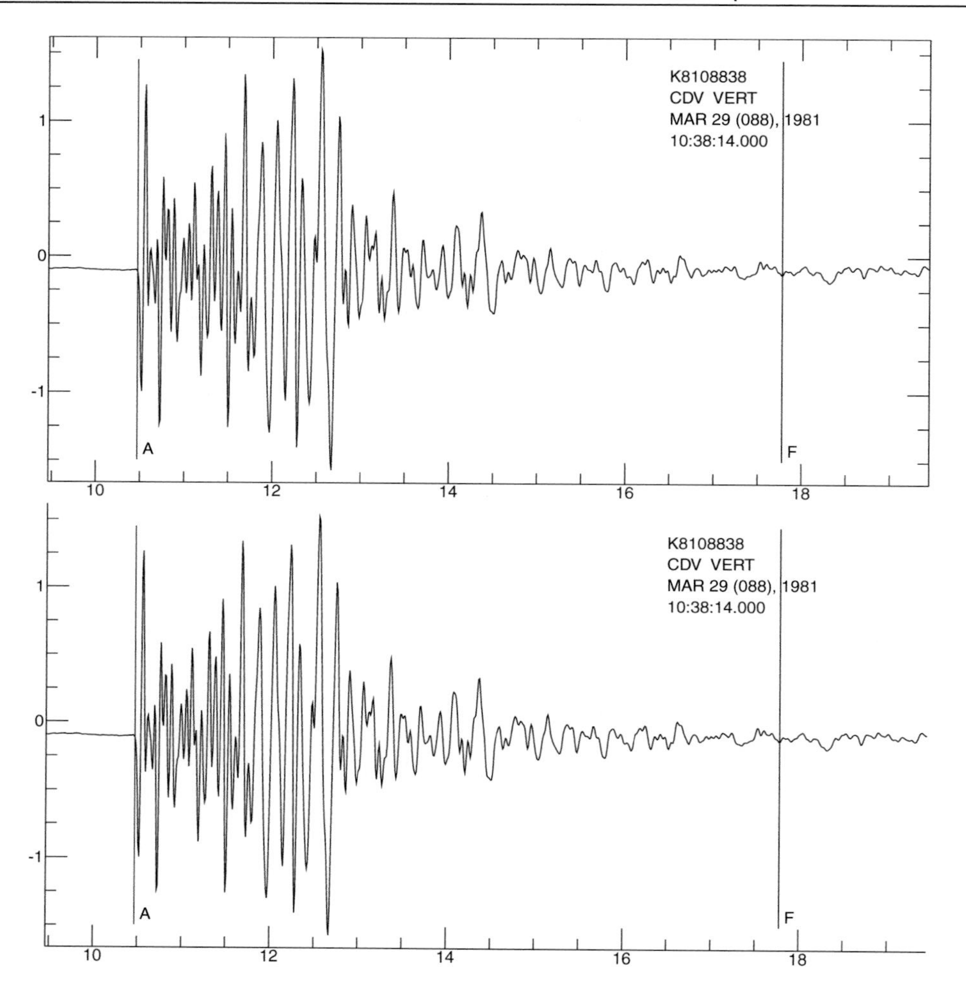

Figure 7.1 A basic data plot by PLOT1. (*top*) The plot contains axes on the left and bottom (numerically labeled) and tick marks on the top and right. (*bottom*) Turning off tick marks eliminates the box around the data. Both panels contain a file ID at the upper right consisting of an event name, a station and component name, and a zero time for the trace. The vertical lines indicate time picks stored in the file header. The labels next to the pick lines indicate the default pick name.

The information, its position and its display format can be changed using the FILEID command. The options ON or OFF turns FILEID on or off. The information TYPE can be either DEFAULT, NAME for the file name, or TYPE <list>, where <list> refers to a set of file header variable names to be included. The location in the trace window is given by a LOCATION of UR, UL, LR, LL, where U and L indicate the upper or lower side of the window, and R and L indicate its left or right side. The keyword FORMAT giving the form of information display is either EQUALS (a name = value format), COLON (a name: value format), or NONAMES (simply the value).

Picks

Picks associated with the trace that are recorded in the header (A, F, O, T0, ..., T9) may also be labeled. Within a seismogram, a pick is typically shown as a vertical line at the pick time (see Fig. 7.1), less commonly as a cross, or rarely as a horizontal line at the level of the data point nearest to the time pick. A label may be associated with a pick (for example, "S-wave" might be associated with the pick T0) and, when the pick is displayed, that label is shown alongside of the pick. If no label is given for a pick, its name is shown instead. Use the CHNHDR command to set the pick labels in the file header (KA, KF, KO, KT0, ...).

Pick display is under control of the PICKS command. The ON and OFF options control whether they are shown on a trace. You can set the height and width of the picks using the WIDTH <v> and HEIGHT <v> options. Here, <v> is a numeric value between 0 and 1 representing the fraction of the plot size. By default, the type of all picks is a vertical line, but this can be changed by naming the pick and following it with a type of VERTICAL, CROSS or HORIZONTAL.

Composite plots

SAC's basic display is a time series plot. Depending on the plotting command used, there might be one or more traces in a plot, but they generally share the same time coordinate axis. Often it is useful to show two traces in the same plot with different time axes, or even with different independent variables. For example, one might simultaneously wish to see a seismogram and its spectrum in the same plot.

SAC provides a facility to combine many different types of plots in a single display. Two new concepts underlie this facility. The first is the idea of a *frame*, which collects a group of plots into a single graphical display. When a frame opens, the display is cleared, and when a frame closes, the plot group is shown on a display. Inside an open frame, a plot is made in a *viewport*. One positions the viewport in a location in the frame and makes a plot using one of SAC's plotting commands. Changing the viewport's position and drawing another plot fills the frame with the desired information. Closing the frame realizes the individual plots on the display.

The commands BEGINFRAME and ENDFRAME open and close frames. Once SAC sees a BEGINFRAME command, no plots will be shown until an ENDFRAME appears. Thus these commands must be used in pairs to delimit a display.

A viewport is defined by giving its extent in the X (horizontal) and Y (vertical) directions. The extents are in fractions of the window size and range from 0 to 1. This convention means that, if a window is shrunk or expanded, the viewport correspondingly shrinks or expands. The X and Y viewport extents are set using the commands XVPORT and YVPORT, respectively.

Example

The following is an example of a composite plot that illustrates the ideas of a frame and a viewport and shows how labels may be used. Figure 7.2 shows the resulting plot.

Ex-7.2

```
SAC> FUNCGEN SEISMOGRAM   ;* create test data
SAC> CUTIM A -0.2 N 512   ;* window data in memory
```

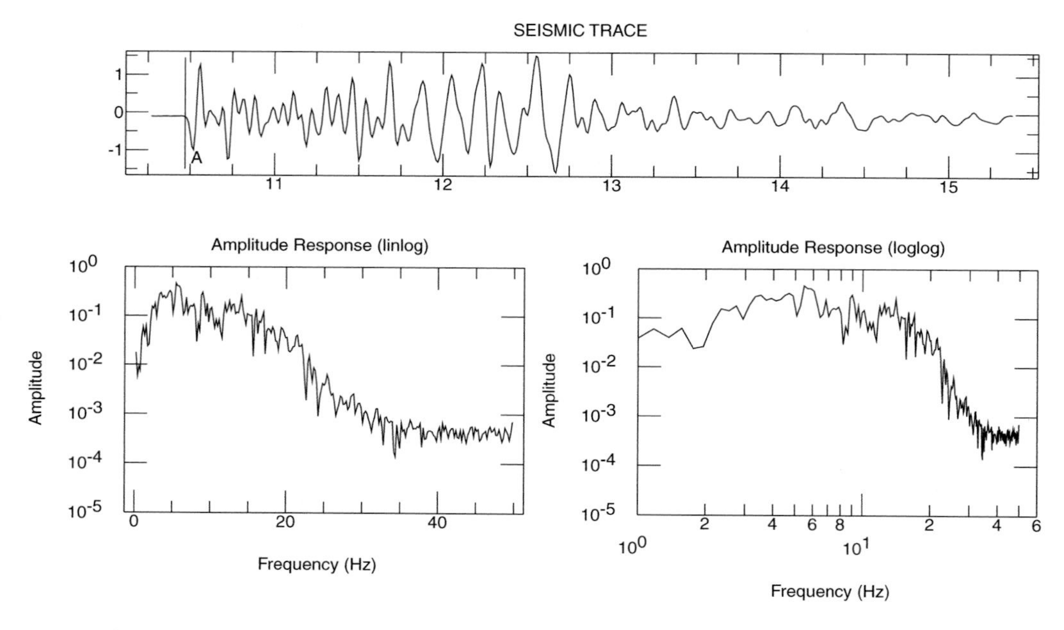

Figure 7.2 A multi-plot display example. The plot consists of three panels, each described with a viewport. Each panel has axis labels associated with it and is displayed inside of its corresponding viewport.

```
SAC> BEGINFRAME              ;* turn off automatic framing
SAC> XVPORT .1 .9            ;* define viewport and options
SAC> YVPORT .7 .9
SAC> TITLE 'SEISMIC TRACE'
SAC> FILEID OFF              ;* turn off fileid and qdp option
SAC> QDP OFF
SAC> PLOT                    ;* plot the trace

SAC> XVPORT PREVIOUS         ;* restore viewport
SAC> YVPORT PREVIOUS
SAC> TITLE PREVIOUS          ;* restore title

SAC> FFT WMEAN               ;* take transform of data
SAC> XVPORT .1 .45           ;* second viewport and options
SAC> YVPORT .15 .55
SAC> TITLE 'Amplitude Response @(linlog@)'
                             ;* use of escape character @ prevents
                             ;* parsing of linlog as a function name
SAC> LINE FILL OFF           ;* suppress any fill of spectral plots
SAC> YLIM 1E-5 1
SAC> PLOTSP AM LINLOG        ;* plot the amplitude

SAC> XVPORT PREVIOUS         ;* restore viewport
SAC> TITLE PREVIOUS          ;* restore title
```

```
SAC> XVPORT .55 .9         ;* third viewport and options
SAC> TITLE 'Amplitude Response @(loglog@)'
SAC> XLIM 1 60
SAC> PLOTSP AM LOGLOG      ;* plot amplitude again
SAC> XVPORT PREVIOUS       ;* restore viewport
SAC> YVPORT PREVIOUS
SAC> TITLE PREVIOUS        ;* restore title
SAC> ENDFRAME              ;* resume automatic framing

SAC> LINE FILL PREVIOUS    ;* reset changes to previous values
SAC> FILEID PREVIOUS
SAC> XLIM PREVIOUS
SAC> YLIM PREVIOUS
```

7.2 ANNOTATING PLOTS WITH GRAPHICAL ELEMENTS

Changing elements of SAC's standard plots and grouping plots together to form a larger display are two ways outlined so far to form specialized displays. SAC provides a more powerful suite of commands for building up elaborate displays that do not necessarily use SAC's data plotting facilities.

Figure 7.3 shows an example of a display produced by SAC. The display, showing a clock, is built up from a series of plotting elements assembled in a particular way. The elements comprising the clock are circles (the face outline and the hand boss), arrows (hands), lines (hour markers) and text (numerals and manufacturer). SAC's plotting facilities include these (and more) elements that may be assembled into more elaborate displays or may be used to annotate any of SAC's basic plots within the same frame. The SAC command that draws these elements is PLOTC. Because the assembly of the plot elements can be quite complex, the description is contained in auxiliary files that are read by PLOTC to create the plot. PLOTC

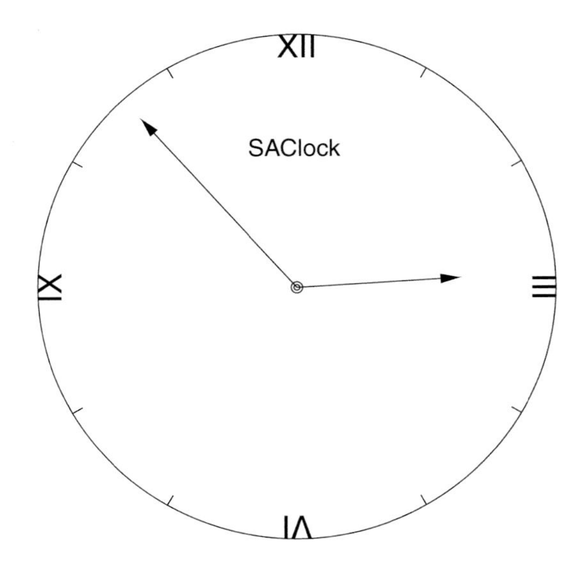

Figure 7.3 An example of a SAC display that is not time series data. The display consists of graphical elements (circles, lines, arrows, text) assembled in a way to form a clock that tells the time.

can be used to create auxiliary files for display annotation, usually at a preliminary stage in the design phase of a complex display when a rough placement of plot elements is made. Subsequently, the file is edited to refine the display before use in routine seismic analysis.

The files used by PLOTC to make displays must have names ending with either .PCF or .PCM. Files with the .PCM suffix are confusingly called macro files, but they are distinct from SAC command macros in that they only contain display descriptions. PLOTC makes the distinction because macros contain groups of graphical elements that may be scaled and rotated as a group and repetitively placed in a display. They only get invoked in the course of reading from a .PCF file while PLOTC makes a display. The suffixes are implied when these files are referred to. For example, when making a display described by the file MYCLOCK.PCF, invoke the PLOTC command via

Ex-7.3

```
SAC> PLOTC REPLAY FILE MYCLOCK
```

Graphical elements

SAC's PLOTC command provides the following graphical elements:

- arrow;
- set of tick marks;
- circle;
- line;
- polygon (2–9 sides);
- sector of a circle;
- rectangle;
- line of text;
- block of text;
- group of other elements (called a macro).

Additional items that appear in display descriptions are comments (which are ignored), parameter assignments (to set line styles and polygon sides, for example), element placement information and the end-of-plot indicator.

SAC's plot windows typically do not have a 1:1 aspect ratio, which distorts the shape of the graphical element. If shape is important, use the VSPACE command to set the view space aspect ratio to 1.0, make the display with PLOTC, and then restore the aspect ratio to its PREVIOUS value.

Assembling graphical elements

PLOTC draws the elements of a display in the order they are met in the description. The coordinate system is a relative one in the plot window with (0,0) at lower left and (1,1) at upper right. Two positions are always kept track of: the *origin* and the *cursor*. Usually the origin is the previous cursor position, but it may be explicitly set. A display description consists of a series of single-letter commands and a cursor position given by two real numbers between 0 and 1. Table 7.1 lists the commands recognized by PLOTC.

Table 7.1 PLOTC commands and graphical elements

Command	Coordinates		Description
*	—	—	Comment-text ignored
Q	—	—	Draw display and quit PLOTC
O	x	y	Set origin to (x,y)
G	x	y	Set origin to (x,y) and make it global
T	x	y	Place line of text (next line in file) at (x,y)
text			
U	x	y	Place multiple lines of text (following lines in file up to the next blank line) at (x,y)
$text_1$			
$text_2$			
...			
L	x	y	Draw a line from the origin to (x,y)
A	x	y	Draw an arrow from the origin to (x,y)
C	x	y	Draw a circle centered at the origin to (x,y)
R	x	y	Draw a rectangle with opposing corners at the origin and at (x,y)
N	x	y	Draw an N-sided polygon centered at the origin with one vertex at (x,y)
S	x	y	Draw a sector of a circle centered at the origin with one vertex at (x,y); following command is either S or C, indicating whether the minor (S) or major arc (C) is drawn to next cursor position
B	x^a	y^a	Draw border tick marks around the plot region (if requested by the BORDER option)
M	x	y	Invoke macro name at (x,y); following line contains macro name, multiplicative scale factor (between 0 and 1), and baseline orientation as clockwise angle from horizontal in degrees
name	*scale*	*angle*	
[	—	—	Set parameters using text on the line up to closing] character

[a]Values ignored but must be present.

A PLOTC display description file typically starts by setting parameters, then sets an origin, and then draws a sequence of graphical elements, moving the origin when necessary. For example, the following file describes the display of a circle in the center of the plot with a rectangle around it:

Ex-7.4

```
* The following line sets the line style to be thicker than default
[W3]
O 0.5 0.5
C 0.5 0.8
O 0.2 0.2
R 0.8 0.8
Q
```

The commands set an origin, draw a circle, reset the origin, and draw a rectangle enclosing the circle.

Macro files are extremely useful for repeating groups of elements at different places in a display. The following commands, placed in a macro file called `ex7-1note.pcm`, draw a small cross labeled "Peak posn." when the macro is invoked.

Ex-7.5

```
* Macro; set text size to medium
[SM]
O -0.01 0.0
L  0.01 0.0
O  0.0 -0.01
L  0.0  0.01
T  0.03 0.0
Peak posn.
```

In macros, the coordinates are relative to the cursor position where the macro is used, and the coordinate axes are rotated by the angle given when the macro is invoked. This is why some of the coordinate values are negative. A Q command does not need to end a macro; the end of the file is sufficient.

The following file, `ex7-1.pcf`, creates the display and uses the macro to place text at various places in it.

Ex-7.6

```
M  0.5 0.5
ex7-1note 1.0 0
M  0.5 0.5
ex7-1note 1.0 90
M  0.5 0.5
ex7-1note 1.0 180
M  0.5 0.5
ex7-1note 1.0 270
M  0.25 0.25
ex7-1note 2.0 -45
M  0.25 0.75
ex7-1note 2.0 45
M  0.75 0.25
ex7-1note 2.0 -135
M  0.75 0.75
ex7-1note 2.0 135
Q
```

Figure 7.4 shows the resulting display. There is a cross in the center labeled with text in each 90° orientation. There are four more crosses centered in each of the four quadrants of the display, with text angled toward the center. The cross orientations are rotated along with the text and are twice as large as the center cross due to the scale factor used when the macro is used. Text is not subject to scaling (see following section).

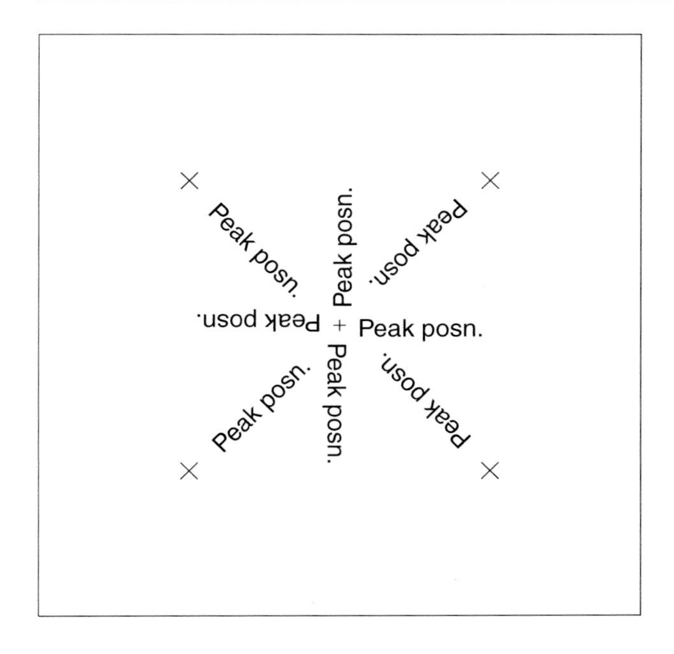

Figure 7.4 Display showing the use of a PLOTC macro in different orientations. The macro draws a cross labeled "Peak posn." The cross size depends on the scale factor used when the macro is invoked to draw the graphical elements within it. The cross and text orientation depend on the angle.

Parameters controlling graphical elements

While the shapes of graphical elements are fixed, SAC allows the way they are displayed to be changed for stylistic purposes. For example, a line might be drawn as dashed or dotted, an arrowhead might be open or filled, a rectangle might be outlined in color or black. These are called parameters by PLOTC. Parameters are set by strings enclosed in square brackets ([. . .]) that begin with the first character on a line of a display description file.

Table 7.2 lists the entire collection of parameters that may be changed in a display. Each parameter consists of a one- or two-letter code followed by an integer argument or character code. Line drawing is one area under parameter control: a line's style and width, whether symbols ornament a line and the symbols' sizes. The graphical element color is also under parameter control. Parameters also affect the way text is rendered: the font, size and justification (both horizontal and vertical). The number of polygon sides is also set by a parameter. Finally, the numbers of tick marks in the horizontal and vertical directions are also under parameter control.

7.3 USING PLOTC

SAC's PLOTC command works in two ways. Typically, it replays an existing display annotation file to annotate a data plot as part of a seismic data analysis procedure. Used this way, it typically appears in a SAC macro that implements the analysis procedure. The command is essentially replaying a pre-existing file to annotate a plot inside a frame consisting of a composite plot. Thus the command in the macro is

```
SAC> plotc replay file XXX
```

where XXX.PCF is the name of the file containing annotation commands (Fig. 7.4 was produced this way). A border is drawn around the annotated area by default. If unwanted, use the BORDER OFF command option.

Table 7.2 PLOTC parameters affecting graphical elements

Code	Argument	Description
H*j*	*j*=L,C,R	Text justification horizontal (L)eft, (C)enter, (R)ight
V*j*	*j*=B,C,T	Text justification vertical (B)ottom, (C)enter, (T)op
S*s*[a]	*s*=T,S,M,L	Text size (T)iny, (S)mall, (M)edium, (L)arge
Q*n*	*n*=1-4	Text quality 1,2=hardware, 3,4=software
Q*q*	*q*=H,S	Text (H)ardware, (S)oftware
F*n*	*n*=1-8	Text Font n
A*s*	*s*=T,S,M,L	Arrowhead size (T)iny, (S)mall, (M)edium, (L)arge
A*t*	*t*=F,U	Arrowhead type (F)illed, (U)nfilled
A*v*	*v*=V,I	Arrowhead shaft (V)isibile, (I)nvisible
L*n*	*n*=1,4	Line type 1=solid, 2-4=other
LN*n*	*n*=1,...	Line type n (can be larger than 4)
W*n*[b]	*n*=1,4	Line width n, 1=thinnest, 4=thickest
N*n*	*n*=2,9	Number of polygon sides
P*n*	*n*=1,4	No-op (ignored)
C*s*	*s*=N, other single character	Set color to N(ormal), or other color in color table identified by its first character; default color table provides R(ed), G(reen), Y(ellow), B(lue), M(agenta), C(yan), W(hite)
BH*n*	*n*=integer	Set number of horizontal border ticks to n
BV*n*	*n*=integer	Set number of vertical border ticks to n
SY*n*	*n*=integer	Set symbol number for line plotting; n=0 turns off symbols
SZ*n*	*n*=integer	Set symbol size; n in milliunits (e.g., 5 is 0.005)
	n=0	scales symbols to current character size
SP*n*	*n*=integer	Set symbol spacing; n in milliunits (e.g., 100 is 0.100)
	n=0	puts symbols at every point on line

[a]Saved and restored when a macro is invoked.
[b]WIDTH ON must be in effect for option to affect widths.

The second use of the PLOTC command is to design annotations that will become part of a data analysis procedure (or, less commonly, a one-off plot). This involves a graphical interaction with the user and a partial plot, similar to the way the PLOTPK command interacts graphically with the user and a data trace. The command to start PLOTC in this way is

```
SAC> plotc create file XXX
```

In this mode, PLOTC takes a sequence of commands and cursor positions and writes them into a display annotation file (in this case, a file called XXX.PCF). When the interaction is finished, the file contains the annotations on the plot. At this point the file can be edited with a text editor to modify the commands, cursor positions, or parameters to refine the annotations for production use.

The interaction is quite simple. When PLOTC is invoked with the CREATE option, a graphical cursor appears on the screen, and SAC waits for you to type a command character (one from Table 7.1). At that point, the command and cursor position are written to the file, and any extra information needed to describe the graphical element (the text, the macro name, etc.) is read and put into the file. SAC displays the graphical element and the cursor reappears awaiting the next command character. The interaction finishes when you type Q.

The Q character is not copied to the annotation description file. It must be added by hand using a text editor. In REPLAY mode, if the Q command is missing, PLOTC assumes that you are continuing to create an annotation and displays the graphical cursor crosshairs. If, after replaying a file, a set of crosshairs unexpectedly appears on the screen, you have forgotten to add the Q to your file.

CHAPTER
EIGHT

Array data handling

8.1 SAC SUBPROCESSES

SAC's commands potentially affect all aspects of a trace. For some types of analysis, however, it is computationally simpler to apply certain assumptions to a collection of traces before they are analyzed. For example, trace stacking is easier when it is assumed that every trace to be stacked has the same sample rate and the same start and end times. To enforce such assumptions, SAC provides the use of a *subprocess*, which, when entered, limits the commands that may be issued. SAC contains two major subprocesses: the signal stacking subprocess (SSS) and the spectral estimation subprocess (SPE). Entry to the subprocess is through the SAC commands `SSS` and `SPE`, respectively, and the return from either is through the `QUITSUB` command (abbreviated `QS`).

Inside the subprocess some commands are restricted and some other commands become available. The rest of SAC's commands continue to operate. To see which commands a subprocess allows, use `HELP COMMANDS` when inside the subprocess:

```
SAC> sss
 Signal Stacking Subprocess.
SAC/SSS> help commands

 Alphabetical list of SAC SSS commands.  Type

    help xxxx

 for information about the xxxx command, and

    syntax xxxx

 ...
```

When looking for a command using `HELP APROPOS`, if the command is restricted to use within a subprocess, `(SPE)` or `(SSS)` will appear after its command name.

In the previous example, note that SAC announces entry to the subprocess, and its command prompt changes. This is a reminder that use of certain commands is restricted inside the subprocess.

8.2 THE SIGNAL STACKING SUBPROCESS

The `SSS` command enters the subprocess, which is designed for combining many traces for graphical display, for trace picking and quality control (QC) and for stacking data to suppress noise and enhance signal. The subprocess works with a trace collection that is initially defined by the traces in memory before entering SSS. The collection may be added to or winnowed within the subprocess.

Trace collections

The trace collection (also known as a *stack*) is the set of files that SSS operates on, either to display them or to stack them. Each trace in the collection has a set of *properties* that may be defaulted or explicitly set. The properties are as follows:

- name - the file name from which the data was read;
- weight - its numerical weight when stacked with other traces;
- polarity - its polarity sense (positive/normal or negative/reversed);
- range - the position of the trace relative to an origin;
- sum - a true or false value indicating whether the trace participates in any stacks calculated with the trace collection;
- delay time and increment - the time shift relative to the trace start for alignment with other traces and an incremental time shift factor;
- sample delay and increment - the shift in samples relative to the trace start for alignment with other traces and an incremental sample shift factor;
- velocity model - a velocity model designation (1 or 2) to calculate time shifts.

The range and polarity affect plotting and picking. The weight and polarity affect stacking. The velocity model and the shifts affect trace plotting and stacking. To view the trace properties, use the `LISTSTACK` command.

A time window must be defined to plot or stack the trace collection. The `TIMEWINDOW` command defines the window, which consists of a start and an end time in seconds. The time window applies to the trace after defining all of its time shifts. There is no default time window.

A range window also must be defined to plot a trace collection. Ranges may be defined in various ways by using SAC's `DISTANCEWINDOW` command:

```
SAC/SSS> help distancewindow

 SUMMARY:
 Controls the distance window properties on subsequent record
 section plots.
```

```
SYNTAX:
DISTANCEWINDOW [{USEDATA|WIDTH <w>|FIXED <lo> <hi>}]
     [UNITS {KILOMETERS|DEGREES}]
```

The range window may be data-based (nearest and farthest trace from the origin), a fixed distance (a range from the nearest trace), or a low and high range value. Ranges may be specified in either kilometers or degrees of arc.

A trace's range value is normally its distance from an event origin. However, it can actually be any value associated with a trace. For example, the natural way to order receiver functions is on the basis of their horizontal slowness, which controls the delays of the primary conversions and the multiples from the parent (main) arrival. Thus a receiver function trace's "range" can actually be a slowness in seconds/degree or seconds/kilometer, explicitly set when the trace is added to the collection.

Adding, deleting and changing traces

The basic command to add a new trace to a collection is the ADDSTACK command:

```
SAC/SSS> help addstack

 SUMMARY:
 Add a new file to the stack file list.

 SYNTAX:
 ADDSTACK <filename> [WEIGHT <v>] [DISTANCE <v>]
      [DELAY <v> [{SECONDS|POINTS}]]
      [INCREMENT <v> [{SECONDS|POINTS}]]
      [{NORMAL|REVERSED}] [SUM {ON|OFF}]
```

This command names a file to be added to the trace collection. All other options act to define the trace's properties. If a property is not defined, it inherits a default property set by the GLOBALSTACK command. If the DISTANCE property is neither defined nor a global default specified, it is taken from the event and station information (EVLA, EVLO, STLA, STLO) if known.

The CHANGESTACK command changes trace collection information:

```
SAC/SSS> help changestack

 SUMMARY:
 Change properties of files currently in the stack file list.

 SYNTAX:
 CHANGESTACK {<name>|<number>|ALL} [WEIGHT <v>] [DISTANCE <v>]
      [DELAY <v> {SECONDS|POINTS}]
      [INCREMENT <v> {SECONDS|POINTS}]
      [{NORMAL|REVERSED}] [SUM {ON|OFF}]
```

A trace may be designated either by its name, by its numerical position in the collection, or by the whole set of traces "ALL". Similarly, a trace may be purged from the collection by using the DELETESTACK command:

```
SAC/SSS> help deletestack

 SUMMARY:
 Deletes one or more files from the stack file list.

 SYNTAX:
 DELETESTACK <name>|<n> ...
```

Plotting record sections

After a collection of traces is defined, they may be plotted by using the PLOTRECORDSECTION command (abbreviated PRS). A basic record section plot of some of SAC's built-in data is shown in Figure 8.1.

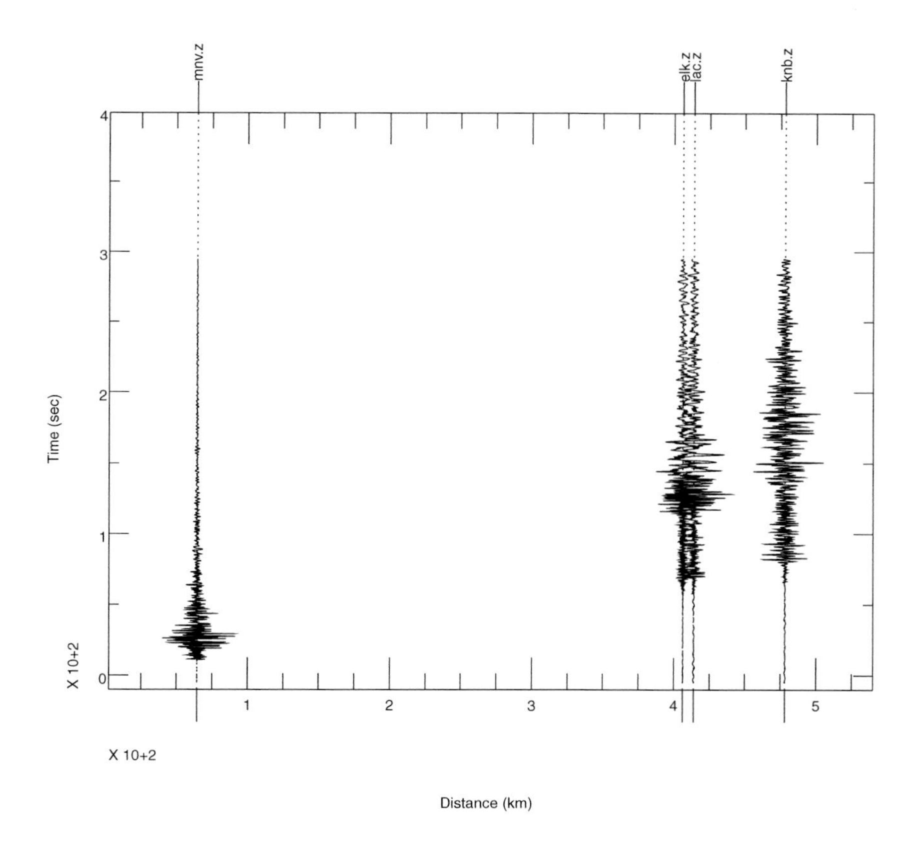

Figure 8.1 This plot shows four vertical component seismogram records from a single event using PLOTRECORDSECTION. The X axis is the range from the event origin in kilometers. The Y axis is the time in seconds from the reference time. Each trace is plotted relative to an amplitude axis anchored at the distance of the station from the range reference (in this case, the event location). The axis is shown with a dotted line and labeled at the top of the figure with the file name. With increasing distance, the P-wave arrival is later in the record section, as expected from normal crustal travel times.

Ex-8.1

```
SAC> datagen sub regional *.z  ;* Built-in data
SAC> sss
SAC/SSS> tw -10 400            ;* Define time window
SAC/SSS> prs                   ;* Default record section plot
```

This is a default plot. Each trace in the collection is plotted with a baseline (zero level) at its range from the reference position. The baseline position is always marked outside of the axis frame by a tab that plays a role in picking traces (see Section 8.2). PLOTRECORDSECTION (abbreviated PRS) has many options to change plot details. Of particular note is the ability to label traces (see the LABEL option) with information other than the file name; any file header variable value may be used as a label. The reference line (the zero amplitude range position) may also be suppressed (see REFERENCELINE). Finally, the orientation of the plot may be changed from landscape (a horizontal range axis) to portrait (a vertical range axis). See the ORIENTATION and ORIGIN options for axis orientation.

Record section plots may be annotated with any of the features described in Chapter 7. It is often desirable to see travel-time curves for particular arrivals from an earthquake. The SSS subprocess provides a simple way to define which travel-time curves to plot using the TRAVELTIME command. The example below shows how to produce a simple record section plot with a set of P and S travel times. Figure 8.2 shows the resulting plot:

Ex-8.2

```
SAC> datagen sub regional *.z          ;* Built-in data
SAC> sss
SAC/SSS> traveltime model iasp91 phase Pn Sn
SAC/SSS> tw -10 400                    ;* Define time window
SAC/SSS> prs ttime on orient portrait  ;* Portrait, travel times
```

Here, the TRAVELTIME command requests that travel-time curves for the regional P and S arrivals (Pn and Sn) be calculated. The PRS TTIME ON option adds the curves to the record section, and the ORIENT PORTRAIT option changes the distance axis to vertical. If you do not like the sense of increase along the range axis (left-to-right in LANDSCAPE orientation, top-to-bottom in PORTRAIT), use the ORIGIN REVERSED option to change it.

Stacking

SAC's SSS subprocess also provides the ability to stack the traces in the collection. The idea underlying stacking is that combining many traces recording the same arrival in a seismic wavefield suppresses noise. The noise is presumably incoherent, and therefore the noisy parts of the traces will sum to a value whose average is zero. Only the coherent parts of the traces will survive the summation to yield a nonzero value. In this way, signal is enhanced over noise.

The SUMSTACK command implements stacking. A collection of traces must be defined along with a time window (see the TIMEWINDOW command). The data in the time window is summed after applying all the time delays associated with the trace properties.

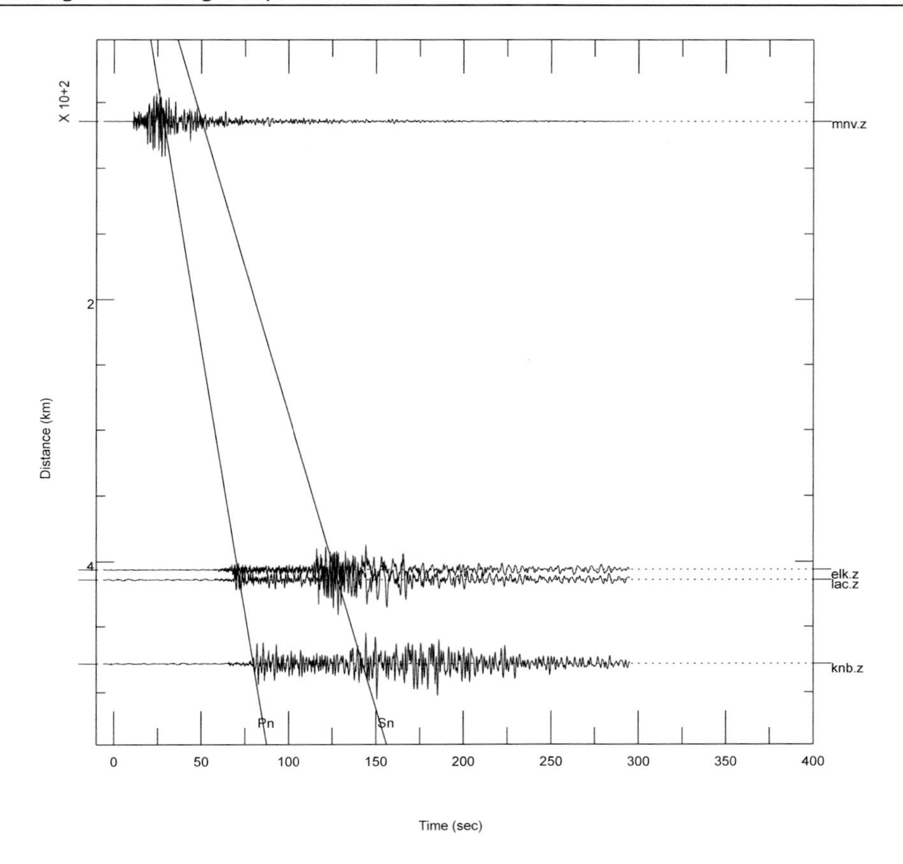

Figure 8.2 This plot is similar to the previous record section plot but differs in the orientation of the range axis (vertical) and added travel-time curves. The regional P- and S-wave arrivals match the first disturbance (Pn) and the largest amplitude arrival (Sn) on almost all of the traces except the closest one.

By default, there is no time delay other than the one defined by the ADDSTACK or CHANGESTACK command. However, further time delays may be specified using the VELOCITYMODEL command.

Velocity models

SAC's VELOCITYMODEL command (abbreviated VM) recognizes two types of velocity models: one for a direct arrival (a so-called "refracted wave" model) and another for a reflected arrival (from a horizontal discontinuity like the Moho, called a "normal moveout" model). The key function of a velocity model is to provide a way to adjust trace delays depending on distance. Adjustments made to the trace delays sharpen waveform features, if the arrival in the stack behaves like one of the models. INCREMENTSTACK successively adjusts the trace delays through increments defined by either the VELOCITYMODEL command or the ADDSTACK or CHANGESTACK commands. INCREMENTSTACK behaves like the stacking control parameter (the relative slowness) in a slant stack.

The refracted wave model defines a time delay t_{delay} as follows:

$$t_{delay} = t_{R,VM} - (t_{0,VM} + \Delta / V_{app}) \tag{8.1}$$

This identifies a lag due to a direct wave propagating a distance Δ (km) divided by an apparent velocity V_{app}. There are two lags, one used to correct for a particular source depth (t_0) and one for alignment at a reference range (t_R). These are set with the `VM T0VM` and `VM TVM` options, respectively, and V_{app} with `VM VAPP`.

The reflected wave model defines a time delay t_{delay} as follows:

$$t_{delay} = t_{R,VM} - \sqrt{t_{0,VM}^2 + (\Delta / V)^2} \tag{8.2}$$

This identifies a lag due to a wave propagating downward from a source and reflecting off a horizontal interface at some level. In this case the delay is a hyperbolic function that consists of a constant factor t_0 representing the arrival time for a reflection from the interface at zero lag (depth/velocity) and a range-dependent factor (Δ/velocity) depending on the constant velocity V in the layer above the interface. An alignment time at a reference range (t_R) is included. These times are set with the `VM T0VM` and `VM TVM` options, respectively, and the velocity is set with `VM VAPP`.

To shift the timings in the stack, increments may be defined for any (or all) t_0 and V_{app} or V. The reference timings t_R are defined once per model with `VM TVM` at the reference range given by `VM DVM`. SAC increments stack control values each time the `INCREMENTSTACK` (abbreviation `IS`) command is used. Use `LISTSTACK` to see the prevailing delays resulting from application of the velocity model. Usually only one item is incremented at a time so that its effect may be understood. In practice, the most common use of velocity models is to use none at all – the static trace delay is enough to define the lags used for trace alignment.

Types of stacks

The goal of stacking is to enhance signal by suppressing noise. The straightforward way to stack is simply to sum each sample at the corresponding time in each trace and to divide each sample sum by the number of traces. This results in a *linear stack*.

A slightly more sophisticated way is to recognize that a signal typically lies at a higher amplitude than noise. By nonlinearly stretching the amplitude axis, the amplitude range is made more nearly constant. Thus the incoherent noise is suppressed more readily through its random phase rather than its lower amplitude relative to the signal (McFadden *et al.*, 1986). After summation, the resulting sum is nonlinearly shrunk back to its original amplitude range. The nonlinear stretching is usually done by taking the Nth root of the sample value, where N is usually a small positive integer. After calculating the sum, the result is raised to the Nth power. This method is called the *Nth root stack* due to the nonlinear mapping.

Another method that recognizes the importance of phase coherence in separating signal from noise is phase-weighted stacking (Schimmel and Paulssen, 1997). This method explicitly computes the phase of each trace point from the analytic signal and weights the point by its phase sum. A coherent phase sum of M point samples has a magnitude of M whereas an incoherent phase sum has a magnitude close to zero. The amplitude sum at each point is therefore multiplied by the Nth power of the phase sum, which suppresses incoherent samples in the stack. It is called a *phase-weighted stack* on this account.

No matter which stacking method is chosen, traces with high signal amplitudes will be emphasized over those with lower amplitudes. Thus the traces can be normalized by their trace weights. If the weight is taken as the reciprocal of the peak trace amplitude, the overemphasis of high-amplitude traces is ameliorated. Use ADDSTACK to provide suitable trace weights.

Stack uncertainty bounds

It is useful to have an estimate of the uncertainty associated with each sample point in a stack. For a linear stack, this may be defined easily through the variance of each sample point and its relationship to a confidence level through the χ^2 distribution. For the other stacking methods, this concept is no longer useful because the variance is of a sum of a distorted sample and the statistical model no longer connects it to a confidence level. Fortunately, jackknife uncertainties may be used to calculate uncertainty in this situation (Efron, 1982). The method calculates uncertainty by forming M subsets of the M traces by leaving out each trace successively. The variance of the sample estimates from each of these subsets of size $M-1$ may be used to estimate the uncertainty. The price of this approach is that when many traces are in a stack it can be slow to calculate.

SAC retains the uncertainty bounds for the stack along with the stack in memory and will display it along with the summation trace using the SUMSTACK command. It will also write it out to a file using WRITESTACK.

Example: 2006 North Korea nuclear test

Data recorded by the Japanese Hi-net array (Obara *et al.*, 2005) of the 9 October 2006 North Korean nuclear test provide a good example of the use of the SSS stacking commands. The following commands download the dataset and show a record section plot:

Ex-8.3

```
SAC> read ds http://www1.gly.bris.ac.uk/MacSAC/korea.sds
SAC> sc curl http://www1.gly.bris.ac.uk/MacSAC/korea.mac \
   > /tmp/korea.mac
SAC> sc awk '/sss/,/quitsub/{print}' /tmp/korea.mac ;* List file
sss
timewindow -5 12
distancewindow fixed 6 12.5 units degrees
changestack flt/KSIH.U delay -123.115800 s n w 1.833053
changestack flt/HIRH.U delay -123.310400 s n w 1.049097
changestack flt/MHSH.U delay -124.211400 s n w 2.782429
...
changestack flt/HNKH.U delay -203.844500 s n w 1.500755
message " 753 traces."
border off
beginframe
plotrecordsection referenceline off labels off size 1 weight off
plotc replay file /tmp/tmp_rs
endframe
border previous
quitsub
SAC> m /tmp/korea.mac                    ;* Produces annotated plot
```

The first command reads a collection of SAC files in dataset format from an internet source. The second command uses the UNIX *curl* command to read a SAC macro from the

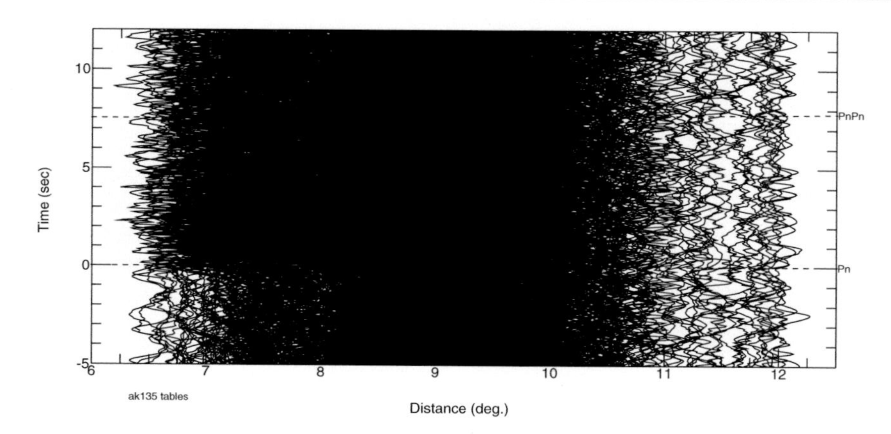

Figure 8.3 This plot shows Japanese network recordings of the 9 October 2006 North Korea nuclear test. The traces, roughly aligned on the predicted Pn arrival from the explosion (horizontal dashed line at relative time 0), do not show a signal that is significantly above the noise level. Its presence is evident, however, by a change in frequency between 6.5 and 10.5°.

same internet source. This file is a SAC macro and contains plot annotation commands followed by SSS commands to align the traces to the Pn arrival from the event. Only the part of the file containing SSS commands is shown (and significantly shortened for this example).

First, a time window is defined using TIMEWINDOW and a distance window by DISTANCEWINDOW. The READ DS command read the traces into memory, so after entering SSS, the traces are already in the collection. However, their properties need to be adjusted for alignment purposes. The series of CHANGESTACK commands sets the lag for each trace in seconds, sets the polarity to NORMAL (N), and sets a weight related to the maximum trace amplitude. Finally, after all the trace properties are set, the macro writes a message giving the number of traces in memory.

A series of plotting commands follow. The border around the plot is temporarily suppressed (the BORDER OFF ... BORDER PREVIOUS pair) and a composite plot constructed (BEGINFRAME ... ENDFRAME) consisting of a record section (PLOTRECORDSECTION) followed by annotation of the record section with the expected Pn arrival time (PLOTC REPLAY). For brevity, the macro commands to set up the PLOTC annotation file (/tmp/tmp_rs.pcf) are omitted here.

The basic record section plot of the data is shown in Figure 8.3. The explosion arrival is not significantly visible above the noise level. Stacking can enhance the signal, however. The commands

Ex-8.4

```
SAC> sss
SAC> cs all sum on                 ;* Add all traces to stack sum
SAC> sumstack type linear
SAC> sumstack type nthroot 2
SAC> sumstack type phaseweight 2
SAC> writestack /tmp/pw-sum.sac unc /tmp/pw-unc.sac ;* Save
```

make a series of stacks with different methodologies, resulting in the plots shown in Figure 8.4. The linear stack is not nearly as good in suppressing noise as is the Nth root or the phase-weighted stack.

Saving stack data and uncertainties

SUMSTACK produces both a stacked signal estimate and the uncertainty of that estimate. The previous example showed the stack and its uncertainty written into the files /tmp/pw-sum.sac and /tmp/pw-unc.sac. To retrieve them as traces in their own right, use an approach as shown in the following commands:

Ex-8.5

```
SAC> read /tmp/pw-unc.sac /tmp/pw-unc.sac /tmp/pw-unc.sac
SAC> mul 1 0 -1            ;* pos., zero and neg. uncertainty
SAC> addf /tmp/pw-sum.sac /tmp/pw-sum.sac /tmp/pw-sum.sac
SAC> ch file 1 kevnm +1_sigma      ;* name traces for P2
SAC> ch file 2 kevnm stack_sum
SAC> ch file 3 kevnm -1_sigma
SAC> fileid type list kevnm        ;* KEVNM identifies trace
SAC> line list 5 1 5 increment on  ;* unc traces dashed
SAC> p2                    ;* overlay three plot traces
```

This reads three copies of the uncertainty and then turns them $+1\sigma$, 0σ, and -1σ values, to which the stack estimate is added. The KEVNM file header identifies the trace's identity, and the FILEID command tells SAC to use it to label the trace in future plots. The PLOT2 command yields a composite plot that shows the $\pm1\sigma$ envelope around the signal estimate (Fig. 8.5).

Picking data in stacks

SAC's PLOTRECORDSECTION command also includes an interactive mode, like PLOTPK, where actions are controlled by the screen's cursor and keyboard. The features provided in this interactive mode include the following:

- zooming in on parts of the record section (either in time or in range);
- amplifying and shrinking traces;
- estimating the apparent velocity of features on the trace;
- selecting traces for further analysis;
- picking trace features (peaks, troughs, or zero crossings).

The PLOTRECORDSECTION CURSOR command enters interactive mode. It draws a record section plot and displays crosshairs on the screen to prompt for graphical interaction. As with PPK, a keyboard character indicates an action to take. Table 8.1 shows a complete list of possible actions. Some actions require use of macro language features to be exploited because they return values in blackboard variables.

The basic action is windowing the trace display. The trace amplitudes may be expanded or shrunk using the + or - keys. A pair of X key presses defines a zoom into a distance

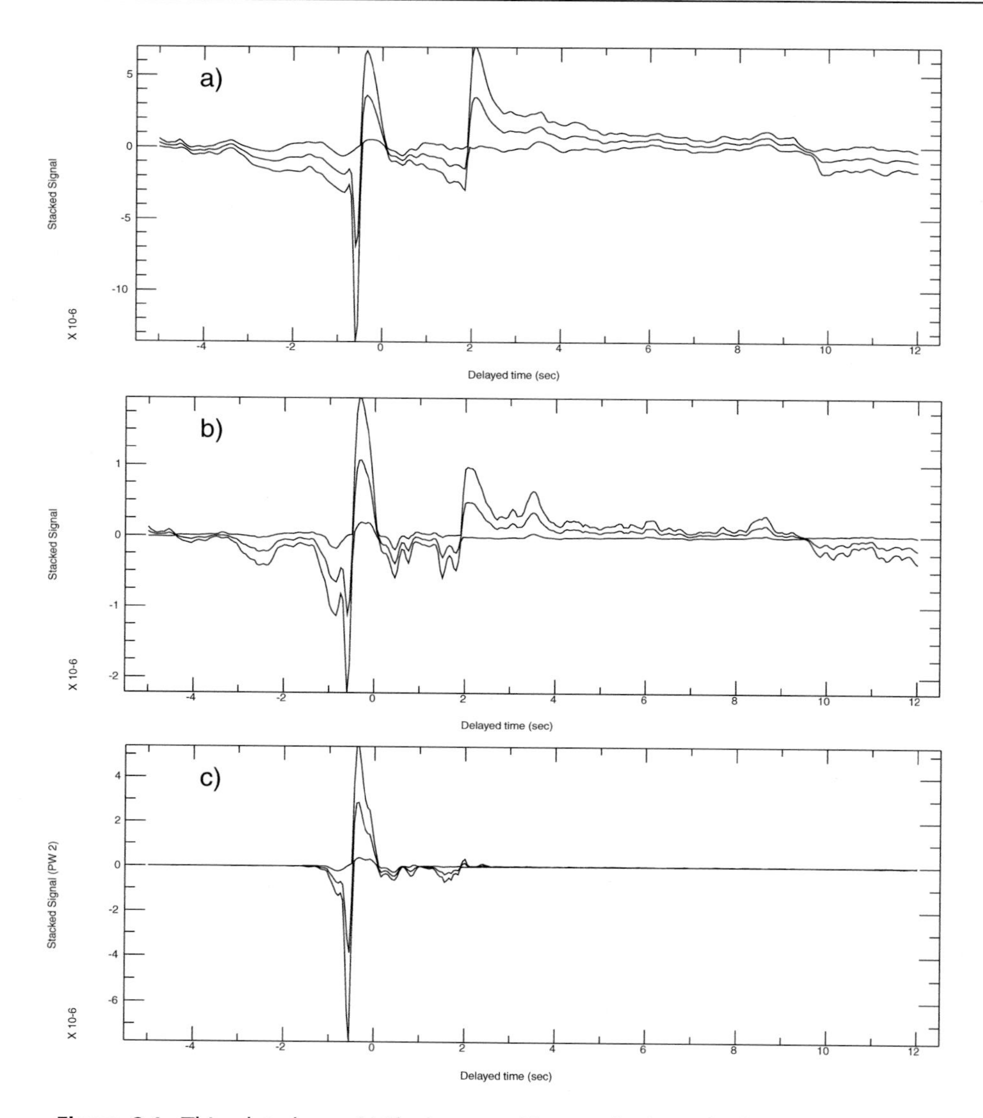

Figure 8.4 This plot shows SAC's three stacking methods applied to Japanese network recordings of the 9 October 2006 North Korea nuclear test. The traces are roughly aligned on the predicted Pn arrival from the explosion. (*a*) Linear stack, (*b*) *N*th root stack ($N = 2$), (*c*) phase-weighted stack ($N = 2$). SUMSTACK produced each of the plots, which consists of a central trace (the stacked estimate) and the $\pm 1\sigma$ jackknife uncertainty on each sample. Note the higher pre-arrival noise levels in the linear stack and the nearly total suppression of the incoherent second arrival at 2 s by the phase-weighted stack. The later arrival is probably due to trace misalignment rather than any property of the source.

Table 8.1 PLOTRECORDSECTION cursor operations

Char.	Description
K	Finish PPK
Q	Finish PPK
X	Define bound of new range window (low then high)
C	Define bound of new crop window (range and time)
O	Return to last range/crop window (the five most recent windows are kept)
R	Redraw current window, clearing messages
S	Select all traces in a distance range (first low range, then high). Sets blackboard variables[a] SSSSELECTED and SSSUNSELECTED
T	Toggle selected/unselected status of trace
+	Increase trace sizes by a factor of 2[b]
-	Decrease trace sizes by a factor of 2[b]
M	Define point on a moveout curve; displays line fixed to cursor leading from previous point on curve. X and O allowed while moveout curve being defined
U	Undo previous point on moveout curve
V	Report slope of moveout curve (s/km or s/deg). Sets blackboard variable VAPP
P	Pick points along moveout curve (at peak P+, trough P-, zero P0). Sets blackboard variable[a] SSSPICKS

[a]Blackboard variable is set or removed after each use of PLOTRECORDSECTION.
[b]Redraws window.

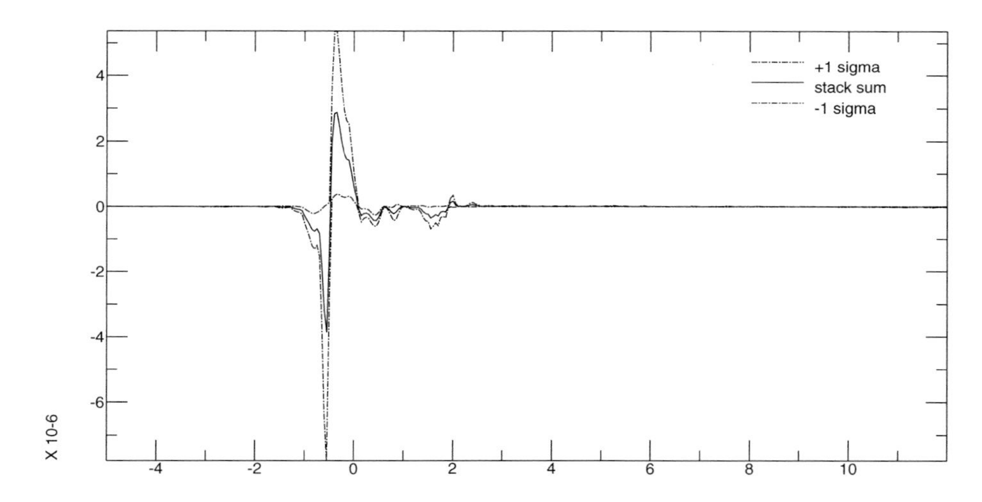

Figure 8.5 Stacked trace from an $N = 2$ phase-weighted stack, saved by WRITE-STACK, re-read and displayed by PLOT2. The stack is shown along with its $\pm 1\sigma$ jackknife uncertainty.

window to expand a set of traces. The O key unzooms to the previous window. Whereas only the cursor's range position is relevant when the X key is used, the C key defines a crop of the data in both time and range. The cursor's position defines the lower left corner of the crop window at the first C key press, and the upper right corner is defined by the second.

Only the selected time and distance window is displayed following the crop. The O key also unzooms to the previous window. R redraws the display.

Trace features may be measured and recorded by using the M key to define anchor points of a moveout curve. Every time the M key is pressed, it defines a new point on the moveout curve. A line leads from the previous moveout curve point to the cursor's position. This defines a slope (in seconds/degree or seconds/kilometer depending on the range unit) related to an apparent velocity of a feature recorded in traces at different ranges. Each time the V key is pressed, SAC reports the slope of the last segment of the moveout curve on the screen (R will erase previous slopes and redraw the screen). The X and C commands may be used to zoom in on portions of the record section even while defining a moveout curve. The moveout curve stays active until `PLOTRECORDSECTION` finishes or until a pick is made.

The points where the moveout curve crosses the trace baselines in the plot provide a time reference where picks on each trace may be made. When a moveout curve is active, the commands P+, P- and P0 cause a pick to be made on every trace crossed by the moveout curve. The nearest peak (+), trough (-) or zero (0) crossing to the intersection defines the pick. SAC saves the pick values in the blackboard variable called SSSPICKS as (trace number, time) pairs. These may be read and handled using the `WHILE READ` command in a loop to act on each pick: write it to a file, change the properties of the trace in the stack, etc.

Groups of traces may also be selected graphically. The range position of the cursor defines the beginning or end of a group of traces after S is typed. To flag the selected traces, the trace's baseline tab changes to the color red. Individual traces also may be selected or deselected by using the T character to toggle its selection status. Similar to the way that trace pick information is returned, SAC sets the trace numbers of the selected traces in the SSSSELECTED blackboard variable, and the trace numbers of the unselected traces in the SSSUNSELECTED blackboard variable. The trace numbers may be read and processed in a loop in a macro using the `WHILE READ` command.

The SSSPICKS, SSSSELECTED and SSSUNSELECTED blackboard variables are set and erased after each `PLOTRECORDSECTION` command. If relevant, they must be processed or saved before viewing the next record section. Use *quotebb* when saving these values to preserve the (number, time pick) association.

Picking examples

The following command examples illustrate the ideas associated with trace selection and time picks in record sections using some of SAC's built-in data. The `DATAGEN SUB LOCAL` dataset is a collection of trace records of a local event. The data contain a few bad traces that will be discarded after visual selection.

Ex-8.6

```
SAC> sc cat ex-prs-remove.m
* Contents of file ex-prs-remove.m
if Y EQ (existbb SSSSELECTED)
   while read SSSSELECTED n
      message "Removing &$n$,filename&"
   enddo
   deletestack %SSSSELECTED%
else
   message "No traces selected for removal"
```

```
endif
SAC> datagen sub local *.z; rmean ;* Remove baseline offset
cal.z cao.z cda.z cdv.z cmn.z cps.z cva.z cvl.z cvy.z
SAC> sss
SAC/SSS> tw 10 50; dw usedata
SAC/SSS> prs c                 ;* Use T to select bad traces
SAC/SSS> m ex-prs-remove.m
 Removing cda.z
 Removing cps.z
SAC/SSS> prs                   ;* Redisplay without bad traces
```

The final example involves making picks using a moveout curve. The P-wave arrival in the same collection of traces is clear and has an apparent velocity that causes it to appear later in time as range increases. In order to change the static shifts of each trace so that the P arrival appears at zero relative time in the window, use the pick option to approximately align to the P-wave peak. Then, adjust the static delay for each trace to remove the moveout with range.

Ex-8.7

```
SAC> sc cat ex-prs-align.m
* Contents of file ex-prs-align.m
if Y EQ (existbb SSSPICKS)
   while read SSSPICKS n t
      message "Offset for &$n$,filename& changed to $t$"
      changestack $n$ delay -$t$ s
   enddo
else
   message "No picks made"
endif
SAC> datagen sub local *.z; rmean
cal.z cao.z cda.z cdv.z cmn.z cps.z cva.z cvl.z cvy.z
SAC> sss
SAC/SSS> tw 10 50; dw usedata
SAC/SSS> prs c                 ;* Use M and P+ for picks
SAC/SSS> m ex-prs-align.m
 Offset for cal.z changed to 20.848566
 Offset for cao.z changed to 23.187908
 Offset for cda.z changed to 23.281334
 Offset for cdv.z changed to 19.336441
 Offset for cmn.z changed to 20.839802
 Offset for cps.z changed to 22.320883
 Offset for cva.z changed to 20.935202
 Offset for cvl.z changed to 22.021576
 Offset for cvy.z changed to 22.235571
SAC/SSS> tw -10 40             ;* New reference time
SAC/SSS> prs                   ;* Aligned to P-wave arrivals
```

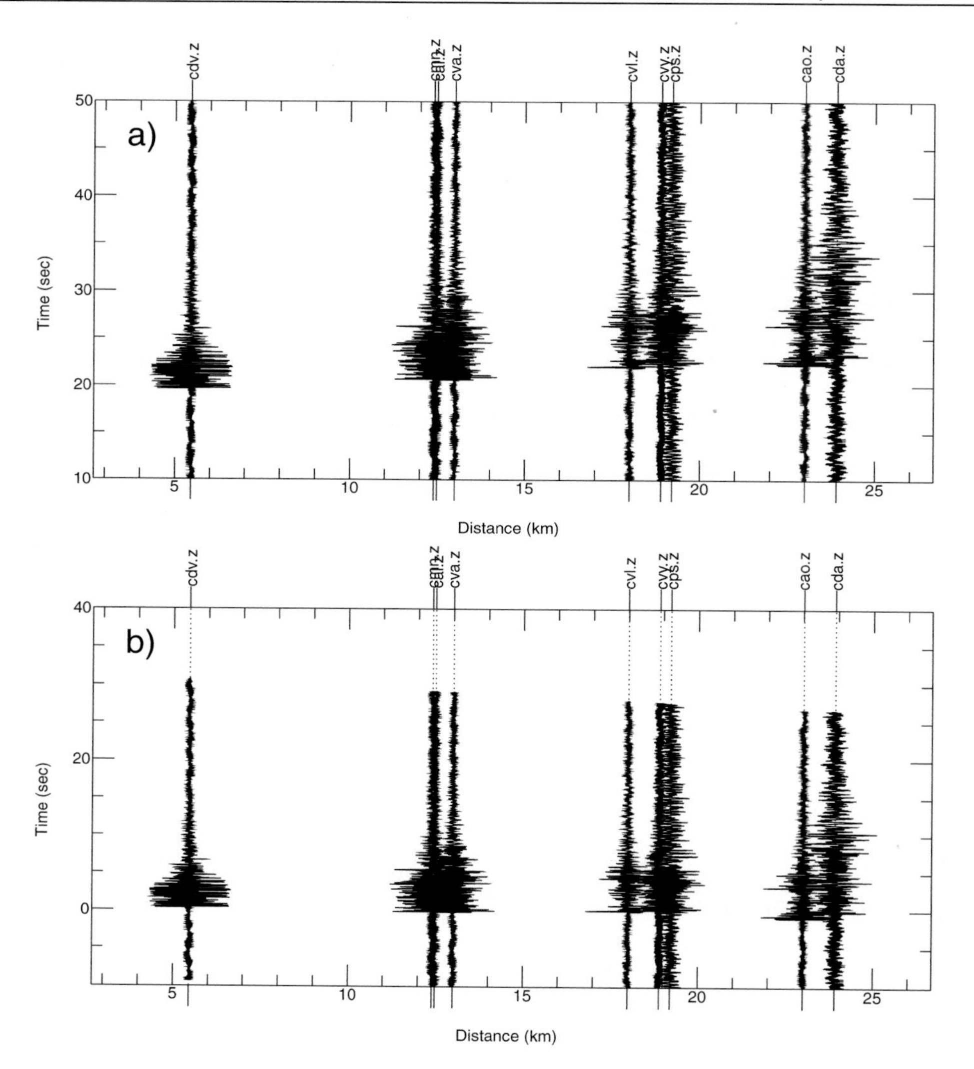

Figure 8.6 Vertical component trace collection of a local event. Time zero is the origin time of the event. Thus, with the default static delays, the P-wave arrival is later with increasing range (*a*). After making a set of picks along a moveout curve, the static delays for each trace are adjusted to bring the P arrival to relative time zero in each trace (*b*). The adjustment is made automatically by a macro processing the picks retained in SAC's memory.

Figure 8.6 shows the trace configuration before and after the picking defines the new static offsets for each trace. The first use of `PRS` involves a graphical interaction. Using M to define a moveout curve extending from (20 s, 5 km) to (22 s, 25 km), and then using P+ to make picks, the offsets are available for the macro `ex-prs-align.m` to make new static shifts. The second record section plot shows the P arrivals aligned at relative time zero in the time window.

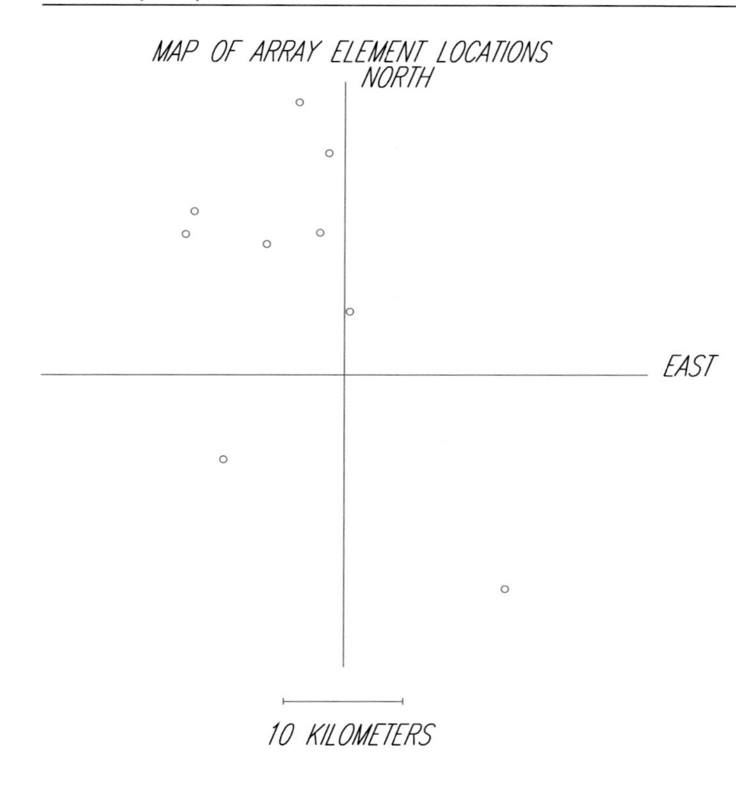

Figure 8.7 This plot shows an X–Y map of stations in a local network recording an earthquake, produced by ARRAYMAP. The origin of the map is the epicenter, and circles show individual station locations. (For color version, see Plates section.)

8.3 ARRAY MAPS

Whenever a trace collection is read in, the geometry of the stations may be relevant. For example, the azimuthal coverage around the event origin might be useful to assess how well the event is located, because gaps along particular azimuths lead to increased epicentral uncertainty in that direction. Another example is: when a regional velocity model is to be derived from the moveout of an arrival recorded by a regional network, path coverage knowledge would help assess the model's applicability.

SAC provides a quick way to make the assessments by producing a station map evoked by the ARRAYMAP command. SAC makes a local Cartesian approximation based on either the first station's location or a fixed point given by a CENTER latitude and longitude and calculates a range and azimuth to every other station that is turned into an (X,Y) coordinate. The commands below produce a map (Fig. 8.7) of a local network centered on the event epicenter:

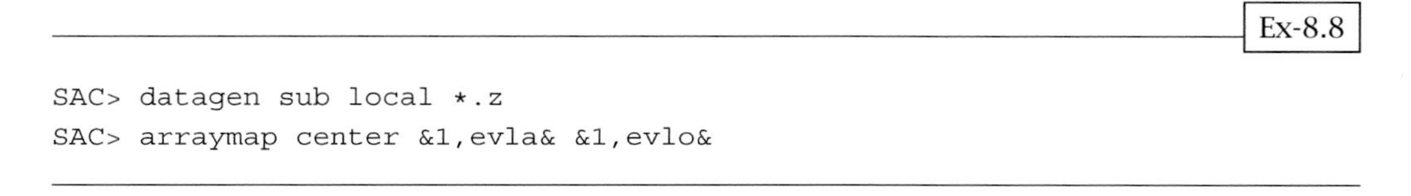

Ex-8.8

```
SAC> datagen sub local *.z
SAC> arraymap center &1,evla& &1,evlo&
```

Array maps are also useful for assessing the geometry of small-aperture arrays used for nuclear discrimination purposes. ARRAYMAP COARRAY will plot the array symmetries that define its angular resolving power.

8.4 BEAMFORMING

BBFK and BEAM are two commands that SAC provides to do array analyses of the approach of a signal that is modeled as a plane wave. The plane wave assumption is clearly better when the array is small-aperture, but over teleseismic distances the approximation holds even for regional seismic arrays (e.g., the national seismic networks of Japan and parts of USArray).

BBFK (broadband f-k) is designed to determine the bearing of approach of a signal toward the array. Under the signal's plane wave approximation, it has a horizontal slowness p. The wavenumber k is related to frequency f and slowness via

$$k = f \times p \tag{8.3}$$

If f is a frequency in Hz and p a slowness in s/km, then k is 1/km. The signal is assumed to be spread across the wavenumber spectrum, which is equivalent to saying that it is a polychromatic signal at a single slowness p (a good model for a seismic wave arrival) or that it is a monochromatic signal at a range of slownesses (unlikely in the seismic case). If an array of sensors is spread over the ground, both the array geometry and the assumption of a wideband signal help to break the spatial aliasing inherent in the approach. In particular, the wideband assumption means that the wavenumber spectrum, plotted in polar form with the angle representing approach bearing, will radiate in a ridge from zero wavenumber (the center) toward the outer circumference of the plot. Spatial aliasing due to the array will also create ridges parallel (in a Cartesian sense) to the true spectral bearing but at different offsets from the center. Once it is identified, the peak amplitude along the ridge will show the dominant wavenumber of the signal. The horizontal slowness may be calculated from Equation 8.3.

A synthetic dataset may be used to illustrate the ideas, based on the macro listed below:

Ex-8.9

```
SAC> sc cat ex-bbfk.m
* BBFK example macro. Usage: m ex-bbfk.m [zero | n-s | e-w | diag]
$default 1 zero
setbb cutew * cutns *               ;* no cuts by default
if $1$ eq diag
   setbb cutew cutim cutns cutim    ;* time shifts N-S and E-W dirs
endif
if $1$ eq n-s
   setbb cutns cutim                ;* time shift on N-S dir
endif
if $1$ eq e-w
   setbb cutew cutim                ;* time shift on E-W dir
endif
cuterr fillz                        ;* Zero exterior trace areas
setbb xo -1 yo 0                    ;* xo and yo are x-y offsets
do f list a b c d e f g h           ;* E-W array arm
   fg seismogram; rmean             ;* Sample data
   %cutew% (&1,b& - %xo%) (&1,e - %xo%)      ;* shift: add/cut 1 s
   ch kuser1 &1,kstnm& user7 %xo% user8 %yo% ;* element coord.
   write /tmp/arr_r$f$.sac                   ;* write data file
```

```
   setbb xo (before . (%xo% + 1))             ;* increment X
enddo
setbb xo 0 yo -1                      ;* Reset offsets
do f list a b c d e f g h             ;* N-S array arm
   fg seismogram; rmean               ;* Sample data
   %cutns% (&1,b& - %yo%) (&1,e - %yo%)       ;* shift: add/cut 1 s
   ch kuser1 &1,kstnm& user7 %xo% user8 %yo% ;* element coord.
   write /tmp/arr_b$f$.sac
   setbb yo (before . (%yo% + 1))             ;* increment Y
enddo
sc rm /tmp/arr_rb.sac                 ;* Repeated array element
read /tmp/arr_[r,b]*.sac              ;* Read traces
bbfk filter off wave 1 pds norm size 180 100
SAC> m ex-bbfk.m zero
SAC> m ex-bbfk.m n-s
SAC> m ex-bbfk.m e-w
SAC> m ex-bbfk.m diag
SAC> arraymap
```

The synthetic data is a built-in seismogram that is replicated at each station in the array with time shifts of 1 s related to either or both of its horizontal and vertical positions. Thus the apparent horizontal velocity is 1 km/s N-S, E-W or 0.707 km/s at 45°. The array form is that of the UK Atomic Energy Agency (UKAEA) arrays, a rectilinear cross with non-symmetric arms (Rost and Thomas, 2002) (Figure 8.8). CUTERR FILLZ adds 1 s to the front or cuts 1 s from the end of the synthetic data trace for every kilometer of offset along the array arms. The CUTIM commands either act on the trace or are comments (depending on the macro parameter), being expanded through the cutew and cutns blackboard variable values, set with SETBB. BBFK expects array offsets to be set in the USER7 and USER8 file header variables for small-aperture arrays. Otherwise they are taken from the STLA and STLO location of the station.

Figure 8.9 shows the results of the BBFK analysis on the synthetic data. With no delays, the signal (plane wave) appears to be emerging vertically and arriving at each sensor simultaneously. The peak is at zero wavenumber (infinite horizontal velocity), and the response function of the array (Aki and Richards, 1980) is evident in the aliasing pattern due to the element separation (Fig. 8.8). With delays along the N-S array arm, the signal appears as a wideband ridge extending N-S. The peak is at $k = 0.12121$ km^{-1}. We know $p = 1$ s/km, so the signal's dominant f must be 8.25 Hz, which an inspection of the spectrum bears out.

The diagonal approach angle shows the aliasing of the signal most clearly. The ridge extending from the origin is the true plane wave representation. The others are aliases of it. With these observations in hand, BBFK analysis of a real signal may be better appreciated.

The built-in DATAGEN SUB ARRAY dataset contains regional array records of a teleseismic arrival. The record section plot in Figure 8.10, constructed using the commands below, shows that the core phases PKPdf, PKPbc and PKPab are fairly clearly recorded by the network. The moveout across the array of the core arrivals (PKPdf horizontal slowness is approximately 1.6 s/deg, PKPbc 2.5 s/deg and PKPab 4.1 s/deg). From PKPab we can estimate that the largest expected wavenumber k will be 0.037 km^{-1}. Therefore, the maximum

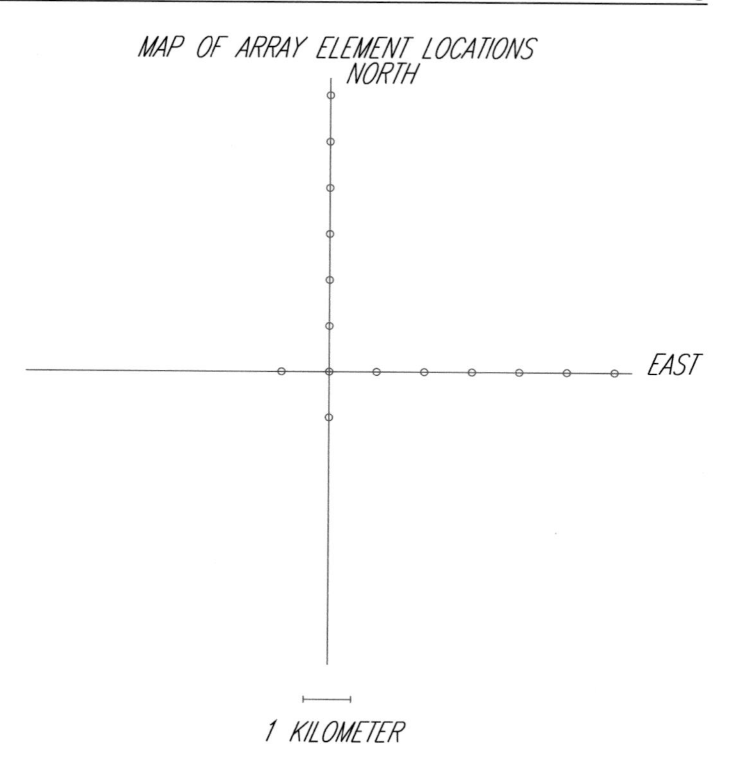

Figure 8.8 This plot shows an X–Y map of the example local array whose synthetic response is explored by the `ex-bbfk.m` macro (see text). The plot is produced by ARRAYMAP. Elements are spaced every kilometer along two asymmetric arms, similar to the UK Atomic Energy Agency design. (For color version, see Plates section.)

k to be explored is 0.045 km^{-1}. BBFK retains the peak position in blackboard variables BBFK_WVNBR and BBFK_BAZIM.

Ex-8.10

```
SAC> datagen sub array *.SHZ
SAC> sss
SAC/SSS> dw usedata units deg; tw 1100 1170
SAC/SSS> traveltime model ak135 phase PKPbc PKPdf PKiKP PKPab
SAC/SSS> line fill yellow/0
SAC/SSS> prs ttime on label kstnm
SAC/SSS> qs
SAC> bbfk wave 0.045 size 180 100 exp 4 ;* Finer az. sampling
SAC> message "k %BBFK_WVNBR%, az %BBFK_BAZIM%"
```

The result is shown in Figure 8.11. The contoured power shows a peak at k = 0.0159 km^{-1} along azimuth 355°. Using SUMSTACK and WRITESTACK, a stack of the traces aligned on PKPbc arrival indicates that its spectral peak lies at 0.7 Hz. At this frequency, the peak wavenumber indicates that the slowness at peak power is p = 0.0227 s/km or 2.526 s/deg. This corresponds to the PKPbc slowness expected for this distance range, which is the major arrival in the record section (Fig. 8.10). The back-azimuth at the

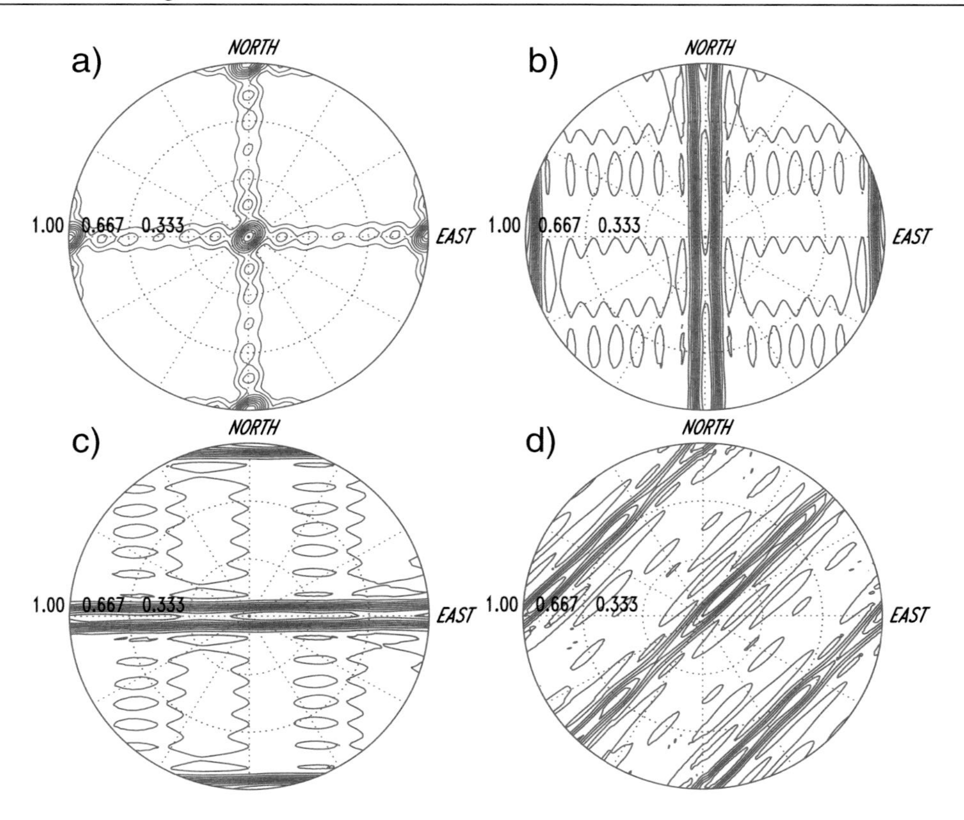

Figure 8.9 Polar plots of contoured beam power as a function of azimuth (angle) and wavenumber (radius) for a synthetic array (Fig. 8.8). The maximum radial wavenumber is 1 km^{-1} in all cases. (*a*) Zero-slowness (vertical) arrival. The cross-shaped pattern is the response function of the array; the power peak is at zero wavenumber. Aliased peaks also exist N, S, E and W bearings at $k = 1\ \mathrm{km}^{-1}$. (*b*) Arrival along a N–S bearing. The ridge extending N–S includes the origin and represents the unaliased signal. (*c*) Arrival along an E–W bearing. Again, the ridge extending E–W includes the origin and represents the unaliased signal. (*d*) Arrival along 45°. Aliasing is more evident in this case because the wavenumber is 0.707 km^{-1}. (For color version, see Plates section.)

approximate center of the network, station ESK, is 354.4°, which confirms the detection made by BBFK.

With this approach information in hand, the BEAM command may be used to form a trace aligned along that azimuth at a reference station located at 55.3167N, 3.2050W,

```
SAC> beam bearing 355 vel (1 / 0.0227) ref 55.3167 -3.2050
```

with the display shown in Figure 8.12. To examine the waveform details, use the BEAM WRITE option to write a separate trace file, read it in, and zoom in on the waveform using PPK.

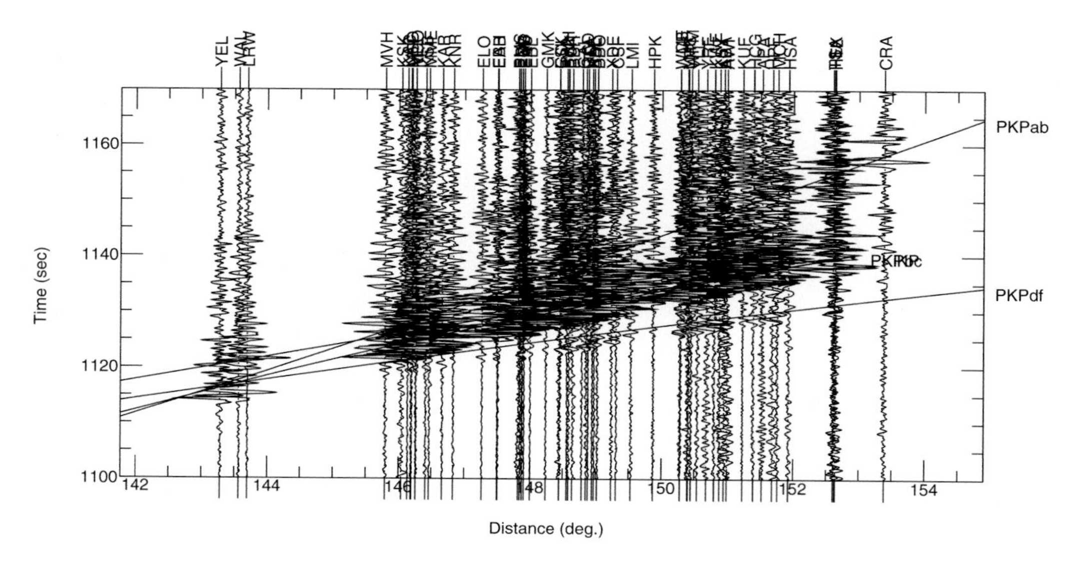

Figure 8.10 The record section shows a teleseismic arrival recorded by a regional seismic network that constitutes an array. The superimposed travel-time curves show that the arrivals PKPdf, PKPbc and PKPab are well recorded across the array. (For color version, see Plates section.)

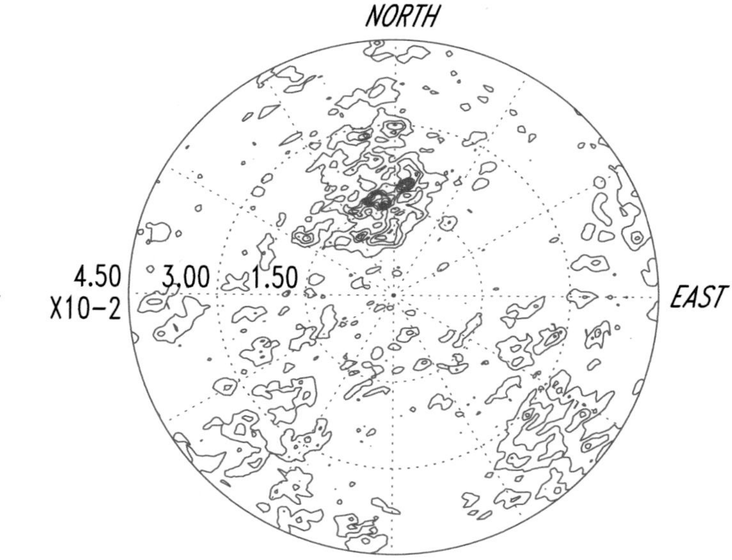

Figure 8.11 BBFK analysis of the array data (see Fig. 8.10) shows a peak along a northerly bearing and at wavenumber slightly larger than 1.5×10^{-2} km^{-1}. (For color version, see Plates section.)

8.5 TRAVEL-TIME ANALYSIS

It is not immediately apparent but SSS also has uses when applied to single station, three component data. The `TRAVELTIME` command is able to add theoretical arrival times to traces. The traces are usually from stations at different ranges from a single event. Thus

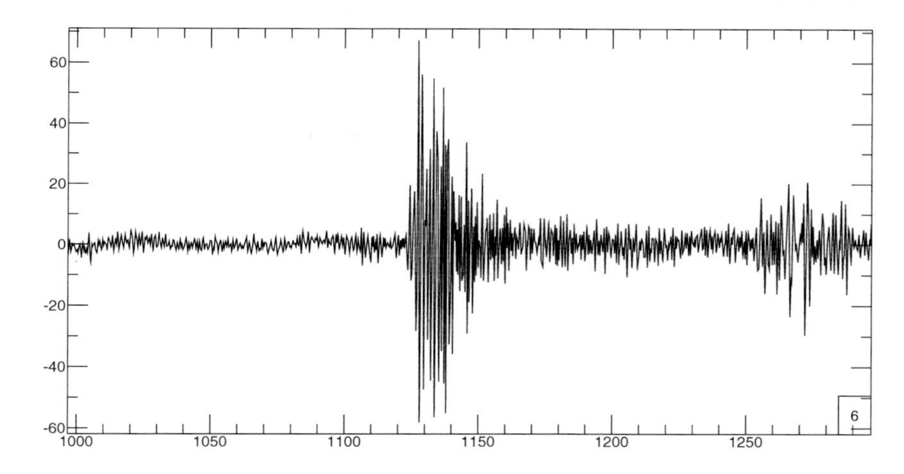

Figure 8.12 Array beam formed from BBFK analysis of the array data (see Fig. 8.10) using the spectral peak's back-azimuth and wavenumber. The entire trace is shown by the `BEAM` display, but it can be saved as a file and its features examined using `READ` followed by `PPK`.

the travel times document the moveout of an arrival with range. If the traces just happen to be three traces at the same station, well, so be it:

```
SAC/SSS> help traveltime

 SUMMARY:
 Computes travel-time curves for pre-defined models or reads
 travel-time curves from ascii text files.

 SYNTAX:
    TRAVELTIME [MORE] MODEL <name> [DEPTH <depth>]
        [PICKS <n>] [PHASE <phs> ...]

    TRAVELTIME [MORE] [DIR CURRENT|<dir>] CONTENT <cont>
        [HEADER <n>] [UNITS DEGREES|KM] [PICKS <n>] <file> ...

    TRAVELTIME [DIR CURRENT|<dir>] [PICKS <n>] TAUP <file>
```

The `PICKS` option is the key one. The various forms of the command will assign arrivals to a sequence of file header variables starting with T<n>. SAC labels the picks with the phase names. When plotted with `PLOT1` as a three-component set, the predicted arrival times will be shown as labeled picks. Up to ten picks may be added to each trace in this way (to `T0` ... `T9`).

The various forms of `TRAVELTIME` allow theoretical travel-time picks to be provided in different ways. Tables for the *iasp91* (Kennett and Engdahl, 1991), *ak135* (Kennett *et al.*, 1995) and *sp6* (Morelli and Dziewonski, 1993) models are built in, and naming the proper model in the first command form supplies their predicted values for the pick values for the chosen phase names. The other forms read travel times from a file, either as a list of range versus time or written by the Tau-p toolkit (Crotwell *et al.*, 1999).

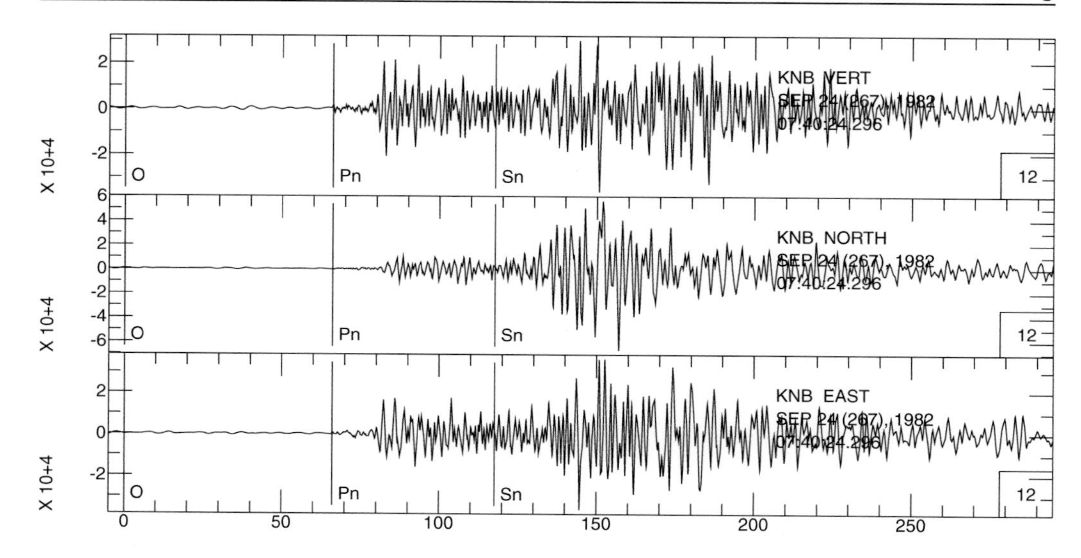

Figure 8.13 Three-component data from a single station with theoretical travel-time picks added by the TRAVELTIME command. PLOT1 adds the picks and their phase name labels to the combined trace plot.

Figure 8.13 shows a plot resulting from a set of picks arranged by the following commands:

Ex-8.11

```
SAC> datagen sub regional knb.[z,n,e]; rmean
SAC> sss
SAC/SSS> traveltime phase Pn Sn picks 0
SAC/SSS> qs
SAC> plot1
```

This is a simple way to add theoretical pick information to a plot. Another way is to use a SAC macro to make a three-component plot annotated (using PLOTC) with predicted arrivals from travel-time models. This way is more complex but does not involve using the scarce resource of file header variables for plot-related information. See Figure 11.2 for an example of this type of display.

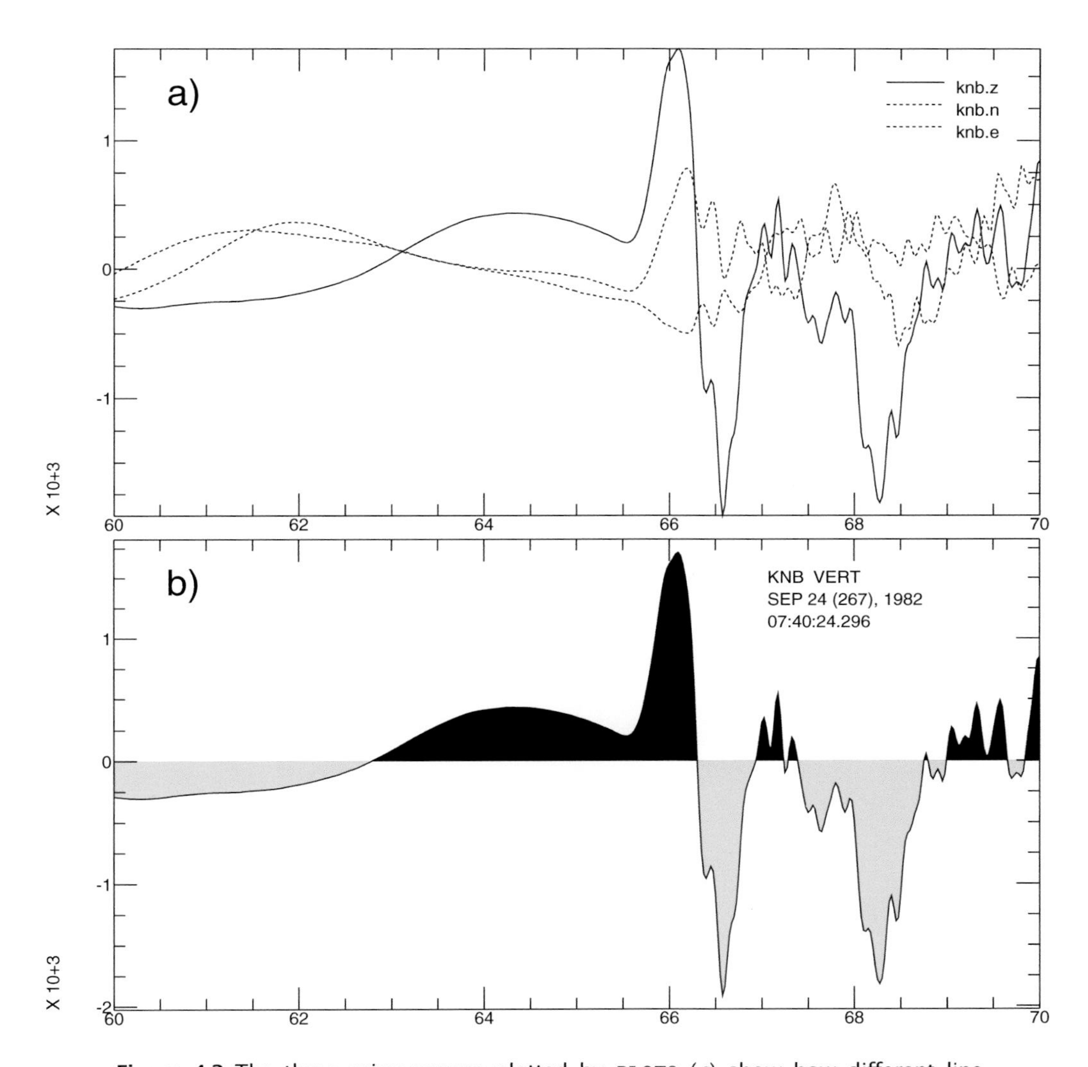

Figure 4.2 The three seismograms plotted by PLOT2 (*a*) show how different line styles enhance comprehension of the display. The line style for the horizontal component traces is different from the vertical component trace. PLOT2 shows this in the legend at top right in the display frame. Another line style feature that SAC provides to improve trace feature visibility is color-filled traces (*b*). The positive and negative areas may be filled with the same color, different colors, or limned selectively.

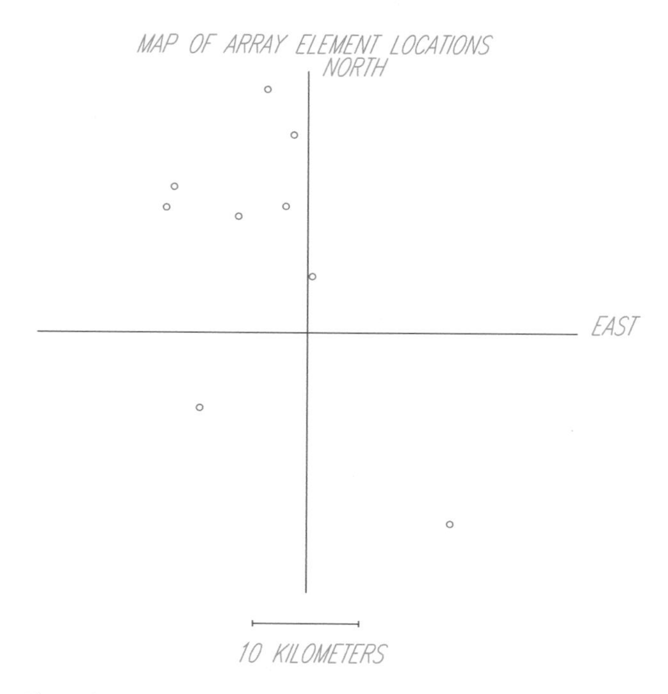

Figure 8.7 This plot shows an X–Y map of stations in a local network recording an earthquake, produced by ARRAYMAP. The origin of the map is the epicenter, and circles show individual station locations.

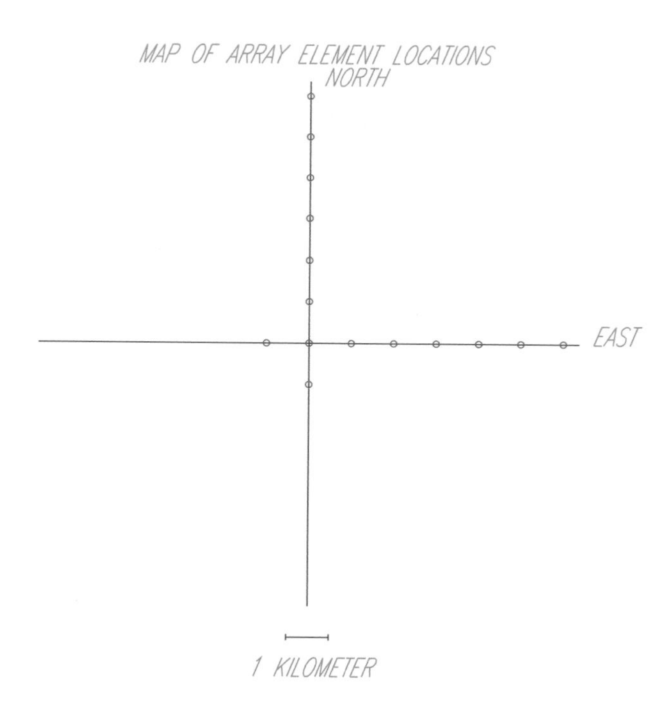

Figure 8.8 This plot shows an X–Y map of the example local array whose synthetic response is explored by the `ex-bbfk.m` macro (see text). The plot is produced by ARRAYMAP. Elements are spaced every kilometer along two asymmetric arms, similar to the UK Atomic Energy Agency design.

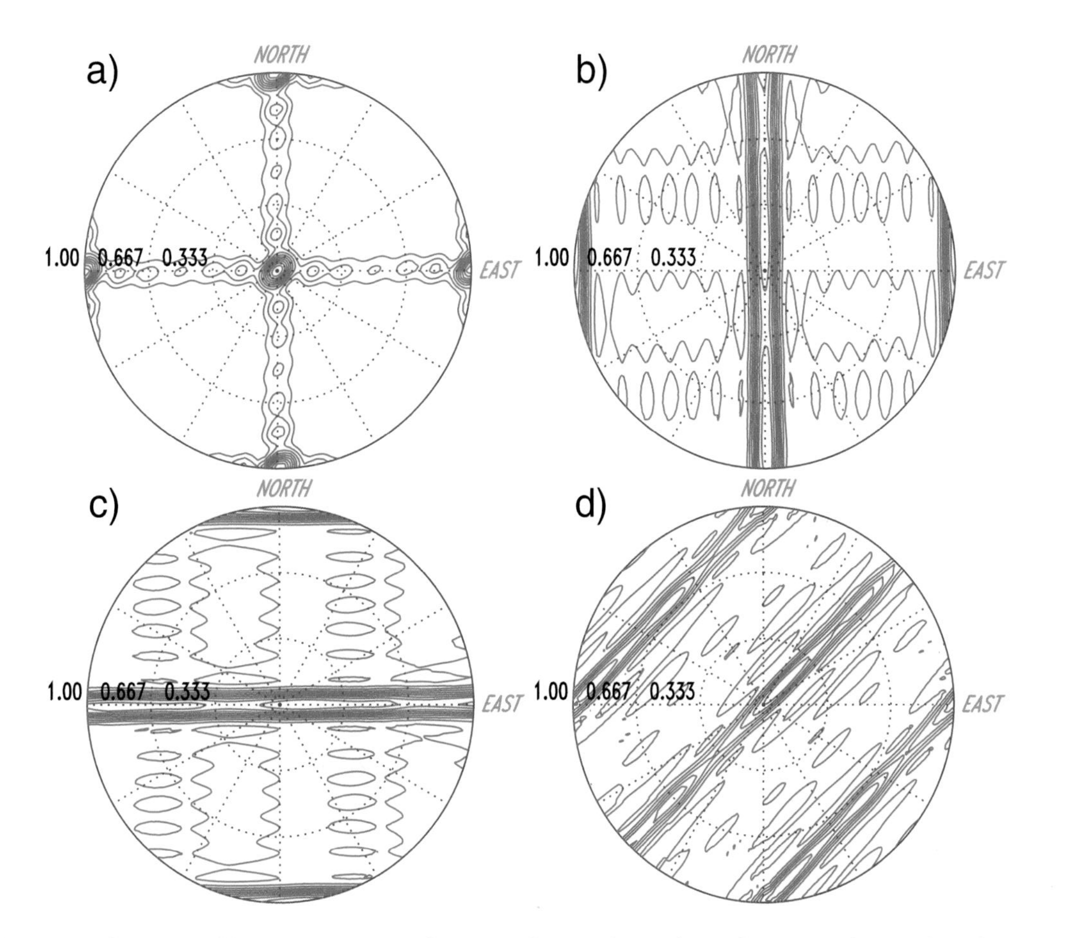

Figure 8.9 Polar plots of contoured beam power as a function of azimuth (angle) and wavenumber (radius) for a synthetic array (Fig. 8.8). The maximum radial wavenumber is 1 km^{-1} in all cases. (*a*) Zero-slowness (vertical) arrival. The cross-shaped pattern is the response function of the array; the power peak is at zero wavenumber. Aliased peaks also exist N, S, E and W bearings at $k = 1$ km^{-1}. (*b*) Arrival along a N–S bearing. The ridge extending N–S includes the origin and represents the unaliased signal. (*c*) Arrival along an E–W bearing. Again, the ridge extending E–W includes the origin and represents the unaliased signal. (*d*) Arrival along 45°. Aliasing is move evident in this case because the wavenumber is 0.707 km^{-1}.

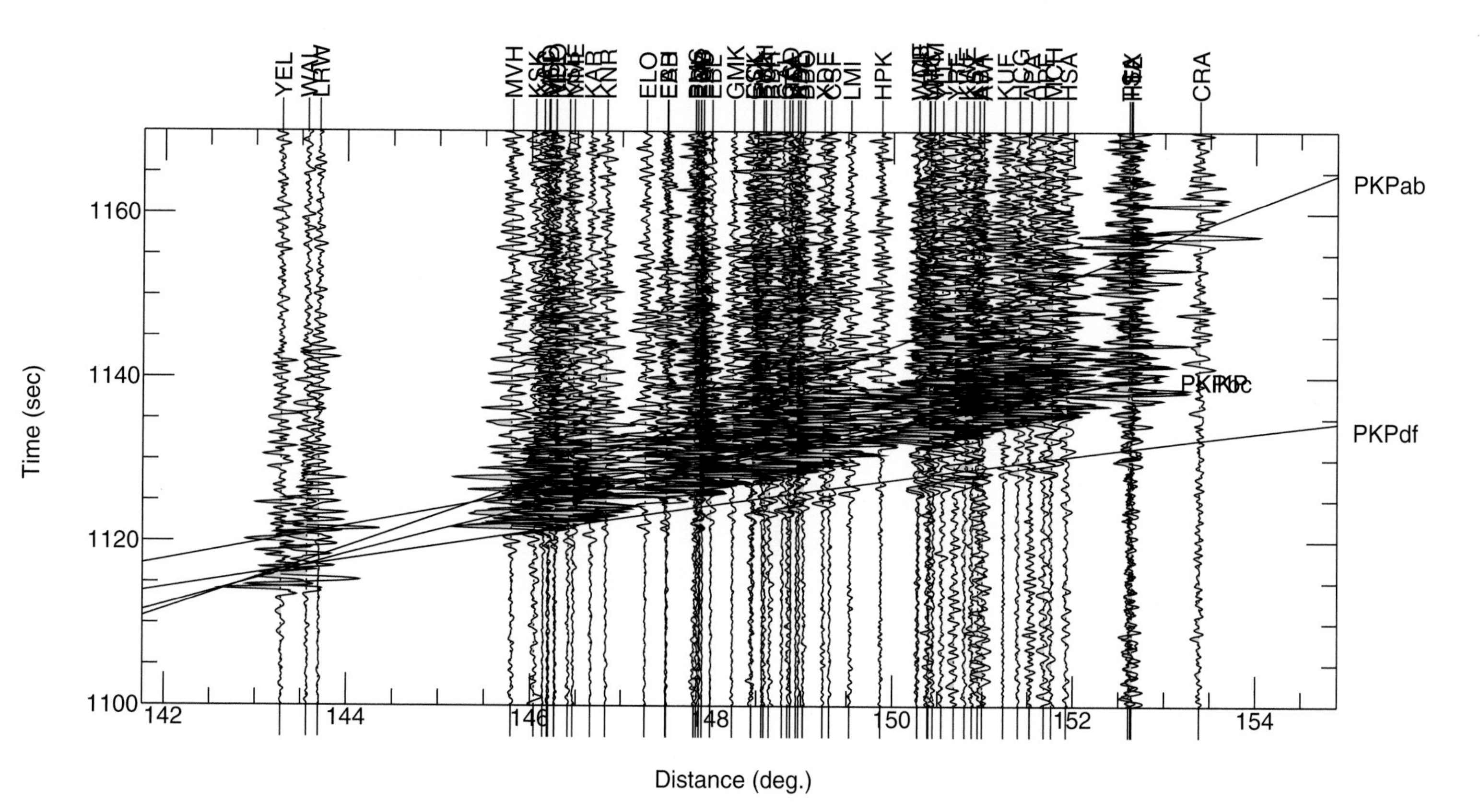

Figure 8.10 The record section shows a teleseismic arrival recorded by a regional seismic network that constitutes an array. The superimposed travel-time curves show that the arrivals PKPdf, PKPbc and PKPab are well recorded across the array.

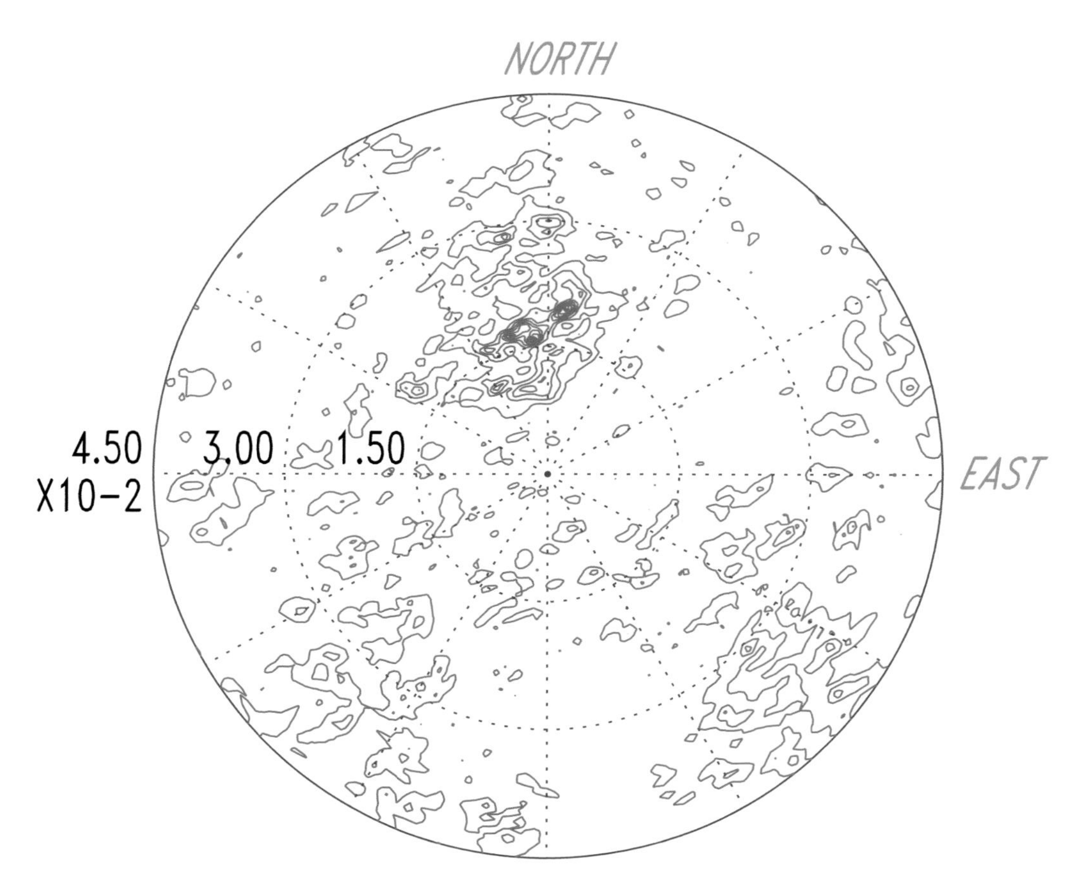

Figure 8.11 BBFK analysis of the array data (see Fig. 8.10) shows a peak along a northerly bearing and at wavenumber slightly larger than 1.5×10^{-2} km^{-1}.

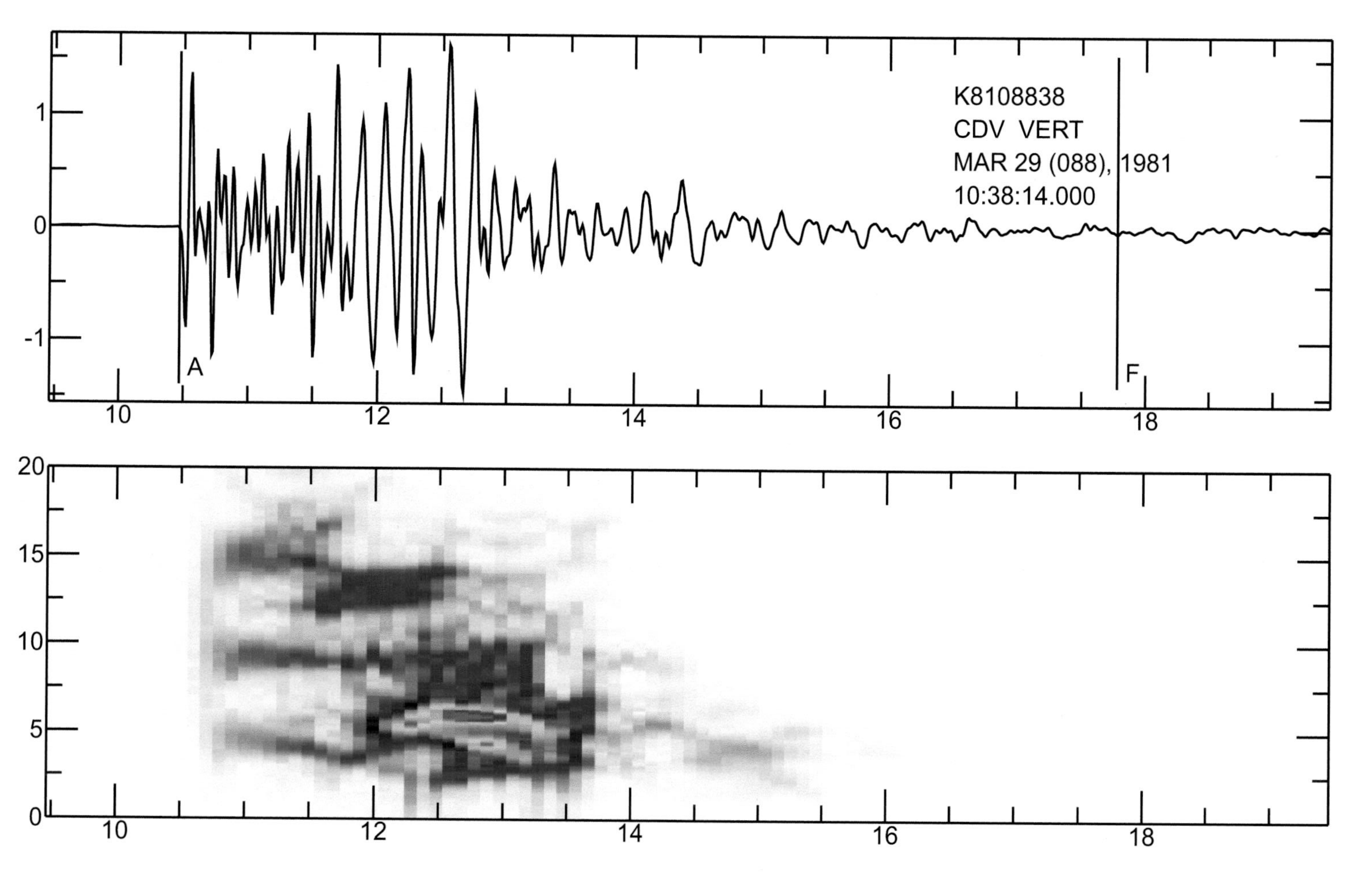

Figure 10.1 Spectrogram of a seismogram: (top) the time series and (bottom) the spectrogram. The X axis is time as it runs through the seismogram. The Y axis is frequency (Hz). In a time window of 2 s recalculated every 0.1 s through the seismogram, the color represents the spectral power at each frequency.

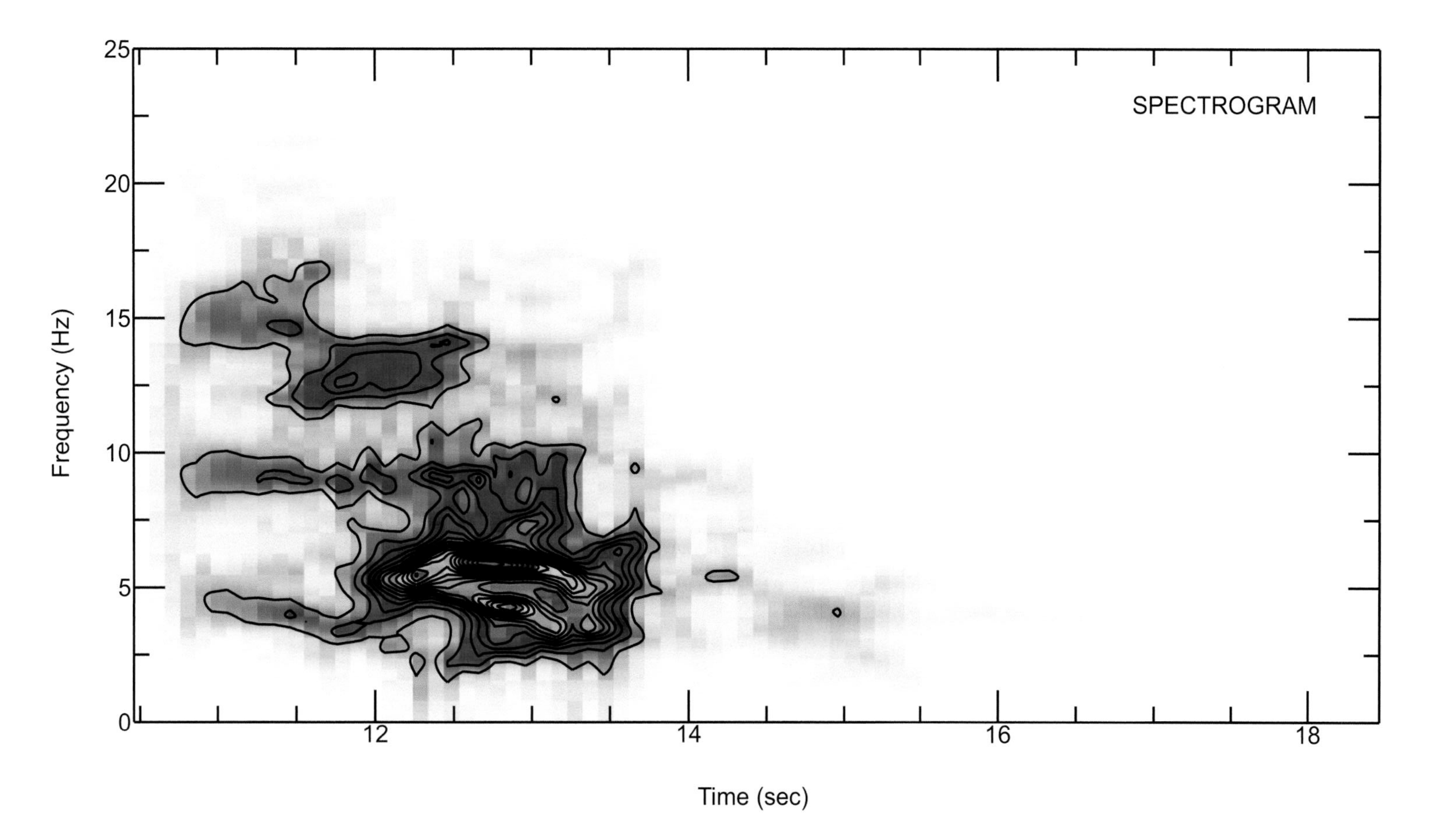

Figure 10.4 XYZ file data may be usefully displayed in alternative forms by SAC. The figure shows a spectrogram, saved as a 3D file, plotted in color with the `GRAYSCALE` command. It is overlaid with contours of spectral power. This provides both a quantitative report on the spectrogram and a qualitative graphical summary.

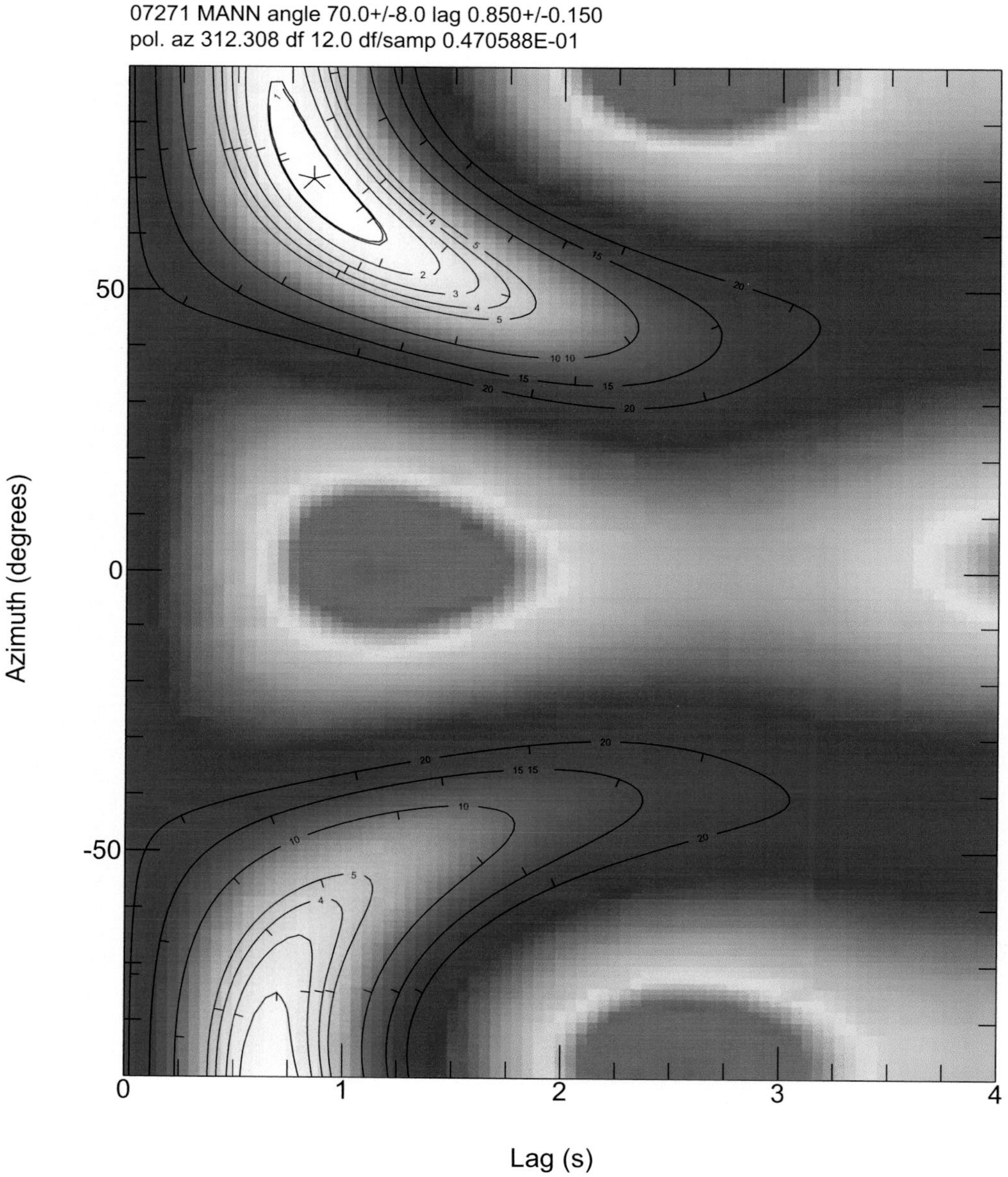

07271 MANN angle 70.0+/-8.0 lag 0.850+/-0.150
pol. az 312.308 df 12.0 df/samp 0.470588E-01

Azimuth (degrees)

Lag (s)

Figure 11.4 Results of the grid search over δt and ϕ. The optimum splitting parameters are shown by the star (top left). The 95% confidence region is shown (double contour), along with multiples of that contour level.

CHAPTER

NINE

Spectral estimation in SAC

9.1 SPECTRAL ESTIMATION

Spectral estimation is the task of taking a time series and decomposing it into its component frequencies. The total length of the time series and, if it has been sampled at a fixed time interval, the inter-sample spacing control the minimum and maximum frequencies that the time series contains. Different methodologies may be used to estimate the power at each frequency at constant intervals between these bounds. This so-called power spectrum provides a way to quantitatively characterize the frequency content of the time series. The information is usually presented in the form of a graph of power versus frequency.

The power spectrum informs further processing avenues for the time series. Usually this involves discrimination of any signal in the spectrum from noise. If the signal is of high quality, the signal's power will dominate the noise. Thus the power spectrum will be peaked in a frequency band containing the signal. If the goal is to design a filtering strategy to minimize the noise, this analysis will suggest the type of filter and the corner frequency to use. Another application might be to seek tidal resonances at a coastal site based on repeated sea level measurements, or a marigram. In this case, the frequencies of the spectral peaks are the desired information, and perhaps their widths or positional uncertainties.

Spectral estimation is technically complex due to the characteristics of the signal under study. Consequently, SAC provides a suite of analysis methods applicable to different signal types (transient versus stationary) and different properties of the spectrum (its broad shape or its narrowband features). These methods, power density (PDS), maximum likelihood (MLM) (Lacoss, 1971) and maximum entropy (MEM) (Burg, 1972), represent the state of the art in about 1985. None of the authors claimed particular expertise in spectral estimation, but in the field of seismology the only later developments unacknowledged by SAC of which we are aware are multi-taper methods (Thomson, 1982), estimation of the spectra of unequally sampled time series (Scargle, 1989), and wavelet methods for decomposing time series (Daubechies, 1992).

9.2 THE SPECTRAL ESTIMATION SUBPROCESS

SAC wraps spectral estimation into a separate subprocess (see Section 8.1). The reason is that only a small set of commands applies to spectral estimation, and those commands must be followed strictly.

The command `SPE` enters the subprocess. In it, SAC makes a spectral estimate for only one trace at a time. If the subprocess is invoked with more than one trace in memory, SAC will complain. Similarly, no trace in memory also elicits a complaint.

The spectral estimate begins with the calculation of the autocorrelation function of the trace. The autocorrelation itself may be examined or saved, but it is only an intermediate step to the spectral estimate itself. In the next step, select and apply one of the three estimation methods, PDS, MEM or MLM, to operate on the stored autocorrelation and produce the spectrum. Finally, view the spectrum by plotting it, or save it as a SAC trace in a file. Saved as files, the results from different estimation methods may be compared by using an overlay plot.

Correlation

The `COR` command calculates the autocorrelation function. SAC assumes that the trace represents a sample of a stationary statistical process. Consequently, any part of the trace is – spectrally – just like any other. The self-similarity means that the trace may be chopped into shorter segments whose spectral estimates may be averaged together to reduce the uncertainty of the whole trace's spectrum. However, this reduced uncertainty trades off with a decreased frequency resolution of the spectrum because the shorter segments have a wider frequency stride in their spectra.

The `COR` command therefore allows control over the number of segments and their length, each called a *window*. Arbitrary segmentation cuts of the trace will not have zero endpoints, which will lead to high-frequency spectral artifacts unless they are tapered. Thus `COR` offers a choice of tapers on each segment: Hamming, Hanning, cosine, triangle (Bartlett) and rectangle (or boxcar, or, equivalently, none). The `COR TYPE` option chooses the window, and the `COR LENGTH` and the `COR NUMBER` select the window length and number in the trace.

Some high-resolution estimation methods, particularly MLM, are sensitive to the high-frequency content of the data where specific frequencies might have near-zero power. This leads to apparent instability in the spectral estimate and is remedied by smoothing out the high frequencies. In the time domain, this corresponds to adding high-frequency noise to the trace and is therefore called *prewhitening*. Use the `COR PREWHITEN ON` option for prewhitening. Unfortunately, this choice actually changes the data in the trace itself, and the data must be re-read to repeat `COR` with prewhitening off.

The normalization for the autocorrelation function depends on whether the signal is stationary or transient. SAC gives a choice of normalization through the `COR TRANSIENT` or the `COR STOCHASTIC` options.

Once the autocorrelation is available, `PLOTCOR` will plot it. In itself, the autocorrelation is not very interesting. It is an oscillatory function whose maximum is at zero lag

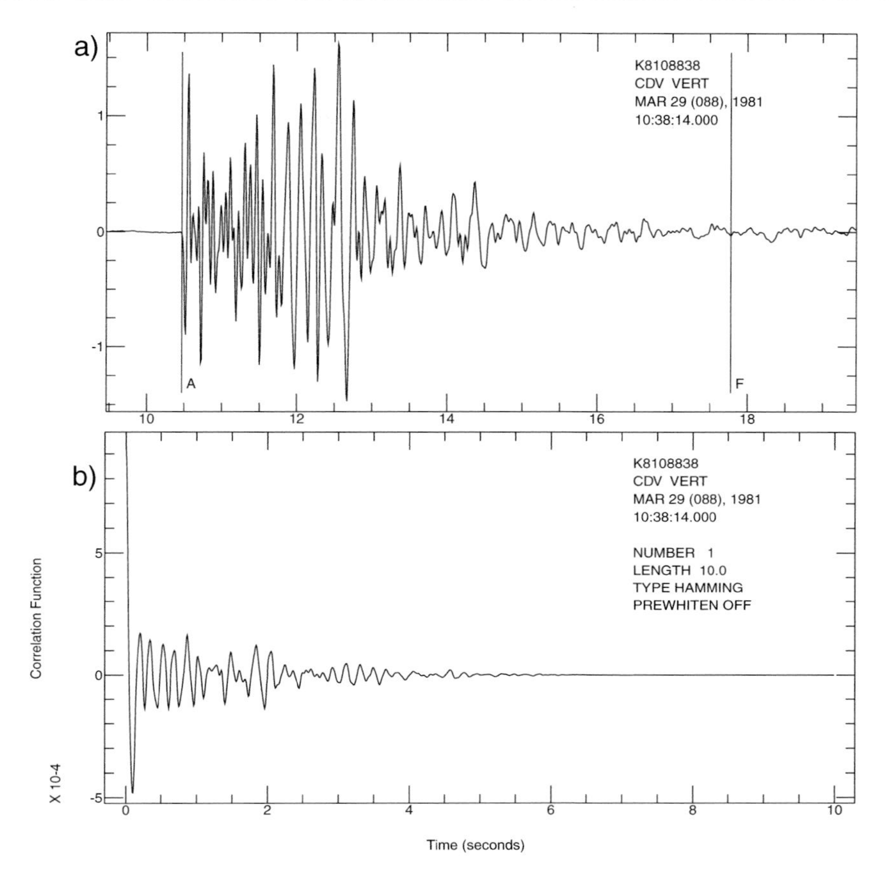

Figure 9.1 (*a*) The now-familiar seismic trace of built-in data provides a time series to show the structure of the autocorrelation function. (*b*) After COR calculates the autocorrelation, a plot by PCOR shows a peak at zero lag and an oscillatory but decreasing amplitude with increasing lag out to the window length of 10 s (the default value). Information about the COR command options appears in the label in the window.

and that decays with increasing lag. Because it is symmetric with respect to lag, positive lags contain all the autocorrelation information (Fig. 9.1). These commands will produce an autocorrelation plot of built-in data in SAC:

Ex-9.1

```
SAC> fg seismogram; rmean    ;* Data
SAC> spe                     ;* Start subprocess
SAC/SPE> cor                 ;* Default parameters - no windowing
SAC/SPE> pcor                ;* plot
```

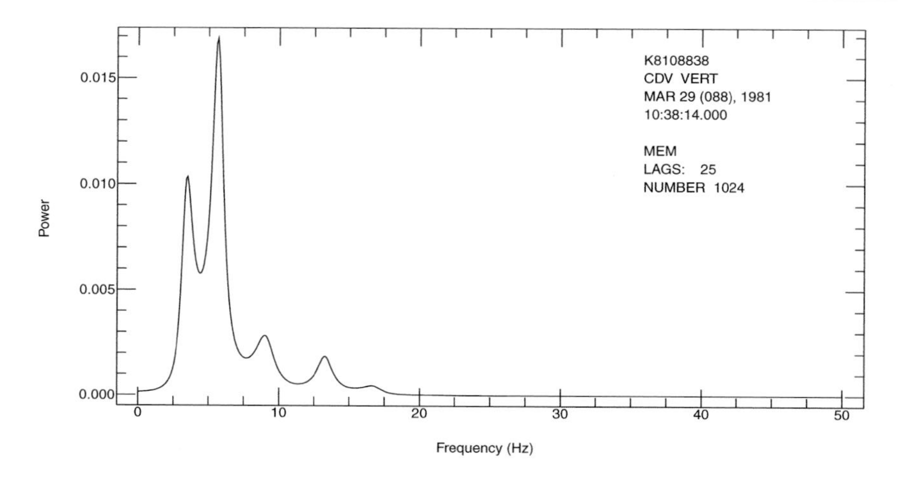

Figure 9.2 The spectrum resulting from a MEM estimate on the trace data shown in Fig. 9.1. The MEM command calculated the maximum entropy spectrum from the correlation function. PSPE plots the spectrum.

Spectrum

Once the autocorrelation is available, the spectrum follows directly from it. SAC makes the spectrum estimate by naming the method to be used: PDS, MEM or MLM. The commands recognize the NUMBER option that gives the number of frequency points in the spectrum. This should be a power of 2 that is ≥ 512. The high-resolution methods require an autoregressive model order as well, but the default is usually satisfactory. To change it, use the ORDER option. The PLOTPE command is a diagnostic to help select the order for the maximum entropy estimate. The prediction error decreases rapidly with increasing order. The plot helps to select the minimum order required to model the features in the data. Using YLOG and XLOG may help to better identify breakpoints in the curve.

PDS provides further controls on the tradeoffs inherent in spectral estimation. This method's estimated spectrum tends to behave like the true spectrum convolved with the Fourier transform of the window used on the correlation function. Thus PDS provides control over the window shape (the same available with COR TYPE) and the window length (the equivalent of the ORDER parameter for the high-resolution methods, measured in either SECONDS or LAGS).

Once the spectrum is available, the PLOTSPE command will plot it. Extending the previous example, the spectral plot shown in Figure 9.2 results from the following commands:

Ex-9.2

```
SAC> fg seismogram; rmean    ;* Data
SAC> spe                     ;* Start subprocess
SAC/SPE> cor                 ;* Default parameters - no windowing
SAC/SPE> mem                 ;* Maximum entropy spectrum
SAC/SPE> pspe                ;* plot
```

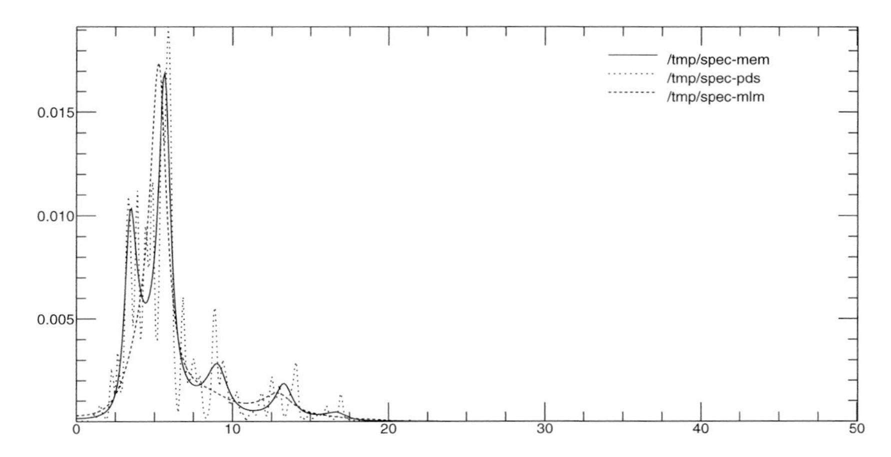

Figure 9.3 The plot shows overlaid spectra resulting from PDS, MEM and MLM estimates on the data trace shown in Figure 9.1. WRITESPE wrote out the individual estimates, which READ and PLOT2 gathered into a single plot. Different line styles denote the different methods.

Estimating the spectrum using any method does not change either the original data or the correlation function (see the caveat about prewhitening, however). Thus different methods may be tried on the same trace and viewed until the result is satisfactory.

Saving the correlation and the spectrum

Both the autocorrelation function and the spectrum may be saved as SAC files. This is principally for plotting use, because the autocorrelation is intrinsically uninteresting. However, because the autocorrelation contains all the information needed to estimate a spectrum, it is a compact way to save spectral information.

Use WRITECOR and READCOR to write and read the autocorrelation function. Autocorrelation information is saved in the file at both positive and negative lags. Thus, if read in as a SAC time series, it will be padded to a power of 2 in length and will appear to have a time-reversed copy of the positive lag autocorrelation at its end, which actually is the negative lag values.

The power spectrum estimate is the more useful quantity for an analyst. The details of these traces might be the subject of further analysis or scrutiny, possibly using PLOTPK, or to which to fit a spectral slope for an attenuation analysis procedure. The WRITESPE command writes out the spectrum. Figure 9.3 is a trivial example of what is possible with WRITESPE. The following commands estimate the spectrum of our favorite time series using each of the three methods provided by SAC. The plot shows a comparison of the estimates overlaid in a single plot using PLOT2 after WRITESPE.

Ex-9.3

```
SAC> fg seismogram; rmean          ;* Data
SAC> spe                           ;* Start subprocess
SAC/SPE> cor                       ;* Default parameters
```

```
SAC/SPE> mem                      ;* Maximum entropy spectrum
SAC/SPE> writespe /tmp/spec-mem   ;* save
SAC/SPE> pds                      ;* Power density spec
SAC/SPE> writespe /tmp/spec-pds   ;* save
SAC/SPE> cor prewhiten on         ;* Stabilize MLM
SAC/SPE> mlm                      ;* Maximum likelihood spec
SAC/SPE> writespe /tmp/spec-mlm   ;* save
SAC/SPE> qs                       ;* Leave subprocess
SAC> read /tmp/spec-[mem,pds,mlm];* re-read spectra
SAC> line list 1 2 3 increment on;* ID by line type
SAC> plot2                        ;* overlay
SAC> line previous                ;* restore line options
```

CHAPTER

TEN

Three-dimensional data in SAC

10.1 THE CONCEPT OF 3D DATA

SAC has facilities for handling a basic type of three-dimensional (3D) data, which are function values evaluated on a regular grid of (X,Y) positions. This type of data can be viewed as evenly sampled in space and fits naturally into SAC file types. The file consists of the string of samples at each grid point, and file header information that flags it as 3D gives the X and Y dimensions of the grid. SAC calls data of this type XYZ.

The simplest visualization of 3D data is map elevation data. Imagine a section of land whose elevation is specified on a grid every 100 m eastwards and 100 m northwards. Over a plot of land 1 km by 1 km, the elevation could be described by a string of 121 values, the first 11 values along the westernmost traverse and the final 11 along the easternmost traverse.

Another type of data that can be represented in 3D is a spectrum as a function of time. If equal lengths of time are cut from a seismogram, the number of points in their FFTs will be the same. If the interval between spectral samples is constant, these values will also form a grid.

A final type of data inherently 3D is a misfit function evaluated over a two-dimensional (2D) grid of points. The misfit minimum within the sampled parameter space yields the optimum combination of parameters that best fit some observations, and the shape of the minimum provides a measure of the joint uncertainty of the parameter estimates.

10.2 SPECTROGRAMS

A spectrogram is a depiction of the time variation of a signal spectrum. In a time window of fixed length, the spectrum of that subset of the time series will have a fixed number

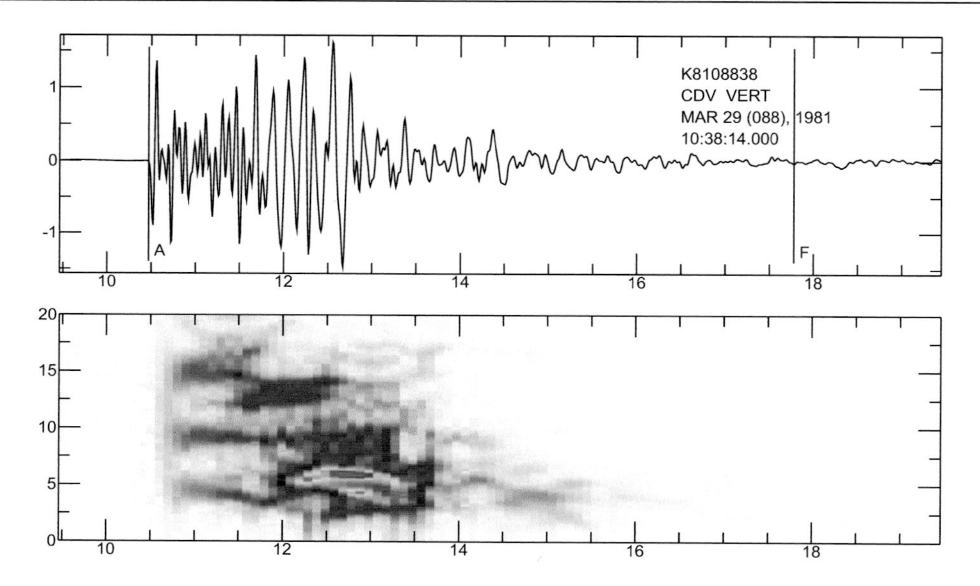

Figure 10.1 Spectrogram of a seismogram: (top) the time series and (bottom) the spectrogram. The X axis is time as it runs through the seismogram. The Y axis is frequency (Hz). In a time window of 2 s recalculated every 0.1 s through the seismogram, the color represents the spectral power at each frequency. (For color version, see Plates section.)

of frequencies. As the time window marches through the time series, the evolution of the spectral content provides a visual assessment of a signal onset and its power at each frequency. This visualization provides a way to graphically assess the dominant frequency of a body wave arrival or the dispersion curve of a surface wave arrival.

SAC provides the SPECTROGRAM command for this use. The command's options control the length of the time window (thus its frequency content), its stride (repetition in time), and the method of spectral estimation (standard, maximum entropy or maximum likelihood). The result is a 3D SAC file that is displayed with color encoding the amplitude at each frequency (Y axis) at each point in time (X axis). The file may be saved using the SPECTROGRAM XYZ option, or, as is commonly done, displayed immediately. Figure 10.1 shows a spectrogram produced using the commands below:

Ex-10.1

```
SAC> fg seismogram; rmean                 ;* Data
SAC> cuterr fillz; cutim (&1,b - 1) (&1,e + 1)    ;* Pad
SAC> bf                                   ;* Multi-plot display
SAC> yvport 0.55 0.95
SAC> xlim (&1,b + 1) (&1,e - 1); p1; xlim previous ;* Data plot
SAC> yvport previous
SAC> yvport 0.1 0.48
SAC> spectrogram slice 0.1 ymax 20 method mlm      ;* Spectrogram
SAC> yvport previous
SAC> ef                                   ;* end display
```

This spectrogram was plotted directly. However, the SPECTROGRAM XYZ option will replace the data in memory with an XYZ-type file that may be written or processed using further SAC commands.

10.3 CONTOUR PLOTS

Any file of 3D data of type XYZ can be treated as elevations over a 2D surface. A natural way to visualize the data is by contouring it, and SAC provides the command CONTOUR to do this. A basic contour plot of SAC's built-in contour data is simple to make

```
SAC> datagen sub xyz volcano.xyz
SAC> contour
```

and results in the plot shown in Figure 10.2. This is a basic plot without any ornamentation on the contours, or even an indication of which contour is high and which is low. SAC provides a plethora of commands to improve contour plot annotation.

In addition to adding axis labels (XLABEL, YLABEL) and a title (TITLE) to a plot, SAC provides ways to affect the contour lines themselves. ZLEVELS sets the values at which to draw the contours. It is possible to set a low and high range for plotting, with either a number of lines in that range or an increment between levels. Alternatively, list the desired levels specifically. After defining a set of levels, associate a drawing color (see ZCOLORS) and a line style (see ZLINES) with each. To attach a label to each level, use ZLABELS. It provides the text or format for each level's label and its size and inter-label spacing rules. (Good labeling on contour plots is an art that requires much experimentation to achieve a pleasing result. The features of ZLABELS will reward close study.) To show the gradient sense, SAC provides a ZTICKS to place ticks perpendicular to the contour in either the uphill or downhill direction.

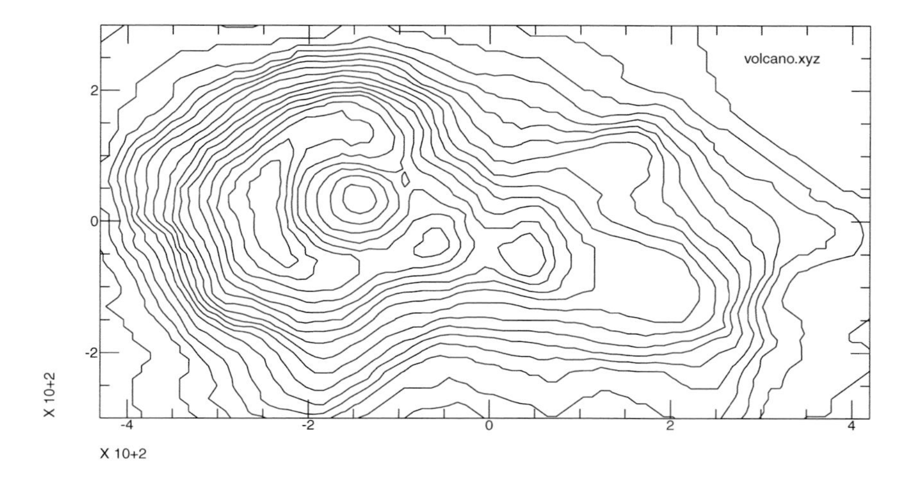

Figure 10.2 Contour plot of a topograpic feature. The basic contour plot is not labeled, and the contour levels are solid lines unlabeled with the contour value. This provides a quick overview of the data, but is not quantitative.

Two of SAC's built-in contouring demonstrations illustrate some of these features; use `ECHO ON` to see details. The command

```
SAC> macro demo contour simple
```

draws the plot shown in Figure 10.3a where `ZLINES` is used to make contour lines solid every 20 m of elevation. The 5 m contour increment (`ZLEVELS INCREMENT 5`) means that

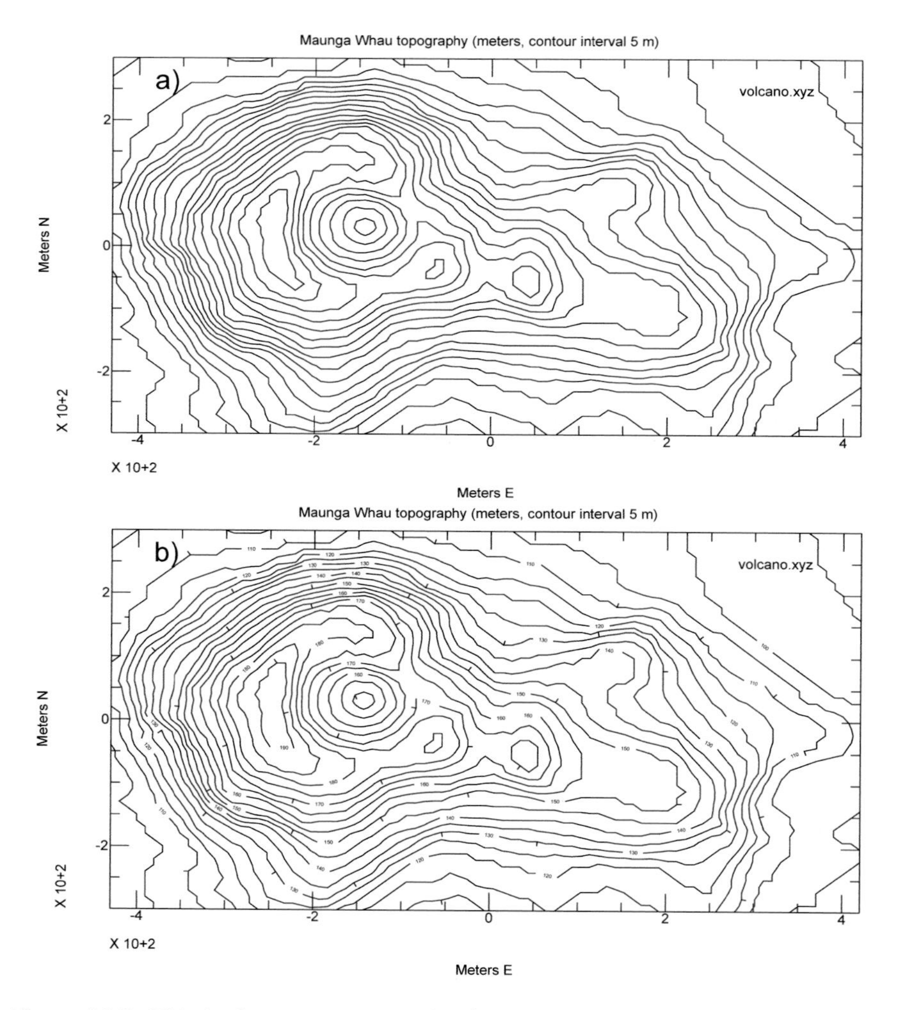

Figure 10.3 This is the same topographic feature seen previously. (*a*) The feature with different line styles that repeat every 20 m. This provides the viewer a visual estimate of the vertical scale of the feature without labeling each contour line. (*b*) The feature with the same line style for all contours. Tick marks on the downhill side of the contours clearly show the pit crater at the summit of the feature and the summit-girdling ridge of the crater. Every other contour line bears a level label.

every fourth line will lie at a multiple of 20 m. Thus, ZLINES LIST 1 2 3 4 returns to a solid line (line style 1) after cycling through styles 2–4.

The command

```
SAC> macro demo contour complex
```

uses more features of CONTOUR. Figure 10.3b shows the result. Here, the same data is displayed with tick marks showing the downhill direction on every fourth level, and every other contour level is labeled with an integer elevation.

10.4 COMPOSITE 3D DATA PLOTS

Once written into an XYZ file, 3D data can be usefully represented in composite forms. The GRAYSCALE command draws a color or grayscale image of an XYZ file using the prevailing color palette to translate Z values into color. A default palette is supplied. To set GRAYSCALE's palette to a custom palette, use the COLOR RASTER LIST command. REPORT COLOR will show the pre-defined color palettes to choose from, or a particular one may be built up from a user's own color list.

A color palette is a file containing lines in the form

```
0.667 0.000 0.000 BRICK-RED
1.000 0.000 0.000 RED
1.000 0.333 0.000 ORANGE
1.000 0.667 0.000 PUMPKIN
1.000 1.000 1.000 WHITE
0.353 1.000 0.118 ACID-GREEN
0.000 0.941 0.431 SEA-GREEN
0.000 0.314 1.000 ROYAL-BLUE
0.000 0.000 0.804 DEEP-BLUE
```

Each line represents a color. The three numbers are a red, green and blue intensity (1 = full, 0 = none) describing the color. The text following the numbers is optional, but provides a way to refer to the color by name in the LINE and COLOR command. The range of Z values encountered in a plot is mapped linearly onto the colors. Linear interpolation of the (red, green, blue) values between bracketing colors provides intermediate color values. The palette here, with white the central color, would be suitable for plotting a zero-mean value.

Color is a qualitative indicator of level because humans respond to it individually. Not only do forms of color blindedness thwart discrimination between some color palette choices, but it is hard to judge the difference between, say, the color indigo and the color violet marking a Z value boundary in a 3D plot. For quantitative use, SAC therefore provides the ability to combine contour plots with color images (or black and white images). The contour lines provide the quantitative levels, and the color provides the visual summary.

A good example of the idea is the composite spectrogram and contour plot shown in Figure 10.4, resulting from the following commands:

Ex-10.2

```
SAC> fg seismogram; rmean                      ;* Data
SAC> spectrogram slice 0.1 method mlm xyz ;* Make XYZ spectrogram
SAC> bf                                        ;* Start composite plot
```

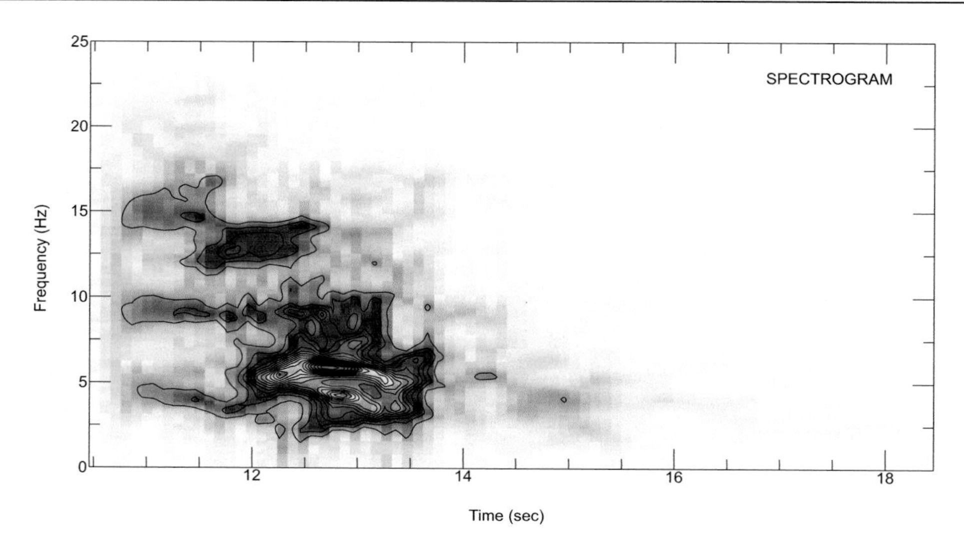

Figure 10.4 XYZ file data may be usefully displayed in alternative forms by SAC. The figure shows a spectrogram, saved as a 3D file, plotted in color with the GRAYSCALE command. It is overlaid with contours of spectral power. This provides both a quantitative report on the spectrogram and a qualitative graphical summary. (For color version, see Plates section.)

```
SAC> ylim 0 25                                  ;* Limits and labels
SAC> ylabel 'Frequency @(Hz@)'; xlabel 'Time @(sec@)'
SAC> grayscale                                  ;* Color spectrogram
SAC> ylabel off; xlabel off                     ;* Only label once
SAC> contour                                    ;* Overlay contours
SAC> ef                                         ;* Finish composite plot
```

The commands calculate a spectrogram of some of SAC's built-in data and save it as an XYZ file. The first plot (GRAYSCALE) is a color realization of the spectrogram inside a BEGIN-FRAME ... ENDFRAME pair. The second CONTOUR plot overlays the contour levels. A useful addition to this basic plot would be labels on the contour lines.

10.5 PROPERTIES OF 3D DATA

Here we turn to the technical details of 3D data. These are unimportant except for writing programs to create or handle 3D data for SAC's use. The details are mercifully brief.

3D data are Z values of function sampled on a regular X–Y grid. There is a compelling analogy between time series and this structure because a time series is also a function that is sampled on a regular time grid. Thus the only values to be stored for either are the sample values from a starting value (time or position) and an interval (time or space).

For an XYZ file, the starting value is the (X,Y) point of the grid specified by the file header variables XMINIMUM and YMINIMUM. The file header variables NXSIZE and NYSIZE give the number of grid elements along the two space axes. The grid point separations

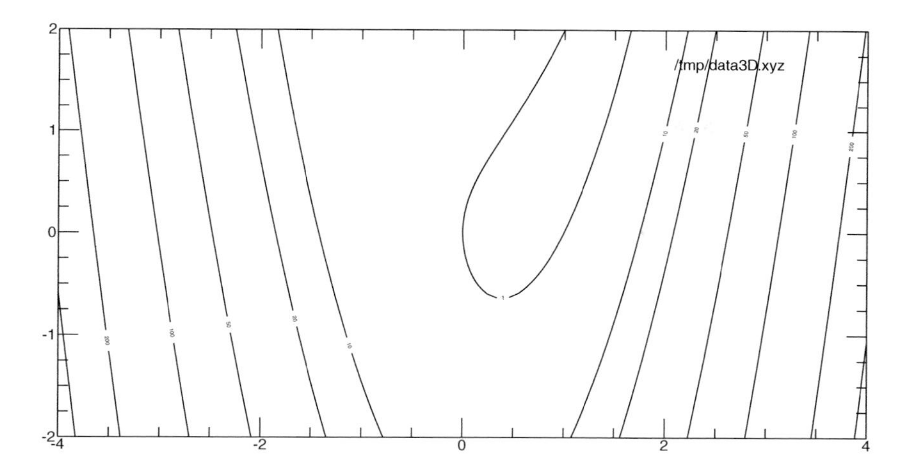

Figure 10.5 Contour plot of an XYZ data file written by the sample program "plot3d" listed in the text. The commands to produce this plot are

```
SAC> read /tmp/data3D.xyz
SAC> zlevels list 0 1 10 20 50 100 200 300
SAC> zlabels on list 0 1 10 20 50 100 200 300
SAC> contour
```

are implicit, defined by XMAXIMUM and YMAXIMUM. Thus the spatial scale is defined by (XMAXIMUM-XMINIMUM)/(NXSIZE-1) rather than as an explicit value like DELTA for a time series. The file type IFTYPE must also be set to IXYZ for these values to be recognized. (Otherwise, SAC reckons the file is a kind of time series. In fact, PLOT will plot an XYZ file – an occasionally useful feature to view the data range.) Finally, the number of points in the file, NPTS, must be set to NXSIZE×NYSIZE.

In the file itself, the data is stored along streaks of Y values with increasing X (see Section 2.4). This is congruent to the way that time series values (Y) are stored with increasing time (X).

10.6 WRITING 3D DATA FILES

Writing 3D data files is simple: first set the file header variables and then call WSAC0. The following Fortran program shows the necessary steps. First, create the data. Then call NEWHDR to create a skeleton file header. Populate the header with values using SETxHV and then use WSAC0. Figure 10.5 shows the resulting figure.

Ex-10.3

```
      program plot3D
C     Grid extends symmetrically from zero
C        for nxgr cells in X and nygr in Y
      parameter (nxgr=50,nygr=50)
      parameter (nx=1+2*nxgr, ny=1+2*nygr)
C     Parameters: depth of valley (fscl), X and Y ranges (4x2)
```

```
      parameter (fscl=1, xscl=4, yscl=2)
C     Data for grid.  Grid layout is
C        (*, 1) - lowest streak of y values
C        (*,ny) - highest streak of y values
C        (1, *) - lowest x value at any y position
C        (nx,*) - highest x value at any y position
      real data(nx,ny)

C     Create data:  Curved valley defined by a Rosenbrock fcn
      f(x,y) = (1-x)**2 + fscl*(y-x**2)**2

C     Run over grid and insert elevation information into it.

      do j=1,ny
         y = yscl*float(j-1-nygr)/nygr
         do i=1,nx
            x = xscl*float(i-1-nxgr)/nxgr
            data(i,j) = f(x,y)
         enddo
      enddo

C     Create default SAC file header; modify to make it 3-D data
      call newhdr

C     File type to XYZ for 3D data
      call setihv('IFTYPE', 'IXYZ', nerr)

C     Grid dimensions: NXSIZE, NYSIZE
      call setnhv('NXSIZE', nx, nerr)
      call setnhv('NYSIZE', ny, nerr)

C     Grid scale:  XMINIMUM, XMAXIMUM, YMINIMUM, YMAXIMUM
      call setfhv('XMINIMUM', -xscl, nerr)
      call setfhv('XMAXIMUM',  xscl, nerr)
      call setfhv('YMINIMUM', -yscl, nerr)
      call setfhv('YMAXIMUM',  yscl, nerr)

C     Number of data points, begin, delta (required but irrelevant)
      call setfhv('B', 0.0, nerr)
      call setfhv('DELTA', 1.0, nerr)
      call setnhv('NPTS', nx*ny, nerr)

C     Write data to file using present header values
      call wsac0('/tmp/data3D.xyz',data,data,nerr)
      end
```

CHAPTER

ELEVEN

Implementation of common processing methodologies using SAC

11.1 SEISMIC ANISOTROPY AND SHEAR WAVE SPLITTING

Overview

The observation of two independent, orthogonally polarized shear waves, one traveling faster than the other, is arguably the most unambiguous indicator of wave propagation through an anisotropic medium. The splitting can be quantified by the time delay (δt) between the two shear waves and the orientation (ϕ) of the fast shear wave (Fig. 11.1).

In this chapter, we review briefly the theory behind shear wave splitting, with particular focus on the popular method of Silver and Chan (1991), for which we provide source code and documentation.

11.2 SHEAR WAVE SPLITTING ANALYSIS

Seismic anisotropy can be studied via shear wave splitting throughout the Earth, ranging in depth from hydrocarbon reservoirs in the shallow crust (e.g., Verdon *et al.*, 2009) to the core and deep mantle (e.g., Wookey and Helffrich, 2008). By selecting earthquakes at angular distances $\geq 88^\circ$ from a recording site, phases such as *SKS*, *PKS* and *SKKS* can be readily isolated for analysis of upper mantle anisotropy (for a review, see Savage, 1999). We will focus on analysis of these core phases here.

Patterns of seismic anisotropy can develop due to the preferential alignment of minerals in the crust and/or mantle, the preferential alignment of fluid or melt, layering of isotropic materials, or some combination thereof (e.g., Blackman and Kendall, 1997). Several

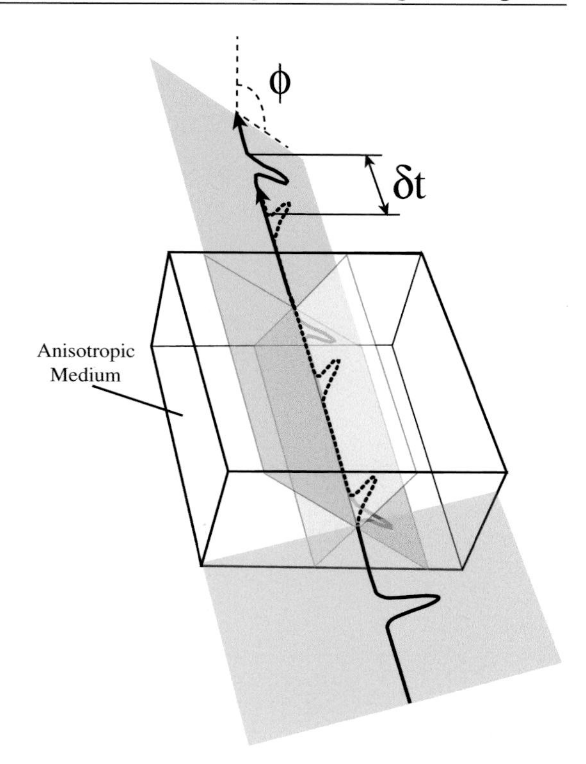

Figure 11.1 When a horizontally polarized shear wave enters an anisotropic medium, it splits into two orthogonally polarized waves. When the two pulses reach a seismometer, one can measure the polarization direction (ϕ) of the fast shear wave and the delay time (δt) between it and the later arriving "slow" shear wave. These splitting parameters can subsequently be used to characterize the anisotropic medium.

processes can develop such anisotropy, including (1) asthenospheric flow parallel to absolute plate motion (e.g., Bokelmann and Silver, 2002), (2) mantle flow around cratonic roots (e.g., Fouch *et al.*, 2000) or subducting slabs (e.g., Di Leo *et al.*, 2012), and (3) pre-existing fossil anisotropy frozen in the lithosphere (e.g., Bastow *et al.*, 2007).

For any S arrival passing through the core, the P-to-S conversion developed at the core–mantle boundary should, in an isotropic Earth, display energy only on the radial component seismogram; its particle motion should thus be linear. In the presence of seismic anisotropy, however, this assumption breaks down and energy can be found on the tangential component seismograms as well. This results in elliptical particle motion.

Parameter estimation methodologies

In principle, the goal of any shear wave splitting analysis methodology is to minimize tangential component energy and to linearize particle motion. This can be achieved in a number of ways (see e.g., Wuestefeld and Bokelmann, 2007) but by far the most common is the method of Silver and Chan (1991). This methodology rotates and time shifts the horizontal components to minimize the second eigenvalue of the covariance matrix for particle motion in a time window around the shear wave arrival. This corresponds to linearizing the particle motion and usually reduces the tangential component energy (assuming the incoming *SKS* wave is radially polarized before entering the anisotropic medium). The data is typically filtered prior to analysis with a zero-phase two-pole Butterworth filter with corner frequencies 0.04–0.3 Hz using SAC's `BANDPASS` command or with an equivalent filtering application.

Macro and auxiliary program design

The implementation of the Silver and Chan (1991) methodology is a program written in Fortran. However, interfacing with the software is required at three principal stages of the analysis:

1. visually inspecting seismic waveforms and picking analysis windows;
2. transferring data to the analysis program;
3. displaying analysis results in a graphical format.

SAC is used in each of these steps. The macro `split` uses keyword parameters (identified with the `$KEYS` command) to specify a file name prefix (`file`) and a component suffix set (`comps`, e.g., bhe, bhn, bhz) on the command line. A `pick` keyword is optional with a default. Further input data is obtained from the SAC file headers. If sufficient information exists in the headers for processing, the macro proceeds, unless `pick yes` indicates that the waveform should be re-picked. If picking is needed, `PLOTPK` starts a graphics interaction to define the start and end times of a waveform in the seismogram through picks of `A` and `F`.

Data is then provided to the program through the command line arguments and via standard input. The macro uses `SYSTEMCOMMAND` to pass the args to the program and to connect standard input to an input file written by the macro using the *echo* command and `SYSTEMCOMMAND`. The program uses the Fortran library subroutine `GETARG` to retrieve the command line inputs and uses Fortran I/O to read text from the standard input. To read header information, the program uses the SAC I/O library routine `RSAC1` to read the file without reading any data. Then the SAC I/O library routine `GETxHV` extracts information from the header. After preliminary checks to ensure that the headers contain event and station information, the program uses `RSAC1` again to read the data for each trace.

After the splitting analysis, the SAC I/O subroutine `WSAC0` is used to write out the original horizontal waveforms (a pair of traces) and the time-shifted waveforms (another pair). The program avoids using `NEWHDR` so that the old header information in the data traces is retained. Calls to `SETxHV` change the header variables that provide the macro with the estimated splitting parameters: ϕ and its uncertainty (`USER0`, `USER1`), δt and its uncertainty (`USER2` and `USER3`). They also return the estimated initial polarization ϕ_0 (`USER4`) and the degrees of freedom in the measurement window (`USER5` and `USER6`). The program writes these files using `WSAC0`. The misfit error surface (an example of a 3D SAC file - see Chapter 10) is similarly output, but the header variables `USER0` ... `USER6` are filled in with values used later in the `split` macro. Upon return from the analysis program, the `split` macro reads these various new SAC files and presents them for QC of the analysis. The particle motion plot (Fig. 11.4) is made with `PLOTPM` and `PLOT2` in different viewports of a composite plot. The misfit surface is displayed using `CONTOUR` to show the optimum value (ϕ and δt) and their joint uncertainty.

In addition to the Silver and Chan (1991) codes, SAC macros and codes to implement the stacking method of Restivo and Helffrich (1999) accompany this text. In this technique, high signal-to-noise ratio measurements are given more weight. Additionally, every individual measurement is scaled to a factor of $1/N$, with its back-azimuth defining a wedge of $\pm 10°$ in which N observations fall. This compensates for the effects of over-represented back-azimuths in the uneven sampling of a station. The stacking procedure is implemented via the SAC macro `splstack`. The macro uses keyword parameters, with the obligatory `file` keyword indicating a file name that contains a list of splitting estimates to combine, one per line. Other keyword parameters are optional with suitable defaults. The macro passes the

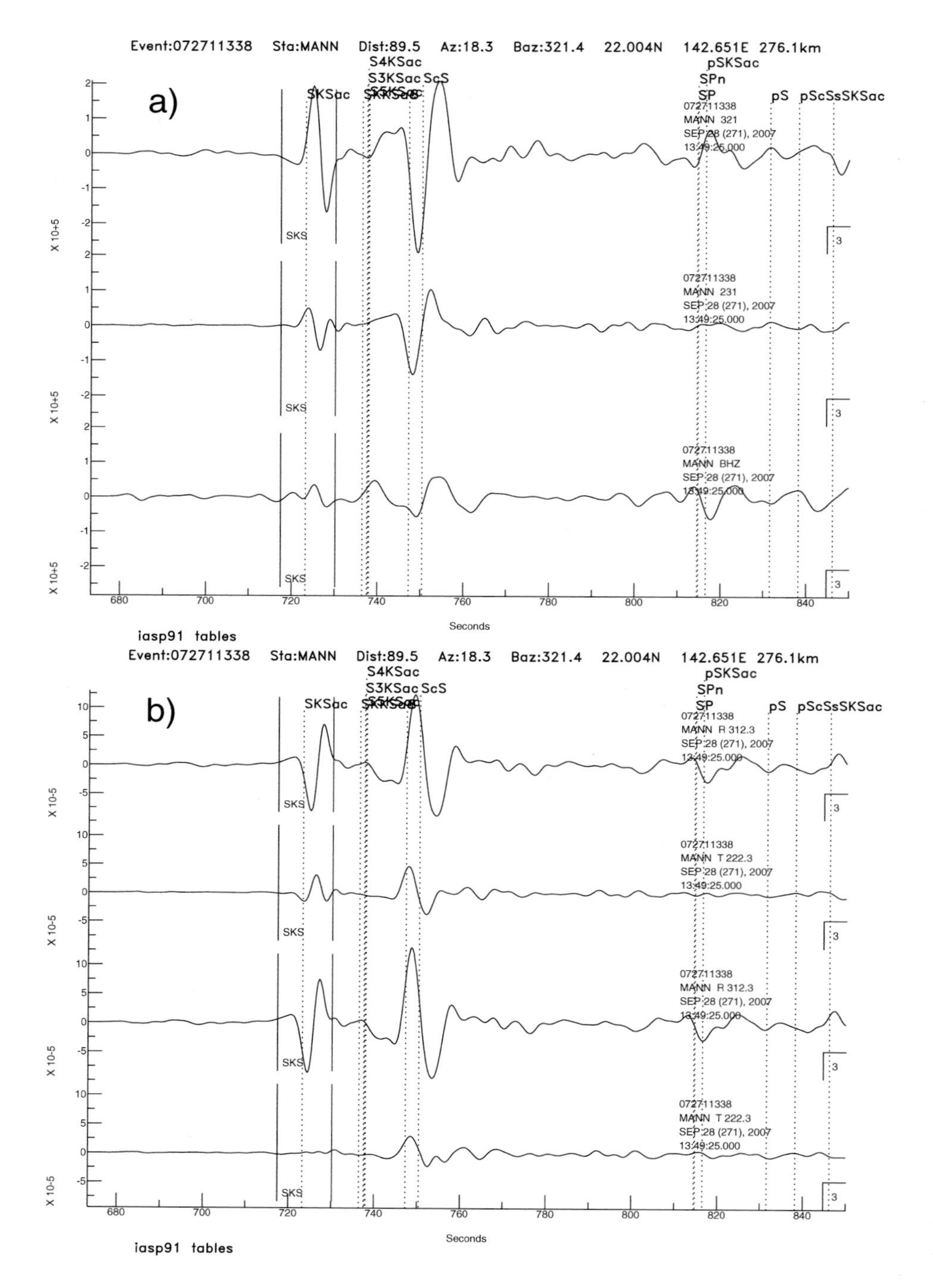

Figure 11.2 Shear wave splitting analysis example using data from station MANN in northern Hudson Bay, Canada (Bastow *et al.*, 2011). (*a*) Three-component broadband seismograms, cut around the *SKS* phase. Time window for splitting analysis indicated with labeled picks around the *SKS* arrival. (*b*) Top two traces are original radial and tangential component seismograms; bottom two traces are radial and tangential traces after correction for splitting. Note that, in the corrected seismogram, tangential component energy has been minimized.

input file to the auxiliary shell script `splitstack.sh` that invokes the program `splstacksac` to average the estimates. It then returns a 3D file to the macro to display by `CONTOUR`, with annotations added by `PLOTC`.

The program reads a list of file names and a command line arg, similar to the way that the splitting analysis program works. Each file header is read (using `RSAC1`), suitable information is loaded, and then data is read for stacking. A 3D file, written using `WSAC0`, is the result of the stack estimation. The stack-estimated ϕ and δt, their uncertainty and the joint degrees of freedom are returned in `USER1` ... `USER5` for macro use. The macro reads the stack result and presents it as a contoured file using the `CONTOUR` command with `PLOTC` annotations.

SAC implementation

To perform the shear wave splitting analysis on a set of three seismograms for station MANN in northern Hudson Bay, Canada (Bastow *et al.*, 2011), where files take the form MANN.133859.sks.bh[e,n,z], type the following:

```
SAC> window 1 x 0.05 0.80 y 0.05 0.74
SAC> m split file MANN.133859.sks comps bhe bhn bhz option e
```

The `split` macro uses SAC graphics to display the seismogram and allows the analyst to define a window with the keys *a* and *f* (to set the `A` and `F` file header variables) around the *SKS* phase (Fig. 11.2a). The program will then display a number of images to document the stages in the shear wave splitting analysis. Radial and tangential component seismograms are shown both before and after the splitting analysis (Fig. 11.2b). Note the energy on the tangential component of Figure 11.2b, minimized after correction.

Next, the windowed data is shown with focus on the fast and slow shear waves before and after correction (Fig. 11.3). Particle motion plots enable the analyst to establish that elliptical particle motion, diagnostic of shear wave splitting, is linearized after correction.

Finally, an uncertainty plot is displayed (Fig. 11.4), showing the results of the grid search over all possible values of ϕ and δt. The bold contour outlines the area of joint uncertainty in ϕ and δt, showing how good the measurement is – a smaller area corresponds to lower uncertainty. The splitting parameters resulting from methods such as Silver and Chan (1991) are generally shown as vectors on a map, with the orientation of the arrow paralleling ϕ, and the length of it proportional to δt (see e.g., figure 1 in Bastow *et al.*, 2011).

11.3 RECEIVER FUNCTION ANALYSIS

Overview

Receiver functions are time series computed from three-component seismic data that show the impulse response of Earth structure beneath a seismometer to an incoming plane wave (e.g., Langston, 1979). The P receiver function waveform is a composite of P-to-S converted waves and subsequent reverberations generated at impedance discontinuities such as the Moho (Fig. 11.5) and mantle transition zone.

Modeling the amplitude and timing of those reverberating waves can supply valuable constraints on the underlying geology and mantle structure. In many cases, the Earth structure can be approximated by a series of roughly horizontal layers. The arrivals generated by

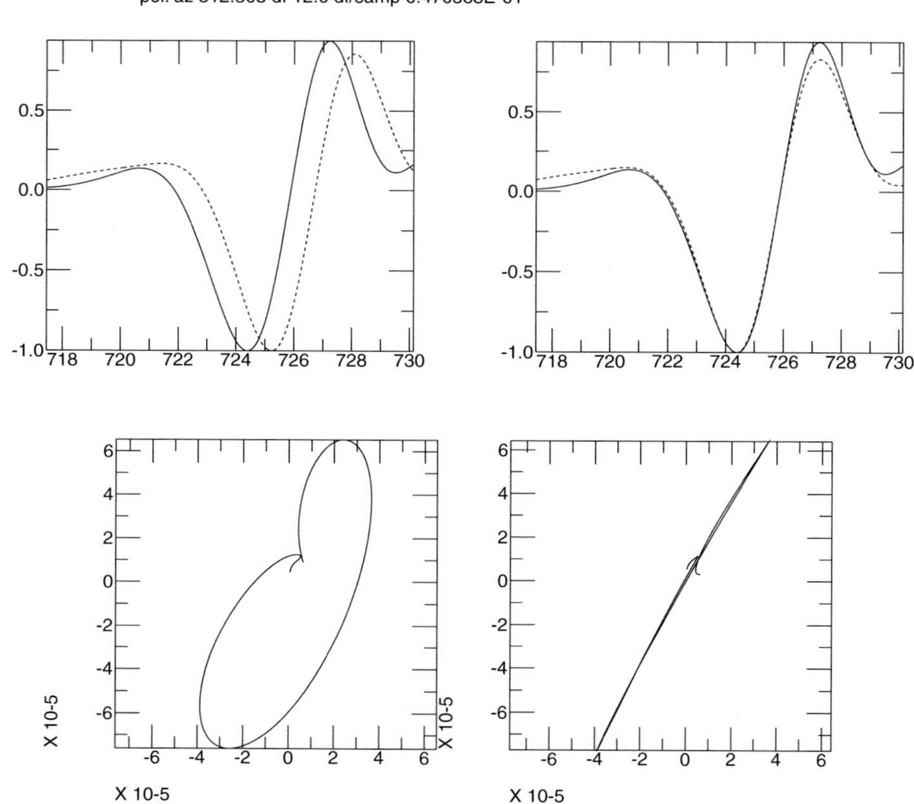

Figure 11.3 (*top*) Superposition of fast (ϕ direction) and slow components ($\phi + 90°$), uncorrected (*left*) and corrected (*right*). (bottom) Particle motion for slow components, uncorrected (*left*) and corrected (*right*). Corrections are made using estimated values of the splitting parameters, ϕ and δt, to assess their validity.

each sharp boundary (that is, sharp relative to the shortest wavelength in the observations) look something like the cartoon seismogram in Figure 11.5.

A radial receiver function can be computed using the source equalization procedure of Langston (1979) by deconvolving the vertical component seismogram from the radial (SV) component (and by following a similar procedure for the tangential component, SH). In the frequency domain, this can be written simply as

$$H(f) = R(f)/Z(f) \tag{11.1}$$

where f is the frequency, $Z(f)$ and $R(f)$ are the Fourier transforms of the vertical and radial seismograms, and $H(f)$ is the Fourier transform of the receiver function.

In the following section, we review briefly the theory behind receiver function analysis, for which we provide source code and operator instructions.

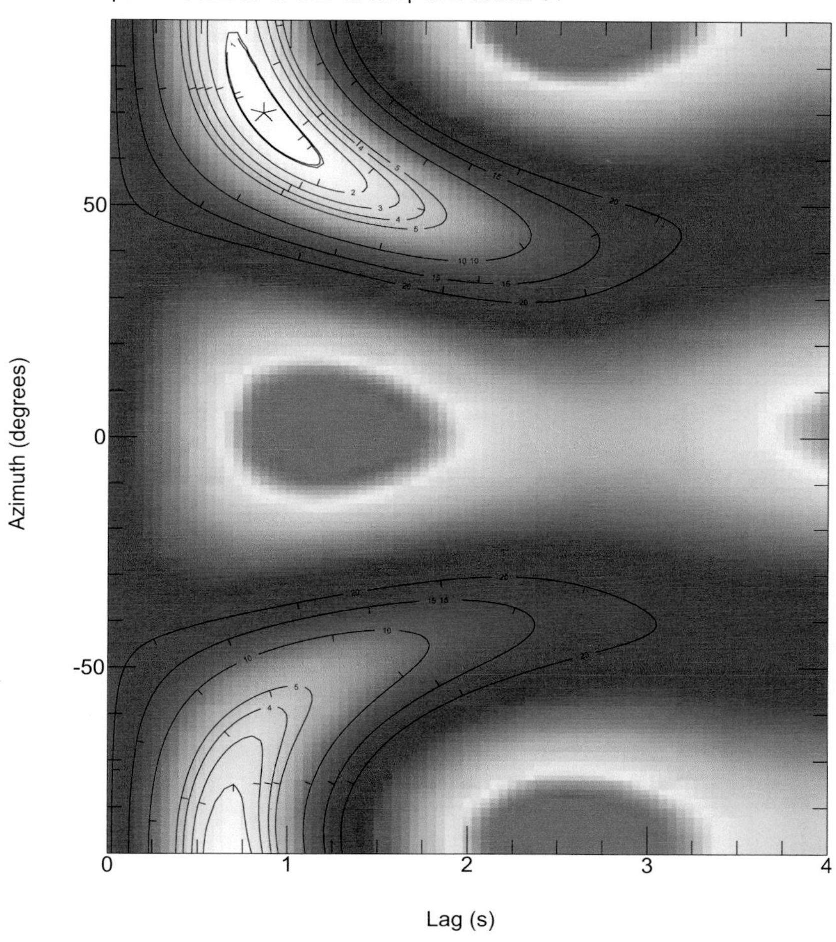

Figure 11.4 Results of the grid search over δt and ϕ. The optimum splitting parameters are shown by the star (top left). The 95% confidence region is shown (double contour), along with multiples of that contour level. (For color version, see Plates section.)

Estimation methodologies

Equation 11.1 is a relatively simple concept but implementation is difficult because of the instability of deconvolution. Thus, a number of different approaches to the computation of receiver functions has been developed over the years:

- water level deconvolution (e.g., Langston, 1979);
- deconvolution in the time domain by least squares (e.g., Abers *et al.*, 1995);
- iterative deconvolution in the time domain (e.g., Ligorrìa and Ammon, 1999);
- multi-taper frequency-domain cross-correlation receiver function (MTRF) (e.g., Park and Levin, 2000); MTRF is more resistant to noise so better for ocean island environments, for

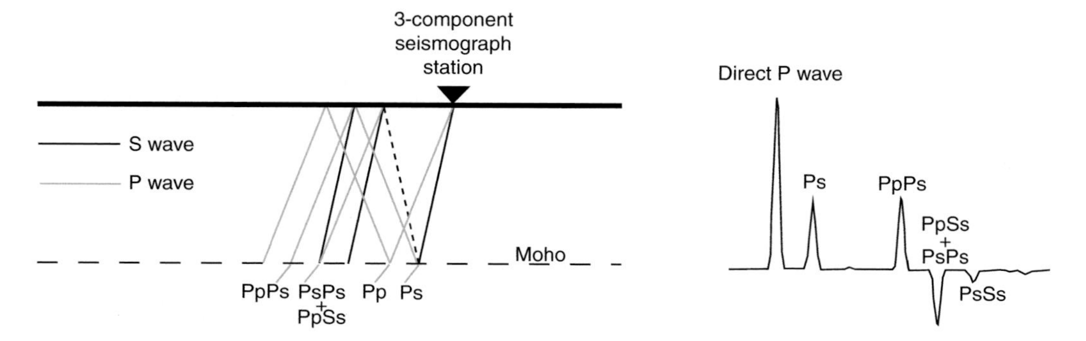

Figure 11.5 Receiver function analysis of Earth structure beneath a seismograph station. (*left*) When a P-wave is incident on a velocity discontinuity such as the Moho, it generates a shear wave at the interface. These arrivals, and their subsequent reverberations, can be isolated by receiver function analysis (*right*). Modified after Ammon (1991).

example, an advantage that is due to the use of multi-tapers to minimize spectral leakage and its frequency-dependent down-weighting in noisy portions of the spectrum.

Macro and auxiliary program design

The implementation of MTRF methodology in SAC is intuitively very similar to `split` code. The `mtrf` macro invokes SAC to display three-component seismogram data such that the analyst can define a measurement interval using picks for the `A` and `F` header variables. SAC rotates the horizontal component traces to the radial and tangential frames for the program. Macro keyword parameters provide user input, including a file name prefix, a component suffix set (e.g., bhe, bhn, bhz), a frequency-domain tapering limit ($f_{\max}$) and phase information. The macro passes the information to the Fortran program *mtdecon* via command line args and the standard input. The program calls `GETARG` and the SAC library subroutine `RSAC1` to extract header information and trace data. `WSAC0` is used to output the results of the analysis (vertical, radial and tangential component receiver functions), which subsequently can be displayed in SAC for quality control (Fig. 11.6).

SAC implementation

To perform the receiver function analysis on a set of three seismograms for permanent station FRB in northern Hudson Bay, Canada, where file names take the form FRB.133859.p.bh[e,n,z], type the following:

```
SAC> m mtrf file FRB.133859.p comps bhe bhn bhz pick yes fmax 1
```

Here, `fmax` governs the frequency content of the resulting receiver function.

The `mtrf` macro uses graphical interactions (`PLOTPK`) to display the seismogram and allow the analyst to define a window through picks of the file header variables `A` and `F` beginning just before the first arriving P-wave phase and ending ~30 s into the P-wave coda

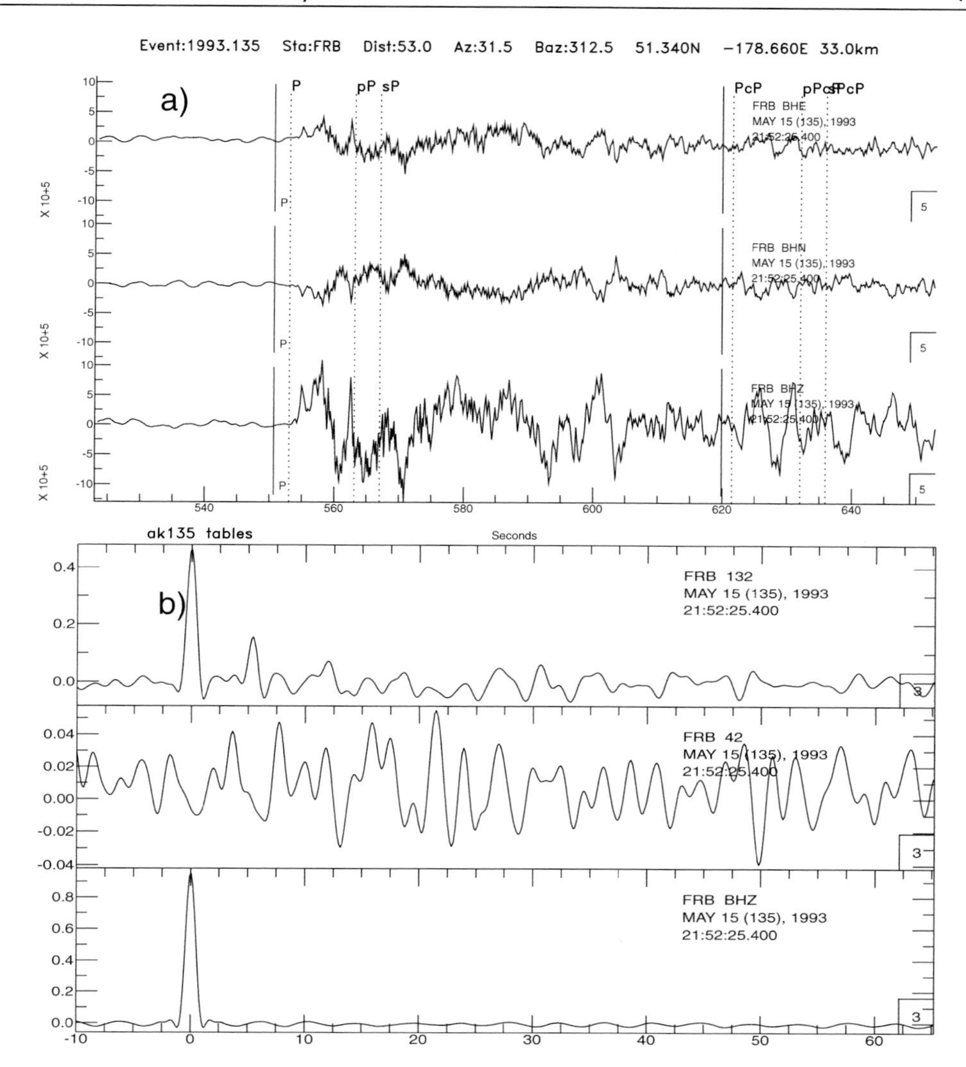

Figure 11.6 Example receiver function analysis using data from station FRB in northern Hudson Bay, Canada. (*a*) Three-component broadband seismograms, windowed around the P phase. (*b*) Radial (*top*) and tangential (*middle*) component receiver functions resulting from the extended-time MTRF method (Helffrich, 2006). Note the identifiable P-to-S conversion arriving at ~5.5 s, which likely developed at the Moho beneath the station.

(Fig. 11.6a). Then `mtrf` displays the results of the receiver function analysis for quality assurance. Figure 11.6b shows the resulting display.

H–κ analysis

After computation, receiver function data is routinely analyzed further by auxiliary SAC programs. One example is the H–κ stacking method of Zhu and Kanamori (2000), which estimates bulk crustal properties (crustal thickness and V_P/V_S ratio) from the receiver function time series. This procedure utilizes SAC commands and subroutines for reading SAC-format data, writing output files and, finally, displaying output graphically.

Crustal thickness (H) and V_P/V_S ratio (κ) are known to trade off strongly (e.g., Ammon *et al.*, 1990; Zandt *et al.*, 1995). In an effort to reduce the ambiguity inherent in this trade-off, Zhu and Kanamori (2000) incorporated the later arriving crustal reverberations PpPs and PpSs + PsPs in a stacking procedure whereby the stacking itself transforms the time-domain receiver functions directly into objective function values in H–κ parameter space. This method, known as the H–κ stacking technique, is used here. The objective function for stacking is

$$s(H,\kappa) = \sum_{j=1}^{N} w_1 r_j(t_1[H,\kappa]) + w_2 r_j(t_2[H,\kappa]) - w_3 r_j(t_3[H,\kappa]) \tag{11.2}$$

where w_1, w_2, w_3 are weights; $r_j(t)$ are the receiver function amplitudes at the predicted arrival times of the direct P-to-S conversion Ps and subsequent reverberations PpPs and PsPs + PpSs, respectively, for the jth receiver function; and N is the number of receiver functions used. The predicted travel times t_i are given by Equations 11.3–11.5:

$$t_1[H,\kappa] = H\left[\sqrt{\frac{\kappa^2}{V_P^2} - p^2} - \sqrt{\frac{1}{V_P^2} - p^2}\right] \tag{11.3}$$

$$t_2[H,\kappa] = H\left[\sqrt{\frac{\kappa^2}{V_P^2} - p^2} + \sqrt{\frac{1}{V_P^2} - p^2}\right] \tag{11.4}$$

$$t_3[H,\kappa] = 2H\sqrt{\frac{\kappa^2}{V_P^2} - p^2} \tag{11.5}$$

where p is the ray parameter (s / km) and V_P is an assumed P-wave speed representing the reverberative interval.

The stacked objective function should attain its maximum value when H and κ are correct. Thus by performing a grid search for a range of plausible H and κ values, its maximum value can be determined (Zhu and Kanamori, 2000). For example, an H–κ grid search might range over $20 \le H \le 50$ km and $1.5 \le \kappa \le 2.3$. The H–κ method provides a better estimate of Moho depth and V_P/V_S ratio than a simple stacking of all receiver functions because it accounts for the receiver function dependence on ray parameter, upon which the objective function sensitively depends.

Once a number of radial receiver functions have been computed for a given seismograph station, the receiver functions can be viewed and stacked in SAC:

```
SAC> read *.nrfr
SAC> sss
SAC/SSS> tw -10 30; dw units degrees usedata
SAC/SSS> prs
SAC/SSS> cs all sum on
SAC/SSS> sumstack
```

This can often result in later reverberant phases such as PpPs and PsPs + PpSs (Fig. 11.5) emerging from below noise level. However, if earthquakes from variable angular distances from the seismograph station are used in the analysis, the expected travel times of Ps and its subsequent reverberant phases change (Equations 11.3–11.5). The controlling parameter is horizontal slowness p, rather than range.

Further analyses like H–κ depend on p, so associating slowness with the trace is beneficial. An auxiliary program *setrfslow* adds the horizontal slowness information to the headers of the radial receiver function files in units of seconds/degree.

```
ls *nrfr | setrfslow -set USER9 -units s/km -verbose
```

In this case, the variable is `USER9`. The following commands display the radial receiver function dependence on slowness:

```
SAC> sc cat doss
* Contents of macro "doss"
read *.nrfr
sss
do i from 1 to (status NFILES)
   changestack $i$ distance &$i$,user9&
enddo
timewindow -10 30 ; distancewindow usedata
prs
qs
SAC> m doss
```

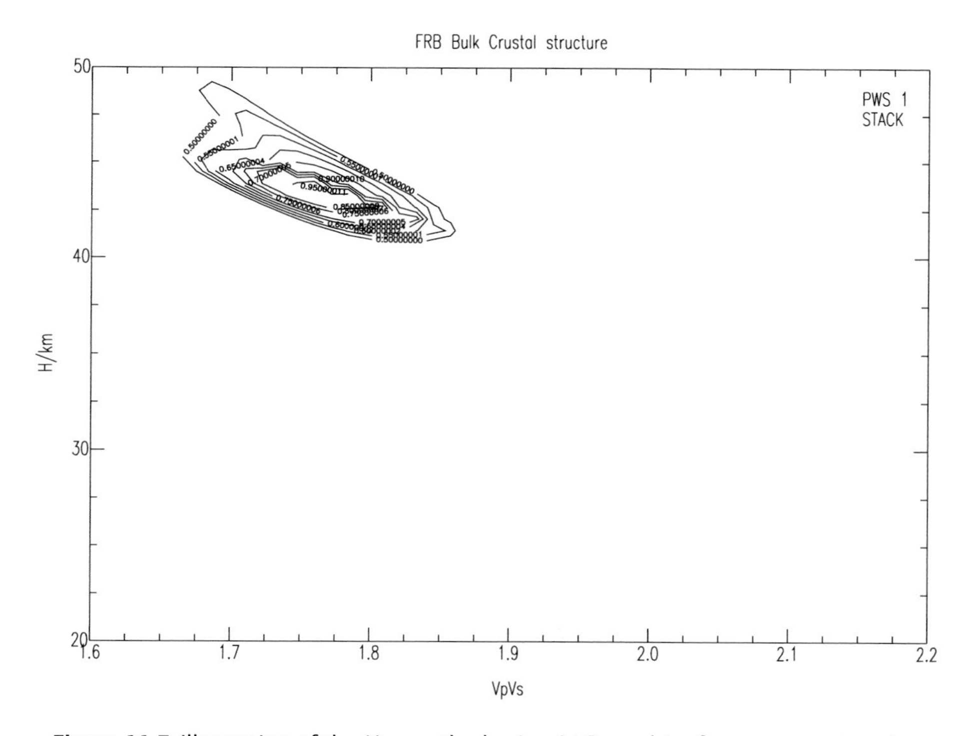

Figure 11.7 Illustration of the H–κ method using SAC graphics for permanent station FRB in Canada. In their receiver function study, Thompson *et al.* (2010) constrained bulk crustal properties for station FRB to be H = 43.5 km, V_P/V_S= 1.75, as is observed here.

In the directory of radial receiver functions, pipe the receiver functions into the H–κ stacking code. This will evaluate the objective function (Equation 11.2) for 50 crustal thickness (H) in the range 20–50 km, V_P/V_S in the range 1.6–2.1 with weights 0.5, 0.4 and 0.1.

```
ls *nrfr | \
   hk -vp 6.5 -h 20 50 50 -k 1.6 2.1 50 \
      -weight 0.5 0.4 0.1 -phaseweight 1 \
      -p USER9 -info HK.sac
```

Then, use SAC graphics to view the H–κ stack as follows (Fig. 11.7):

```
SAC> r HK.sac
SAC> zlevels range 0.5 1 increment 0.05
SAC> xlabel "VpVs"
SAC> ylabel "H / km"
SAC> title "FRB Bulk Crustal Structure"
SAC> zlabels on
SAC> contour
```

It is easy to envisage developing a macro to aid in the display of H–κ stacks that might even include annotating the plot with PLOTC, showing the peak using a crosshair and overlaying the contoured data onto a color background.

Appendix

A

Alphabetical list of SAC commands

absolutevalue	Takes the absolute value of each data point.
add	Adds a constant to each data point.
addf	Adds a set of data files to data in memory.
addstack	Adds a new file to the stack file list.
apk	Applies an automatic event picking algorithm.
arraymap	Produces a map of the array or "coarray" using all files in SAC memory.
axes	Controls the location of annotated axes.
bandpass	Applies an infinite-impulse-response (IIR) bandpass filter.
bandrej	Applies an IIR bandreject filter.
bbfk	Computes the broadband frequency–wavenumber (FK) spectral estimate, using all files in SAC memory.
beam	Computes the beam using all data files in SAC memory.
begindevices	Begins plotting to one or more graphics devices.
beginframe	Turns off automatic new frame actions between plots.
beginwindow	Begins plotting to a new graphics window.
benioff	Applies a Benioff filter to the data.
binoperr	Controls errors that can occur during binary file operations.
border	Controls the plotting of a border around plots.
break	Exits early from a sequence of repeated commands inside of a macro.
changestack	Changes properties of files currently in the stack file list.
chnhdr	Changes the values of selected header fields.
chpf	Closes the currently open HYPO pick file.
color	Controls color selection for color graphics devices.

comcor	Controls SAC's command correction option.
comment	Allows a comment in a SAC macro or on a SAC command line.
contour	Produces contour plots of data in memory.
convert	Converts data files from one format to another.
convolve	Computes the convolution of a master signal with one or more other signals; computes the auto- and cross-convolution functions.
copyhdr	Copies header variables from one file in memory to all others.
copyright	Displays copyright.
cor	Computes the correlation function.
correlate	Computes the auto- and cross-correlation functions.
cut	Defines how much of a data file is to be read.
cuterr	Controls errors due to bad cut parameters.
cutim	Cuts files in memory; can cut multiple segments from each file.
datagen	Generates sample data files and stores them in memory.
decimate	Decimates (downsamples) data, including an optional anti-aliasing finite-impulse-response (FIR) filter.
deletechannel	Deletes one or more files from the file list.
deletestack	Deletes one or more files from the stack file list.
deltacheck	Changes the sampling rate checking option.
dif	Differentiates data in memory.
distanceaxis	Defines the record section plot distance axis parameters.
distancewindow	Controls the distance window properties on subsequent record section plots.
div	Divides each data point by a constant.
divf	Divides data in memory by a set of data files.
divomega	Performs integration in the frequency domain.
do	Repeats a sequence of commands inside of a macro.
echo	Controls echoing of input and output to the terminal.
else	Conditionally processes a sequence of commands inside of a macro.
elseif	Conditionally processes a sequence of commands inside of a macro.
enddevices	Terminates one or more graphics devices.
enddo	Ends a sequence of repeated commands inside of a macro.
endframe	Resumes automatic new frame actions between plots.
endif	Conditionally processes a sequence of commands inside of a macro.
endwindow	Closes a graphics window.
envelope	Computes the envelope function using a Hilbert transform.
erase	Erases the graphics display area.
evaluate	Evaluates simple arithmetic expressions.
exp	Computes the exponential of each data point.
exp10	Computes the base 10 exponential (10**y) of each data point.
expressions	Overview of expressions in SAC
fft	Performs a discrete Fourier transform.
fileid	Controls the file id display found on most SAC plots.
filenumber	Controls the file number display found on most SAC plots.
filterdesign	Produces a graphic display of a filter's digital versus analog characteristics for amplitude, phase and impulse response curves, and the group delay.
fir	Applies an FIR filter.

floor	Puts a minimum value on logarithmically scaled data.
funcgen	Generates a function and stores it in memory.
getbb	Gets (prints) values of blackboard variables.
getfhv	Returns a real-valued header variable from the present SAC file.
getihv	Returns an enumerated header variable from the present SAC file.
getkhv	Returns a character-valued header variable from the present SAC file.
getlhv	Returns a logical-valued header variable from the present SAC file.
getnhv	Returns an integer-valued header variable from the present SAC file.
globalstack	Sets global stack properties.
grayscale	Produces grayscale or color images of data in memory.
grid	Controls the plotting of grid lines in plots.
gtext	Controls the quality and font of text used in plots.
hanning	Applies a "Hanning" window to each data file.
help	Displays information about SAC commands and features on the screen.
highpass	Applies an IIR highpass filter.
hilbert	Applies a Hilbert transform.
history	Prints a list of the recently issued SAC commands
if	Conditionally processes a sequence of commands inside of a macro.
ifft	Performs an inverse discrete Fourier transform.
image	Produces color sampled image plots of data in memory.
incrementstack	Increments properties for files in the stack file list.
inicm	Reinitializes all of SAC's common blocks.
installmacro	Installs macro files in the global SAC macro directory.
int	Performs integration using the trapezoidal or rectangular rule.
interpolate	Interpolates evenly or unevenly spaced data to a new sampling rate.
keepam	Keep amplitude component of spectral files (of either the AMPH or RLIM format) in SAC memory.
khronhite	Applies a Khronhite filter to the data.
line	Controls the linestyle selection in plots.
linefit	Computes the best straight-line fit to the data in memory and writes the results to header blackboard variables.
linlin	Turns on linear scaling for the x and y axes.
linlog	Turns on linear scaling for x axis and logarithmic for y axis.
listhdr	Lists the values of selected header fields.
liststack	Lists the properties of the files in the stack file list.
load	Load an external command (not available in MacSAC).
loadctable	Allows the user to select a new color table for use in image plots.
log	Takes the natural logarithm of each data point.
log10	Takes the base 10 logarithm of each data point.
loglab	Controls labels on logarithmically scaled axes.
loglin	Turns on logarithmic scaling for x axis and linear for y axis.
loglog	Turns on logarithmic scaling for the x and y axes.
lowpass	Applies an IIR lowpass filter.
macro	Executes a SAC macro file and the startup/init commands when invoking SAC.

map	Generate a Generic Mapping Tools (GMT) map with station/event symbols topography and station names using all the files in SAC memory and an optional event file specified on the command line.
markptp	Measures and marks the maximum peak-to-peak amplitude of each signal within the measurement time window.
marktimes	Marks files with travel times from a velocity set.
markvalue	Searches for and marks values in a data file.
mem	Calculates the spectral estimate using the maximum entropy method.
merge	Merges (concatenates) a set of files to data in memory.
message	Sends a message to the user's terminal.
mlm	Calculates the spectral estimate using the maximum likelihood method.
mtw	Determines the measurement time window for use in subsequent measurement commands.
mul	Multiplies each data point by a constant.
mulf	Multiplies a set of files by the data in memory.
mulomega	Performs differentiation in the frequency domain.
news	Prints current news concerning SAC.
null	Sets the undefined sample value for plotting data.
oapf	Opens an alphanumeric pick file.
ohpf	Opens a HYPO formatted pick file.
pause	Sends a message to the terminal and pauses.
pds	Calculates the spectral estimate using the power density spectrum method.
phase	Specifies phase names to include in record section plots.
picks	Controls the display of time picks on most SAC plots.
plabel	Defines general plot labels and their attributes.
plot	Generates a single-trace single-window plot.
plot1	Generates a multi-trace multi-window plot.
plot2	Generates a multi-trace single-window (overlay) plot.
plotalpha	Reads alphanumeric data from disk into memory and plots the data, optionally labeling each data point, to the current output device.
plotcor	Plots the correlation function.
plotc	Annotates SAC plots and creates figures using cursor.
plotdy	Plots one or more data files versus another data file with vertical error bars around each data value.
plotpe	Plots the root mean square (RMS) prediction error function.
plotpk	Produces a plot for the picking of arrival times.
plotpm	Generates a "particle-motion" plot of pairs of data files.
plotrecordsection	Plots a record section of the files in the stack file list.
plotsp	Plots spectral data in several different formats.
plotspe	Plots the spectral estimate.
plotstack	Plots the files in the stack file list.
plotxy	Plots one or more data files versus another data file.
production	Controls the production mode option.
qdp	Controls the "quick and dirty plot" option.

qsacxml	Queries a SAC XML dataset into a program and returns summary information about it.
quantize	Converts continuous data into its quantized equivalent.
quit	Terminates SAC.
quitsub	Terminates the currently active subprocess.
read	Reads data from SAC, SEGY, GCF or MSEED data files on disk into memory.
readbbf	Reads a blackboard variable file into memory.
readcor	Reads a SAC file containing the correlation function.
readcss	Reads data files in CSS external format from disk into memory.
readerr	Controls errors that occur during the READ command.
readgse	Reads data files in GSE 2.0 format from disk into memory.
readhdr	Reads headers from SAC data files into memory.
readsp	Reads spectral files written by WRITESP and WRITESPE.
readtable	Reads alphanumeric data files in column format on disk into memory.
report	Informs the user about the current state of SAC.
reverse	Reverses the order of data points.
rglitches	Removes glitches and timing marks.
rmean	Removes the mean.
rms	Computes the RMS of the data within the measurement time window.
rotate	Rotates a pair of data components through an angle.
rq	Removes the seismic Q factor from spectral data.
rsac1	Reads evenly spaced files into a program.
rsac2	Reads unevenly spaced or spectral files into a program.
rsach	Reads the header from a SAC file into a program.
rsacxml	Reads a trace from an XML SAC dataset into a program.
rtrend	Removes the linear trend.
setbb	Sets (defines) values of blackboard variables.
setdevice	Defines a default graphics device to use in subsequent plots.
setfhv	Sets a real-valued header variable in the present SAC file.
setihv	Sets an enumerated header variable in the present SAC file.
setkhv	Sets a character-valued header variable in the present SAC file.
setlhv	Sets a logical header variable in the present SAC file.
setmacro	Defines a set of directories to search when executing a SAC macro file.
setnhv	Sets an integer-valued header variable in the present SAC file.
sgf	Controls the SAC Graphics File (SGF) device options.
smooth	Applies an arithmetic smoothing algorithm to the data.
spe	Activates the spectral estimation subprocess.
spectrogram	Calculates a spectrogram using all of the data in memory.
speid	Controls annotation of plots from the spectral estimation subprocess.
speread	Reads data from a SAC data file into memory.
sqr	Squares each data point.
sqrt	Takes the square root of each data point.
sss	Activates the signal stacking subprocess.

stretch	Stretches (upsamples) data, including an optional interpolating FIR filter.
sub	Subtracts a constant from each data point.
subf	Subtracts a set of data files from data in memory.
sumstack	Sums the files in the stack file list.
symbol	Controls the symbol plotting attributes.
synchronize	Synchronizes the reference times of all files in memory.
syntax	Prints basic information about SAC commands, the command summary and syntax.
systemcommand	Executes system commands from SAC.
taper	Applies a symmetric taper to each end of data.
ticks	Controls the location of tick marks on plots.
timeaxis	Controls the time axis properties on subsequent record section plots.
timewindow	Sets the time window limits for subsequent stack summations.
title	Defines the plot title and attributes.
trace	Controls the tracing of blackboard and header variables.
transcript	Controls output to the transcript files.
transfer	Performs deconvolution to remove an instrument response and convolution to apply another instrument response.
traveltime	Computes travel-time curves for pre-defined models or reads travel-time curves from ASCII text files.
tsize	Controls the text size attributes.
unsetbb	Unsets (deletes) blackboard variables.
unwrap	Computes amplitude and unwrapped phase.
utilities	Overview of utility programs for SGF file plots, header information and file format swapping.
velocitymodel	Sets stack velocity model parameters for computing dynamic delays.
velocityroset	Controls the placement of a velocity rosette on subsequent record section plots.
view	Changes the set of files that will be affected by SAC commands.
vspace	Changes the maximum size and shape of plots.
wait	Tells SAC whether or not to pause between plots or during text output.
while	Repeats a sequence of commands inside of a macro.
whiten	Flattens the spectrum of the input time series.
whpf	Writes auxiliary cards into the HYPO pick file.
width	Controls line widths for plotting.
wiener	Designs and applies an adaptive Wiener filter.
wild	Sets wildcard characters used in read commands to expand file lists.
window	Sets the location and shape of graphics windows.
write	Writes data in memory to disk.
writebbf	Writes a blackboard variable file to disk.
writecor	Writes a SAC file containing the correlation function.
writegse	Writes data files in GSE 2.0 format from memory to disk.
writehdr	Overwrites the headers on disk with those in memory.
writespe	Writes a SAC file containing the spectral estimate.
writesp	Writes spectral files to disk as "normal" data files.

writestack	Writes the stack summation to disk.
wsac0	Writes out a SAC file, using the current header variables, from a program.
wsac1	Writes out an evenly spaced SAC file from a program.
wsac2	Writes out an unevenly spaced SAC time series file from a program.
wsach	Writes the header in the program's memory to a SAC file.
wsacxml	Writes out an XML SAC dataset from a program.
xapiir	Subroutine access to SAC time series filtering capabilities.
xdiv	Controls the x axis division spacing.
xfudge	Changes the x axis "fudge factor."
xfull	Controls plotting of x axis full logarithmic decades.
xgrid	Controls plotting of grid lines in the x direction.
xlabel	Defines the x axis label and attributes.
xlim	Determines the plot limits for the x axis.
xlin	Turns on linear scaling for the x axis.
xlog	Turns on logarithimic scaling for the x axis.
xvport	Defines the viewport for the x axis.
ydiv	Controls the y axis division spacing.
yfudge	Changes the y axis "fudge factor."
yfull	Controls plotting of y axis full logarithmic decades.
ygrid	Controls plotting of grid lines in the y direction.
ylabel	Defines the y axis label and attributes.
ylim	Determines the plot limits for the y axis.
ylin	Turns on linear scaling for the y axis.
ylog	Turns on logarithimic scaling for the y axis.
yvport	Defines the viewport for the y axis.
zcolors	Controls the color display of contour lines.
zerostack	Zeroes or reinitializes the signal stack.
zlabels	Controls the labeling of contour lines with contour level values.
zlevels	Controls the contour line spacing in subsequent contour plots.
zlines	Controls the contour line styles in subsequent contour plots.
zticks	Controls the labeling of contour lines with directional tick marks.

Appendix

B

Keyword in context for SAC command descriptions

The entries in this table are sorted based on a selected set of keywords drawn from the command descriptions in SAC's online help information. The entries are sorted based on the keyword in the middle of each text column and its context (four words to either side of the keyword) is provided along with the name of the command at the left side of each entry. Equivalent information may also be obtained by the `HELP APROPOS KKKK` command where the keyword is `KKKK`.

absolutevalue	Takes the	absolute value of each...
beginframe	...off automatic new frame	actions between plots.
endframe	Resumes automatic new frame	actions between plots.
spe		Activates the spectral estimation...
sss		Activates the signal stacking...
quitsub	Terminates the currently	active subprocess.
wiener	Designs and applies an	adaptive Wiener filter.
addstack		Adds a new file...
add		Adds a constant to...
addf		Adds a set of...
view	...files that will be	affected by SAC commands.
apk	...an automatic event picking	algorithm.
smooth	Applies an arithmetic smoothing	algorithm to the data.
oapf	Opens an	alphanumeric pick file.
plotalpha	Reads	alphanumeric data from disk...
readtable	Reads	alphanumeric data files in...
keepam	...files (of either the	AMPH or RLIM format)...

filterdesign	...vs. analog characteristics for	amplitude, phase and impulse...
keepam	Keeps	amplitude component of spectral...
markptp	...maximum peak to peak	amplitude of each signal...
unwrap	Computes	amplitude and unwrapped phase.
filterdesign	...a filter's digital versus	analog characteristics for amplitude,...
rotate	...data components through an	angle.
axes	Controls the location of	annotated axes.
speid	Controls	annotation of plots from...
decimate	...data, including an optional	anti-aliasing FIR filter.
erase	Erases the graphics display	area.
evaluate	Evaluates simple	arithmetic expressions.
smooth	Applies an	arithmetic smoothing algorithm to...
arraymap	...a map of the	array or "coarray" using...
plotpk	...for the picking of	arrival times.
traveltime	...reads travel-time curves from	ASCII text files.
plabel	...plot labels and their	attributes.
symbol	Controls the symbol plotting	attributes.
title	...the plot title and	attributes.
tsize	Controls the text size	attributes.
xlabel	...x axis label and	attributes.
ylabel	...y axis label and	attributes.
convolve	...other signals. Computes the	auto- and cross-convolution functions.
correlate	Computes the	auto- and cross-correlation...
apk	Applies an	automatic event picking algorithm.
beginframe	Turns off	automatic new frame actions...
endframe	Resumes	automatic new frame actions...
whpf	Writes	auxiliary cards into the...
axes	...the location of annotated	axes.
linlin	...the x and y	axes.
loglab	...labels on logarithmically scaled	axes.
loglog	...the x and y	axes.
distanceaxis	...record section plot distance	axis parameters.
linlog	...and logarithmic for y	axis.
linlog	...linear scaling for x	axis and logarithmic for...
loglin	...and linear for y	axis.
loglin	...logarithmic scaling for x	axis and linear for...
timeaxis	Controls the time	axis properties on subsequent...
xdiv	Controls the x	axis division spacing.
xfudge	Changes the x	axis "fudge factor."
xfull	Controls plotting of x	axis full logarithmic decades.
xlabel	Defines the x	axis label and attributes.
xlim	...limits for the x	axis.
xlin	...scaling for the x	axis.
xlog	...scaling for the x	axis.
xvport	...viewport for the x	axis.
yvport	...viewport for the y	axis.
bandpass	Applies an IIR	bandpass filter.
bandrej	Applies an IIR	bandreject filter.

plotdy	... file with vertical error	bars around each data...
exp10	Computes the	base 10 exponential (10**y)...
log10	Takes the	base 10 logarithm of...
syntax	Prints	basic information about SAC...
beam	Computes the	beam using all data...
benioff	Applies a	Benioff filter to the...
binoperr	... that can occur during	binary file operations.
getbb	Gets (prints) values of	blackboard variables.
linefit	... the results to header	blackboard variables.
readbbf	Reads a	blackboard variable file into...
setbb	Sets (defines) values of	blackboard variables.
trace	Controls the tracing of	blackboard and header variables.
unsetbb	Unsets (deletes)	blackboard variables.
writebbf	Writes a	blackboard variable file to...
inicm	... all of SAC's common	blocks.
border	... the plotting of a	border around plots.
bbfk	Computes the	broadband frequency–wavenumber...
whpf	Writes auxiliary	cards into the HYPO...
filterdesign	... filter's digital versus analog	characteristics for: amplitude, phase,...
wild	Sets wildcard	characters used in read...
getkhv	Returns a	character-valued header variable from...
setkhv	Sets a	character-valued header variable in...
deltacheck	Changes the sampling rate	checking option.
chpf		Closes the currently open...
endwindow		Closes a graphics window.
arraymap	... of the array or	"coarray" using all files...
color	Controls	color selection for color...
color	Controls color selection for	color graphics devices.
grayscale	Produces grayscale or	color images of data...
image	Produces	color sampled image plots...
loadctable	... to select a new	color table for use...
zcolors	Controls the	color display of contour...
readtable	... alphanumeric data files in	column format on disk...
comcor	Controls SAC's	command correction option.
comment	... or on a SAC	command line.
load	Loads an external	command (not available in...
map	... file specified on the	command line.
readerr	... occur during the READ	command.
syntax	... about SAC commands, the	command summary and syntax.
break	... a sequence of repeated	commands inside of a...
do	Repeats a sequence of	commands inside of a...
else	... processes a sequence of	commands inside of a...
elseif	... processes a sequence of	commands inside of a...
enddo	... a sequence of repeated	commands inside of a...
endif	... processes a sequence of	commands inside of a...
help	Displays information about SAC	commands and features on...
history	... the recently issued SAC	commands.
if	... processes a sequence of	commands inside of a...

macro	...file and the startup/init	commands when invoking SAC.
mtw	...use in subsequent measurement	commands.
syntax	...basic information about SAC	commands, the command summary...
systemcommand	Executes system	commands from SAC.
view	...be affected by SAC	commands.
while	Repeats a sequence of	commands inside of a...
wild	...characters used in read	commands to expand filelists.
comment	Allows a	comment in a SAC...
keepam	Keeps amplitude	component of spectral files...
rotate	...a pair of data	components through an angle.
merge	Merges	(concatenates) a set of...
else		Conditionally processes a sequence...
elseif		Conditionally processes a sequence...
endif		Conditionally processes a sequence...
if		Conditionally processes a sequence...
add	Adds a	constant to each data...
div	...data point by a	constant.
mul	...data point by a	constant.
sub	Subtracts a	constant from each data...
quantize	Converts	continuous data into its...
contour	Produces	contour plots of data...
zcolors	...the color display of	contour lines.
zlabels	Controls the labeling of	contour lines with contour...
zlabels	...of contour lines with	contour level values.
zlevels	Controls the	contour line spacing in...
zlevels	...line spacing in subsequent	contour plots.
zlines	Controls the	contour line styles in subsequent...
zlines	...contour linestyles in subsequent	contour plots.
zticks	Controls the labeling of	contour lines with directional...
convolve	Computes the	convolution of a master...
transfer	...an instrument response and	convolution to apply another...
copyhdr		Copies header variables from...
copyright	Displays	copyright.
comcor	Controls SAC's command	correction option.
cor	Computes the	correlation function.
correlate	...the auto- and cross-	correlation functions.
plotcor	Plots the	correlation function.
readcor	...SAC file containing the	correlation function.
writecor	...SAC file containing the	correlation function.
correlate	Computes the auto- and	cross-correlation functions.
convolve	...Computes the auto- and	cross-convolution functions.

readcss	Reads data files in	CSS external format from...
plotc	...and creates figures using	cursor.
filterdesign	...phase, and impulse response	curves, and the group...
traveltime	Computes travel-time	curves for pre-defined models...
traveltime	...models or reads travel-time	curves from ASCII text...
cuterr	...errors due to bad	cut parameters.
cutim	...files in memory. Can	cut multiple segments from...
cutim		Cuts files in memory....
absolutevalue	...absolute value of each	data point.
add	...a constant to each	data point.
addf	Adds a set of	data files to data...
addf	...of data files to	data in memory.
beam	...the beam using all	data files in SAC...
benioff	...Benioff filter to the	data.
contour	Produces contour plots of	data in memory.
convert	Converts	data files from one...
cut	...how much of a	data file is to...
datagen	Generates sample	data files and stores...
decimate	Decimates (downsamples)	data, including an optional...
dif	Differentiates	data in memory.
div	Divides each	data point by a...
divf	Divides	data in memory by...
divf	...by a set of	data files.
exp	...the exponential of each	data point.
exp10	...exponential (10**y) of each	data point.
floor	...value on logarithmically scaled	data.
grayscale	...or color images of	data in memory.
hanning	..."Hanning" window to each	data file.
image	...sampled image plots of	data in memory.
interpolate	...evenly or unevenly spaced	data to a new...
khronhite	...Khronhite filter to the	data.
linefit	...line fit to the	data in memory and...
log	...natural logarithm of each	data point.
log10	...10 logarithm of each	data point.
markvalue	...marks values in a	data file.
merge	...set of files to	data in memory.
mul	Multiplies each	data point by a...
mulf	...of files by the	data in memory.
null	...sample value for plotting	data.
plotalpha	...memory and plots the	data, optionally labeling each...
plotalpha	Reads alphanumeric	data from disk into...
plotalpha	...data, optionally labeling each	data point, to the...
plotdy	Plots one or more	data files versus another...
plotdy	...data files versus another	data file with vertical...
plotdy	...error bars around each	data value.
plotpm	...plot of pairs of	data files.
plotsp	Plots spectral	data in several different...
plotxy	Plots one or more	data files versus another...

plotxy	... data files versus another	data file.
quantize	Converts continuous	data into its quantized...
read	Reads	data from SAC, SEGY,...
read	... SEGY, GCF or MSEED	data files on disk...
readcss	Reads	data files in CSS...
readgse	Reads	data files in GSE...
readhdr	Reads headers from SAC	data files into memory.
readtable	Reads alphanumeric	data files in column...
reverse	Reverses the order of	data points.
rms	... mean square of the	data within the measurement...
rotate	Rotates a pair of	data components through an...
rq	... Q factor from spectral	data.
smooth	... smoothing algorithm to the	data.
spectrogram	... using all of the	data in memory.
speread	Reads	data from a SAC...
speread	... data from a SAC	data file into memory.
sqr	Squares each	data point.
sqrt	... square root of each	data point.
stretch	Stretches (upsamples)	data, including an optional...
sub	... a constant from each	data point.
subf	Subtracts a set of	data files from data...
subf	... of data files from	data in memory.
taper	... to each end of	data.
write	Writes	data in memory to...
writegse	Write	data files in GSE...
writesp	... to disk as "normal"	data files.
qsacxml	Queries a SAC XML	dataset into a program...
rsacxml	... from an XML SAC	dataset into a program.
wsacxml	... out an XML SAC	dataset from a program.
xfull	... x axis full logarithmic	decades.
yfull	... y axis full logarithmic	decades.
decimate		Decimates (downsamples) data, including...
transfer	Performs	deconvolution to remove an...
setdevice	Defines a	default graphics device to...
setbb	Sets	(defines) values of blackboard...
filterdesign	... curves, and the group	delay.
velocitymodel	... parameters for computing dynamic	delays.
deletechannel		Deletes one or more...
deletestack		Deletes one or more...
unsetbb	Unsets	(deletes) blackboard variables.
pds	... estimate using the power	density spectrum method.
wiener		Designs and applies an...
plotalpha	... to the current output	device.
setdevice	Defines a default graphics	device to use in...
sgf	... SAC Graphics File (SGF)	device options.
begindevices	... one or more graphics	devices.
color	... selection for color graphics	devices.

enddevices	...one or more graphics	devices.
plotsp	...spectral data in several	different formats.
dif		Differentiates data in memory.
mulomega	Performs	differentiation in the frequency...
filterdesign	...display of a filter's	digital versus analog characteristics...
xgrid	...lines in the x	direction.
ygrid	...lines in the y	direction.
zticks	...of contour lines with	directional tick marks.
setmacro	Defines a set of	directories to search when...
installmacro	...the global SAC macro	directory.
qdp	Controls the "quick and	dirty plot" option.
fft	Performs a	discrete Fourier transform.
ifft	Performs an inverse	discrete Fourier transform.
plotalpha	Reads alphanumeric data from	disk into memory and...
read	...MSEED data files on	disk into memory.
readcss	...CSS external format from	disk into memory.
readgse	...GSE 2.0 format from	disk into memory.
readtable	...in column format on	disk into memory.
write	...data in memory to	disk.
writebbf	...blackboard variable file to	disk.
writegse	...format from memory to	disk.
writehdr	Overwrites the headers on	disk with those in...
writesp	Writes spectral files to	disk as "normal" data...
writestack	...the stack summation to	disk.
copyright		Displays copyright.
erase	Erases the graphics	display area.
fileid	Controls the file id	display found on most...
filenumber	Controls the file number	display found on most...
filterdesign	Produces a graphic	display of a filter's...
picks	Controls the	display of time picks...
zcolors	Controls the color	display of contour lines.
help		Displays information about SAC...
distanceaxis	...the record section plot	distance axis parameters.
distancewindow	Controls the	distance window properties on...
div		Divides each data point...
divf		Divides data in memory...
xdiv	Controls the x axis	division spacing.
ydiv	Controls the y axis	division spacing.
divomega	...integration in the frequency	domain.
mulomega	...differentiation in the frequency	domain.
decimate	Decimates	(downsamples) data, including an...
velocitymodel	...model parameters for computing	dynamic delays.
echo	Controls	echoing of input and...
taper	...symmetric taper to each	end of data.
enddo		Ends a sequence of...
mem	...estimate using the maximum	entropy method.

getihv	Returns an	enumerated header variable from...
setihv	Sets an	enumerated header variable in...
envelope	Computes the	envelope function using a...
quantize	...data into its quantized	equivalent.
erase		Erases the graphics display...
plotdy	...data file with vertical	error bars around each...
plotpe	Plots the RMS prediction	error function.
binoperr	Controls	errors that can occur...
cuterr	Controls	errors due to bad...
readerr	Controls	errors that occur during...
getihv	...from the present SAC	file.
bbfk	...-wavenumber (FK) spectral	estimate, using all files...
mem	Calculates the spectral	estimate using the maximum...
mlm	Calculates the spectral	estimate using the maximum...
pds	Calculates the spectral	estimate using the power...
plotspe	Plots the spectral	estimate.
writespe	...file containing the spectral	estimate.
spe	Activates the spectral	estimation subprocess.
speid	...plots from the spectral	estimation subprocess.
interpolate	Interpolates	evenly or unevenly spaced...
rsac1	Reads	evenly spaced files into a...
wsac1	Writes out an	evenly spaced SAC file from...
apk	Applies an automatic	event picking algorithm.
map	...memory and an optional	event file specified on...
macro		Executes a SAC macro...
systemcommand		Executes system commands from...
setmacro	...directories to search when	executing a SAC macro...
break		Exits early from a...
wild	...in read commands to	expand file lists.
exp	Computes the	exponential of each data...
exp10	Computes the base 10	exponential (10**y) of each...
evaluate	Evaluates simple arithmetic	expressions.
expressions	Overview of	expressions in SAC.
load	Loads an	external command (not available...
readcss	...data files in CSS	external format from disk...
rq	Removes the seismic Q	factor from spectral data.
xfudge	...the x axis “fudge	factor.”
yfudge	...the y axis “fudge	factor.”
help	...about SAC commands and	features on the screen.
chnhdr	...values of selected header	fields.
listhdr	...values of selected header	fields.
plotc	...SAC plots and creates	figures using cursor.
addstack	Adds a new	file to the stack...
addstack	...file to the stack	file list.
binoperr	...can occur during binary	file operations.
changestack	...currently in the stack	file list.

chpf	...currently open HYPO pick	file.
copyhdr	...header variables from one	file in memory to...
cut	...much of a data	file is to be...
cutim	...multiple segments from each	file.
deletechannel	...more files from the	file list.
deletestack	...files from the stack	file list.
fileid	Controls the	file id display found...
filenumber	Controls the	file number display found...
getfhv	...from the present SAC	file.
getkhv	...from the present SAC	file.
getlhv	...from the present SAC	file.
getnhv	...from the present SAC	file.
hanning	...window to each data	file.
incrementstack	...files in the stack	file list.
liststack	...files in the stack	file list.
macro	Executes a SAC macro	file and the startup/init...
map	...and an optional event	file specified on the...
markvalue	...values in a data	file.
oapf	Opens an alphanumeric pick	file.
ohpf	...a HYPO formatted pick	file.
plotdy	...files versus another data	file with vertical error...
plotrecordsection	...files in the stack	file list.
plotstack	...files in the stack	file list.
plotxy	...files versus another data	file.
readbbf	Reads a blackboard variable	file into memory.
readcor	Reads a SAC	file containing the correlation...
rsach	...header from a SAC	file into a program.
setfhv	...in the present SAC	file.
setihv	...in the present SAC	file.
setkhv	...in the present SAC	file.
setlhv	...in the present SAC	file.
setmacro	...executing a SAC macro	file.
setnhv	...in the present SAC	file.
sgf	Controls the SAC Graphics	File (SGF) device options.
speread	...from a SAC data	file into memory.
sumstack	...files in the stack	file list.
utilities	...plots, header information and	file format swapping.
utilities	...utility programs for SGF	file plots, header information...
whpf	...into the HYPO pick	file.
writebbf	Writes a blackboard variable	file to disk.
writecor	Writes a SAC	file containing the correlation...
writespe	Writes a SAC	file containing the spectral...
wsac0	Writes out a SAC	file, using the current...
wsac1	...out an evenly spaced SAC	file from a program.
wsac2	...unevenly spaced SAC time series	file from a program.
wsach	...memory to a SAC	file.
wild	...read commands to expand	filelists.
addf	...a set of data	files to data in...

arraymap	...or "coarray" using all	files in SAC memory.
bbfk	...spectral estimate, using all	files in SAC memory.
beam	...beam using all data	files in SAC memory.
changestack	Changes properties of	files currently in the...
convert	Converts data	files from one format...
cutim	Cuts	files in memory; can...
datagen	Generates sample data	files and stores them...
deletechannel	Deletes one or more	files from the file...
deletestack	Deletes one or more	files from the stack...
divf	...a set of data	files.
incrementstack	Increments properties for	files in the stack...
installmacro	Installs macro	files in the global...
keepam	...amplitude component of spectral	files (of either the...
liststack	...the properties of the	files in the stack...
map	...names using all the	files in SAC memory...
marktimes	Marks	files with travel times...
merge	...(concatenates) a set of	files to data in...
mulf	Multiplies a set of	files by the data...
plotdy	...one or more data	files versus another data...
plotpm	...of pairs of data	files.
plotrecordsection	...record section of the	files in the stack...
plotstack	Plots the	files in the stack...
plotxy	...one or more data	files versus another data...
read	...GCF or MSEED data	files on disk into...
readcss	Reads data	files in CSS external...
readgse	Reads data	files in GSE 2.0...
readhdr	...headers from SAC data	files into memory.
readsp	Reads spectral	files written by WRITESP...
readtable	Reads alphanumeric data	files in column format...
rsac1	Reads evenly spaced	files into a program.
rsac2	Reads unevenly spaced or spectral	files into a program.
subf	...a set of data	files from data in...
sumstack	Sums the	files in the stack...
synchronize	...reference times of all	files in memory.
transcript	...output to the transcript	files.
traveltime	...curves from ASCII text	files.
view	Changes the set of	files that will be...
writegse	Writes data	files in GSE 2.0...
writesp	Writes spectral	files to disk as...
writesp	...disk as "normal" data	files.
bandpass	Applies an IIR bandpass	filter.
bandrej	Applies an IIR bandreject	filter.
benioff	Applies a Benioff	filter to the data.
decimate	...an optional anti-aliasing FIR	filter.
fir	Applies a finite-impulse-response	filter.
highpass	Applies an IIR highpass	filter.
khronhite	Applies a Khronhite	filter to the data.
lowpass	Applies an IIR lowpass	filter.

stretch	...an optional interpolating FIR	filter.
wiener	...applies an adaptive Wiener	filter.
xapiir	...to SAC time series	filtering capabilities.
filterdesign	...graphic display of a	filter's digital versus analog...
fir	Applies a	finite-impulse-response filter.
decimate	...including an optional anti-aliasing	FIR filter.
stretch	...including an optional interpolating	FIR filter.
linefit	...the best straight-line	fit to the data...
bbfk	...broadband frequency–wavenumber	(FK) spectral estimate, using...
whiten		Flattens the spectrum of...
gtext	Controls the quality and	font of text used...
convert	...data files from one	format to another.
keepam	...the AMPH or RLIM	format) in SAC memory.
readcss	...files in CSS external	format from disk into...
readgse	...files in GSE 2.0	format from disk into...
readtable	...data files in column	format on disk into...
utilities	...header information and file	format swapping.
writegse	...files in GSE 2.0	format from memory to...
plotsp	...data in several different	formats.
ohpf	Opens a HYPO	formatted pick file.
fileid	...the file id display	found on most SAC...
filenumber	...the file number display	found on most SAC...
fft	Performs a discrete	Fourier transform.
ifft	Performs an inverse discrete	Fourier transform.
beginframe	Turns off automatic new	frame actions between plots.
endframe	Resumes automatic new	frame actions between plots.
bbfk	Computes the broadband	frequency–wavenumber (FK)...
divomega	Performs integration in the	frequency domain.
mulomega	Performs differentiation in the	frequency domain.
xfudge	Changes the x axis	"fudge factor."
yfudge	Changes the y axis	"fudge factor."
cor	Computes the correlation	function.
envelope	Computes the envelope	function using a Hilbert...
funcgen	Generates a	function and stores it...
plotcor	Plots the correlation	function.
plotpe	...the RMS prediction error	function.
readcor	...file containing the correlation	function.
writecor	...file containing the correlation	function.
convolve	...the auto- and cross-convolution	functions.
correlate	...auto- and cross-correlation	functions.
read	...data from SAC, SEGY,	GCF or MSEED data...
plabel	Defines	general plot labels and...
map	Generate a GMT	(Generic Mapping Tools) map...
getbb		Gets (prints) values of...
rglitches	Removes	glitches and timing marks.
globalstack	Sets	global stack properties.
installmacro	...macro files in the	global SAC macro directory.
map	Generate a	GMT (Generic Mapping Tools)...

filterdesign	Produces a	graphic display of a...
begindevices	...to one or more	graphics devices.
beginwindow	...plotting to a new	graphics window.
color	...color selection for color	graphics devices.
enddevices	Terminates one or more	graphics devices.
endwindow	Closes a	graphics window.
erase	Erases the	graphics display area.
setdevice	Defines a default	graphics device to use...
sgf	Controls the SAC	Graphics File (SGF) device...
window	...location and shape of	graphics windows.
grayscale	Produces	grayscale or color images...
grid	Controls the plotting of	grid lines in plots.
xgrid	Controls plotting of	grid lines in the...
ygrid	Controls plotting of	grid lines in the...
filterdesign	...response curves, and the	group delay.
readgse	Reads data files in	GSE 2.0 format from...
writegse	Writes data files in	GSE 2.0 format from...
hanning	Applies a	"Hanning" window to each...
chnhdr	...the values of selected	header fields.
copyhdr	Copies	header variables from one...
getfhv	Returns a real-valued	header variable from the...
getihv	Returns an enumerated	header variable from the...
getkhv	Returns a character-valued	header variable from the...
getlhv	Returns a logical-valued	header variable from the...
getnhv	Returns an integer-valued	header variable from the...
linefit	...writes the results to	header blackboard variables.
listhdr	...the values of selected	header fields.
rsach	Reads the	header from a SAC...
setfhv	Sets a real-valued	header variable in the...
setihv	Sets an enumerated	header variable in the...
setkhv	Sets a character-valued	header variable in the...
setlhv	Sets a logical	header variable in the...
setnhv	Sets an integer-valued	header variable in the...
trace	...tracing of blackboard and	header variables.
utilities	...for SGF file plots,	header information and file...
wsac0	...file, using the current	header variables, from a...
wsach	Writes the	header in the program's...
readhdr	Reads	headers from SAC data...
writehdr	Overwrites the	headers on disk with...
highpass	Applies an IIR	highpass filter.
envelope	...envelope function using a	Hilbert transform.
hilbert	Applies a	Hilbert transform.
chpf	Closes the currently open	HYPO pick file.
ohpf	Opens a	HYPO formatted pick file.
whpf	...auxiliary cards into the	HYPO pick file.
fileid	Controls the file	id display found on...
bandpass	Applies an	IIR bandpass filter.
bandrej	Applies an	IIR bandreject filter.

highpass	Applies an	IIR highpass filter.
lowpass	Applies an	IIR lowpass filter.
image	Produces color sampled	image plots of data...
loadctable	...table for use in	image plots.
grayscale	Produces grayscale or color	images of data in...
filterdesign	...for amplitude, phase and	impulse response curves, and...
incrementstack		Increments properties for files...
help	Displays	information about SAC commands...
qsacxml	...program and returns summary	information about it.
syntax	Prints basic	information about SAC commands,...
utilities	...SGF file plots, header	information and file format...
report		Informs the user about...
echo	Controls echoing of	input and output to...
whiten	...the spectrum of the	input time series.
transfer	...convolution to apply another	instrument response.
transfer	...deconvolution to remove an	instrument response and convolution...
getnhv	Returns an	integer-valued header variable from...
setnhv	Sets an	integer-valued header variable in...
divomega	Performs	integration in the frequency...
int	Performs	integration using the trapezoidal...
interpolate		Interpolates evenly or unevenly...
stretch	...data, including an optional	interpolating FIR filter.
ifft	Performs an	inverse discrete Fourier transform.
macro	...the startup/init commands when	invoking SAC.
history	...list of the recently	issued SAC commands.
khronhite	Applies a	Khronhite filter to the...
xlabel	Defines the x axis	label and attributes.
ylabel	Defines the y axis	label and attributes.
plotalpha	...plots the data, optionally	labeling each data point,...
zlabels	Controls the	labeling of contour lines...
zticks	Controls the	labeling of contour lines...
loglab	Controls	labels on logarithmically scaled...
plabel	Defines general plot	labels and their attributes.
zlabels	...contour lines with contour	level values.
mlm	...estimate using the maximum	likelihood method.
timewindow	Sets the time window	limits for subsequent stack...
xlim	Determines the plot	limits for the x...
ylim	Determines the plot	limits for the y...
comment	...on a SAC command	line.
linefit	Computes the best straight	line fit to the...
map	...specified on the command	line.
width	Controls	line widths for plotting.
zlevels	Controls the contour	line spacing in subsequent...
linlin	Turns on	linear scaling for the...

linlog	Turns on	linear scaling for x...
loglin	...for x axis and	linear for y axis.
rtrend	Removes the	linear trend.
xlin	Turns on	linear scaling for the...
ylin	Turns on	linear scaling for the...
grid	...the plotting of grid	lines in plots.
xgrid	Controls plotting of grid	lines in the x...
ygrid	Controls plotting of grid	lines in the y...
zcolors	...color display of contour	lines.
zlabels	...the labeling of contour	lines with contour level...
zticks	...the labeling of contour	lines with directional tick...
line	Controls the	line style selection in plots.
zlines	Controls the contour	line styles in subsequent contour...
addstack	...to the stack file	list.
changestack	...in the stack file	list.
deletechannel	...files from the file	list.
deletestack	...from the stack file	list.
history	prints a	list of the recently...
incrementstack	...in the stack file	list.
liststack	...in the stack file	list.
plotrecordsection	...in the stack file	list.
plotstack	...in the stack file	list.
sumstack	...in the stack file	list.
listhdr		Lists the values of...
liststack		Lists the properties of...
load		Load an external command...
axes	Controls the	location of annotated axes.
ticks	Controls the	location of tick marks...
window	Sets the	location and shape of...
xlog	Turns on	logarithimic scaling for the...
ylog	Turns on	logarithimic scaling for the...
log	Takes the natural	logarithm of each data...
log10	Takes the base 10	logarithm of each data...
linlog	...for x axis and	logarithmic for y axis.
loglin	Turns on	logarithmic scaling for x...
loglog	Turns on	logarithmic scaling for the...
xfull	...of x axis full	logarithmic decades.
yfull	...of y axis full	logarithmic decades.
floor	...a minimum value on	logarithmically scaled data.
loglab	Controls labels on	logarithmically scaled axes.
setlhv	Sets a	logical header variable in...
getlhv	Returns a	logical-valued header variable from...
lowpass	Applies an IIR	lowpass filter.
break	...commands inside of a	macro.
comment	...comment in a SAC	macro or on a...
do	...commands inside of a	macro.
else	...commands inside of a	macro.

elseif	... commands inside of a	macro.
enddo	... commands inside of a	macro.
endif	... commands inside of a	macro.
if	... commands inside of a	macro.
installmacro	Installs	macro files in the...
installmacro	... in the global SAC	macro directory.
macro	Executes a SAC	macro file and the...
setmacro	... when executing a SAC	macro file.
while	... commands inside of a	macro.
load	... command (not available in	MacSAC).
arraymap	Produces a	map of the array...
map	... GMT (Generic Mapping Tools)	map with station/event symbols...
map	Generate a GMT (Generic	Mapping Tools) map with...
markptp	Measures and	marks the maximum peak...
marktimes		Marks files with travel...
markvalue	Searches for and	marks values in a...
rglitches	Removes glitches and timing	marks.
ticks	... the location of tick	marks on plots.
zticks	... lines with directional tick	marks.
convolve	... the convolution of a	master signal with one...
markptp	Measures and marks the	maximum peak to peak...
mem	... spectral estimate using the	maximum entropy method.
mlm	... spectral estimate using the	maximum likelihood method.
vspace	Changes the	maximum size and shape...
rmean	Removes the	mean.
rms	Computes the root	mean square of the...
markptp	... each signal within the	measurement time window.
mtw	Determines the	measurement time window for...
mtw	... for use in subsequent	measurement commands.
rms	... the data within the	measurement time window.
markptp		Measures and marks the...
cutim	Cuts files in	memory; can cut multiple...
merge		Merges (concatenates) a set...
message	Sends a	message to the user's...
pause	Sends a	message to the terminal...
mem	... using the maximum entropy	method.
mlm	... using the maximum likelihood	method.
pds	... the power density spectrum	method.
floor	Puts a	minimum value on logarithmically...
production	Controls the production	mode option.
velocitymodel	Sets stack velocity	model parameters for computing...
traveltime	... travel-time curves for pre-defined	models or reads travel-time...
read	... SAC, SEGY, GCF or	MSEED data files on...
cutim	... in memory; can cut	multiple segments from each...
mul		Multiplies each data point...
mulf		Multiplies a set of...
plot1	Generates a	multi-trace multi-window plot.
plot2	Generates a	multi-trace single-window (overlay) plot.

plot1	Generates a multi-trace	multi-window plot.
map	... symbols topography and station	names using all the ...
phase	Specifies phase	names to include in ...
log	Takes the	natural logarithm of each ...
news	Prints current	news concerning SAC.
writesp	... files to disk as	"normal" data files.
load	Loads an external command	(not available in MacSAC).
filenumber	Controls the file	number display found on ...
keepam	... component of spectral files	(of either the AMPH ...
chpf	Closes the currently	open HYPO pick file.
oapf		Opens an alphanumeric pick ...
ohpf		Opens a HYPO formatted ...
binoperr	... occur during binary file	operations.
comcor	Controls SAC's command correction	option.
deltacheck	... the sampling rate checking	option.
production	Controls the production mode	option.
qdp	... "quick and dirty plot"	option.
plotalpha	... and plots the data,	optionally labeling each data ...
sgf	... Graphics File (SGF) device	options.
reverse	Reverses the	order of data points.
echo	... echoing of input and	output to the terminal.
plotalpha	... point, to the current	output device.
transcript	Controls	output to the transcript ...
wait	... plots or during text	output.
plot2	Generates a multi-trace single-window	(overlay) plot.
expressions		Overview of expressions in ...
utilities		Overview of utility programs ...
writehdr		Overwrites the headers on ...
rotate	Rotates a	pair of data components ...
plotpm	... a "particle-motion" plot of	pairs of data files.
cuterr	... due to bad cut	parameters.
distanceaxis	... section plot distance axis	parameters.
velocitymodel	Sets stack velocity model	parameters for computing dynamic ...
plotpm	Generates a	"particle-motion" plot of pairs ...
wait	... whether or not to	pause between plots or ...
pause	... to the terminal and	pauses.
markptp	... and marks the maximum	peak-to-peak amplitude ...
markptp	... the maximum peak to	peak amplitude of each ...
filterdesign	... analog characteristics for amplitude,	phase and impulse response ...
phase	Specifies	phase names to include ...
unwrap	Computes amplitude and unwrapped	phase.
chpf	... the currently open HYPO	pick file.
oapf	Opens an alphanumeric	pick file.
ohpf	Opens a HYPO formatted	pick file.
whpf	... cards into the HYPO	pick file.
apk	Applies an automatic event	picking algorithm.
plotpk	... a plot for the	picking of arrival times.

picks	...the display of time	picks on most SAC...
velocityroset	Controls the	placement of a velocity...
distanceaxis	Defines the record section	plot distance axis parameters.
plabel	Defines general	plot labels and their...
plot	Generates a single-trace single-window	plot.
plot1	Generates a multi-trace multi-window	plot.
plot2	...a multi-trace single-window (overlay)	plot.
plotpk	Produces a	plot for the picking...
plotpm	Generates a “particle-motion”	plot of pairs of...
qdp	...the “quick and dirty	plot” option.
title	Defines the	plot title and attributes.
xlim	Determines the	plot limits for the...
ylim	Determines the	plot limits for the...
beginframe	...new frame actions between	plots.
border	...of a border around	plots.
contour	Produces contour	plots of data in...
distancewindow	...on subsequent record section	plots.
endframe	...new frame actions between	plots.
fileid	...found on most SAC	plots.
filenumber	...found on most SAC	plots.
grid	...of grid lines in	plots.
gtext	...of text used in	plots.
image	Produces color sampled image	plots of data in...
line	...the line style selection in	plots.
loadctable	...for use in image	plots.
phase	...include in record section	plots.
picks	...picks on most SAC	plots.
plotalpha	...disk into memory and	plots the data, optionally...
plotc	Annotates SAC	plots and creates figures...
plotcor		Plots the correlation function.
plotdy		Plots one or more ...
plotpe		Plots the RMS prediction...
plotrecordsection		Plots a record section...
plotsp		Plots spectral data in...
plotspe		Plots the spectral estimate.
plotstack		Plots the files in...
plotxy		Plots one or more...
setdevice	...to use in subsequent	plots.
speid	Controls annotation of	plots from the spectral...
ticks	...of tick marks on	plots.
timeaxis	...on subsequent record section	plots.
utilities	...programs for SGF file	plots, header information and...
velocityroset	...on subsequent record section	plots.
vspace	...size and shape of	plots.
wait	...not to pause between	plots or during text...
zlevels	...spacing in subsequent contour	plots.
zlines	...line styles in subsequent contour	plots.

begindevices	Begins	plotting to one or...
beginwindow	Begins	plotting to a new...
border	Controls the	plotting of a border...
grid	Controls the	plotting of grid lines...
null	...undefined sample value for	plotting data.
symbol	Controls the symbol	plotting attributes.
width	Controls line widths for	plotting.
xfull	Controls	plotting of x axis...
xgrid	Controls	plotting of grid lines...
yfull	Controls	plotting of y axis...
ygrid	Controls	plotting of grid lines...
plotalpha	...optionally labeling each data	point, to the current...
pds	...spectral estimate using the	power density spectrum method.
traveltime	Computes travel-time curves for	pre-defined models or reads...
plotpe	Plots the RMS	prediction error function.
getbb	Gets	(prints) values of blackboard...
history		Prints a list of...
news		Prints current news concerning...
syntax		Prints basic information about...
production	Controls the	production mode option.
production	Controls the	production mode option.
qsacxml	...XML dataset into a	program and returns summary...
rsac1	...evenly spaced files into a	program.
rsac2	...spectral files into a	program.
rsach	...SAC file into a	program.
rsacxml	...SAC dataset into a	program.
wsac0	...header variables, from a	program.
wsac1	...SAC file from a	program.
wsac2	...series file from a	program.
wsacxml	...SAC dataset from a	program.
utilities	Overview of utility	programs for SGF file...
wsach	...the header in the	program's memory to a...
changestack	Changes	properties of files currently...
distancewindow	Controls the distance window	properties on subsequent record...
globalstack	Sets global stack	properties.
incrementstack	Increments	properties for files in...
liststack	Lists the	properties of the files...
timeaxis	Controls the time axis	properties on subsequent record...
rq	Removes the seismic	Q factor from spectral...
gtext	Controls the	quality and font of...
quantize	...continuous data into its	quantized equivalent.
qsacxml		Queries a SAC XML...
qdp	Controls the	"quick and dirty plot"...
deltacheck	Changes the sampling	rate checking option.
interpolate	...to a new sampling	rate.
cut	...file is to be	read.
readcss		Reads data files in...
readerr	...that occur during the	READ command.

readgse		Reads data files in...
wild	...wildcard characters used in	read commands to expand...
plotalpha		Reads alphanumeric data from...
read		Reads data from SAC,...
readbbf		Reads a blackboard variable...
readcor		Reads a SAC file...
readhdr		Reads headers from SAC...
readsp		Reads spectral files written...
readtable		Reads alphanumeric data files...
rsac1		Reads evenly spaced files into...
rsac2		Reads unevenly spaced or spectral...
rsach		Reads the header from...
rsacxml		Reads a trace from...
speread		Reads data from a...
traveltime	...for pre-defined models or	reads travel-time curves from...
getfhv	Returns a	real-valued header variable from...
setfhv	Sets a	real-valued header variable in...
history	...a list of the	recently issued SAC commands.
distanceaxis	Defines the	record section plot distance...
distancewindow	...window properties on subsequent	record section plots.
phase	...names to include in	record section plots.
plotrecordsection	Plots a	record section of the...
timeaxis	...axis properties on subsequent	record section plots.
velocityroset	...velocity rosette on subsequent	record section plots.
int	...using the trapezoidal or	rectangular rule.
synchronize	Synchronizes the	reference times of all...
inicm		Reinitializes all of SAC's...
zerostack	Zeroes or	reinitializes the signal stack.
transfer	Performs deconvolution to	remove an instrument response...
rglitches		Removes glitches and timing...
rmean		Removes the mean.
rq		Removes the seismic Q...
rtrend		Removes the linear trend.
break	...from a sequence of	repeated commands inside of...
enddo	Ends a sequence of	repeated commands inside of...
do		Repeats a sequence of...
while		Repeats a sequence of...
filterdesign	...amplitude, phase, and impulse	response curves, and the...
transfer	...to apply another instrument	response.
transfer	...to remove an instrument	response and convolution to...
linefit	...memory and writes the	results to header blackboard...
endframe		Resumes automatic new frame...
getfhv		Returns a real-valued header...
getihv		Returns an enumerated header...

getkhv		Returns a character-valued header...
getlhv		Returns a logical-valued header...
getnhv		Returns an integer-valued header...
qsacxml	...into a program and	returns summary information about...
reverse		Reverses the order of...
keepam	...either the AMPH or	RLIM format) in SAC...
plotpe	Plots the	RMS prediction error function.
rms	Computes the	root mean square of...
sqrt	Takes the square	root of each data...
velocityroset	...placement of a velocity	rosette on subsequent record...
rotate		Rotates a pair of...
int	...the trapezoidal or rectangular	rule.
datagen	Generates	sample data files and...
null	Set the undefined	sample value for plotting...
image	Produces color	sampled image plots of...
deltacheck	Changes the	sampling rate checking option.
interpolate	...data to a new	sampling rate.
floor	...minimum value on logarithmically	scaled data.
loglab	Controls labels on logarithmically	scaled axes.
linlin	Turns on linear	scaling for the x...
linlog	Turns on linear	scaling for x axis...
loglin	Turns on logarithmic	scaling for x axis...
loglog	Turns on logarithmic	scaling for the x...
xlin	Turns on linear	scaling for the x...
xlog	Turns on logarithimic	scaling for the x...
ylin	Turns on linear	scaling for the y...
ylog	Turns on logarithimic	scaling for the y...
help	...and features on the	screen.
setmacro	...set of directories to	search when executing a...
markvalue		Searches for and marks...
distanceaxis	Defines the record	section plot distance axis...
distancewindow	...properties on subsequent record	section plots.
phase	...to include in record	section plots.
plotrecordsection	Plots a record	section of the files...
timeaxis	...properties on subsequent record	section plots.
velocityroset	...rosette on subsequent record	section plots.
cutim	...memory; can cut multiple	segments from each file.
read	Reads data from SAC,	SEGY, GCF or MSEED...
rq	Removes the	seismic Q factor from...
loadctable	Allows the user to	select a new color...
color	Controls color	selection for color graphics...
line	Controls the line style	selection in plots.
message		Sends a message to...
pause		Sends a message to...

plotsp	Plots spectral data in	several different formats.
sgf	...the SAC Graphics File	(SGF) device options.
utilities	...of utility programs for	SGF file plots, header...
vspace	...the maximum size and	shape of plots.
window	Sets the location and	shape of graphics windows.
convolve	...convolution of a master	signal with one or...
markptp	...peak amplitude of each	signal within the measurement...
sss	Activates the	signal stacking subprocess.
zerostack	Zeroes or reinitializes the	signal stack.
convolve	...one or more other	signals; computes the auto-...
plot	Generates a	single-trace single-window plot.
plot	Generates a single-trace	single-window plot.
plot2	Generates a multi-trace	single-window (overlay) plot.
tsize	Controls the text	size attributes.
vspace	Changes the maximum	size and shape of...
smooth	Applies an arithmetic	smoothing algorithm to the...
interpolate	Interpolates evenly or unevenly	spaced data to a...
xdiv	...the x axis division	spacing.
ydiv	...the y axis division	spacing.
zlevels	Controls the contour line	spacing in subsequent contour...
bbfk	...frequency–wavenumber (FK)	spectral estimate, using all...
keepam	Keeps amplitude component of	spectral files (of either...
mem	Calculates the	spectral estimate using the...
mlm	Calculates the	spectral estimate using the...
pds	Calculates the	spectral estimate using the...
plotsp	Plots	spectral data in several...
plotspe	Plots the	spectral estimate.
readsp	Reads	spectral files written by...
rq	...seismic Q factor from	spectral data.
rsac2	Reads unevenly spaced or	spectral files into a...
spe	Activates the	spectral estimation subprocess.
speid	...of plots from the	spectral estimation subprocess.
writesp	Writes	spectral files to disk...
writespe	...SAC file containing the	spectral estimate.
spectrogram	Calculates a	spectrogram using all of...
pds	...using the power density	spectrum method.
whiten	Flattens the	spectrum of the input...
rms	Computes the root mean	square of the data...
sqrt	Takes the	square root of each...
sqr		Squares each data point.

addstack	...new file to the	stack file list.
changestack	...files currently in the	stack file list.
deletestack	...more files from the	stack file list.
globalstack	Sets global	stack properties.
incrementstack	...for files in the	stack file list.
liststack	...the files in the	stack file list.
plotrecordsection	...the files in the	stack file list.
plotstack	...the files in the	stack file list.
sumstack	...the files in the	stack file list.
timewindow	...window limits for subsequent	stack summations.
velocitymodel	Sets	stack velocity model parameters...
writestack	Writes the	stack summation to disk.
zerostack	...or reinitializes the signal	stack.
sss	Activates the signal	stacking subprocess.
macro	...macro file and the	startup/init commands when invoking...
report	...user about the current	state of SAC.
map	...station/event symbols topography and	station names using all...
map	...Mapping Tools) map with	station/event symbols topography and...
linefit	Computes the best	straight-line fit to...
stretch		Stretches (upsamples) data, including...
quitsub	Terminates the currently active	subprocess.
spe	Activates the spectral estimation	subprocess.
speid	...from the spectral estimation	subprocess.
sss	Activates the signal stacking	subprocess.
xapiir		Subroutine access to SAC...
sub		Subtracts a constant from...
subf		Subtracts a set of...
qsacxml	...a program and returns	summary information about it.
syntax	...SAC commands, the command	summary and syntax.
writestack	Writes the stack	summation to disk.
timewindow	...limits for subsequent stack	summations.
sumstack		Sums the files in...
utilities	...information and file format	swapping.
symbol	Controls the	symbol plotting attributes.
map	...Tools) map with station/event	symbols topography and station...
taper	Applies a	symmetric taper to each...
synchronize		Synchronizes the reference times...
syntax	...the command summary and	syntax.
systemcommand	Executes	system commands from SAC.
loadctable	...select a new color	table for use in...
taper	Applies a symmetric	taper to each end...

echo	... and output to the	terminal.
message	... message to the user's	terminal.
pause	... a message to the	terminal and pauses.
enddevices		Terminates one or more ...
quit		Terminates SAC.
quitsub		Terminates the currently active ...
gtext	... quality and font of	text used in plots.
traveltime	... travel-time curves from ASCII	text files.
tsize	Controls the	text size attributes.
wait	... between plots or during	text output.
ticks	Controls the location of	tick marks on plots.
zticks	... contour lines with directional	tick marks.
markptp	... signal within the measurement	time window.
mtw	Determines the measurement	time window for use ...
picks	Controls the display of	time picks on most ...
rms	... data within the measurement	time window.
timeaxis	Controls the	time axis properties on ...
timewindow	Sets the	time window limits for ...
whiten	... spectrum of the input	time series.
wsac2	... out an unevenly spaced SAC	time series file from ...
xapiir	Subroutine access to SAC	time series filtering capabilities.
marktimes	Marks files with travel	times from a velocity ...
plotpk	... the picking of arrival	times.
synchronize	Synchronizes the reference	times of all files ...
rglitches	Removes glitches and	timing marks.
title	Defines the plot	title and attributes.
map	... a GMT (Generic Mapping	Tools) map with station/event ...
map	... map with station/event symbols	topography and station names ...
rsacxml	Reads a	trace from an XML ...
trace	Controls the	tracing of blackboard and ...
transcript	Controls output to the	transcript files.
envelope	... function using a Hilbert	transform.
fft	Performs a discrete Fourier	transform.
hilbert	Applies a Hilbert	transform.
ifft	... an inverse discrete Fourier	transform.
int	Performs integration using the	trapezoidal or rectangular rule.
marktimes	Marks files with	travel times from a ...
traveltime	Computes	travel-time curves for pre-defined ...
traveltime	... pre-defined models or reads	travel-time curves from ASCII ...
rtrend	Removes the linear	trend.
null	Set the	undefined sample value for ...
interpolate	Interpolates evenly or	unevenly spaced data to ...
rsac2	Reads	unevenly spaced or spectral files ...
wsac2	Writes out an	unevenly spaced SAC time series ...
unsetbb		Unsets (deletes) blackboard variables.
unwrap	Computes amplitude and	unwrapped phase.
stretch	Stretches	(upsamples) data, including an ...

loadctable	Allows the	user to select a...
report	Informs the	user about the current...
message	...a message to the	user's terminal.
utilities	Overview of	utility programs for SGF...
absolutevalue	Takes the absolute	value of each data...
floor	Puts a minimum	value on logarithmically scaled...
null	Sets the undefined sample	value for plotting data.
plotdy	...bars around each data	value.
chnhdr	Changes the	values of selected header...
getbb	Gets (prints)	values of blackboard variables.
listhdr	Lists the	values of selected header...
markvalue	Searches for and marks	values in a data...
setbb	Sets (defines)	values of blackboard variables.
zlabels	...lines with contour level	values.
getfhv	Returns a real-valued header	variable from the present...
getihv	Returns an enumerated header	variable from the present...
getkhv	Returns a character-valued header	variable from the present...
getlhv	Returns a logical-valued header	variable from the present...
getnhv	Returns an integer-valued header	variable from the present...
readbbf	Reads a blackboard	variable file into memory.
setfhv	Sets a real-valued header	variable in the present...
setihv	Sets an enumerated header	variable in the present...
setkhv	Sets a character-valued header	variable in the present...
setlhv	Sets a logical header	variable in the present...
setnhv	Sets an integer-valued header	variable in the present...
writebbf	Writes a blackboard	variable file to disk.
copyhdr	Copies header	variables from one file...
getbb	...(prints) values of blackboard	variables.
linefit	...results to header blackboard	variables.
setbb	...(defines) values of blackboard	variables.
trace	...of blackboard and header	variables.
unsetbb	Unsets (deletes) blackboard	variables.
wsac0	...using the current header	variables, from a program.
marktimes	...travel times from a	velocity set.
velocitymodel	Sets stack	velocity model parameters for...
velocityroset	...the placement of a	velocity rosette on subsequent...
plotdy	...another data file with	vertical error bars around...
xvport	Defines the	viewport for the x...
yvport	Defines the	viewport for the y...
filterdesign	...of a filter's digital	versus analog characteristics for...
bbfk	...the broadband frequency–	wavenumber (FK) spectral estimate,...
width	Controls line	widths for plotting.
wiener	...and applies an adaptive	Wiener filter.
wild	Sets	wildcard characters used in...
beginwindow	...to a new graphics	window.
distancewindow	Controls the distance	window properties on subsequent...
endwindow	Closes a graphics	window.

hanning	Applies a “Hanning”	window to each data...
markptp	...within the measurement time	window.
mtw	Determines the measurement time	window for use in...
rms	...within the measurement time	window.
timewindow	Sets the time	window limits for subsequent...
window	...and shape of graphics	windows.
writegse		Writes data files in...
linefit	...data in memory and	writes the results to...
whpf		Writes auxiliary cards into...
write		Writes data in memory...
writebbf		Writes a blackboard variable...
writecor		Writes a SAC file...
writesp		Writes spectral files to...
writespe		Writes a SAC file...
writestack		Writes the stack summation...
wsac0		Writes out a SAC...
wsac1		Writes out an evenly spaced...
wsac2		Writes out an unevenly spaced...
wsach		Writes the header in...
wsacxml		Writes out an XML...
readsp	...spectral files written by	WRITESP and WRITESPE.
readsp	...written by WRITESP and	WRITESPE.
readsp	Reads spectral files	written by WRITESP and...
qsacxml	Queries a SAC	XML dataset into a...
rsacxml	...a trace from an	XML SAC dataset into...
wsacxml	Writes out an	XML SAC dataset from...
zerostack		Zeroes or reinitializes the...

References

Abers, G.A., Hu, X., and Sykes, L.R. 1995. Source scaling of earthquakes in the Shumagin region, Alaska: time-domain inversions of regional waveforms. *Geophys. J. Int.*, **123**(1), 41–58.

Aki, K., and Richards, P.G. 1980. *Quantitative Seismology – Theory and Methods.* San Francisco: W.H. Freeman.

Ammon, C.J. 1991. The isolation of receiver effects from teleseismic *P* waveforms. *Bull. Seism. Soc. Am.*, **81**(6), 2504–2510.

Ammon, C.J., Randall, G.E., and Zandt, G. 1990. On the nonuniqueness of receiver function inversions. *J. Geophys. Res.*, **95**(B10), 15303–15318.

Anderson, J., Farrell, W., Garcia, K., Given, J., and Swanger, H. 1990. *CSS Version 3 Database: Schema Reference Manual.* Tech. Rep. C90-01. Science Applications International Corporation.

Barry, K.M., Cavers, D.A., and Kneale, C.W. 1975. Recommended standards for digital tape formats. *Geophysics*, **40**, 344–352.

Bastow, I.D., Owens, T.J., Helffrich, G., and Knapp, J.H. 2007. Spatial and temporal constraints on sources of seismic anisotropy: evidence from the Scottish highlands. *Geophys. Res. Lett.*, **34**(5), doi:10.1029/2006GL028911.

Bastow, I.D., Thompson, D.A., Wookey, J., Kendall, J., Helffrich, G., Snyder, D., Eaton, D., and Darbyshire, F. 2011. Precambrian plate tectonics: seismic evidence from northern Hudson Bay, Canada. *Geology*, **39**(1), 91–94.

Blackman, D., and Kendall, J.M. 1997. Sensitivity of teleseismic body waves to mineral texture and melt in the mantle beneath a mid-ocean ridge. *Phil. Trans. R. Soc. Lond.*, **355**, 217–231.

Bokelmann, G.H.R., and Silver, P.G. 2002. Shear stress at the base of shield lithosphere. *Geophys. Res. Lett.*, **29**(23), doi:10.1029/2002GL015925.

Burg, J.P. 1972. A new analysis technique for time series data. In: Childers, D.G. (ed.), *Modern Spectrum Analysis.* New York: IEEE Press. Originally published in NATO Advanced Study Institute: Signal processing with emphasis on underwater acoustics, Enschede, the Netherlands, 12–23 August 1968.

Crotwell, H.P., Owens, T.J., and Ritsema, J. 1999. The TauP toolkit: flexible seismic travel-time and ray-path utilities. *Seism. Res. Lett.*, **70**, 154–160.

Daubechies, I. 1992. *Ten Lectures on Wavelets.* CBMS-NSF Regional Conference Series in Applied Mathematics, vol. 61. Philadelphia, PA: SIAM.

Di Leo, J.F., Wookey, J., Hammond, J.O.S., Kendall, J.M., Kaneshima, S., Inoue, H., Yamashina, T., and Harjadi, P. 2012. Deformation and mantle flow beneath the Sangihe subduction zone from seismic anisotropy. *Phys. Earth Planet. Inter.*, **194**, 38–54.

Efron, B. 1982. *The Jackknife, the Bootstrap and Other Resampling Plans.* CBMS-NSF Regional Conference Series in Applied Mathematics, vol. 38. Philadelphia, PA: SIAM.

Fouch, M.J., Fischer, A.M., Parmentier, E.M., Wysession, M.E., and Clarke, T.J. 2000. Shear wave splitting, continental keels, and patterns of mantle flow. *J. Geophys. Res.*, **105**, 6255–6276.

GSETT-3. 1995. *GSETT-3 Documentation, Conference Room Paper Series.* Tech. Rep. 243. United Nations, Comprehensive Test-ban Treaty Office.

Hamming, R.W. 1989. *Digital Filters.* Englewood Cliffs, NJ: Prentice-Hall.

Harris, D. 1990. *XAPiir: A Recursive Digital Filtering Package.* Tech. Rep. UCRL-ID-106005. Lawrence Livermore National Laboratory.

Helffrich, G. 2006. Extended-time multitaper frequency domain cross-correlation receiver-function estimation. *Bull. Seism. Soc. Am.*, **96**(1), 344–347.

IRIS. 2006. *SEED Reference Manual*, 3rd edn. Incorporated Research Institutions for Seismology.

Kennett, B.L.N., and Engdahl, E.R. 1991. Traveltimes for global earthquake and phase identification. *Geophys. J. Int.*, **105**, 429–465.

Kennett, B.L.N., Engdahl, E.R., and Buland, R. 1995. Constraints on seismic velocities in the Earth from traveltimes. *Geophys. J. Int.*, **122**, 108–124.

Kradolfer, U. 1996. AutoDRM: the first five years. *Seism. Res. Lett.*, **67**, 30–33.

Lacoss, R.T. 1971. Data adaptive spectral analysis methods. *Geophysics*, **36**(4), 661–675.

Langston, C.A. 1979. Structure under Mount Rainer, Washington, inferred from teleseismic body waves. *J. Geophys. Res.*, **84**, 4749–4762.

Ligorrìa, J., and Ammon, C.J. 1999. Iterative deconvolution and receiver-function estimation. *Bull. Seism. Soc. Am.*, **89**(5), 1395–1400.

McFadden, P.L., Drummond, B.J., and Kravis, S. 1986. The *N*th-root stack: theory, applications, and examples. *Geophysics*, **51**, 1879–1892.

Morelli, A., and Dziewonski, A.M. 1993. Body-wave traveltimes and a spherically symmetric P- and S-wave velocity model. *Geophys. J. Int.*, **112**, 178–194.

Obara, K., Kasahara, K., Hori, S., and Okada, Y. 2005. A densely distributed high-sensitivity seismograph network in Japan: Hi-net by National Research Institute for Earth Science and Disaster Prevention. *Rev. Sci. Instrum.*, **76**, doi:10.1063/1.1854197.

Park, J., and Levin, V. 2000. Receiver functions from multiple-taper spectral correlation estimates. *Bull. Seism. Soc. Am.*, **90**(6), 1507–1520.

Restivo, A., and Helffrich, G. 1999. Teleseismic shear-wave splitting measurement in noisy environments. *Geophys. J. Int.*, **137**, 821–830.

Rost, S., and Thomas, C. 2002. Array seismology: methods and applications. *Rev. Geophys.*, **40**, doi:10.1029/2000RG000100.

Savage, M.K. 1999. Seismic anisotropy and mantle deformation: what have we learned from shear wave splitting? *Rev. Geophys.*, **37**, 65–106.

Scargle, J.D. 1989. Studies in astronomical time series analysis. III. Fourier transforms, autocorrelation functions, and cross-correlation functions of unevenly spaced data. *Astrophys. J.*, **343**, 874–887.

Schimmel, M., and Paulssen, H. 1997. Noise reduction and detection of weak, coherent signals through phase-weighted stacks. *Geophys. J. Int.*, **130**, 497–505.

Silver, P., and Chan, G. 1991. Shear wave splitting and subcontinental mantle deformation. *J. Geophys. Res.*, **96**, 16429–16454.

Thompson, D.A., Bastow, I.D., Helffrich, G., Kendall, J.-M., Wookey, J., Snyder, D.B., and Eaton, D.W. 2010. Precambrian crustal evolution: seismic constraints from the Canadian Shield. *Earth Planet. Sci. Lett.*, **297**, 655–666.

Thomson, D.J. 1982. Spectrum estimation and harmonic analysis. *Proc. IEEE*, **70**(9), 1055–1096.

Verdon, J.P., Kendall, J.M., and Wuestefeld, A. 2009. Imaging fractures and sedimentary fabrics using shear wave splitting measurements made on passive seismic data. *Geophys. J. Int.*, **179**(2), 1245–1254.

Vergino, E., and Snoke, A. 1993. SAC and MAP. *IRIS Newsletter*, **XII**(2), 7–8.

Wiggins, R.A. 1976. Interpolation of digitized curves. *Bull. Seism. Soc. Am.*, **66**, 2077–2081.

Wookey, J., and Helffrich, G. 2008. Inner-core shear-wave anisotropy and texture from an observation of PKJKP waves. *Nature*, **454**(7206), 873–876.

Wuestefeld, A., and Bokelmann, G. 2007. Null detection in shear-wave splitting measurements. *Bull. Seism. Soc. Am.*, **97**(4), 1204–1211.

Zandt, G., Myers, S.C., and Wallace, T.C. 1995. Crust and mantle structure across the Basin and Range Colorado Plateau boundary at 47°N latitude and implications for Cenozoic extensional mechanism. *J. Geophys. Res.*, **100**(B6), 10529–10548.

Zhu, L., and Kanamori, H. 2000. Moho depth variation in southern California from teleseismic receiver functions. *J. Geophys. Res.*, **105**(B2), 2969–2980.

Index

AH, 3
aliasing, 31, 102, 103
array
  map, 101
  response function, 103
aspect ratio, 80
attenuation, 113

blackboard variable, 41–42, 49, 54, 104

C, 3
causality, 34
cepstral analysis, 34
color, 23, 116, 117, 119
commands
  $DEFAULT, 58, 59
  $ENDRUN, 15, 39, 52
  $KEYS, 59, 125
  $KILL, 50
  $RESUME, 50, 51
  $RUN, 15, 39, 52
  $TERMINAL, 50, 51
  ADDSTACK, 88, 91, 93
  APK, 15, 26
  ARRAYMAP, 101, 104
  AXES, 14, 15, 75
  BANDPASS, 14, 35, 59, 124
  BANDREJ, 14, 35
  BBFK, 102–105
  BD, 19
  BEAM, 102, 105, 107
  BEGINDEVICES, 19, 20, 24
  BEGINFRAME, 77, 94, 120
  BEGINWINDOW, 20
  BORDER, 75, 94
  BREAK, 54
  CH, 28
  CHANGESTACK, 88, 91, 94
  CHNHDR, 14, 28, 41, 77
  CHNHDR ALLT, 14
  COLOR, 14, 15, 23, 36, 119
  CONTOUR, 117, 119, 120, 125, 127
  COR, 110–112
  CUT, 14, 24
  CUTERR, 24, 103
  CUTIM, 14, 24, 103
  DATAGEN, 98, 103
  DECIMATE, 31
  DELETESTACK, 88
  DISTANCEWINDOW, 87, 94
  DO, 41, 49, 52, 53, 55, 56
  ECHO, 39–41, 118
  ELSE, 41
  ELSEIF, 41, 49
  ENDDO, 41, 53–56
  ENDFRAME, 77, 94, 120
  ENDIF, 41
  ENDWINDOW, 20
  EVALUATE, 42
  FD, 35
  FFT, 32
  FILEID, 76, 95
  FILTERDESIGN, 17, 35, 36
  FUNCGEN, 57
  GLOBALSTACK, 88
  GRAYSCALE, 119, 120
  GRID, 74, 75
  GTEXT, 36
  HELP, 15, 18, 22, 24, 27, 37, 40, 69, 72, 86, 87, 142
  HELP APROPOS, 15
  HELP COMMANDS, 15
  HELP LIBRARY, 8
  HELP UTILITIES, 8

HIGHPASS, 14, 35
IF, 41, 49
IFFT, 32
INCREMENTSTACK, 91, 92
INTERPOLATE, 31
IS, 92
LH, 27
LINE, 14, 15, 22, 23, 36, 119
LISTHDR, 27, 36
LISTSTACK, 87, 92
LOWPASS, 14, 35
M, 38
MACRO, 38, 39
MEM, 112
MERGE, 14
MESSAGE, 14, 15, 39, 56
MLM, 112
MTW, 24
P1, 21
PCOR, 111
PDS, 112
PICKS, 77
PLABEL, 75
PLOT, 21, 32, 74, 121
PLOT1, 14, 21, 22, 74, 76, 107, 108
PLOT2, 21-23, 74, 95, 97, 113, 125
PLOTC, 14, 15, 79-81, 83-85, 94, 108, 127, 134
  commands and graphical elements, 81
  graphical elements, 80
  parameters, 84
PLOTCOR, 110
PLOTPE, 112
PLOTPK, 14, 15, 25, 26, 84, 95, 113, 125, 130
  cursor commands, 26
PLOTPM, 125
PLOTRECORDSECTION, 14, 15, 89, 90, 94, 95, 97, 98
  cursor commands, 97
PLOTSP, 32
PLOTSPE, 112
PPK, 95, 105, 107
PRS, 89, 90, 100
PSPE, 112
QDP, 23, 36
QS, 86
QUIT, 18, 64
QUITSUB, 86
R, 18
READ, 6, 7, 14, 17-19, 34, 94, 107, 113
READCOR, 113
READCSS, 8, 14
READGSE, 6, 14, 15
READSP, 34
REPORT, 23, 119
RGLITCHES, 14, 24, 29, 30
RMEAN, 14, 30
RMS, 24
ROTATE, 32
RTREND, 14, 30
SC, 51
SETBB, 14, 15, 41, 47, 103
SETDEVICE, 20, 36
SETMACRO, 36, 40
SGF, 25
SPE, 86, 110
SPECTROGRAM, 116, 117
SSS, 86, 87, 94
SUMSTACK, 90, 93, 95, 96, 104
SYNCHRONIZE, 14, 28
SYNTAX, 16
SYSTEMCOMMAND, 15, 51, 52, 61, 125
TAPER, 14, 32
TICKS, 14, 15, 74
TIMEWINDOW, 87, 90, 94
TITLE, 14, 15, 75, 117
TRANSCRIPT, 18, 36
TRANSFER, 14
TRAVELTIME, 14, 90, 106-108
TSIZE, 36
UNSETBB, 41, 42
UNWRAP, 34
VELOCITYMODEL, 91
VM, 91, 92
VSPACE, 80
W, 19
WH, 29
WHILE, 41, 49, 52-55, 98
WHILE READ, 61
WIENER, 24
WINDOW, 20, 36
WRITE, 14, 17, 19, 29, 34
WRITECOR, 113
WRITEGSE, 15
WRITEHDR, 24, 29
WRITESP, 34
WRITESPE, 113
WRITESTACK, 93, 97, 104
XDIV, 74
XGRID, 74, 75
XLABEL, 14, 15, 75, 117
XLIM, 22, 24, 25
XLOG, 112
XVPORT, 77
YDIV, 74
YGRID, 74, 75
YLABEL, 14, 15, 75, 117
YLIM, 22, 25
YLOG, 112
YVPORT, 77
ZCOLORS, 117
ZLABELS, 117
ZLEVELS, 117, 118
ZLINES, 117-119
ZTICKS, 117
components, 31
coordinate system, 32

digital Fourier transform, *see* Fourier transform

elevation data, 115

Encapsulated PostScript, *see* EPS
EPS, 20, 24, 37
escape character, 48, 52, 62
expressions
  before, 56
  built-in functions
    character, 46
    numerical, 44
    system, 48
  existbb, 45
  existmv, 45
  getval, 43
  hdrnum, 45, 47
  hdrval, 47
  integer, 55
  itemcount, 47
  quotebb, 98
  reply, 14, 15, 44, 45, 50, 57
  status, 48

FFT, *see* Fourier transform
file header, 14
file header variable, 37, 41, 49
  A, 25, 27, 69, 77, 125, 127, 130
  AZ, 26, 27
  B, 24, 25, 27
  BAZ, 26, 27
  CMPINC, 32
  DELTA, 9, 10, 25, 121
  DIST, 26
  E, 24, 25, 27
  EVDP, 26, 47, 57, 69
  EVLA, 26, 27, 32, 57, 69, 88
  EVLO, 26, 27, 32, 57, 69, 88
  F, 25, 27, 69, 77, 125, 127, 130
  GCARC, 26, 27
  IFTYPE, 10, 121
  KA, 77
  KCMPNM, 45, 47
  KEVNM, 26, 37, 95
  KF, 77
  KHOLE, 26
  KO, 77
  KSTNM, 26
  KT0, 77
  KUSER0, 70, 71
  LCALDA, 27
  NPTS, 9, 10, 25, 71, 121
  NXSIZE, 9, 10, 120, 121
  NYSIZE, 9, 10, 120, 121
  NZJDAY, 25
  NZYEAR, 25
  O, 26, 77
  SCALE, 57
  STEL, 26
  STLA, 26, 27, 32, 57, 88, 103
  STLO, 26, 27, 32, 57, 88, 103
  T0, 25, 69, 77, 107
  T1, 67
  T9, 25, 69, 77, 107
  USER0, 125
  USER1, 125, 127
  USER2, 125
  USER3, 125
  USER4, 125
  USER5, 125, 127
  USER6, 125
  USER7, 103
  USER8, 103
  USER9, 133
  XMAXIMUM, 121
  XMINIMUM, 120, 121
  YMAXIMUM, 121
  YMINIMUM, 120
Fortran, 3, 68–70, 72, 121
Fourier transform, 32, 112, 115, 128
frame, 77, 83

gSAC, 2

IEEE floating point, 27
input redirection, 36

library routine
  GETFHV, 69
  GETKHV, 69
  GETLHV, 69
  GETNHV, 69
  GETxHV, 69, 125
  NEWHDR, 125
  RSAC1, 68, 69, 125, 127, 130
  RSAC2, 68
  RSACH, 68, 69
  sacio, 68–70, 72
  sacio90, 70
  SETFHV, 69
  SETKHV, 69
  SETLHV, 69
  SETNHV, 69
  SETxHV, 125
  WSAC0, 69, 121, 125, 127, 130
  WSAC1, 69
  WSAC2, 69
  xapiir, 72
linear phase, 34
linear stack, 92, 95

macro, 35
  keyword parameters, 57, 59
  positional parameters, 57–59
macro variable, 41, 49, 54
map, 115
marigram, 109
MATLAB, 3, 71, 72

*N*th root stack, 92, 95

palette, *see* color
partial data window, 24
PDF, 24, 37

phase-weighted stack, 92, 95, 97
PITSA, 3
plot
  axes, 75
  border, 75
  grid lines, 74
  labels, 75
  picks, 77
  tick marks, 74
Portable Document Format, *see* PDF
PostScript, *see* PS
PS, 20, 24, 37
Python, 3, 71

R, 3
ray parameter, *see* slowness
receiver function, 88

SAC Graphics File, *see* SGF file
SAC utilities, 24
SEISAN, 3
Seismic UNIX, 3
Seismic-Handler, 3
SGF file, 20, 23, 37
shell script, 37, 65–68
signal stacking subprocess, 86–100, 106
slowness, 88, 102
spectral estimation, 109–113
  prediction error, 112
  prewhitening, 110, 113
  window, 110
spectral estimation subprocess, 86, 109–114
spectrogram, 115
spectrum
  amplitude, 32
  phase, 34
stack
  *N*th root, 92, 95
  linear, 92, 95
  phase-weighted, 92, 95, 97
  properties, 87, 88
  uncertainty, 93, 95
standard input, 36, 37, 52, 125
standard output, 37, 65
subprocess
  signal stacking, *see* signal stacking subprocess
  spectral estimation, *see* spectral estimation subprocess

time, 27

UNIX, 3, 12, 15, 18, 20, 25, 28, 36–38, 41, 51, 64, 65, 93
UNIX commands
  awk, 52
  bash, 36, 64
  cal, 28
  cat, 52
  csh, 37
  curl, 93
  date, 51, 61
  echo, 125
  evalresp, 1
  f2c, 2
  jweed, 1
  ls, 12, 51, 61
  nedit, 38
  pico, 38
  rseed, 1, 11
  sh, 12, 36, 54
  tar, 11
  tcsh, 37, 64
  vi, 38
  xfig, 37
  xterm, 20
utilities
  sachdrinfo, 37
  sactosac, 8
  sgftoeps, 25, 37
  sgftopdf, 25, 37
  sgftops, 25, 37
  sgftoxfig, 37

viewport, 77

wavenumber, 102, 103
window
  autocorrelation, 110
  plot, 20
  spectrogram, 116
  taper, 32

XYZ data, 9, 10, 115

Printed in the United States
by Baker & Taylor Publisher Services

Practical Management of Bipolar Disorder

Practical Management of Bipolar Disorder

Edited by

Allan H. Young
Department of Psychiatry, University of British Columbia, Vancouver, Canada

I. Nicol Ferrier
Institute of Neuroscience, University of Newcastle, Newcastle upon Tyne, UK

Erin E. Michalak
Department of Psychiatry, University of British Columbia, Vancouver, Canada

CAMBRIDGE
UNIVERSITY PRESS

University Printing House, Cambridge CB2 8BS, United Kingdom

One Liberty Plaza, 20th Floor, New York, NY 10006, USA

477 Williamstown Road, Port Melbourne, VIC 3207, Australia

314-321, 3rd Floor, Plot 3, Splendor Forum, Jasola District Centre, New Delhi - 110025, India

79 Anson Road, #06-04/06, Singapore 079906

Cambridge University Press is part of the University of Cambridge.

It furthers the University's mission by disseminating knowledge in the pursuit of education, learning and research at the highest international levels of excellence.

www.cambridge.org
Information on this title: www.cambridge.org/9780521734899

First published 2010

A catalogue record for this publication is available from the British Library

ISBN 978-0-521-73489-9 Paperback

Contents

List of contributors *page* vi

1. **Reflections and insights from the 'lived' experience** 1
   Victoria Maxwell
2. **Overview of bipolar disorder** 9
   Allan H. Young, I. Nicol Ferrier and Erin E. Michalak
3. **The treatment of mania** 10
   Heinz Grunze
4. **Pharmacological treatment of bipolar depression** 24
   Allan H. Young and Charles B. Nemeroff
5. **Practical treatment guidelines for management of bipolar disorder** 33
   Lakshmi N. Yatham
6. **Psychosocial interventions in bipolar disorder: theories, mechanisms and key clinical trials** 44
   Sagar V. Parikh and Vytas Velyvis
7. **Physical treatments in bipolar disorder** 62
   Marisa Le Masurier, Lucie L. Herrmann, Louisa K. Coulson and Klaus P. Ebmeier
8. **Treating bipolar disorder in the early stages of illness** 73
   E. Jane Garland and Anne Duffy
9. **Special populations: the elderly** 84
   Alan J. Thomas
10. **Special populations: women and reproductive issues** 93
    Karine A.N. Macritchie and Carol Henshaw
11. **Physical health issues** 106
    Chennattucherry John Joseph and Yee Ming Mok
12. **Anxiety associated with bipolar disorder: clinical and pathophysiological significance** 120
    Pratap R. Chokka and Vikram K. Yeragani
13. **Bipolar disorder co-morbid with addictions** 129
    Jose M. Goikolea and Eduard Vieta
14. **Practical management of cyclothymia** 139
    Giulio Perugi and Dina Popovic
15. **Circadian and sleep/wake considerations in the practical management of bipolar disorder** 152
    Greg Murray
16. **A clinician's guide to psychosocial functioning and quality of life in bipolar disorder** 163
    Erin E. Michalak and Greg Murray
17. **Service delivery of integrated care for bipolar disorder** 175
    Richard Morriss
18. **Training and assessment issues in bipolar disorder: a clinical perspective** 184
    Sagar V. Parikh and Erin E. Michalak
19. **Practical guide to brain imaging in bipolar disorder** 194
    I. Nicol Ferrier and Adrian J. Lloyd
20. **From neuroscience to clinical practice** 199
    Annie J. Kuan, Vivienne A. Curtis and Allan H. Young

Index 210

The colour plates can be found between pages 184 and 185.

Contributors

Pratap R. Chokka
Department of Psychiatry, University of Alberta and Grey Nuns Hospital, Edmonton, Alberta, Canada

Louisa K. Coulson
Department of Psychiatry, University of Oxford, Oxford, UK

Vivienne A. Curtis
Maudsley Hospital and Institute of Psychiatry, London, UK

Anne Duffy
Department of Psychiatry, Dalhousie University, Halifax, Nova Scotia, Canada

Klaus P. Ebmeier
Department of Psychiatry, University of Oxford, Oxford, UK

I. Nicol Ferrier
Institute of Neuroscience, Newcastle University, Newcastle upon Tyne, UK

E. Jane Garland
University of British Columbia and British Columbia's Children's Hospital,Vancouver, British Columbia, Canada

Jose M. Goikolea
Bipolar Disorders Program, Institute of Clinical Neuroscience, University of Barcelona, Barcelona, Spain

Heinz Grunze
Institute of Neuroscience, Newcastle University, Newcastle upon Tyne, UK

Carol Henshaw
Staffordshire University and Liverpool Women's NHS Foundation Trust, Liverpool, UK

Lucie L. Herrmann
Department of Psychiatry, University of Oxford, Oxford, UK

Chennattucherry John Joseph
Department of Liaison Psychiatry, Royal Victoria Infirmary, Newcastle upon Tyne, UK

Annie J. Kuan
Department of Psychiatry, University of British Columbia, Vancouver, British Columbia, Canada

Adrian J. Lloyd
Institute of Neuroscience, Newcastle University, Newcastle upon Tyne, UK

Karine A.N. Macritchie
Institute of Mental Health, Department of Psychiatry, University of British Columbia, Vancouver, British Columbia, Canda

Marisa Le Masurier
Department of Psychiatry, University of Oxford, Oxford, UK

Victoria Maxwell
Crazy for Life Co., Sechelt, Canada

Erin E. Michalak
Mood Disorders Centre, Department of Psychiatry, University of British Columbia, Vancouver, British Columbia, Canada

Yee Ming Mok
Institute of Mental Health, Woodbridge Hospital, Singapore

Richard Morriss
University of Nottingham, Nottingham, UK

Greg Murray
Swinburne University of Technology, Melbourne, Australia

Charles B. Nemeroff
Department of Psychiatry and Behavioral Sciences, Emory University School of Medicine, Atlanta, USA

Sagar V. Parikh
University Health Network, Toronto Western Hospital, Toronto, Ontario, Canada

Giulio Perugi
University of Pisa, Pisa, Italy

Dina Popovic
Institute of Behavioural Science 'G De Lisio', Pisa, Italy

Alan J. Thomas
Wolfson Research Centre Institute for Ageing and Health, Newcastle University Campus for Ageing and Vitality, Newcastle upon Tyne, UK

Vytas Velyvis
Ontario Shores Centre for Mental Health Sciences, Ontario, Canada

Eduard Vieta
Bipolar Disorders Program, Institute of Clinical Neuroscience, University of Barcelona, Barcelona, Spain

Lakshmi N. Yatham
Mood Disorders Centre, Department of Psychiatry, University of British Columbia, Vancouver, British Columbia, Canada

Vikram K. Yeragani
Wayne State University School of Medicine, Detroit, USA

Allan H. Young
Institute of Mental Health, Department of Psychiatry, University of British Columbia, Vancouver, British Columbia, Canada

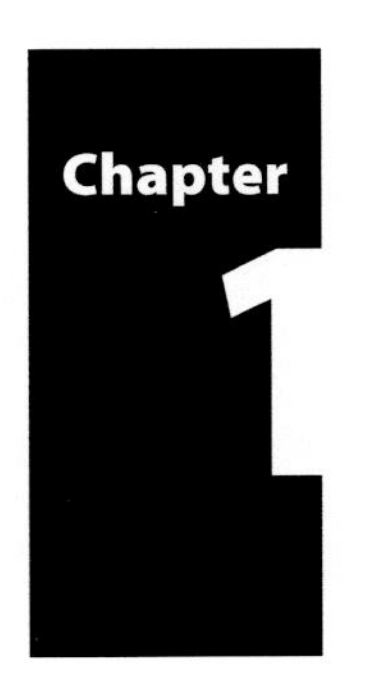

Reflections and insights from the 'lived' experience

Victoria Maxwell

Introduction: practical for both clinician and consumer

Bipolar disorder (BD) destabilises more than just a person's mood, thinking and physiology. It subverts one's identity, life and dreams. It is a disease that ferociously and expertly destroys lives and the sense of meaning, not just of the individual who has the illness but those loved ones around that person. I know. After what was thought to be a brief reactive psychosis (by the way: nothing is really brief when it comes to psychosis), I was eventually diagnosed with rapid-cycling, mixed state BD with psychotic features, mild temporal lobe epilepsy and generalised anxiety disorder. The regrettable (and sadly common) fact is that it took many more psychotic, manic and depressive episodes over a four-year period before I would accept I had BD. And it took an additional five years to regain some semblance of self-esteem and confidence and return to the workforce and independent living.

Prior to and during those first four years, my career careened out of control, my significant other left, my money evaporated, my friends were scared off, my car and home were no longer affordable and my idea of myself permanently changed. The assistance I received from the health professionals who worked with me (more accurately who *tried* to work with me – I wasn't exactly a willing participant for those first few years) included potentially helpful treatment plans and medication regimes. However, I didn't care about these plans – I cared about gathering what was left of my life (and my dignity) and envisioning something beyond attending self-help meetings and cognitive behavioural therapy sessions and cashing welfare cheques and living with my parents.

Understandably, wellness is defined in various ways by different groups. Decreasing the occurrence of disabling symptoms is not to be dismissed. Those living with BD know the importance of being able to get out of bed and to be free of the relentless fatigue that dogs us during depression. But to have a reason and a life to get out of bed *for* is at the heart of wellbeing. It is something that must be taken into consideration when creating treatment plans – it encompasses wholly different elements but is as essential as reducing episodes of psychosis, mania and depression. To be truly practical, the management of BD must go beyond the reduction of symptoms. It is about understanding what we, as your patients, want in specifics. Does your patient want to work full time, have children, go back to school, afford a car, learn to cook, have a circle of friends, quit smoking, live downtown but not on the Eastside? To be valuable to us, your help must be relevant to our daily lives. That is why including the consumer in the planning of treatment and in the process of delivery of those treatments is essential. Shared decision-making is pivotal. It is also why I am grateful and honoured to be contributing to this guide as one voice of the consumer. In order for this guide to be practical, the information and tactics lying within these pages must in the end benefit not only you, but the patients you serve.

In this chapter I discuss, from the perspective of someone who lives with BD, what makes a guide like this practical to those of us with the illness, and what recovery means and how it relates to the chapters that follow; I also introduce the concept of 'personal medicine' and the critical connection treatment must make with it.

Recovery, personal medicine and treatment: the critical connection

> Instead of focusing primarily on symptom relief, as the medical model dictates, recovery casts a much wider spotlight on restoration of self-esteem and identity and on attaining meaningful roles in society. (US Department of Health & Human Services, 1999)

As mentioned previously, symptom abatement alone is not synonymous with recovery. A life defined by decreased hospital days or better medication adherence is only part of the formula indicating recovery and one that is unsatisfactory at best. Dr Patricia Deegan, a

Practical Management of Bipolar Disorder, eds. Allan H. Young, I. Nicol Ferrier and Erin E. Michalak. Published by Cambridge University Press.

clinical psychologist and pioneer in the mental illness recovery movement and the study of resilience (who also happens to live with schizophrenia), eloquently states: 'recovery is about changing our lives, not just our biochemistry' (Deegan, 2005). Indeed, having a sense of purpose, feeling like a valued part of the community despite experiencing symptoms of BD, is at the core of what it means to be in recovery. Whether we experience intense symptoms or few at all, we as consumers – like all of us really – need to feel our lives have significance, that our lives have meaning, or that at the very least we are creating meaning.

All of us, regardless of our state of health, utilise what Dr Deegan terms 'personal medicine' (PM), which she defines as the things that give life meaning and 'make life worth living' (Deegan, 2004; Deegan, 2005). This idea of personal medicine is integral to recovery and must be taken into account when creating and executing recovery plans for an individual.

Personal medicine defined

The term coined by Deegan refers to activities or wellness strategies we do, rather than things we ingest. They are not goals we set or feelings we have. They are not over-the-counter mixtures of our own making. They are behaviours that enhance our sense of purpose, raise our self-esteem, engage our deepest values and beliefs and improve our wellbeing. At first glance, PM may seem overly simplistic and self-evident, but these behaviours and the ability to engage and align treatment with them have profound impact on patients' recovery and their adherence to treatment plans. Crucial to note is the delicate and often underestimated interplay between PM and treatment. In studies carried out by Deegan, discovering, understanding and aligning medical treatment with a patient's PM can increase medication adherence. She states that 'medication adherence may be improved when clinicians inquire about patients' personal medicine and use pharmaceuticals to support, rather than interfere with, these self-assessed health resources' (Deegan, 2005).

As we explore PM further it will also become apparent that this element in a person's life may be one of the most cost-effective and efficient means of engaging that individual in treatment, as long as this one guideline is followed: medical treatment must align with and complement PM, not the other way around. That is, in order for adherence to improve, treatment and medication must support, not interfere with, PM. Therefore the continued maintenance of PM in a patient's life must be the priority. Treatment and medications are to be adjusted to the PM in order to interfere as little as possible with the ability to engage in the individual's PM. I know this to be true from my own experience. I have seen and continue to see this phenomenon in action.

Categories, subtypes and unique features of PM

Personal medicine is not instead of, but in addition to, medication and other forms of treatment. It expands and challenges our idea of traditional medicine, entailing a broader definition of how an illness can be treated.

All of us have examples of PM in our lives. For me, these include: window shopping for shoes (note the operative word, 'window' – remember I have bipolar disorder); listening to almost any song of Joni Mitchell or Emmylou Harris; running in trails near my home; watching reruns of 'Friends'; being able to work and perform full time; soaking in our hot tub; spending time with my husband; reading fiction by my favourite authors; and prayer. Personal medicine works simultaneously, on multiple levels, is interdependent and rests on a spectrum. Deegan cites two categories of PM: self-care strategies and activities that give meaning and purpose (Deegan, 2005). Within each of those two classes of PM, I distinguish two further types (Figure 1.1): (1) self-care strategies that soothe the self and decrease the intensity of and occurrence of symptoms; and (2) activities that engage and enhance core ethics and identity.

Dual functionality of a PM measure

Within these two groups, the actions can either contain or activate energy (Erin Michalak, personal communication, 2009). Paradoxically, the same behaviour can accomplish both activation and containment. For example, running in trails for me: when hypomanic, running serves to contain and channel my extraneous energy; when experiencing depression, running and the outdoor environment stimulate dormant energy and counteract inertia. Another example that echoes this dual function is spending one-on-one time with my husband: when I am cycling through mild mania, my time with him acts as a grounding agent, calming and directing my frenetic energy; when I am depressed, his presence soothes and stimulates my enervated sense of aliveness.

Interplay between PM and treatment adherence

How does this alignment of PM and treatment and medication work to increase treatment adherence?

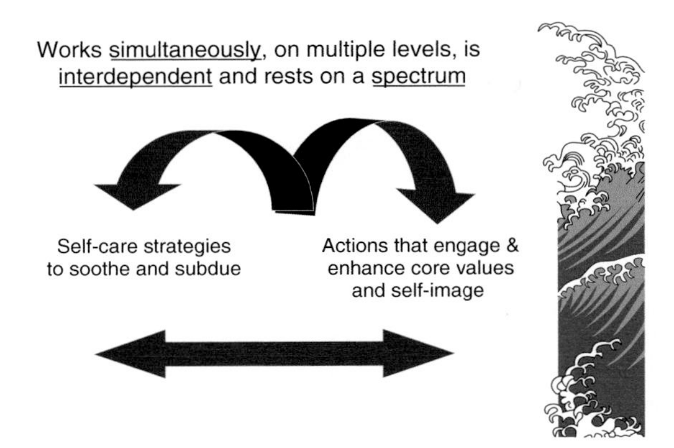
Works simultaneously, on multiple levels, is interdependent and rests on a spectrum
Self-care strategies to soothe and subdue
Actions that engage & enhance core values and self-image

Figure 1.1 Types of personal medicine.

Most PM is present before the onset of the illness and certainly precedes the diagnosis (Deegan, 2005). In addition, PM is actively pursued despite, or perhaps because of, the illness, in order to create some familiarity. Just as personality traits and interests generally remain constant throughout an individual's life (Costa & McCrae, 1997; Soldz & Vaillant, 1999), so too those activities considered to be PM remain stable and persistent (Low & Rounds, 2007). We are intrinsically motivated to continue doing these actions because they are highly valued and connected to our core principles. Because there is a pre-existing commitment to engage in the PM, the actions are most often self-initiated.

These behaviours that help create meaning in a patient's life, that help restore esteem, are also at times compromised and limited by treatment and the side effects of medication. And frequently, a patient is encouraged to tolerate these impositions rather than being given the opportunity, no less the power, to explore options to better integrate the suggested recovery plans into existing and personally important PM.

Discussing PM with patients to avoid non-adherence

Why must health providers make discussing and identifying current and potential PM with their patients a priority? Part of treatment needs to be working with the patient to identify current PM and discover new or latent PM. As Rapp describes in his strengths approach (Rapp, 1998), it is imperative and more effective to concentrate on what is already working in an individual's life or what has worked in the past. The congruency between treatment and PM will provide a vehicle to engage the patient proactively and to benchmark

Box 1.1. Personal medicine (PM) primer

- All of us practise PM, regardless of wellbeing
- PM pre-exists illness and diagnosis
- PM is as crucial to recovery as medication and other treatment plans
- Harmony between PM and treatment and medication is key to improving adherence and outcomes
- Patients do not commonly disclose PM to their clinicians
- Clinicians do not often enquire about patients' PM
- If clinicians begin a dialogue with patients about PM, medication adherence is likely to improve, in addition to quality of life
- Providers can be of most benefit by helping patients to identify current PM and discover new PM
- Ideally this discussion is done before or alongside the prescription of medication or other treatments.

significant barriers to taking medication or accepting certain treatment plans. The most effective management of BD ensures that prescribed treatment and medication supports rather than encumbers individualised life goals, current 'meaning-making' activities and already existing valued aspects and coping tools within your patient's life. This may seem like common sense, but unfortunately, common sense is not always common practice. More accurately even, practitioners often believe they understand this interplay and already align PM with medication and treatment. But I would dare to differ. The subtlety of PM is underestimated, as is the hesitation – and quite frankly fear – that we as patients have in challenging our health-care provider's mandates.

Non-adherence to medication and other treatment elements is a common issue in BD (Lingam & Scott, 2002). Studies indicate that over 60% of patients with BD have some adherence difficulties with medication (Gaudiano *et al.*, 2008). But the real reasons for the resistance are often misunderstood. Exploring a patient's PM and how that patient personally defines quality of life are keys to removing barriers to treatment cooperation (Box 1.1).

PM and removing barriers to treatment and medication adherence: real-life examples

As Deegan asserts, 'non-adherence with prescribed psychiatric medications was found to occur when pharmaceuticals interfered with personal medicine resulting in a diminished quality of life' (Deegan,

2005). I have acutely experienced how the failure to enquire about my PM resulted in my refusal to take medication, even to accept the diagnosis of BD, and resulted in much more time being sick. When I had my first psychotic experience, I was in a meditation group and involved in an intense spiritual practice. My manic psychosis was, in fact, very important to me. Although obviously many aspects were delusional, some also held valuable and cherished spiritual insights for me. But when I landed in the hospital, what I underwent was pathologised, and medication was prescribed so that it never happened again. One well-meaning, but misdirected doctor emphasising the benefits of pharmacology, explained to me: 'We don't want you to go down that path again...' What he failed to realise is that the 'path' that he saw only from a medical model, I held in a much wider framework. Like many in Deegan's study, I didn't discuss my interest in pursuing meditation and my practice further, as I feared, understandably, that it would not meet with the approval of my doctors (Deegan, 2005). In my mind I was being prescribed mood stabilisers, antipsychotics and antidepressants to eradicate the very thing that gave meaning to my life. So I refused to take them. Rather than a discussion taking place about my PM and how the medication and diagnosis undermined it, a power struggle between me, my family and my psychiatrist began. When faced with the choice of either taking pharmaceuticals and denying my spiritual life or refusing medication but embracing my relationship with my spirituality regardless of the symptoms I might have to endure – it was obvious to me which I was going to choose. It wasn't until further manias, several additional psychoses and more trips to the hospital, where I met a different psychiatrist, that a breakthrough occurred. In my initial meeting with this doctor he asked me to explain what led to my first psychosis, and what the meditation I was engaged in meant to me. In essence, he was exploring my PM in order to understand the vital role it played in my life.

This astute doctor worked with me and discussed how to incorporate my desire, and in fact my need, to follow my spiritual path with the goals of medication and his treatment plans. Our time together became a process of shared decision-making (Deegan & Drake, 2006). We agreed that temporarily I would put meditation on the 'back burner', and instead I would find a spiritual counsellor to help make sense of what happened to me and put it in a context that would be meaningful. This wise doctor told me that having a spiritual experience (which is how I defined what I went through, at least in part) and a mental illness are not mutually exclusive. It is not an 'either/or' scenario. Most importantly we discussed the fact that if I accepted I had the mental disorders I had been diagnosed with and took the medication prescribed, this in no way had to minimise the impact of what I went through or render it meaningless and merely a result of the illnesses. I see in hindsight that without either of us knowing it, he was ensuring that the treatment and medication goals were in harmony and alignment with one of my most important self-assessed health resources, or my PM.

Other examples can be less complex but no less pivotal. As one woman diagnosed with anxiety and BD explained to me:

> When I am as anxious and depressed as I am, it's important to be able to enjoy the small things. So when I get drowsy too early to watch my favorite TV show or not be able to get up in the morning and be able to be clear headed to write my to do list...I don't want to take my medication...it's not the kind of life I want (Velma, personal communication, 2008)

The ability to watch a much-loved television programme can be an important ingredient in a patient's recovery plan – and if it is one of the few activities a person is willing or able to partake in, it is essential that medication and treatment support these endeavours. Although hesitant to 'complain' about side effects and ask for the pills to be adjusted, Velma eventually did speak with her doctor to find a better match of medication to her PM.

Although I am speaking from my 'lived' experience and with anecdotal evidence, I posit that the more that prescribed pharmaceuticals and treatments plans align with an individual's already established and strongly embraced PM, the more a person will comply. And the more likely a person's quality of life will improve, even if symptom abatement does not occur fully. And if this discussion happens early on, during the initial stages of intervention with the health-care provider and patient, the better the rapport and the less likely non-adherence will take place.

PM as priority

Self-perpetuating reinforcement phenomenon of alignment

When clinicians I work with make my PM the benchmark from which to develop my treatment and set the goals of my medication, an interesting process begins to occur. I have also witnessed this sequence of events in others as they work with doctors who make PM the starting point. Deegan & Drake describe it as the

- The more my doctor helps me to align my treatment with my existing PM…
- The more valued and respected I feel…
- The better the rapport I have with my doctor; the more I trust him/her…
- The more willingly I embrace treatment plans…
- The more my wellness improves…
- The more my self-esteem improves…
- The more I practise my PM…
- The more I adhere to treatment…
- The more my wellness improves…
- The more I adhere to treatment and practise my PM…

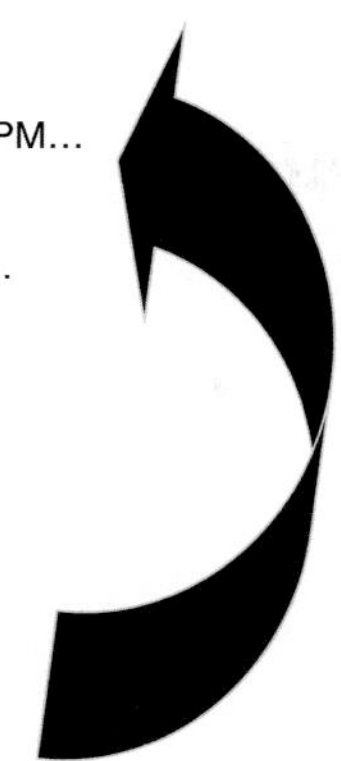

Figure 1.2 Self-perpetuating reinforcement phenomenon of alignment, or Deegan's 'cascading effect of personal medicine and psychiatric medication'.

'cascading effect of personal medicine and psychiatric medication' (Deegan & Drake, 2006). I see it in a more comprehensive light, enlarging the focus to highlight the potent effect the congruency of treatment with PM has on the recovery process. I call it the 'self-perpetuating reinforcement phenomenon of alignment'.

Figure 1.2 outlines what occurred and continues to occur when my psychiatrist takes into account my PM first rather than focusing solely on symptom abatement. This is not to say that the reduction of symptoms is not important, but if it comes at the cost of cherished wellness activities and beliefs, then any decrease in symptoms may not be seen as important from a patient's perspective.

I experience this process to this day, and I see others experience it on their journey to recovery. As with any other course of action, there may be setbacks, but the overall effect is one of growth and improvement.

The interrelatedness of respect, rapport and recovery

A myriad studies reveal the same conclusion: rapport of the clinician with the patient is one of the most powerful agents of change when it comes to positive outcomes and recovery (Luborsky *et al.*, 1985; Truax & Carkhuff, 1967). And this is even more salient when it comes to blending treatment with PM. Within this sequence of events, rapport and trust with the clinician prove to be essential components. Without strong therapeutic alliance and attunement, this 'self-perpetuating reinforcement phenomenon of alignment' will not easily occur. When we are treated as partners in the process of our recovery, where opportunities are set up for shared decision-making and to discuss our PM, this is when true healing can occur. Even in the acute stages of our illness – mania, depression and psychosis – in which our ability to contribute and make sound decisions is limited, if we are treated with respect and given an opportunity to contribute in any way, no matter how small, this is never forgotten. This is healing in and of itself.

When the opposite occurs, when I am almost literally locked out of discussing my own recovery process, not only does healing not occur, but in fact damage is also done. It is close to impossible to tell how aware a person is when they are actively manic or psychotic and just what will be remembered. But I'd advise you to err on the side of caution. Awareness is a powerful and mysterious element that should not be underestimated. Each time I entered into psychosis and mania, I was acutely conscious of how I was being treated, even if it was impossible for me to express that. I clearly recall – although floridly psychotic and struggling to free myself from the straps that held my wrists in place – the words that spouted from an emergency room worker as I made a run for a nearby washroom: 'Hey…stop the *crazy* woman!'. I can laugh at it now, but days later in the hospital and months later in my life, those words scalded me to the core when I thought of them. However, I also recollect just the opposite experience when the female police officer and two ambulance attendants caught up with me after I had decided (in another manic psychosis) to take an ecstatic skip through my neighbourhood 'sans clothes'. I recall the kind way the officer reassured me and offered me a blanket, how the EMS workers gently guided me into the ambulance and sat me down. It was the look in their eyes, the language of their body and face – you can't fake compassion. I never met the officer or those attendants again; I wish I had (under different circumstances, mind you). I would thank them for giving me hope and respect. Their behaviour registered within me (perhaps my unconsciousness, initially). But that experience stays with me. And for years it was a beacon of encouragement and flame of promise that I might indeed recover. That brief encounter with them gave me a palpable and lasting sense that I was something, someone other than my illness; the way they treated me indicated they saw who I was – not what I had become. It is not just what my health-care providers say, or even whether what my clinicians suggest as a treatment works (although this is important); more vital is how they make me feel when I leave their offices.

It is exciting to see the emphasis on making the data in each of the following chapters relevant to the lives of individuals like myself who live with the illness – the consumers. Your rapport with your patients is one of those invisible factors that will determine the level of wellness they will reach, and whether or not adherence continues to be an obstacle. I invite you to listen, not just to what your patients are telling you, but also to what they aren't telling you – and in fact why they don't.

Summary

When clinicians inquire about a patient's PM, and align treatment and pharmaceutical goals with it, non-adherence is likely to be diminished. This is because important activities and core values of the patient are not disrupted, and the patient's quality of life remains a priority, not just the reduction of symptoms.

Accurate reflection of bipolar disorder in the chapters that follow

It is both eerily and excitingly surreal to see my life distilled into precise clinical depictions. Seeing the words 'anxiety' and 'lethality' bookending the illness I live with is at once both frightening and encouraging. Encouraging? Yes, because I have been fortunate to receive and eventually actively participate in treatment for BD that has allowed me to regain a meaningful life. I missed falling into the crevasse of completed suicide, active addiction and chronic anxiety. This is because, like the researchers and practitioners who are the authors of the words that follow, those of you reading these pages are clinicians who care deeply about helping the people you work with.

As I review the various chapters, I can't help but be stunned at how my personal experience is reflected, and faithfully so, in the wide array of scientific research. Far from feeling like a number in a study, I feel somehow validated and comforted. My 'lived' experience is 'normal' (for a person with bipolar disorder, that is). It's good to be normal sometimes. I am not alone in what I have undergone – repeated hospitalisations, denial of the diagnosis, resistance to take prescribed medication or in the length of time or the challenges I faced – it took over four years for me to accept the illness and another five to live independently and regain employment.

Apparently I am a textbook case of BD (much to the chagrin of my ego). In Chapter 8, Drs E. Jane Garland and Anne Duffy cite among other conclusions that 'at least one-quarter of patients with bipolar disorder experience the onset of their first major mood episode during adolescence…which is usually depressive and (have) a positive family history of bipolar disorder.' With hindsight I recognise that my first major depressive episode, which went undetected and was chalked up to adolescent angst, was around the age of 17. Anxiety, an aspect that Garland explores, was present as early as 8 years old but again went undetected, manifesting itself in a refusal to go to school, social awkwardness and home sickness (which lasted until I was almost 13). The generalised anxiety, however, persisted throughout and into my adult years, co-occurring with my mood disorder.

My mother was hospitalised several times before being diagnosed with BD when I was 8 and relapsed into depression several times over the course of my childhood. I had what I kindly call a 'double-barrelled' genetic vulnerability: my mother has BD and anxiety; my father was never diagnosed with but dealt with anxiety and depression; and many of my relatives on my father's side make up a motley crew of individuals with different psychiatric disorders. So goes the luck of the draw. I doubt I will ever wear a pink ribbon and run for breast cancer – my genetics are not stacked in that direction – but I am already hoping I can wear a coloured ribbon and do a 10 k for my wonky brain in the not-so-distant future.

Again my profile is unmistakably reflected in other chapters, such as Chapter 6 with a review and comparisons of various psychosocial interventions by Drs Parikh and Velyvis. They describe the study of Lam and colleagues (Lam *et al.*, 2004) in which traits of 'perfectionism and excessive striving for achievement may confer increased vulnerability to experience mood relapse'. All through high school, in university and as I pursued my acting career, I set high – and frequently unrealistic – goals, fuelling a sense of failure when these targets weren't met. And perfectionism was learned early from my mother – again feeding a sense of anxiety, insecurity and helplessness as I unsuccessfully met my self-imposed goals. All of this contributed to feelings of low mood and anxiousness – emotions that eventually developed into painful clinical depression and a nerve-racking anxiety disorder. Although my pursuit of achievement is still strong, it is manageably so. The perfectionism I wrestled with for years thankfully left after a good amount of time in psychotherapy – in large part due to the cognitive therapy. Paradoxically, as my perfectionism faded into the background, my ability to reach my goals successfully increased.

In addition, Parikh and Velyvis explain that psychoeducation is loosely based on the age-old adage that 'knowledge is power'. And as such, patients make better decisions about their illness based on accurate information offered in such interventions. This is evident in my own journey to wellness – once the myths of mental illness and psychiatric medications I held were exposed, the easier it was to accept treatment as it was prescribed. As I gained correct information about how and why doctors believed pharmaceuticals would work – and that indeed BD is as biological as it is environmental and how those elements interact – the less problematic it was for me to admit I had a mental illness (several actually) and that the antidepressant and mood stabiliser I was prescribed were necessary. I recognised that my personal determination to get well would not be enough. In doing so, my adherence to medication and treatment improved.

In Chapter 16, Drs Erin Michalak and Greg Murray beautifully articulate what I feel must be foremost in the minds of practitioners when treating BD: that functioning and quality of life are as significant, if not more so, than improvements in symptoms. In addition, I echo the sentiment that there has been a misplaced assumption that symptoms and functioning are highly related.

Conclusion

This guide offers concrete tools and data that can result in better and more positive outcomes for those of us living with BD – not necessarily quickly or easily, but over time and with effort and patience. And when we work as partners, this knowledge can be harnessed into power. The alignment of treatment with a patient's PM also plays an integral role in the effectiveness of the treatment you offer. Ours is a symbiotic relationship. I rely on your expertise, and you rely on my feedback to inform the application of your expertise.

I still face bouts of depression (milder ones mind you, but disruptive all the same) and mini hypomanias, yet I live with a strong sense of recovery and self – a purposeful existence that incorporates meaningful relationships, a rewarding career, a robust identity and goals I actively pursue. This is a result of the tools I use to stay mentally well and manage my illnesses. Key reasons I am willing to be proactive and take charge of my mental health are the relationships my health professionals have established with me and the respect they have of my personal medicine. They work with me as partners, reminding me my insight into my illnesses is more crucial than their and is equally as important and is equally as important as their clinical expertise. This dynamic is what keeps me well. And this I am grateful for – just as I am grateful to contribute a voice to this text.

References

Costa, P.T. & McCrae, R.R. (1997). Stability and change in personality assessment: the revised NEO Personality Inventory in the year 2000. *Journal of Personality Assessment*, **68**, 86–94.

Deegan, P.E. (2004). *Recovery journal: the importance of personal medicine.* www.patdeegan.com/blog/archives/000013.php [accessed 1 March 2009].

Deegan, P.E. (2005). The importance of personal medicine: a qualitative study of resilience in people with psychiatric disabilities. *Scandinavian Journal of Public Health Supplement*, **66**, 29–35.

Deegan, P.E. & Drake, R.E. (2006). Shared decision making and medication management in the recovery process. *Psychiatric Services (Washington, D.C.)*, **57**, 1636–9.

Gaudiano, B.A., Weinstock, L.M., Miller, I.W. (2008). Improving treatment adherence in bipolar disorder: a review of current psychosocial treatment efficacy and recommendations for future treatment development. *Behavior Modification*, **32**, 267–301.

Lam, D.H., Wright, K., Smith, N. (2004). Dynsfunctional assumptions in bipolar disorder. *Journal of Affective Disorders*, **79**, 193–9.

Lingam R. & Scott, J. (2002). Treatment non-adherence in affective disorders. *Acta Psychiatrica Scandinavica*, **105**, 164–72.

Low, K.S. & Rounds, J. (2007). Interest change and continuity from early adolescence to middle adulthood. *International Journal of Educational and Vocational Guidance*, **7**, 23–36.

Luborsky L., McLellan A.T., Woody G.E., O'Brien C.P., Auerbach A. (1985). Therapist success and its determinants. *Archives of General Psychiatry*, **42**, 602–11.

Rapp C.A. (1998). *The Strengths Model*. New York: Oxford University Press.

Soldz, S. & Vaillant, G. (1999). The big five personality traits and the life course: a 45-year longitudinal study. *Journal of Research in Personality*, **33**, 208–32.

Truax C.B. & Carkhuff R.R. (1967). *Toward Effective Counseling and Psychotherapy*. Chicago: Aldine Publishing.

US Department of Health & Human Services (1999). *Mental Health: a Report of the Surgeon General*, Chapter 2, Section 10. www.surgeongeneral.gov/library/mentalhealth/chapter2/sec10.html [accessed 1 March 2009].

Chapter 2

Overview of bipolar disorder

Allan H. Young, I. Nicol Ferrier and Erin E. Michalak

Bipolar disorder (BD) is a complex mood disorder that is typically characterised by recurring episodes of depression and mania (a distinct period of abnormally elevated, expansive or irritable mood) called BD type I, or depression and hypomania (a less severe form of mania) called BD type II.

The past two decades have witnessed a burgeoning of interest in BD. This has been fuelled in part by a wealth of new research evidence but also by an increasing understanding of the significant impact that this severe illness has, not only for sufferers and their loved ones, but also society more broadly. Bipolar disorder is surprisingly common; BD type I and type II subtypes have a lifetime prevalence of between 1 and 2%, and the bipolar spectrum, a category about which much remains unknown, is found in up to an additional 5% (Angst, 2007). The World Health Organization (WHO) has estimated that BD is the sixth leading cause of disability worldwide among young adults (Murray & Lopez, 1997), underlining the idea that BD is a serious public health concern. Advances are being made in the understanding of the aetiopathogenesis of BD, with certain genes being strongly implicated for the first time and other modalities, such as neuroimaging, identifying fruitful leads. Research interest is also being focused upon issues such as physical morbidity and neurocognition. An avalanche of publications has reported new data on treatment modalities, including psychosocial, somatic and pharmacological interventions. The impact of BD on outcomes such as psychosocial functioning and quality of life is being explored. It is within the context of this sometimes overwhelming and confusing proliferation of new knowledge that we frame this book.

Our objective is to provide a practical and pragmatic guide to the management of BD for the busy clinician. To this end we have sought contributions from acknowledged experts in BD that provide contemporary evidence-based reviews and insights into areas of prime clinical relevance. We have chosen to begin the book from the perspective of an individual who is living with BD (Maxwell), underscoring our belief in the importance of maintaining a patient-centred stance for the successful management of this complex condition. Subsequent chapters deal with the pharmacological treatment of mania (Grunze) and depression (Young and Nemeroff). A summary of treatment guidelines is provided by Yatham and a guide to somatic treatments (electroconvulsive therapy, transcranial stimulation, etc.) by Le Masurier, Herrmann, Coulson and Ebmeier. Psychosocial treatments are ably detailed by Parikh and Velyvis. A number of chapters deal with management in special populations: the elderly (Thomas); women and reproductive issues (Macritchie and Henshaw); and those with physical morbidity (Joseph and Mok). The treatment of BD in the early stages of the illness is reviewed by Garland and Duffy. Anxiety disorders are now recognised as an important feature of BDs, and Chokka and Yeragani expertly review this complex subject. Similarly, substance abuse is reviewed by Goikolea and Vieta. Further chapters cover service delivery (Morriss), psychosocial functioning and quality of life (Michalak and Murray), brain imaging (Ferrier and Lloyd), the practical management of cyclothymia (Perugi and Popovic) and, crucial to the optimal management of this condition, circadian and sleep/wake issues (Murray). There is a review of training issues by Parikh and Michalak, and the book ends with a chapter by Kuan, Curtis and Young, who attempt to integrate the path from neuroscience to clinical practice. We are grateful to all the contributors and thank them all for sharing their expertise with us. This is therefore first and foremost a 'practical guide' to the management of BD; we trust that you will find it a useful aid for this fascinating, complex and ever-challenging illness.

References

Angst J. (2007). The bipolar spectrum. *British Journal of Psychiatry*, **190**, 189–91.

Murray, C.J. & Lopez, A.D. (1997). Global mortality, disability, and the contribution of risk factors: Global Burden of Disease Study. *Lancet*, **349,** 1436–42.

Practical Management of Bipolar Disorder, eds. Allan H. Young, I. Nicol Ferrier and Erin E. Michalak. Published by Cambridge University Press.

Chapter 3

The treatment of mania

Heinz Grunze

Introduction

Mania is the hallmark of bipolar disorder (BD) and differentiates it from unipolar depression. *The Diagnostic and Statistical Manual of Mental Disorders*, 4th edition (DSM-IV-TR; APA, 2000), characterises bipolar I disorder by the occurrence of at least one manic or mixed episode; Table 3.1 and Table 3.2 summarise diagnostic criteria for mania and mixed episodes according to DSM-IV-TR (APA, 2000). By contrast, the *International Classification of Diseases*, 10th edition (ICD-10; WHO, 2007), differentiates unipolar mania (ICD-10, F30) from bipolar disorder (ICD-10, F31), the latter requiring not only a manic episode, but also at least one episode of major depression. However, as unipolar mania is a rare disorder and is also conceptualised as part of the bipolar spectrum, treatment recommendations given in this chapter imply the assumption of a long-term course with opposing mood deflections.

The true complexity of mania, however, is hardly captured by any current categorical classification. Manic states are not monolithic, nor do they always fit into clear clinical distinctions. A wide range of symptoms may occur in an acute manic episode (Table 3.3), with some symptom clusters being more likely than others. The European mania in bipolar longitudinal evaluation of medication (EMBLEM) study (Goetz *et al.*, 2007) found the presence of three major clinical subtypes of patient with acute mania using latent class analysis to define discrete groups of patients. The authors identified: 'typical mania' (59% of patients); 'psychotic mania' (27%) with more severe mania and presence of psychotic symptoms; and 'dual mania' (13%) with a high proportion of substance abuse. These patient groups differed in age of onset, social functioning, service needs and outcome (van Rossum *et al.*, 2008). However, controlled clinical studies usually exclude at least the group of dual mania, which makes it difficult to give substantiated treatment recommendations for all subgroups.

General principles when treating manic patients

Treatment of mania requires hospitalisation in most cases, often even emergency treatment and involuntary detention. There may be a few exceptions where manic patients still have sufficient insight and behavioural control to allow treatment in the community, but ensuring sufficient guarding of the patient to minimise risk is often the limiting factor. In addition, having to cope with a manic relative at home may easily devastate any relationship. Advantages of inpatient treatment include risk limitation in a secure environment, professional care around the clock and the opportunity to create a non-stimulating environment. It also allows for more aggressive pharmacotherapy as vital parameters can be monitored and side effects more rapidly recognised and counteracted.

Treatment should involve the same humane principles as for any other patient, respecting the patient's dignity and rights, and hostility as part of a manic syndrome should lead to adaequate safety measures but never provoke a similar response from professionals. Responsible treatment of a manic patient includes the physician taking full responsibility for diagnosis, physical examination, other investigations and explanation of the medical plan of management to the patient and their relatives. The physician should always communicate clearly, understandably and honestly, but calmly, what he or she thinks. Mania is often accompanied by anxiety, and giving the patient the feeling that they are not threatened or getting tricked, but that they are safe, is essential for cooperation and future compliance.

Ideally, the manic patient will then consent to treatment. If agreement cannot be reached, it is the task of the responsible psychiatrist to estimate the risks of non-treatment. If there are no other means to avert acute risks to the patient or others, and once a decision has been made to initiate treatment against the patient's

Practical Management of Bipolar Disorder, eds. Allan H. Young, I. Nicol Ferrier and Erin E. Michalak. Published by Cambridge University Press.

Table 3.1 Diagnostic criteria for acute mania according to DSM-IV-TR (APA, 2000)

- Criteria for manic episode (DSM-IV-TR, p. 332)
  - A. A distinct period of abnormally and persistently elevated, expansive, or irritable mood, lasting at least 1 week (or any duration if hospitalisation is necessary).
  - B. During the period of mood disturbance, three (or more) of the following symptoms have persisted (four if the mood is only irritable) and have been present to a significant degree:
    - inflated self-esteem or grandiosity
    - decreased need for sleep (e.g. feels rested after only 3 hours of sleep)
    - more talkative than usual, or pressure to keep talking
    - flight of ideas or subjective experience that thoughts are racing
    - distractibility (i.e. attention too easily drawn to unimportant or irrelevant external stimuli)
    - increase in goal-directed activity (either socially, at work or school, or sexually) or psychomotor agitation
    - excessive involvement in pleasurable activities that have a high potential for painful consequences (e.g. engaging in unrestrained buying sprees, sexual indiscretions or foolish business investments).
  - C. The symptoms do not meet criteria for a Mixed Episode.
  - D. The mood disturbance is sufficiently severe to cause marked impairment in occupational functioning or in usual social activities or relationships with others, or to necessitate hospitalisation to prevent harm to self or others, or there are psychotic features.
  - E. The symptoms are not due to the direct physiological effects of a substance (e.g. a drug of abuse, a medication or other treatment) or a general medical condition (e.g. hyperthyroidism).

Table 3.2 Diagnostic criteria for a mixed episode according to DSM-IV-TR (APA, 2000)

- A. The criteria are met both for a Manic Episode and for a Major Depressive Episode (except for duration) nearly every day during at least a 1-week period.
- B. The mood disturbance is sufficiently severe to cause marked impairment in occupational functioning or in usual social activities or relationships with others, or to necessitate hospitalisation to prevent harm to self or others, or there are psychotic features.
- C. The symptoms are not due to the direct physiological effects of a substance (e.g. a drug of abuse, a medication or other treatment) or a general medical condition (e.g. hyperthyroidism).

will, it is important for the staff to act both determinedly and in a skilled way. Every restriction imposed on this occasion on the patient, e.g. constraining mobility or treatment with injections, needs to be constantly scrutinised and to be eased as soon as is justifiable. As soon as a therapeutic alliance has been established, it is the duty of the treating physician to explain to the patient the necessity of these measures.

Therapy of mania means primarily treatment with medication in a relaxed, non-stimulating environment. A formal psychotherapeutic approach may be tried additionally when patients have largely recovered and are cooperative, but so far there is little evidence for its efficacy (Basco & Rush, 1996). Therefore, the focus of the following chapters will be the evidence-based pharmacological treatment modalities of mania.

Dosing and duration of treatment

Table 3.4 shows recommended dosages for different antimanic medication in monotherapy, categorised according to their evidence for antimanic efficacy. The dosages are derived from studies in acute mania and do not necessarily reflect the whole dosage range that is approved for a given medication. In the case of combination treatment, a reduction of dosage may be necessary, especially when side effects of two medications are additive or potentiating. In fact, most combination treatment trials used lower dosage of the investigational drug than in the corresponding monotherapy studies.

Antimanic treatment should be commenced at least until full remission, syndromal and functional, has been achieved. Persistence of subsyndromal mania is associated with a significantly increased risk of relapse (Tohen *et al.*, 2006). Unless there is doubt regarding whether the manic episode may have been consequent to an external perturbant, e.g. steroids, alcohol or other abusable drugs, all patients should be treated in continuation and maintenance regimens. Most guidelines recommend continuation therapy for 6–12 months

Table 3.3 Frequency of symptoms observed clinically during acute manic episodes (adapted from Goodwin and Jamison, 2007)

Symptom	Weighted mean (%)
Mood symptoms	
Irritability	71
Euphoria	63
Depression	46
Lability	49
Expansiveness	60
Cognitive symptoms	
Grandiosity	73
Flight of ideas, racing thoughts	76
Distractibility, poor concentration	75
Confusion	29
Psychotic symptoms	
Any delusions	53
Grandiose delusions	31
Persecutory/paranoid delusions	29
Passivity delusions	12
Any hallucinations	23
Auditory hallucinations	18
Visual hallucinations	12
Olfactory hallucinations	15
Presence or history of psychotic symptoms	61
Thought disorder	19
First rank Schneiderian symptoms	18
Activity and behaviour during mania	
Hyperactivity	90
Decreased sleep	83
Violent, assaultive behavior	47
Rapid, pressured speech	88
Hyperverbosity	89
Nudity, sexual exposure	29
Hypersexuality	51
Extravagance	32
Religiosity	39
Head decoration	34
Regression (pronounced)	28
Catatonia	24
Faecal incontinence (smearing)	13

after remission from an acute mood episode has been achieved; however, this recommendation is based upon expert advice and clinical wisdom, and not derived from controlled studies. Therefore, when selecting a regimen for treatment of acute mania one consideration should be its overall efficacy and tolerability in long-term treatment, thereby minimising switches of medication that may be associated with an increased relapse risk.

Treatment should generally be initiated with a medication with a high degree of evidence and good tolerability and safety profile. Previous experience of the patient is helpful in guiding the choice. If this first choice of medication is inefficacious or leads only to partial response, it is unclear how long clinicians should wait before changing or amending medication. In controlled studies, most successful investigational drugs start to separate from placebo within 1 week, and early partial response can predict later response at study endpoint (Pappadopulos *et al.*, 2008), although response may be delayed with some medications that need titration (e.g. lithium) or are used in lower dosages for tolerability or safety reasons. However, as acute mania constitutes a significant burden to patients and everyone involved, clinicians may not want to wait for too long to tap the last potential of a medication. Hence, in the absence of firm evidence, it may be a rule of thumb that a treatment trial should not last more than 2 weeks, and its continuation or discontinuation should then be decided upon the basis of full, partial or no response (see Figure 3.1).

Monotherapy or skilful polypharmacy?

In clinical practice, fewer than 10% of acutely manic patients receive monotherapy; the average number of medications in acutely manic patients is approximately

Table 3.4 Pharmacological and non-pharmacological treatments used in acute mania (in alphabetical order within one category of evidence for efficacy)

Medication	Typically recommended daily dose for adults (variations may occur due to different approvals)
Medication with good evidence of efficacy	
Aripiprazole	15–30 mg
Asenapine	10–20 mg
Carbamazepine	600–1200 mg (serum level 4–15 mg/l)
Haloperidol	5–20 mg
Lithium	600–1200 mg (serum level 0.8–1.3 mmol/l)
Olanzapine	10–20 mg[a]
Quetiapine	400–800 mg
Risperidone	2–6 mg
Valproate	1200–3000 mg (loading dose 20–30 mg/kg body weight; serum level 75–100 mg/l)
Ziprasidone	80–160 mg
Medication with reasonable evidence of efficacy	
Chlorpromazine	300–1000 mg
Paliperidone	3–12 mg
Phenytoin	300–400 mg
Pimozide	2–16 mg
Tamoxifen	40–80 mg
Medication with some evidence of efficacy	
Amisulpride	400–800 mg
Clonazepam	2–8 mg
Clozapine	100–300 mg
Levetiracetam	500–1500 mg
Lorazepam	4–8 mg
Nimodipine	240–480 mg
Oxcarbazepine	900–1800 mg
Retigabine	600–1200 mg
Verapamil	450 mg
Zonisamide	100–500 mg
Zotepine	200–400 mg
Other modalities	
ECT	Reserved for treatment-refractory mania and special issues (e.g. as alternative option in pregnancy)

[a] A fixed dose of 20 mg olanzapine was sufficient to demonstrate significant antimanic effects in females with moderate to severe mania (Bech *et al*., 2006). However, females achieve significantly higher plasma concentrations of olanzapine than males (Kelly *et al*., 2006). This may imply that higer doses are needed in males with moderate to severe mania (Goodwin & Jamison, 2007).

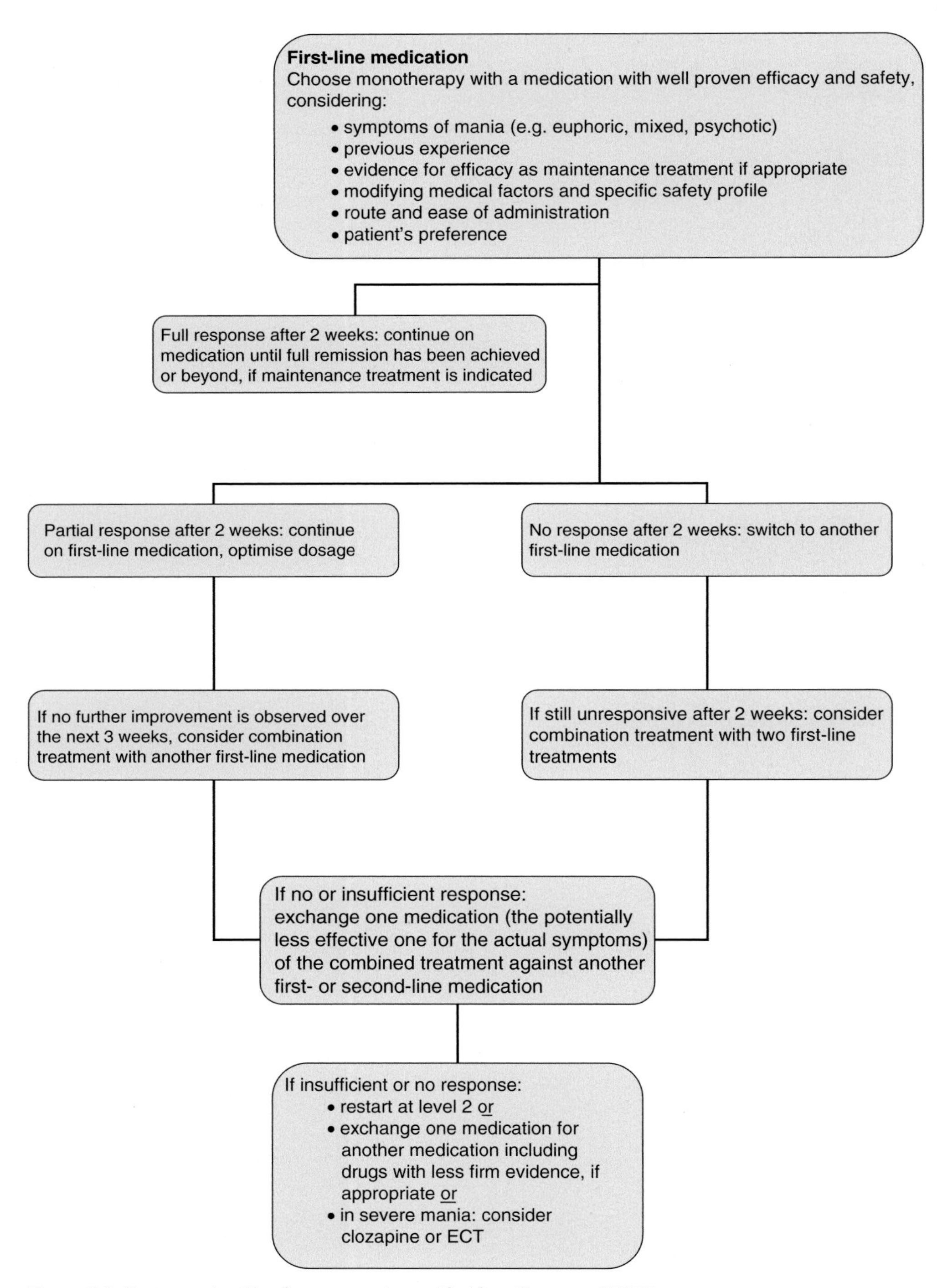

Figure 3.1 Treatment algorithm for acute mania, modified from Grunze *et al.* (2009).

three (Wolfsperger *et al.*, 2007). This surely underlines the difficulties in treating 'real world' patients with severe mania and manifold co-morbidities compared with selected samples in clinical studies – less than 20% of a screened naturalistic patient cohort fulfils all inclusion criteria for entering a randomised, controlled trial. In line with clinical practice results from randomised, controlled trials that show that addition of an antipsychotic drug to patients with manic symptoms despite treatment with lithium or valproate show greater rates of acute efficacy than continuation of lithium or valproate as monotherapy (Smith *et al.*, 2007). Similar beneficial results occurred with a combination of valproate and haloperidol versus haloperidol monotherapy (Müller-Oerlinghausen *et al.*, 2000).

So why not recommend combination therapy as a general first-line treatment? Safety and practicability issues clearly favour a first-line approach with monotherapy, making best use of the dosage range available for a given medication. Combined treatments are potentially associated with higher frequency or greater severity of side effects, putting patients at a potentially unnecessary risk and perhaps disrupting the therapeutic alliance. Thus, best use of a diligently chosen monotherapy should be tried before switching to combinatons in order to minimise side effects and medical risks.

Subtype and severity of mania as treatment modifiers

Clinical experience with various tentative antimanic agents over recent years has suggested that a drug that is efficacious in one subtype of mania is not necessarily the treatment of choice for the overall spectrum of mania.

Controlled studies on antimanic agents have been focused mainly on patients with BD type I disorder and 'euphoric mania', resulting in firm evidence that several agents are effective in this type of mania.

'Dysphoric mania' describes mania with some depressed and dysphoric features that are either not pronounced enough or insufficiently long lasting to fulfil the criteria for a major depressive episode, whereas 'mixed states' imply that diagnostic criteria for a manic episode and a depressive episode (except for the duration criterion) are fulfilled simultaneously. As dysphoric mania and mixed states have not been the sole subjects of prospective controlled trials so far, we have only a limited amount of evidence for efficacy, and even less for the superiority of one drug over another. Secondary analysis of trials also including mixed patients indicates, for example, that lithium may not be as effective as valproate, carbamazepine, olanzapine or risperidone under these conditions. Although there is no direct evidence for lack of efficacy, the use of typical neuroleptics, especially at a higher dose, may exacerbate dysphoric or depressive symptoms and should probably be avoided in mixed/dysphoric manic states (Tohen *et al.*, 2003).

'Psychotic mania' has only recently been arbitrarily designated as a subtype of bipolar mania. It is unclear whether secondary grandiose delusions – the commonest clinical manifestation of 'psychosis'– merits qualitative distinction, as it looks much more like an expression of severity. However, first-rank symptoms also occur in mania and confuse the distinction from schizophrenia. Psychotic mania has been so little studied in clinical trials that recommendations regarding drug regimens are based principally on inferential criteria. Some evidence suggests higher efficacy of typical neuroleptics than lithium (Johnstone *et al.*, 1988). With the emergence of atypical antipsychotics, monotherapy options may increase, but unambiguous prospective, controlled trials in psychotic mania are still missing. However, post-hoc analysis of Phase III studies of olanzapine, risperidone and ziprasidone showed similar response rates in psychotic versus nonpsychotic mania.

Besides symptomatology expressed as (partially aribitrary) subtypes of mania, the severity of behavioural disturbance determines the first-line treatment in acute mania. In clinical practice, severity of mania and speed of onset of action are more likely to be the primary arguments in favour of a particular drug. But unfortunately, no prospective controlled trials are available that specifically look into severly manic patients, defined by, for example, a high threshold on a mania rating scale. Some secondary evidence favours haloperidol over several atypical antipsychotics and lithium in ultra-short-term treatment, but post-hoc subgroup analyses of severly manic patients in randomised, controlled trials also show that risperidone, ziprasidone, aripiprazole and olanzapine are effective in this patient group. Dose-loading with valproate or carbamazepine are suggested alternatives, whereas lithium loading is effective, but associated with higher risks of accidental overdosing (Keck *et al.*, 2000). Finally, electroconvulsive therapy (ECT) is still a valuable last resource in severe delirious mania.

On the other end of the severity spectrum, hypomania has also received little attention as far as controlled treatment studies are concerned. Thus, most recommendations are based on clinical wisdom, but not on evidence. If hypomania is well known to be the prelude to mania in an individual patient, treatment should be as for mania. If no further prophylaxis is planned, short-term treatment with either valproate or an atypical antipsychotic may be the best choice, as both are well tolerated, have a good safety profile and a relatively rapid onset of action, minimising the danger that hypomania develops into mania within the next few days. Hypomania may still be approachable to some extent by behavioural interventions in combination with pharmacotherapy. These inventions may centre on modifications of daily routines, e.g. maintaining a natural sleep/wake cycle, stress avoidance and some elements of cognitive behavioural therapy (CBT) (Basco & Rush, 1996).

Medications and other modalities in the treatment of mania

The following review of different medications is based on a literature search, including the FDA Clinical trial database (www.clinicaltrials.gov]), up to January 2009. However, with the rapid progress in this area, there is always a chance that some evidence – especially negative evidence from failed and hitherto unpublished trials – may have been missed.

Lithium

Lithium has clearly the largest pool of studies in mania, with more than 20 published or presented studies of varying quality, starting with Shou's work in 1954 (Schou *et al.*, 1954). But only more recent studies, beginning with a three-arm study comparing valproate and lithium against placebo (Bowden *et al.*, 1994), can be considered to reach the current methodological standards for a drug-approval study. Since then, several well-designed and placebo-controlled studies have repeatably confirmed lithium's antimanic efficacy. A recent meta-analysis of six randomised, controlled trials with lithium in acute mania (four of them published, and two as part of a registration dossier) revealed an overall standardised effect size of 0.40 (95% CI: 0.28, 0.53) and an overall 'numbers needed to treat' (NNT) for response of 6 (95% CI: 4, 13) (Storosum *et al.*, 2007). The antimanic efficacy of lithium may be higher in pure (euphoric) mania than in mania with concomitant dysphoric or depressive features (Swann *et al.*, 1997). In methodologically less sophisticated comparator studies, lithium was tested against various antipsychotics, including chlorpromazine, haloperidol, olanzapine and risperidone, and against carbamazepine. Overall, these studies showed largely similar efficacy of lithium and the respective comparator.

Target plasma levels for lithium in recent controlled studies were usually in the range of 0.6 to 1.3 mmol/l. In clinical practice, adolescents and young adults may require and tolerate plasma levels at the higher end of this range, whereas elderly patients may tolerate dosages only at the lower end of this range. A slower onset of action of lithium, relative to the investigational drug, has been observed in some studies (e.g. Keck *et al.*, 2009), but not in others (e.g. Bowden *et al.*, 2005). This potentially slower onset of action together with the absence of sedative properties often makes it necessary to combine it with a tranquilising agent at treatment initiation.

Lithium is not suitable with certain medical conditions, which should be excluded before treatment initiation, e.g. renal problems or thyroid dysfunction. The usefulness of lithium in acute mania may also be limited by the need for regular plasma level checks to avoid toxicity, its side-effect profile and other contraindications. These limitations have been dealt with extensively in standard textbooks.

Anticonvulsants

Carbamazepine

Starting with the first studies of Okuma *et al.* (1973), carbamazepine has been tested in acute mania in several small comparator studies against the conventional antipsychotics, lithium and valproate. The impression of these studies was that carbamazepine was overall equally effective as comparators, with a probably slightly slower onset of response compared with antipsychotics and valproate, but slightly faster acting than lithium (Grunze, 2006).

Two large randomised, placebo-controlled mania studies with carbamazepine as the investigational drug were published just a few years ago (Owen, 2006), proving the significant superiority of carbamazepine. Looking into specific subgroups of patients, carbamazepine may be especially helpful in patients with incomplete response to lithium in acute mania, rapid-cycling and co-morbid organic (neurological) disorders, and schizoaffective patients (Grunze, 2006).

Common side effects of carbamazepine include oversedation and blurred vision, especially with high dosages and rapid titration. Tolerability isssues may be less problematic with extended-release formulations.

Rare, but potentially severe, side effects include allergic reactions, lupus erythematosus, agranulocytosis and hyponatraemia. Detailed information on the tolerability and safety profile of carbamazepine is available in recent reviews. In addition, carbamazepine is associated with an increased risk of birth defects (Morrow *et al.*, 2006). The main shortcoming of carbamazepine, however, is its manifold interactions with other psychotropic medication, including several antipsychotics, antidepressants and anticonvulsants (Spina *et al.*, 1996). As most patients with acute mania are treated with several medications, this clearly limits the utility of carbamazepine.

Valproate

This chapter uses 'valproate' as a common generic name for the different preparations tested in acute mania, e.g. valproic acid, sodium valproate, divalproate, divalproex sodium and valpromide. As far as pharmacokinetics and pharmacodynamics are concerned, only valproic acid finally reaches and penetrates the blood–brain barrier. Although tolerability is enhanced with extended-release preparations, the differences do not warrant grouping valproic acid derivatives as different medications.

Several placebo-controlled studies have provided consistent evidence that valproate is an efficacious treatment for acute mania (Macritchie *et al.*, 2003). Valproate also showed similar efficacy in comparator trials with lithium, carbamazepine, haloperidol, and in one study against olanzapine, but not in two others.

Dose loading with 20–30 mg/kg body weight seems to be more effective in mania than slower titration schemes (Hirschfeld *et al.*, 2003), and plasma levels of 75–99 mg/l seem to be associated with the best efficacy/tolerability ratio (Allen *et al.*, 2006). The safety margin of valproate is relatively large, allowing rapid titration and a subsequent earlier onset of action. Tolerability of valproate appears fair across trials. Gastointestinal discomfort, sedation and tremor are more frequently observed with valproate than with placebo in most trials, but usually do not result in higher discontinuation rates. For rare, but severe, complications such as hepatic failure, pancreatitis or hyperammonaemic coma and precautionary measures please refer to the pertinent reviews (e.g. Haddad *et al.*, 2009).

Valproate is not appropriate in some medical conditions, e.g. liver disease, and in combination with some medications, e.g. warfarin. Caution should be used in women of childbearing age, not only because of teratogenicity, but also because of the supposed increased risk of polycystic ovary syndrome (Rasgon *et al.*, 2005).

Other anticonvulsants with potential antimanic properties

Several medications have been proposed as having antimanic properties, but none of them has been studied enough to allow the conclusion that efficacy and tolerability were within the same range as the anticonvulsants previously reviewed in detail.

Phenytoin has demonstrated antimanic properties in a small, double-blind, placebo-controlled add-on study to haloperidol (Mishory *et al.*, 2000). The side effect profile of phenytoin, especially cognitive side effects and cerebellar atrophy, however, makes it a medication of subordinate choice for acute mania.

For oxcarbazepine, a recent review of several small, underpowered or not placebo-controlled studies came to the conclusion that it may be useful in treating manic symptoms (Popova *et al.*, 2007), but conclusive evidence is lacking. Other anticonvulsants with evidence only from open studies, but not from controlled trials, include levetiracetam, zonisamide and retigabine. Negative evidence – failure to separate from placebo in controlled trials – exists for topiramate, gabapentin and lamotrigine. For tiagabine, open studies were suggestive of no efficacy together with an increased risk of epileptiform seizures.

Atypical antipsychotics

The treatment choice in acute mania has significantly widened with the emergence of the atypical antipsychotics. In this chapter we first list the different antipsychotics that were approved for the treatment of mania at the end of 2008, and then we will briefly touch on other atypicals that may be of clinical interest in this indication. Medications are listed in alphabetical order; it does not reflect true or assumed superiority of one atypical over another.

Aripiprazole

Four placebo-controlled acute mania studies have been published so far, one of them also including lithium (Keck *et al.*, 2009) and another one haloperidol as a comparator (Young *et al.*, 2009). In a placebo-controlled combination treatment study with either valproate or lithium, aripiprazole was also superior to valproate or lithium alone (Vieta *et al.*, 2008c). In addition, an intramuscular injectable preparartion of aripiprazole has demonstrated antimanic efficacy (Sanford & Scott, 2008).

Secondary analyses confirmed the broad-spectrum of efficacy of aripiprazole across subtypes of mania.

Headache, somnolence and dizziness were the most frequently reported side effects, but none was significantly more frequent than with placebo. However, akathisia appears to be more frequent with aripiprazole than with placebo. No significant QTc changes or elevations in prolactin, cholesterol and fasting blood glucose levels were reported for aripiprazole.

Olanzapine

Together with lithium, olanzapine is probably the most extensively studied medication for acute mania. It has shown significant superiority over placebo in five double-blind placebo-controlled monotherapy studies, including one in adolescent mania (Tohen *et al.*, 2007). When olanzapine was combined with lithium or valproate, these combinations were also more effective than valproate or lithium alone (Tohen *et al.*, 2004). In all these studies, olanzapine showed comparable efficacy across subtypes of mania.

In addition, a randomised, controlled trial of injectable olanzapine in agitated mania has demonstrated significant superiority of olanzapine against placebo and lorazepam after 2 h (Meehan *et al.*, 2001). Intramuscular injections of olanzapine, however, bear an increased risk of respiratory arrest when patients are on concomitant benzodiazepines.

Several head-to-head comparison studies with olanzapine have also been conducted. Whereas efficacy of olanzapine appeared slightly higher than that of valproate or lithium, olanzapine-treated subjects were also more prone to side effects, limiting clinical utility. Compared with risperidone, no difference in antimanic efficacy was observed (Perlis *et al.*, 2006), wheras in another trial haloperidol was significantly more efficacious than olanzapine after 6 weeks (Tohen *et al.*, 2003).

These positive results in controlled clinical studies are also consistent with a large, pan-European naturalistic mania study (EMBLEM), which reported efficacy both for olanzapine monotherapy and olanzapine in combination with other medications in a broad spectrum of manic patients (Vieta *et al.*, 2008b).

As far as tolerability and safety are concerned, oral olanzapine showed a favourable profile regarding short-term treatment. Somnolence and dizziness were associated significantly more frequently with olanzapine treatment than with placebo. Anticholinergic side effects such as dry mouth, constipation or extrapyramidal side effects (EPS) occurred rarely in the controlled studies. However, already in short-term trials the mean weight gain from baseline to endpoint ranged from 1.65 to 4 kg, and metabolic issues clearly complicate the intermediate and long-term treatment with olanzapine (Franciosi *et al.*, 2005).

Quetiapine

Four randomised, placebo-controlled acute mania monotherapy studies have been conducted (Bowden *et al.*, 2005; McIntyre *et al.*, 2005; Cutler *et al.*, 2008; Vieta *et al.*, 2008a) demonstrating antimanic efficacy of quetiapine. Whereas quetiapine was in one of these studies as effective as lithium, haloperidol showed a faster onset of action and better efficacy than quetiapine in another trial (Scherk *et al.*, 2007). These controlled studies concentrated only on euphoric mania, so it is not possible to make conclusions about mixed or psychotic mania.

Quetiapine was also tested in two placebo-controlled combination studies as an add-on to lithium or valproate, with one of them demonstrating superiority of the combination quetiapine/lithium or valproate.

Quetiapine also appears to be effective and safe in adolescents, as demonstrated by two controlled studies.

The tolerability of quetiapine throughout studies appears good, with the exception of somnolence, which was two- to sixfold greater than with placebo. In addition, mean weight gain was consistently higher in quetiapine-treated patients compared with placebo, haloperidol or lithium. As with olanzapine, metabolic issues may become more prominent when quetiapine is taken as long-term medication but appear not to be of significance for short-term use.

Risperidone

Three double-blind, placebo-controlled monotherapy trials have been published so far, one of them also having a comparator arm with haloperidol. Risperidone was in all studies significantly better than placebo, and comparison against haloperidol showed no difference in antimanic efficacy (Scherk *et al.*, 2007). There is evidence for good efficacy in severe and psychotic mania from one study (Khanna *et al.*, 2005), but a lack of controlled data supporting the use of risperidone in mixed mania.

Two placebo-controlled studies investigated risperidone as an add-on to valproate or lithium and as an add-on to lithium, valproate or carbamazepine. Whereas the first study demonstrated superiority of the risperidone add-on, the second one failed due to lack of response in the patients receiving carbamazepine as primary treatment. This illustrates the problematic issues associated with carbamazepine as an inducer of the cytochrome P450 enzyme in combination treatment.

Higher dosages of risperidone (6 mg/day) were associated in one study with an increased rate of EPS, whereas in lower dosages, as applied in the other studies, discontinuation of risperidone due to EPS was similar to placebo. Two studies reported higher rates of somnolence, and also dizziness occurred slightly more often with risperidone than with placebo. The mean weight gain ranged from 1.7 to 2.4 kg at endpoint, and the observed elevation of prolactin in blood was even higher than with haloperidol in one study (Smulevich *et al.*, 2005). This may be because of the relatively low brain penetration of risperidone and the relatively high plasma levels required for efficacy: these preferentially elevate prolactin because the pituitary lies outside the blood–brain barrier.

Ziprasidone

Ziprasidone monotherapy demonstrated antimanic efficacy in three double-blind placebo-controlled studies. Post-hoc analyses of these trials also supplied evidence that ziprasidone is effective in dysphoric/mixed states and psychotic mania (Greenberg & Citrome, 2007). Ziprasidone is also available as intramuscular injectable solution. In a placebo-controlled add-on study to lithium or valproate, ziprasidone did not separate from placebo at endpoint, although it enhanced antimanic response at the beginning.

No significant differences compared with placebo were observed in EPS-related scales, but cases of akathisia were numerically more frequent in one study. In addition, initial somnolence was reported about three to four times as often for ziprasidone when compared with placebo.

Of some concern is the cardiovascular safety profile of ziprasidone, although no QTc prolongation beyond 500 ms was observed in the cited mania studies, and the main safety concern – torsades de pointes – has also not been reported in post-marketing surveillance.

Ziprasidone and aripiprazole are the two atypical antipsychotics licensed for bipolar disorder (but not in all countries) that may be largely weight neutral.

Other atypical antipsychotics used off label in mania

Amisulpride is a frequently used antimanic medication in some parts of the world, but evidence from controlled studies is still vague and disputable (Thomas *et al.*, 2008). Amisulpride is not associated with weight gain. However, high dosages of amisulpride usully administered in acute mania do cause hyperprolactinaemia.

For asenapine, two placebo-controlled acute mania monotherapy trials and one add-on study to lithium or valproate have been presented in scientific meetings as posters so far. In all three studies, asenapine demonstrated significant superiority over placebo and was well tolerated; in particular, the incidence of EPS was low. Limited data, however, suggest that asenapine may be associated with weight gain and metabolic issues. Licensing for mania is still pending but may have been granted by the time this book is published.

Numerous case reports and several small investigator-initiated trials support the antimanic as well as mild antidepressive and good prophylactic efficacy of clozapine in bipolar patients, making it a last-resort drug in treatment-refractory bipolar patients (Gitlin, 2006). Large-scale methodologically unambiguous studies, however, are missing because of the lack of commercial interest and the potentially life-threatening side effects of clozapine.

More recently, paliperidone (an active metabolite of risperidone) has been tested in two placebo-controlled monotherapy trials, one of them positive and another one reaching significance only for the highest dose of paliperidone (12 mg/day). Similarly to asenapine, a licence for mania had not been granted by the end of 2008.

The situation with zotepine is similar to that of clozapine. At least two open studies are in line with antimanic efficacy, but a lack of commercial interest has prohibited further evaluation in randomised, controlled studies. As zotepine is capable of causing significant weight gain, its value may be limited with the emergence of weight-neutral atypicals with proven antimanic efficacy.

Typical neuroleptics

Haloperidol

Haloperidol has been the primary clinical choice in severe mania for decades, but a sound scientific basis for this use has been produced only recently, when it has served as a comparator in randomised, placebo-controlled studies of atypical antipsychotics (Cipriani *et al.*, 2006; Scherk *et al.*, 2007). The usefulness of haloperidol, however, is clearly limited by its high propensity to induce acute extrapyramidal motor symptoms at established doses and – probably even more importantly – tardive dyskinesia. Nevertheless, at least in the emergency treatment of severe mania or in patients who do not respond to other therapies, haloperidol has its place and justification. Doses of haloperidol chosen in the past may have been unnecessarily high, and lower dosages may have the same efficacy with less side-effect burden (Rifkin *et al.*, 1994). Low doses of haloperidol still lead to sufficient D2 receptor occupancy which, in turn, predicts antimanic efficacy (Harrison-Read, 2009). Thus, low-dose typical neuroleptics may still be a sensible alternative to atypical antipsychotics in selected patients, and this may apply as much to mania as it does to schizophrenia.

Chlorpromazine

Only one placebo-controlled randomised trial for chlorpromazine has been reported in acutely manic patients. Further evidence stems from head-to-head comparisons versus lithium, carbamazepine, pimozide, thiothixene and ECT. Doses established for use in acute mania (200–800 mg/day) are associated with a high risk of EPS, and similar considerations apply to the use of chlorpromazine as to haloperidol. Other frequent side effects with chlorpromazine include pronounced sedation, tardive dyskinesia, hypersensitivity of the skin to sunlight and hepatotoxicity.

Benzodiazepines

Clonazepam and lorazepam are quite frequently used in bipolar disorder – lorazepam particularly as a rescue medication. However, they are usually not considered as primary mood-stabilising agents. Nevertheless, there are some studies supporting antimanic effects of these two drugs (Grunze, 2004), although their addictive potential should deter psychiatrists from using them as primary and long-term treatment (Chouinard, 2004).

Experimental medications

Three placebo-controlled studies gave evidence for the antimanic efficacy of the protein kinase C inhibitor tamoxifen, although some methodological issues may impact on the generalisability of the result (Tohen, 2008). Tamoxifen was well tolerated, but its nature as an anti-oestrogen may raise some concerns for longer-term treatment. The evidence for the calcium antagonists nimodipine and verapamil is limited and partly contradictory. The viability of these calcium antagonists is also limited because of the effect on blood pressure (verapamil) and a short half-life necessitating multiple dosing per day (nimodipine).

Electroconvulsive therapy (ECT)

No randomised, controlled studies have been conducted for ECT in mania, but numerous case reports and chart reviews support the utility of ECT in severe mania. A comprehensive review of open studies and case reports describes improvement in approximately 80% of manic patients (Mukherjee *et al.*, 1994) – thus greater than for any pharmacological intervention. ECT is still a valuable last resource in severe mania that is otherwise treatment refractory (Karmacharya *et al.*, 2008).

Recent work suggests that bifrontal ECT is at least as efficacious as bitemporal ECT in severe mania and, at the same time, better tolerated (Barekatain *et al.*, 2008).

Conclusions

Skilful treatment of mania will remain a clinical challenge, although treatment options have significantly increased over recent years, and new targets for drug development will emerge. Protein kinase C inhibition is one example of a mechanism with some recent evidence of efficacy. However, with the substantial number of medications available, it will become more essential that any new medication shows additional benefits besides being a pure antimanic agent. Most clinicians are likely to prefer antimanic drugs that also have established long-term prophylactic efficacy not only against manic relapse, but also against depressive episodes, or – even more challenging – substances that also have antidepressant activity. With the expanding range of drugs with evidence of efficacy in mania, psychiatrists as well as patients may reasonably place safety, tolerability and evidence of good persistence over time on equal footing with efficacy in selecting and continuing a regimen. Similarly, tolerability and ease of adhering to the prescribed dosage can benefit from selection of drug formulations with extended-release properties and/or once daily dosing. In a highly competitive field, future research and development will have to take that into account at an earlier stage than in the past; it will not be enough if you have 'just another antimanic drug' to be clinically accepted.

References

Allen, M.H., Hirschfeld, R.M., Wozniak, P.J., *et al.* (2006). Linear relationship of valproate serum concentration to response and optimal serum levels for acute mania. *American Journal of Psychiatry*, **163**, 272–5.

APA (2000). *The Diagnostic and Statistical Manual of Mental Disorders*, 4th edition. Arlington, VA: American Psychiatric Association. www.psychiatryonline.com/resourceTOC.aspx?resourceID=1 [accessed 23 December 2009].

Barekatain, M., Jahangard, L., Haghighi, M., *et al.* (2008). Bifrontal versus bitemporal electroconvulsive therapy in severe manic patients. *Journal of ECT*, **24**, 199–202.

Basco, M.R. & Rush, A.J. (1996). *Cognitive Behavioral Therapy for Bipolar Disorder*. New York: Guildford Press.

Bech, P., Gex-Fabry, M., Aubry, J.M., *et al.* (2006). Olanzapine plasma level in relation to antimanic effect in the acute therapy of manic states. *Nordic journal of psychiatry*, **60**, 181–2.

Bowden, C.L., Brugger, A.M., Swann, A.C., *et al.* (1994). Efficacy of divalproex vs lithium and placebo in the treatment of mania. The Depakote Mania Study Group. *JAMA: Journal of the American Medical Association*, **271**, 918–24.

Bowden, C.L., Grunze, H., Mullen, J., *et al.* (2005). A randomized, double-blind, placebo-controlled efficacy and safety study of quetiapine or lithium as monotherapy for mania in bipolar disorder. *Journal of Clinical Psychiatry*, **66**, 111–21.

Chouinard, G. (2004). Issues in the clinical use of benzodiazepines: potency, withdrawal, and rebound. *Journal of Clinical Psychiatry*, **65** (Suppl. 5), 7–12.

Cipriani, A., Rendell, J.M. Geddes, J.R. (2006). Haloperidol alone or in combination for acute mania. *Cochrane Database of Systematic Reviews*, **3**, CD004362.

Cutler, A.J., Datto, C., Nordenhem, A., *et al.* (2008). Effectiveness of extended release formulation of quetiapine as montherapy for the treatment of acute bipolar mania (trial D144CC00004). *International Journal of Neuropsychopharmacology / Official Scientific Journal of the Collegium Internationale Neuropsychopharmacologicum (CINP)*, **11** (Suppl. 1), 184.

Franciosi, L.P., Kasper, S., Garber, A.J., *et al.* (2005). Advancing the treatment of people with mental illness: a call to action in the management of metabolic issues. *Journal of Clinical Psychiatry*, **66**, 790–8.

Gitlin, M. (2006). Treatment-resistant bipolar disorder. *MolecularPsychiatry*, **11**, 227–40.

Goetz, I., Tohen, M., Reed, C., *et al.* (2007). Functional impairment in patients with mania: baseline results of the EMBLEM study. *Bipolar Disorders*, **9**, 45–52.

Goodwin, F.K. & Jamison, K.R. (2007). *Manic-depressive illness*, 2nd edition. New York: Oxford University Press.

Greenberg, W. M. & Citrome, L. (2007). Ziprasidone for schizophrenia and bipolar disorder: a review of the clinical trials. *CNS.Drug Reviews*, **13**, 137–77.

Grunze, H. (2004). Other agents for bipolar disorder. In: Akiskal, H.S. & Tohen, M. (Eds). *Innovative Bipolar Pharmacotherapy*. Philadelphia, PE: Lippincott.

Grunze, H. (2006). Carbamazepine, other anticonvulsants and augmenting agents. In: Akiskal, H.S. & Tohen, M. (Eds). *Bipolar Psychopharmacotherapy: Caring for the Patient*. Chichester: John Wiley & Sons.

Grunze, H., Vieta, E., Goodwin, G.M., *et al.* (2009). The World Federation of Societies of Biological Psychiatry (WFSBP) guidelines for the biological treatment of bipolar disorders: update 2009 on the treatment of acute mania. *World Journal of Biological Psychiatry*, **10**, 85–116.

Haddad, P.M., Das, A., Ashfaq, M., *et al.* (2009). A review of valproate in psychiatric practice. *Expert Opinion on Drug Metabolism & Toxicology*, **5**, 539–51.

Harrison-Read, P. (2009). Antimanic potency of typical neuroleptic drugs and affinity for dopamine D2 & serotonin 5-HT2A receptors – a new analysis of data from the archives and implications for improved antimanic drug treatments. *Journal of Psychopharmacology*, **8**, 899–907. Epub 2008 Jul 17.

Hirschfeld, R.M., Baker, J.D., Wozniak, P., *et al.* (2003). The safety and early efficacy of oral-loaded divalproex versus standard-titration divalproex, lithium, olanzapine, and placebo in the treatment of acute mania associated with bipolar disorder. *Journal of Clinical Psychiatry*, **64**, 841–6.

Johnstone, E.C., Crow, T.J., Frith, C.D., *et al.* (1988). The Northwick Park "functional" psychosis study: diagnosis and treatment response. *Lancet*, **2**, 119–25.

Karmacharya, R., England, M.L., Ongur, D. (2008). Delirious mania: clinical features and treatment response. *Journal of Affective Disorders*, **109**, 312–16.

Keck, P.E., McElroy, S.L., Bennett, J.A. (2000). Pharmacologic loading in the treatment of acute mania. *Bipolar Disorders*, **2**, 42–6.

Keck, P.E., Orsulak, P.J., Cutler, A.J., *et al.* (2009). Aripiprazole monotherapy in the treatment of acute bipolar I mania: a randomized, double-blind, placebo- and lithium-controlled study. *Journal of Affective Disorders*, **112**, 36–49.

Kelly, D.L., Richardson, C.M., Yu, Y., *et al.* (2006). Plasma concentrations of high-dose olanzapine in a double-blind crossover study. *Human Psychopharmacology*, **21**, 393–8.

Khanna, S., Vieta, E., Lyons, B., *et al.* (2005). Risperidone in the treatment of acute mania: double-blind, placebo-controlled study. *British Journal of Psychiatry*, **187**, 229–34.

Macritchie, K., Geddes, J.R., Scott, J., *et al.* (2003). Valproate for acute mood episodes in bipolar disorder. *Cochrane Database of Systematic Reviews*, **1**, CD004052.

McIntyre, R.S., Brecher, M., Paulsson, B., *et al.* (2005). Quetiapine or haloperidol as monotherapy for bipolar mania – a 12-week, double-blind, randomised, parallel-group, placebo-controlled trial. *European Neuropsychopharmacology*, **15**, 573–85.

Meehan, K., Zhang, F., David, S., *et al.* (2001). A double-blind, randomized comparison of the efficacy and safety of intramuscular injections of olanzapine, lorazepam, or placebo in treating acutely agitated patients diagnosed with bipolar mania. *Journal of Clinical Psychopharmacology*, **21**, 389–97.

Mishory, A., Yaroslavsky, Y., Bersudsky, Y., *et al.* (2000). Phenytoin as an antimanic anticonvulsant: a controlled study. *American Journal of Psychiatry*, **157**, 463–5.

Morrow, J., Russell, A., Guthrie, E., *et al.* (2006). Malformation risks of antiepileptic drugs in pregnancy: a prospective study from the UK Epilepsy and Pregnancy Register. *Journal of Neurology, Neurosurgery, and Psychiatry*, **77**, 193–8.

Mukherjee, S., Sackeim, H.A., Schnur, D.B. (1994). Electroconvulsive therapy of acute manic episodes: a review of 50 years' experience. *American Journal of Psychiatry*, **151**, 169–76.

Müller-Oerlinghausen, B., Retzow, A., Henn, F., *et al.* (2000). Valproate as an adjunct to neuroleptic medication for the treatment of acute episodes of mania. A prospective, randomized, double-blind, placebo-controlled multicenter study. *Journal of Clinical Psychopharmacology*, **20**, 195–203.

Okuma, T., Kishimoto, A., Inoue, K., *et al.* (1973). Anti-manic and prophylactic effects of carbamazepine (Tegretol) on manic depressive psychosis. A preliminary report. *Folia Psychiatrica et Neurologica Japonica*, **27**, 283–97.

Owen, R.T. (2006). Extended-release carbamazepine for acute bipolar mania: a review. *Drugs of Today*, **42**, 283–9.

Pappadopulos, E., Vieta, E., Mandel, F. (2008). Day 4 partial resonse to ziprasidone predicts later treatment response in patients with bipolar disorder. *European Neuropsychopharmacology*, **18** (Suppl. 4), 443.

Perlis, R.H., Baker, R.W., Zarate Jr, C.A., *et al.* (2006). Olanzapine versus risperidone in the treatment of manic or mixed states in bipolar I disorder: a randomized, double-blind trial. *Journal of Clinical Psychiatry*, **67**, 1747–53.

Popova, E., Leighton, C., Bernabarre, A., *et al.* (2007). Oxcarbazepine in the treatment of bipolar and schizoaffective disorders. *Expert Review of Neurotherapeutics*, **7**, 617–26.

Rasgon, N.L., Altshuler, L.L., Fairbanks, L., *et al.* (2005). Reproductive function and risk for PCOS in women treated for bipolar disorder. *Bipolar Disorders*, **7**, 246–59.

Rifkin, A., Doddi, S., Karajgi, B., *et al.* (1994). Dosage of haloperidol for mania. *British Journal of Psychiatry*, **165**, 113–16.

Sanford, M. & Scott, L.J. (2008). Intramuscular aripiprazole: a review of its use in the management of agitation in schizophrenia and bipolar I disorder. *CNS Drugs*, **22**, 335–52.

Scherk, H., Pajonk, F.G., Leucht, S. (2007). Second-generation antipsychotic agents in the treatment of acute mania: a systematic review and meta-analysis of randomized controlled trials. *Archives of General Psychiatry*, **64**, 442–55.

Schou, M., Juel-Nielsen, N., Strömgren, E., *et al.* (1954). The treatment of manic psychoses by the administration of lithium salts. *Journal of Neurology, Neurosurgery, and Psychiatry*, **17**, 250–60.

Smith, L.A., Cornelius, V., Warnock, A., *et al.* (2007). Acute bipolar mania: a systematic review and meta-analysis of co-therapy vs. monotherapy. *Acta Psychiatrica Scandinavica*, **115**, 12–20.

Smulevich, A.B., Khanna, S., Eerdekens, M., *et al.* (2005). Acute and continuation risperidone monotherapy in bipolar mania: a 3-week placebo-controlled trial followed by a 9-week double-blind trial of risperidone and haloperidol. *European Neuropsychopharmacology*, **15**, 75–84.

Spina, E., Pisani, F., Perucca, E. (1996). Clinically significant pharmacokinetic drug interactions with carbamazepine. An update. *Clinical pharmacokinetics*, **31**, 198–214.

Storosum, J.G., Wohlfarth, T., Schene, A., *et al.* (2007). Magnitude of effect of lithium in short-term efficacy studies of moderate to severe manic episode. *Bipolar Disorders*, **9**, 793–8.

Swann, A.C., Bowden, C.L., Morris, D., *et al.* (1997). Depression during mania. Treatment response to lithium or divalproex. *Archives of General Psychiatry*, **54**, 37–42.

Thomas, P., Vieta, E.: SOLMANIA study group (2008). Amisulpride plus valproate vs haloperidol plus valproate in the treatment of acute mania of bipolar I patients: a multicenter, open-label, randomized, comparative trial. *Neuropsychiatric disease and treatment*, **4**, 675–86.

Tohen, M. (2008). Clinical trials in bipolar mania: implications in study design and drug development. *Archives of General Psychiatry*, **65**, 252–3.

Tohen, M., Bowden, C.L., Calabrese, J.R., *et al.* (2006). Influence of sub-syndromal symptoms after remission from manic or mixed episodes. *British Journal of Psychiatry*, **189**, 515–19.

Tohen, M., Chengappa, K., Suppes, T., *et al.* (2004). Relapse prevention in bipolar I disorder: 18-month comparison of olanzapine plus mood stabiliser v. mood stabiliser alone. *British Journal of Psychiatry*, **184**, 337–45.

Tohen, M., Goldberg, J.F., Gonzalez-Pinto Arrillaga, A.M., *et al.* (2003). A 12-week, double-blind comparison of olanzapine vs haloperidol in the treatment of acute mania. *Archives of General Psychiatry*, **60**, 1218–26.

Tohen, M., Kryzhanovskaya, L., Carlson, G., *et al.* (2007). Olanzapine versus placebo in the treatment of adolescents with bipolar mania. *American Journal of Psychiatry*, **164**, 1547–56.

van Rossum, I., Haro, J.M., Tenback, D., *et al.* (2008). Stability and treatment outcome of distinct classes of mania. *European Psychiatry*, **23**, 360–7.

Vieta, E., Berwaerts, J., Nuamah, I., *et al.* (2008a). Randomised, placebo, active-controlled study of paliperidone extended-release (ER) for acute manic and mixed episodes in bipolar I disorder. *European Neuropsychopharmacology*, **18** (Suppl. 4), S369.

Vieta, E., Panicali, F., Goetz, I., *et al.* (2008b). Olanzapine monotherapy and olanzapine combination therapy in the treatment of mania: 12-week results from the European Mania in Bipolar Longitudinal Evaluation of Medication (EMBLEM) observational study. *Journal of Affective Disorders*, **106**, 63–72.

Vieta, E., Tjoen, C., McQuade, R.D., *et al.* (2008c). Efficacy of adjunctive aripiprazole to either valproate or lithium in bipolar mania patients partially nonresponsive to valproate/lithium monotherapy: a placebo-controlled study. *American Journal of Psychiatry*, **165**, 1316–25.

WHO (2007). *International Classification of Diseases*, 10th edition. Geneva: World Health Organization. http://apps.who.int/classifications/apps/icd/icd10online/ [accessed 14 December 2009].

Wolfsperger, M., Greil, W., Rossler, W., *et al.* (2007). Pharmacological treatment of acute mania in psychiatric in-patients between 1994 and 2004. *Journal of Affective Disorders*, **99**, 9–17.

Young, A.H., Oren, D.A., Lowy, A., *et al.* (2009). Aripiprazole monotherapy in acute mania: 12-week randomised placebo- and haloperidol-controlled study. *British Journal of Psychiatry*, **194**, 40–8.

Chapter 4

Pharmacological treatment of bipolar depression

Allan H. Young and Charles B. Nemeroff

Introduction

The practical management of bipolar depression represents one of the greatest challenges for the clinician concerned with the treatment of this illness. As discussed elsewhere in this book, bipolar disorder (BD) is common, and much of the considerable disability associated with this illness is due to the depressed phase (Post *et al.*, 2003); patients with bipolar depression have significant psychosocial impairment (Michalak *et al.*, 2008). In addition to this predominant depression, patients suffer higher mortality rates than other bipolar patients, and depression is a major risk factor for suicide in BD (Angst *et al.*, 2005). Pharmacotherapy is the mainstay of treatment, although both psychological and somatic treatment strategies may also be utilised. This chapter will briefly outline the evidence base for current pharmacotherapeutic practice. For a more extensive discussion and specific treatment recommendations, the reader is referred to more comprehensive reviews (Calabrese *et al.*, 2009; Young & Nemeroff, 2009).

Pharmacotherapy: focus on bipolar disorder type I depression

Lithium

A review of the older literature revealed that lithium was significantly more effective than placebo for the treatment of bipolar depression, and approximately half of the patients experienced a relapse of depressive symptoms when lithium was substituted by placebo (Srisurapanont *et al.*, 1995). However, a more recent double-blind, randomised, placebo-controlled study of quetiapine and lithium as acute monotherapy treatment for bipolar depression found no statistically significant difference between lithium and placebo (Young *et al.*, 2010). The mean serum lithium level in this study was 0.61 mEq/l (mmol/l), and it is unknown if response rates would be improved with higher serum lithium levels for a longer duration than the 8 weeks' duration of this trial.

Meta-analyses of trials of lithium in patients with BD include further data concerning the use of lithium as a maintenance treatment (Geddes *et al.*, 2004; Smith *et al.*, 2007). Lithium was shown in these analyses to be more effective than placebo in preventing any new mood episodes. However, although lithium was superior to placebo in preventing manic episodes, this was not clearly shown for depressive episodes (Geddes *et al.*, 2004; Smith *et al.*, 2007). Overall, therefore, the evidence for lithium being antidepressant in BD seems less clear cut than that for its antimanic actions.

Lamotrigine monotherapy

Lamotrigine was initially examined in a double-blind, placebo-controlled study of lamotrigine monotherapy for the acute treatment of BD type I depression (Calabrese *et al.*, 2000). More recently, this study was included in a review of five double-blind, placebo-controlled trials, all of which assessed the efficacy of lamotrigine in the acute treatment of bipolar depression (Calabrese *et al.*, 2008). Although there is a widespread belief that lamotrigine is antidepressant in BD, the overall pooled effect was modest at best. The advantage of lamotrigine over placebo was greater in the more severely depressed participants, and the slow dose titration necessary to minimise the risk of dangerous rashes may mitigate against an antidepressant effect fully manifesting in the time frame of acute treatment studies.

Two longer-term studies compared the effectiveness of lithium and lamotrigine monotherapy over a period of 18 months in 638 patients with BD type I and recent episodes of mania or depression (Bowden *et al.*, 2003; Calabrese *et al.*, 2003). Bowden and colleagues demonstrated that lamotrigine was significantly more effective than placebo at prolonging the time to intervention for a depressive episode (Bowden *et al.*, 2003). Calabrese and co-workers also found that lamotrigine significantly prolonged time to a depressive episode (Calabrese *et al.*, 2003). A pooled analysis of data from

Practical Management of Bipolar Disorder, eds. Allan H. Young, I. Nicol Ferrier and Erin E. Michalak. Published by Cambridge University Press.

these trials showed that lamotrigine, but not lithium, was superior to placebo at delaying the time to intervention for a depressive episode (Goodwin *et al.*, 2004). Lamotrigine is therefore typically used for prevention of depressive relapse in BD, although this is likely to be most effective in patients who have some history of lamotrigine responsiveness.

Adjunctive lamotrigine

The use of adjunctive lamotrigine for bipolar depression was recently evaluated in an 8-week, double-blind, randomised, placebo-controlled trial; the results showed that lamotrigine in combination with lithium was superior to lithium monotherapy (van der Loos *et al.*, 2009). However, a Systematic Treatment Enhancement Program for Bipolar Disorder (STEP-BD) study of open-label lamotrigine, inositol or risperidone as adjuncts to a mood stabiliser for up to 16 weeks in patients (n=66) with treatment-resistant BD type I or type II depression found no significant between-group differences (Nierenberg *et al.*, 2006).

A comparison of the adjunctive use of lamotrigine and lithium in the long-term treatment of BD was carried out in an open, randomised trial (Licht, 2008). Patients were randomised to receive lithium (dosed to attain serum levels of 0.5–0.8 mmol/l, n=78) or lamotrigine (up to 400 mg/d, n=77) for up to 6 years, with concomitant pharmacologic therapy allowed for the first 6 months of the study period. No differences were noted between lamotrigine and lithium when used as adjunctive therapy with respect to the primary outcome measure (Licht, 2008).

Valproate

A small 8-week, double-blind, placebo-controlled, randomised study that evaluated the clinical efficacy of valproate (as divalproex; up to 2500 mg/d) in 25 outpatients with BD type I depression found that the active treatment was significantly more effective than placebo in improving symptoms of depression (Davis *et al.*, 2005). Further evidence is provided by a very small double-blind, randomised study in which patients with BD type I depression (n=9) received divalproex or placebo for 6 weeks (Ghaemi *et al.*, 2007). Divalproex was associated with a significantly greater reduction in Montgomery–Asberg depression rating scale (MADRS) total score, the primary efficacy measure, from baseline to Week 6 compared with placebo (P<0.001). Further studies are required to verify the antidepressant efficacy of valproate in BD.

Carbamazepine

A double-blind study evaluating the acute effects of carbamazepine monotherapy in 35 patients with depression (16 patients with BD type I depression and 8 with BD type II depression) over a median treatment duration of 45 d found that 62% of patients receiving carbamazepine monotherapy (mean dose of 971 mg/d; achieving mean ± SD blood levels of 9.3 ± 1.9 µg/ml; range, 3–12.5 µg/ml) experienced a response (mean improvement of ≥1 point on the Bunney–Hamburg scale) (Post *et al.*, 1986).

Quetiapine monotherapy

Evidence for the acute efficacy of quetiapine monotherapy in patients with BD type I or type II depression is provided by the results of two large, 8-week, randomised, double-blind, placebo-controlled studies that evaluated quetiapine monotherapy (300 and 600 mg/d) (Calabrese *et al.*, 2005a; Thase *et al.*, 2006).

In both studies both doses of quetiapine were significantly better than placebo, with moderate-to-large effects. Importantly, treatment-emergent mania rates did not differ from placebo (Calabrese *et al.*, 2005a; Thase *et al.*, 2006).

The efficacy of quetiapine or lithium as acute monotherapy was further evaluated in a double-blind, placebo-controlled study in patients with BD type I and type II depression. The study consisted of an initial acute phase lasting 8 weeks, during which patients were randomised to receive quetiapine 300 mg/d, 600 mg/d, lithium or placebo. This was followed by a continuation phase lasting between 26 and 52 weeks. The primary endpoint of the acute phase of the study was the change from baseline to Week 8 in MADRS total score. In the BD type I subgroup of patients (n=487), the mean change in MADRS total score at Week 8 was −14.8 with quetiapine 300 mg/d (P<0.05 vs placebo) and −16.5 with quetiapine 600 mg/d (P<0.05 vs placebo) compared with −11.2 for placebo (Young *et al.*, 2010). Notably, lithium did not separate from placebo.

These findings are consistent with a double-blind, placebo-controlled study of similar design that evaluated the efficacy of quetiapine (300 mg/d and 600 mg/d) and paroxetine (20 mg/d) as monotherapy in patients with BD type I and type II depression (McElroy *et al.*, 2010). In the BD type I subgroup (n=448), quetiapine 600 mg/d significantly reduced MADRS total score from baseline to Week 8 (−16.2, quetiapine 300 mg/d (95% CI −5.34 to −0.22; P<0.05); −16.4, quetiapine 600 mg/d (95% CI −5.60 to −0.49; P<0.05) compared with placebo (−13.4)). Both of the above studies were sufficiently large (n=328 and n=256, respectively) to provide adequate evidence for the short- and long-term use of quetiapine monotherapy for the treatment of BD type I or II depression (Young *et al.*, 2010; McElroy *et al.*, 2010). These two studies were powered to allow for a combined continuation phase, during which patients who remitted on quetiapine 300 mg/d or 600 mg/d were randomly reassigned to either continued treatment on quetiapine 300 mg/d or placebo and were studied for an additional 26–52 weeks. Quetiapine significantly increased the time to recurrence of depression compared with placebo (HR 0.48, 95% CI 0.29 to 0.77) (Young *et al.*, 2010) (HR 0.36, 95% CI 0.21 to 0.63) (McElroy *et al.*, 2010) in patients with BD type I or II depression.

Analysis of the individual items on the MADRS revealed significant improvement of virtually all of them in the quetiapine group, indicating that the overall effect was not solely secondary to improvement in sleep. The two comparator agents in these studies, lithium and paroxetine, did not separate from placebo.

Adjunctive quetiapine

To date, no randomised, controlled trials have yet examined the efficacy of adjunctive quetiapine treatment for acute bipolar depression; however, evidence for long-term efficacy in patients with BD type I depression is provided by two randomised, double-blind, parallel group studies that investigated the use of quetiapine in combination with lithium or divalproex (Li/DVP) (Suppes *et al.*, 2008; Vieta *et al.*, 2008). In both studies a 12- to 36-week open-label stabilisation phase was followed by a randomised treatment phase of up to 104 weeks. The primary efficacy endpoint for both studies was the time to recurrence of any mood event (mixed, mania or depression). In one study, quetiapine in combination with Li/DVP was found to be significantly more effective than placebo and Li/DVP in preventing the recurrence of any mood event, and in particular a depressive episode (Suppes *et al.*, 2008). In the other study, quetiapine in combination with Li/DVP was also significantly more effective than placebo and Li/DVP in preventing both the recurrence of any mood event and a depression event (Vieta *et al.*, 2008).

Olanzapine and olanzapine/fluoxetine

Combination

The olanzapine/fluoxetine combination (OFC) was compared to lamotrigine in a 7-week, randomised, double-blind, parallel-group study in patients with BD type I depression. The combination was associated with significantly greater improvement in clinical global impression–severity (CGI-S) rating from baseline to Week 7 compared with lamotrigine (−1.43, OFC; −1.18, lamotrigine; P<0.01). Patients had statistically significantly greater improvement in MADRS total score with OFC than lamotrigine at Week 7 (−14.91, OFC; −12.92, lamotrigine; P<0.01) (Brown *et al.*, 2006).

Tohen *et al.* (1999) carried out a subanalysis of data from an 8-week, placebo-controlled, randomised study that investigated the efficacy of olanzapine and OFC in patients with BD type I depression. The analysis compared the rates of treatment-emergent mania in patients receiving olanzapine or placebo. During the 8-week study, olanzapine and OFC were not associated with a greater risk of treatment-emergent mania compared with placebo (Keck, Jr. *et al.*, 2005).

To date, the long-term use of olanzapine or OFC in patients with bipolar depression has not been evaluated in placebo-controlled trials. The only long-term, controlled data for olanzapine plus fluoxetine in bipolar depression derive from a study evaluating the efficacy of OFC. A 25-week, randomised, double-blind study compared the efficacy of OFC and lamotrigine in patients with BD type I depression (Brown *et al.*, 2009; see above). OFC was associated with significantly greater improvements in CGI-S and MADRS total scores than lamotrigine from baseline to Week 25.

It is important to note the increased risk for the development of metabolic syndrome, diabetes and cardiovascular disease incumbent in long-term treatment with olanzapine and other atypical antipsychotics. This is an important consideration in patients with BD, a patient population already vulnerable to these major causes of morbidity and mortality. However, an 11-year follow-up of mortality in patients with schizophrenia has recently shown that life expectancy has increased since the introduction of atypical antipsychotic drugs (Tiihonen *et al.*, 2009), suggesting that overall health outcomes may not be made worse by the use of this class of drug.

Antidepressants

Despite their common use there is little evidence to support the use of antidepressant monotherapy in patients with BD (Hirschfeld *et al.*, 2003). In line with this, results from the McElroy study that investigated the efficacy of quetiapine and paroxetine for the acute treatment of bipolar depression showed that paroxetine did not significantly improve MADRS total scores at Week 8 from baseline in patients with bipolar depression (–14.9, paroxetine; –13.4, placebo; $P=0.313$) (McElroy *et al.*, 2010), whereas quetiapine did separate from placebo as discussed above.

The adjunctive use of antidepressants is a common approach to the treatment of bipolar depression. A systematic review and meta-analysis of five acute, randomised, double-blind controlled trials ($n=779$) compared the use of antidepressants or placebo as adjuncts to a mood stabiliser in patients with BD and a current depressive or mixed episode. The authors concluded that antidepressants were a more effective adjunctive therapy than placebo, and, moreover, were not associated with a higher incidence of switching to mania (Gijsman *et al.*, 2004).

In some contrast to the Gijsman analysis, the long-term use of adjunctive antidepressants in patients with BD type I or type II depression was evaluated in a large, 26-week, double-blind, randomised, placebo-controlled study (STEP-BD). The primary outcome was durable recovery, defined as euthymia for at least eight consecutive weeks. Adjunctive treatment with paroxetine or bupropion did not significantly increase the rate of durable recovery compared with the use of mood stabilisers alone (23.5 and 27.3%, respectively; $P=0.4$). Notably, the rate of treatment-emergent affective switch in the two groups was not significantly different (Sachs *et al.*, 2007). For a further review of antidepressant use see Young & Nemeroff (2009).

Modafinil

Modafinil is potentially an efficacious adjunctive treatment for patients with BD type I depression who respond inadequately to monotherapy with a mood stabiliser (Frye *et al.*, 2007). In this study patients with BD type I ($n=64$) or BD type II ($n=21$) depression were randomised to receive modafinil (200 mg/d) or placebo in combination with a mood stabiliser for 6 weeks. Significant reductions in inventory of depressive symptomatology (IDS) score ($P=0.047$, effect size 0.47) and the clinical global impression as modified for bipolar illness (CGI-BP) depression severity item ($P=0.009$, effect size 0.63) were seen in the modafinil group compared with placebo. Furthermore, response (>50% improvement in IDS total score) and remission (final IDS total score <12) rates were significantly higher in the modafinil group compared with placebo (43.9 vs 22.7% ($P<0.05$) and 39 vs 18% ($P=0.033$), respectively) (Frye *et al.*, 2007).

Pramipexole

A 6-week, randomised, placebo-controlled trial investigated the efficacy of pramipexole monotherapy (up to 5 mg/d) in patients with treatment-resistant BD type I ($n=15$) and BD type II depression ($n=7$). Of the patients who received pramipexole, 67% responded to treatment, compared with 20% of patients who received placebo ($P<0.05$). The change in mean Hamilton depression rating (HAM-D) scores was greater ($P=0.05$) for pramipexole (48%) compared with placebo (21%). Pramipexole also significantly improved mean CGI-S score from baseline to Week 6 compared with placebo (–2.4 and –0.30, respectively; $P=0.01$) (Goldberg *et al.*, 2004).

Mifepristone

A recent Cochrane systematic review has found some evidence supporting the use of glucocorticoid antagonists in the treatment of mood disorders (Gallagher *et al.*, 2008). Of these, the glucocorticoid receptor antagonist mifepristone has the best, albeit extremely preliminary, evidence in bipolar depression. Improvements in mood and neurocognitive function were reported in a small proof-of-concept study in bipolar depression (Young *et al.*, 2004), effects that were not replicated in a similar study in schizophrenia (Gallagher *et al.*, 2005). A larger study is currently under way. Mifepristone has relatively persistent effects on cortisol levels in bipolar patients, which raises the possibility of similarly persistent beneficial effects on mood and cognition (Gallagher *et al.*, 2008).

Ethyl-eicosapentaenoic acid

Adjunctive ethyl-eicosapentaenoic acid (EPA) (1 g/d and 2 g/d) was evaluated in a 12-week randomised, double-blind, placebo-controlled study in 65 patients with BD type I depression. Both doses of adjunctive EPA (combined data) significantly improved both HAM-D (–3.3 points, 95% CI –6.1 to –0.2; $P<0.05$; effect size 0.34) and CGI (–0.79 points, 95% CI –1.27 to –0.25; $P<0.05$) scores compared with placebo from baseline to the end of the study (Frangou *et al.*, 2006).

Aripiprazole

Aripiprazole is an atypical antipsychotic drug that possesses relatively novel pharmacological properties, including partial agonism at dopamine D_2/D_3 receptor as well as full agonism at $5\text{-}HT_{1A}$ receptors.

Two, identically designed, 8-week, randomised, double-blind, placebo-controlled studies in patients with BD type I depression found that aripiprazole (flexible dose 5–30 mg/d) demonstrated a rapid onset of action (from Week 1), with significant reductions in MADRS total score compared with placebo. However, this effect was lost in the final 2 weeks of the trials (Thase *et al.*, 2008). Subsequent analysis of data from the two trials suggests that the aripiprazole doses used were not adequately determined in advance, leading to high patient withdrawal, which probably contributed to the loss of statistical significance towards the end of the trials (Thase *et al.*, 2008). Aripiprazole has demonstrable efficacy as an adjunctive treatment to antidepressants in unipolar depression (Simon & Nemeroff, 2005; Marcus *et al.*, 2008).

Pharmacotherapy: focus on bipolar disorder type II depression

The trials discussed above mostly provide evidence for depression in BD type I. Contrastingly, there is a relative dearth of clinical evidence from studies in patients with BD type II depression (Vieta & Suppes, 2008). This may be a consequence of BD type I being regarded as a more severe form of illness than BD type II depression, particularly regarding length and severity of individual depressive episodes (Coryell *et al.*, 1985; Vieta *et al.*, 1997; Benazzi, 1999); however, this view may be erroneous, as patients with BD type II experience a greater frequency of episodes and a longer overall time spent in depression (Vieta *et al.*, 1997; Judd *et al.*, 2003). BD type II depression is also poorly recognised by both patients and clinicians.

Quetiapine

The efficacy of quetiapine monotherapy (300 and 600 mg/d) for the acute treatment of patients with BD type II depression was evaluated as part of two 8-week, randomised, double-blind, placebo-controlled studies (Calabrese *et al.*, 2005a; Thase *et al.*, 2006). In the BD type II subgroup in the first study (n=182), quetiapine monotherapy was associated with a statistically significant improvement in mean MADRS total score at most assessments during the study, compared with placebo. However, the difference in MADRS total score was not significant at final assessment (Week 8) with either dose of quetiapine compared with placebo. Effect sizes were 0.39 for quetiapine 600 mg/d and 0.28 for quetiapine 300 mg/d (Calabrese *et al.*, 2005a). In the second study, a significant improvement in mean MADRS total score compared with placebo was sustained from Week 1 to final assessment with quetiapine 300 mg/d ($P \leq 0.05$) and from Week 3 to final assessment with quetiapine 600 mg/d ($P \leq 0.05$) in the BD type II subgroup (n=152) (Thase *et al.*, 2006).

A post-hoc analysis of pooled data from both studies of BD type II depression has recently been published. Quetiapine monotherapy significantly improved mean MADRS total score from the first assessment (Week 1) and at each subsequent assessment (Suppes *et al.*, 2007). At Week 8, mean change from baseline in MADRS total score was −17.1 for quetiapine 300 mg/d ($P<0.01$) and −17.9 for quetiapine 600 mg/d ($P<0.01$) compared with −13.3 for placebo. Effect sizes were calculated as 0.54 for quetiapine 600 mg/d and 0.45 for quetiapine 300 mg/d.

Additional data regarding the use of quetiapine for the acute treatment of BD type II depression derive from a randomised, placebo-controlled study that evaluated the acute (8 weeks') use of quetiapine monotherapy (300 and 600 mg/d) in this patient group (n=252) (McElroy *et al.*, 2010). At the end of the study the investigators reported a mean change in MADRS total score of −16.5 points for quetiapine 300 mg/d, −16.3 points for quetiapine 600 mg/d and −11.53 points for placebo. Differences in mean MADRS total score were significant for quetiapine 300 mg/d vs placebo (95% CI −7.93 to −1.67; $P<0.05$) and for quetiapine 600 mg/d vs placebo (95% CI −8.10 to −1.85; $P<0.05$).

Evidence for the long-term use of quetiapine monotherapy in the treatment of BD type II depression derives from the 26–52-week continuation phases of the Young *et al.* (2010) and McElroy *et al.* (2010) studies. Quetiapine significantly increased the time to recurrence of depression compared with placebo during the continuation phases of both studies in patients with BD type I or type II depression.

Pramipexole

A very small 6-week, double-blind, placebo-controlled study investigated the efficacy of pramipexole (up to 4.5 mg/d) in patients with BD type II depression (n=10). The results of the study revealed a significant treatment effect with pramipexole, as shown by an improvement in total MADRS score compared with placebo at Week 6 (P=0.03, 95% CI 0.104 to 2.27). Furthermore, response (defined as a >50% decrease in MADRS score from baseline) was experienced by 60% of patients in the pramipexole group compared with 9% in the placebo group (P=0.02) (Zarate, Jr. *et al.*, 2004).

Antidepressants

Another very small, 9-month, randomised, placebo-controlled, cross-over study reported significant improvement in depression severity, measured by

HAM-D score and percentage of days impaired (effect sizes 1.07 and 0.85, respectively; $P<0.05$), in patients with BD type II disorder ($n=10$) receiving selective serotonin reuptake inhibitor (SSRI) monotherapy compared with placebo (Parker *et al.*, 2006). However, this study needs to be replicated in a larger sample before any conclusions regarding the efficacy of SSRIs in patients with BD type II depression can be drawn.

Pharmacotherapy: focus on rapid cyclers

Divalproex and lithium

A 20-month, double-blind, parallel-group study comparing the efficacy of divalproex and lithium for the long-term treatment of rapid-cycling BD has been conducted (Calabrese *et al.*, 2005b). Following a 6-month, acute stabilisation phase, during which patients received open-label lithium and divalproex in combination, 60 patients were randomised to receive lithium monotherapy (mean dose 1359 mg/d) or divalproex monotherapy (mean dose 1571 mg/d) for up to 20 months. No statistically significant difference between the lithium and divalproex groups was observed for the primary efficacy measure of time to treatment for a mood episode.

Lamotrigine

The use of lamotrigine as a maintenance treatment in rapid-cycling BD was investigated in a double-blind, placebo-controlled prophylaxis study (Calabrese *et al.*, 2000). Patients ($n=324$) received lamotrigine or placebo as monotherapy for 6 months following a preliminary, open-label stabilisation phase. The primary efficacy measure was time to additional pharmacotherapy for emerging mood symptoms. No significant difference between the lamotrigine and placebo groups with respect to the primary measure was observed. However, lamotrigine was associated with a significantly greater time to premature discontinuation compared with placebo ($P<0.05$). Furthermore, significantly more patients receiving lamotrigine (41%) were stable without relapse for the duration of the study compared with placebo (26%; $P<0.05$) (Calabrese *et al.*, 2000).

Quetiapine

Evidence for the use of quetiapine monotherapy in patients with a rapid-cycling disease course has been provided by a subanalysis of an 8-week, randomised, double-blind, placebo-controlled study (Calabrese *et al.*, 2005a). Quetiapine (600 and 300 mg/d) provided significantly greater mean reductions from baseline to Week 8 in MADRS total score than placebo ($P<0.001$ for both doses) in patients with a rapid-cycling disease course (Vieta *et al.*, 2007).

Antidepressants

A 10-week, randomised, flexible-dose study evaluating sertraline, bupropion and venlafaxine as adjuncts to mood stabilisers investigated the impact of a patient's rapid-cycling status on the relative risk of switching into mania or hypomania (Post *et al.*, 2006). In patients without rapid cycling, the risk of switching was no different with the three study medications ($P=0.55$); however, in patients with a rapid-cycling disease course, bupropion was associated with a significantly lower risk of switching than was venlafaxine ($P<0.01$).

Thyroxine and clozapine

High doses of thyroxine have been shown to have efficacy in rapid-cycling BD (Bauer and Whybrow, 1990). Open studies suggest the efficacy of clozapine in treatment-refractory rapid cyclers (McElroy *et al.*, 1991; Suppes *et al.*, 1992).

Pharmacotherapy: conclusions

Bipolar depression is a common and disabling condition, with highly significant costs to patients, their families and society. Increasing evidence is available to support the efficacy of certain pharmacological agents for this mood disorder. Lithium is the most established medication; however, the evidence base supporting the use of lithium (at least as monotherapy) is not strong, and important negative studies have recently been reported. Although antidepressant drugs have long been used in bipolar depression, evidence of their efficacy is somewhat lacking, and some recent large studies have been negative. However, recent large studies do support the use of some drugs, including lamotrigine and quetiapine. Much further research is needed. Larger trials could examine the efficacy of older agents (such as valproate), as well as newer agents (for example, mifepristone), and proof-of-concept studies may usher in newer agents yet, particularly 'orphan' drugs from other areas of medicine. The requirement for effective pharmacotherapy for bipolar depression is unlikely to lessen in the near future, and clinicians who treat this disorder should be aware of the current evidence in order to facilitate the best clinical practice.

Disclosures

Charles B. Nemeroff

Scientific advisory board: AFSP; AstraZeneca; NARSAD; Quintiles; PharmaNeuroboost.

Stockholder or equity: Corcept; Revaax; NovaDel Pharma; CeNeRx, PharmaNeuroboost.

Board of directors: American Foundation for Suicide Prevention (AFSP); George West Mental Health Foundation; NovaDel Pharma, Mt. Cook Pharma, Inc.

Patents: method and devices for transdermal delivery of lithium (US6 375 990 B1); method to estimate serotonin and norepinephrine transporter occupancy after drug treatment using patient or animal serum (provisional filing April, 2001).

Allan H. Young

Recent honoraria: Astra-Zeneca, Bristol-Myers Squibb, Eli Lilly, Sanofi-Aventis, Servier.

References

Angst, J., Angst, F., Gerber-Werder, R., Gamma, A. (2005). Suicide in 406 mood-disorder patients with and without long-term medication: a 40 to 44 years' follow-up. *Archives of Suicide Research*, **9**, 279–300.

Bauer, M.S. & Whybrow, P.C. (1990). Rapid cycling bipolar affective disorder. II. Treatment of refractory rapid cycling with high-dose levothyroxine: a preliminary study. *Archives of General Psychiatry*, **47**, 435–40.

Benazzi, F. (1999). A comparison of the age of onset of bipolar I and bipolar II outpatients. *Journal of Affective Disorders*, **54**, 249–53.

Bowden, C.L., Calabrese, J.R., Sachs, G., *et al.* (2003). A placebo-controlled 18-month trial of lamotrigine and lithium maintenance treatment in recently manic or hypomanic patients with bipolar I disorder. *Archives of General Psychiatry*, **60**, 392–400.

Brown, E.B., McElroy, S.L., Keck Jr, P.E., *et al.* (2006). A 7-week, randomized, double-blind trial of olanzapine/fluoxetine combination versus lamotrigine in the treatment of bipolar I depression. *Journal of Clinical Psychiatry*, **67**, 1025–33.

Brown, E.B., Dunner, D.L., Adams, D.H., *et al.* (2009). Olanzapine/fluoxetine combination versus lamotrigine in the long-term treatment of bipolar I depression. *Neuropsychopharmacology*, **12**, 773–82.

Calabrese, J.R., Suppes, T., Bowden, C.L., *et al.***:** Lamictal 614 Study Group (2000). A double-blind, placebo-controlled, prophylaxis study of lamotrigine in rapid-cycling bipolar disorder. *Journal of Clinical Psychiatry*, **61**, 841–50.

Calabrese, J.R., Bowden, C.L., Sachs, G.S., *et al.* (2003). A placebo-controlled 18-month trial of lamotrigine and lithium maintenance treatment in recently depressed patients with bipolar I disorder. *Journal of Clinical Psychiatry*, **64**, 1013–24.

Calabrese, J.R., Keck Jr, P.E., Macfadden, W., *et al.*: BOLDER Study Group (2005a). A randomized, double-blind, placebo-controlled trial of quetiapine in the treatment of bipolar I or II depression. *American Journal of Psychiatry*, **162**, 1351–60.

Calabrese, J.R., Shelton, M.D., Rapport, D.J., *et al.* (2005b). A 20-month, double-blind, maintenance trial of lithium versus divalproex in rapid-cycling bipolar disorder. *American Journal of Psychiatry*, **162**, 2152–61.

Calabrese, J.R., Huffman, R.F., White, R.L., *et al.* (2008). Lamotrigine in the acute treatment of bipolar depression: results of five double-blind, placebo-controlled clinical trials. *Bipolar Disorders*, **10**, 323–33.

Calabrese, J.R., Kasper, S., Johnson, G., *et al.* (2009). International Consensus Group on Bipolar Depression treatment guidelines (updated and revised). *Journal of Clinical Psychiatry*, in press.

Coryell, W., Endicott, J., Andreasen, N., Keller, M. (1985). Bipolar I, bipolar II, and nonbipolar major depression among the relatives of affectively ill probands. *American Journal of Psychiatry*, **142**, 817–21.

Davis, L.L., Bartolucci, A., Petty, F. (2005). Divalproex in the treatment of bipolar depression: a placebo-controlled study. *Journal of Affective Disorders*, **85**, 259–66.

Frangou, S., Lewis, M., McCrone, P. (2006). Efficacy of ethyl-eicosapentaenoic acid in bipolar depression: randomised double-blind placebo-controlled study. *British Journal of Psychiatry*, **188**, 46–50.

Frye, M.A., Grunze, H., Suppes, T., *et al.* (2007). A placebo-controlled evaluation of adjunctive modafinil in the treatment of bipolar depression. *American Journal of Psychiatry*, **164**, 1242–9.

Gallagher, P., Watson, S., Smith, M.S., Ferrier, I.N., Young, A.H. (2005). Effects of adjunctive mifepristone (RU-486) administration on neurocognitive function and symptoms in schizophrenia. *Biological Psychiatry*, **57**, 155–61.

Gallagher, P., Malik, N., Newham, J., Young, A.H., Ferrier, I.N., Mackin, P. (2008). Antiglucocorticoid treatments for mood disorders. *Cochrane Database of Systematic Reviews*, **1**, CD005168.

Geddes, J.R., Burgess, S., Hawton, K., Jamison, K., Goodwin, GM. (2004). Long-term lithium therapy for bipolar disorder: systematic review and meta-analysis of randomized controlled trials. *American Journal of Psychiatry*, **161**, 217–22.

Ghaemi, S.N., Gilmer, W.S., Goldberg, J.F., *et al.* (2007). Divalproex in the treatment of acute bipolar depression: a preliminary double-blind, randomized, placebo-controlled pilot study. *Journal of Clinical Psychiatry*, **68**, 1840–4.

Gijsman, H.J., Geddes, J.R., Rendell, J.M., Nolen, W.A., Goodwin, G.M. (2004). Antidepressants for bipolar depression: a systematic review of randomized, controlled trials. *American Journal of Psychiatry*, **161**, 1537–47.

Goldberg, J.F., Burdick, K.E., Endick, C.J. (2004). Preliminary randomized, double-blind, placebo-controlled trial of pramipexole added to mood stabilizers for treatment-resistant bipolar depression. *American Journal of Psychiatry*, **161**, 564–6.

Goodwin, G. M., Bowden, C.L., Calabrese, J.R., *et al.* (2004). A pooled analysis of 2 placebo-controlled 18-month trials of lamotrigine and lithium maintenance in bipolar I disorder. *Journal of Clinical Psychiatry*, **65**, 432–41.

Hirschfeld, R.M., Lewis, L., Vornik, L.A. (2003). Perceptions and impact of bipolar disorder: how far have we really come? Results of the national depressive and manic-depressive association 2000 survey of individuals with bipolar disorder. *Journal of Clinical Psychiatry*, **64**, 161–74.

Judd, L.L., Akiskal., H.S., Schettler, P.J., *et al.* (2003). The comparative clinical phenotype and long term longitudinal episode course of bipolar I and II: a clinical spectrum or distinct disorders? *Journal of Affective Disorders*, **73**, 19–32.

Keck Jr, P.E., Corya, S.A., Altshuler, L.L., *et al.* (2005). Analyses of treatment-emergent mania with olanzapine/fluoxetine combination in the treatment of bipolar depression. *Journal of Clinical Psychiatry*, **66**, 611–16.

Licht, R.W. (2008). Lamotrigine versus lithium in prophylaxis of bipolar disorder: a randomised study mimicking clinical practice. *Bipolar Disorders*, **10** (Suppl. 1), 27.

Marcus, R. N., McQuade, R.D., Carson, W.H., *et al.* (2008). The efficacy and safety of aripiprazole as adjunctive therapy in major depressive disorder: a second multicenter, randomized, double-blind, placebo-controlled study. *Journal of Clinical Psychopharmacology*, **28**, 156–65.

McElroy, S.L., Dessain, E.C., Pope Jr, H.G., *et al.* (1991). Clozapine in the treatment of psychotic mood disorders, schizoaffective disorder, and schizophrenia. *Journal of Clinical Psychiatry*, **52**, 411–14.

McElroy, S., Young, A.H., Carlsson, A., *et al.* (2010). Double-blind, randomized, placebo-controlled study of quetiapine and paroxetine in adults with bipolar depression (EMBOLDEN II). *Journal of Clinical Psychiatry*, in press.

Michalak, E.E., Murray, G., Young, A.H., Lam, R.W. (2008). Burden of bipolar depression: impact of disorder and medications on quality of life. *CNS Drugs*, **22**, 389–406.

Nierenberg, A.A., Ostacher, M.J., Calabrese, J.R., *et al.* (2006). Treatment-resistant bipolar depression: a STEP-BD equipoise randomized effectiveness trial of antidepressant augmentation with lamotrigine, inositol, or risperidone. *American Journal of Psychiatry*, **163**, 210–16.

Parker, G., Tully, L., Olley, A., Hadzi-Pavlovic, D. (2006). SSRIs as mood stabilizers for Bipolar II Disorder? A proof of concept study. *Journal of Affective Disorders*, **92**, 205–14.

Post, R.M., Uhde, T.W., Roy-Byrne, P.P., Joffe, R.T. (1986). Antidepressant effects of carbamazepine. *American Journal of Psychiatry*, **143**, 29–34.

Post, R.M., Denicoff, K.D., Leverich, G.S., *et al.* (2003). Morbidity in 258 bipolar outpatients followed for 1 year with daily prospective ratings on the NIMH life chart method. *Journal of Clinical Psychiatry*, **64**, 680–90.

Post, R.M., Altshuler, L.L., Leverich, G.S., *et al.* (2006). Mood switch in bipolar depression: comparison of adjunctive venlafaxine, bupropion and sertraline. *British Journal of Psychiatry*, **189**, 124–31.

Sachs, G.S., Nierenberg, A., Calabrese, J.R., *et al.* (2007). Effectiveness of adjunctive antidepressant treatment for bipolar depression. *New England Journal of Medicine*, **356**, 1711–22.

Simon, J.S. & Nemeroff, C.B. (2005). Aripiprazole augmentation of antidepressants for the treatment of partially responding and non-responding patients with major depressive disorder. *Journal of Clinical Psychiatry*, **66**, 1216–20.

Smith, L.A., Cornelius, V., Warnock, A., Bell, A., Young, A.H. (2007). Effectiveness of mood stabilizers and antipsychotics in the maintenance phase of bipolar disorder: a systematic review of randomized controlled trials. *Bipolar Disorders*, **9**, 394–412.

Srisurapanont, M., Yatham, L.N., Zis, A.P. (1995). Treatment of acute bipolar depression: a review of the literature. *Canadian Journal of Psychiatry*, **40**, 533–44.

Suppes, T., McElroy, S.L., Gilbert, J., Dessain, E.C., Cole, J.O. (1992). Clozapine in the treatment of dysphoric mania. *Biological Psychiatry*, **32**, 270–80.

Suppes, T., Hirschfeld, R.M., Vieta, E., Raines, S., Paulsson, B. (2007). Quetiapine for the treatment of bipolar II depression: analysis of data from two randomized, double-blind, placebo-controlled studies. *World Journal of Biological Psychiatry*, **9**, 1–14.

Suppes, T., Liu, S., Paulsson, B., Brecher, M. (2008). Maintenance treatment in bipolar I disorder with quetiapine concomitant with lithium or divalproex: a North American placebo-controlled, randomized multicenter trial. *European Psychiatry*, **23** (Suppl. 2), S237.

Thase, M.E., Macfadden, W., Weisler, R.H., *et al.* (2006). Efficacy of quetiapine monotherapy in bipolar I and II depression: a double-blind, placebo-controlled study (the BOLDER II study). *Journal of Clinical Psychopharmacology*, **26**, 600–9.

Thase, M.E., Jonas, A., Khan, A., *et al.* (2008). Aripiprazole monotherapy in nonpsychotic bipolar I depression: results of 2 randomized, placebo-controlled studies. *Journal of Clinical Psychopharmacology*, **28**, 13–20.

Tiihonen, J., Lönnqvist, J., Wahlbeck, K., *et al.* (2009). 11-year follow-up of mortality in patients with schizophrenia: a population-based cohort study (FIN11 study). *Lancet*, **374**, 620–7.

Tohen, M., Sanger, T.M., McElroy, S.L., *et al.*: The Olanzapine HGEH Study Group (1999). Olanzapine versus placebo in the treatment of acute mania. *American Journal of Psychiatry*, **156**, 702–9.

van der Loos, M.L., Mulder, P.G., Hartong, E.G., *et al.*: LamLit Study Group (2009). Efficacy and safety of lamotrigine as add-on treatment to lithium in bipolar depression: a multicenter, double-blind, placebo-controlled trial. *Journal of Clinical Psychiatry*, **70**, 223–31.

Vieta, E., Gasto, C., Otero, A., Nieto, E., Vallejo, J. (1997). Differential features between bipolar I and bipolar II disorder. *Comprehensive Psychiatry*, **38**, 98–101.

Vieta, E., Calabrese, J.R., Goikolea, J.M., Raines, S., Macfadden, W. (2007). Quetiapine monotherapy in the treatment of patients with bipolar I or II depression and a rapid-cycling disease course: a randomized, double-blind, placebo-controlled study. *Bipolar Disorders*, **9**, 413–25.

Vieta, E. , Suppes, T., Eggens, I., Persson, I., Paulsson, B., Brecher, M. (2008). Efficacy and safety of quetiapine in combination with lithium or divalproex for maintenance of patients with bipolar I disorder (international trial 126). *Journal of Affective Disorders*, **109**, 251–63.

Vieta, E. & Suppes, T. (2008). Bipolar II disorder: arguments for and against a distinct diagnostic entity. *Bipolar Disorders*, **10**, 163–78.

Young, A.H., Gallagher, P., Watson, S., Del-Estal, D., Owen, B.M., Ferrier, I.N. (2004). Improvements in neurocognitive function and mood following adjunctive treatment with mifepristone (RU-486) in bipolar disorder. *Neuropsychopharmcology*, **29**, 1538–45.

Young, A.H., McElroy, S., Chang, W., Olausson, B., Paulsson, B., Brecher, M. (2010). A double-blind, placebo-controlled study with acute and continuation phase of quetiapine in adults with bipolar depression (EMBOLDEN I). *Journal of Clinical Psychiatry*, in press.

Young, A.H. & Nemeroff, C.B. (2009). Pharmacological treatment of bipolar depression. In: Yatham, L.N. & Maj, M. (Eds). *Bipolar Disorder: Clinical and Neurobiological Foundations*. London and New York: Wiley Blackwell.

Zarate Jr, C.A., Payne, J.L., Singh, J., *et al.* (2004). Pramipexole for bipolar II depression: a placebo-controlled proof of concept study. *Biological Psychiatry*, **56**, 54–60.

Chapter 5

Practical treatment guidelines for management of bipolar disorder

Lakshmi N. Yatham

Introduction

Since the publication of the first treatment guidelines for bipolar disorder (BD) by the American Psychiatric Association (APA, 1994), several regional, national and international groups have published treatment guidelines for this condition. The intention of these treatment guidelines is to provide an evidence-based standard of care for clinical treatment recommendations for BD. The process by which clinical recommendations were arrived at by various treatment guidelines groups varied, ranging from a consensus process based on the opinion of experts to some combination of evidence for efficacy, expert opinion, support for clinical efficacy and safety considerations. Despite significant variations in methodology to arrive at treatment recommendations, there is in general a consensus with regard to recommendations for pharmacological treatment for acute mania and for maintenance treatment of BD between the various treatment guidelines groups. The most contentious area, however, is with regard to the recommendations for the management of acute bipolar depression and for prophylaxis of depressive episodes, and this is primarily owing to a dearth of research evidence in this area. Therefore, expert opinion and clinical practice experience have heavily influenced treatment recommendations for the management of bipolar depression in various treatment guideline recommendations.

The treatment guidelines for management of BD recommended in this chapter are based primarily on the Canadian Network for Mood and Anxiety Treatments (CANMAT) guidelines for bipolar disorder (Yatham *et al.*, 2005, 2006). These guidelines were chosen primarily because they are comprehensive, practical and incorporate quality of evidence, clinical experience and safety in providing recommendations. The treatments in the CANMAT guidelines are categorised for each phase of BD as being first-line, second-line or third-line treatments or treatments not recommended. The criteria for rating the strength of evidence for a particular treatment are listed in Table 5.1.

Table 5.1 Evidence criteria

1	Meta-analysis or replicated DB-RCT that includes a placebo condition
2	At least 1 DB-RCT with placebo or active comparison condition
3	Prospective uncontrolled trial with at least 10 or more subjects
4	Anecdotal reports or expert opinion

DB-RCT, double-blind randomised controlled trial.

Foundations of management

Diagnosis

The optimal management of BD begins with early and accurate diagnosis of this condition. Previous studies suggest that up to two-thirds of patients with BD are misdiagnosed, and the most common misdiagnosis in patients with BD is, in fact, unipolar depression. There is also evidence that patients with BD suffer from symptoms for an average of 10 to 15 years before they receive the correct diagnosis. Therefore clinicians should be on a high alert for diagnosing BD in any patients who present with depressive symptoms or vague, non-specific somatic symptoms or reverse vegetative symptoms. The guidelines for who to screen, how to screen and what alternative diagnosis should be considered in those presenting with depressive symptoms or vague somatic symptoms are listed in Table 5.2.

Clinicians are aware that there is no diagnostic test or an ideal screening tool for BD. Of those available, the mood disorder questionnaire (MDQ) (Hirschfeld *et al.*, 2000) is a useful screening tool, but it should not be used as a diagnostic instrument. We suggest that each new patient be asked to complete the MDQ before seeing a clinician, and any patient who answered 'yes' to

Practical Management of Bipolar Disorder, eds. Allan H. Young, I. Nicol Ferrier and Erin E. Michalak. Published by Cambridge University Press.

Table 5.2 Screening for bipolar disorder

Who to screen?
Screen patients who present with depressive symptoms for a history of hypomanic or manic symptoms.
Consider an underlying mood disorder in patients presenting with unexplained vague/non-specific somatic symptoms or reverse vegetative symptoms (e.g. hypersomnia and hyperphagia).
How to screen?
Listen to the patient's unprompted presenting complaints. Ask open-ended and non-leading general questions about the common symptoms of depression and mania. Ask questions about specific symptoms of depression and mania, including how long the symptoms have been present during the current episode, how long they lasted during prior episodes (if applicable) and whether they have caused problems in social relationships or work. Always ask about suicidal ideation. Ask about psychotic symptoms. Consider asking the patient to complete the MDQ. Ask about a family history of bipolar disorder. Interview family or friends regarding previous episodes of mania or hypomania. If unclear, ask patients to do prospective mood ratings and assess when patients are rating symptoms in manic or hypomanic range.
Consider alternative diagnoses
General medical conditions that may produce similar symptoms. Alcohol and other substance abuse. Medications that may produce similar symptoms.

MDQ, mood disorder questionnaire.

two or more items should be assessed in greater detail to either exclude or confirm a previous hypomanic or manic episode and thus the diagnosis of BD type I or type II. Patients often have difficulty recalling hypomanic symptoms, particularly when they are questioned during an acute depressive episode, as they may have difficulty recalling symptoms because of problems with concentration and memory. Therefore, it is important to ask several screening questions for hypomania/mania and, when possible, a collateral history from friends or family should be obtained to ascertain the diagnosis. Patients with a family history of BD, psychotic symptoms during a depressive episode, reverse vegetative symptoms such as hypersomnia and hyperphagia as a part of depression and younger age at onset of depressive symptoms, as well as those with multiple episodes of depression, are more likely to have a BD; hence the index of suspicion for a BD should be higher in such patients. For cases in which the diagnosis is still uncertain after a comprehensive assessment, prospective use of a mood diary would be very useful. In such situations, it is prudent that clinicians take the time to assess patients on the days when patients are rating their symptoms in the mood diary in the hypomanic/manic range, as this would help to confirm or refute the diagnosis.

Goals of treatment

Once the diagnosis is confirmed, the optimal management of BD includes treatment of acute episodes to full remission and implementing strategies to minimise the risk of relapse. There is evidence that subsyndromal symptoms predict the relapse of new mood episodes, and the time to the next mood episode is significantly shorter in people who had incompletely recovered from the index episode with subsyndromal symptoms (Judd *et al.*, 2008). Therefore all acute episodes should be treated aggressively to full remission, and both pharmacological and psychosocial maintenance strategies should be applied to prevent relapse of mood episode. The CANMAT guidelines strongly endorse a stepped-care approach, which includes a chronic disease management model. All patients with BD should be offered psychoeducation as well as pharmacotherapy and should be monitored by a physician and at least one health-care professional, such as a nurse or other mental health professional. The psychoeducation should focus on helping the patient to become involved in self-management, to collaborate effectively with health-care providers, to learn early warning signs of relapse as well as key stress management techniques, and to seek help early and as necessary to minimise the risk of relapse. Psychoeducation will also involve helping patients to acknowledge the risks associated with substance abuse and paying attention to biological as well as social rhythms.

The CANMAT guidelines also recommend, as stated above, psychosocial strategies in conjunction with pharmacotherapy for preventing relapses of mood episode. There is evidence from controlled trials suggesting that adjunctive psychoeducation improves outcomes for patients with BD (Colom & Lam, 2005). There is also evidence that other forms of psychotherapy, such as cognitive behavioural therapy, family-focused management and interpersonal social rhythm therapy all improve outcomes when used as adjunctive treatments to pharmacotherapy (Miklowitz, 2008). Given that the cost for psychoeducation would be substantially cheaper than other therapies, and given that a more recent Canadian study has indicated that five group sessions of psychoeducation are as effective as 20 individual sessions of cognitive behavioural therapy as adjuncts to pharmacotherapy in prophylaxis of BD (unpublished observation), we recommend that psychoeducation be offered to all patients with BD before considering other more costly psychological treatments.

Pharmacological management

Emergency management of acute mania

Patients with acute mania – particularly those with moderate to severe symptoms and those with psychotic features – often require admission to hospital for management of mania. The decision to admit the patient to hospital or manage the patient on an outpatient basis must be based on the safety considerations and the psychosocial support that the patient may have at his or her disposal. A patient with mild manic symptoms with good insight and good psychosocial support could be effectively managed on an outpatient basis. Those that are admitted to hospital and refuse to take medications may need to be committed for involuntary admission under the relevant mental health act and may need to be administered medication against their will if there is a potential risk to the patient or others as a result of mania. Treatments that may be considered include an injectable atypical antipsychotic such as intramuscular olanzapine or ziprasidone or a combination of a small dose of an injectable conventional antipsychotic such as haloperidol and a benzodiazepine such as lorazepam. Benzodiazepines should not be used as monotherapy in patients with BD. They should primarily be used in conjunction with either a conventional or an atypical antipsychotic to control acute agitation as well as promote sleep. Patients may need to be kept in a 'locked quiet' room for their safety until the behaviour improves, and in some cases, patients may also need physical restraints to ensure their safety and that of those around them.

Pharmacological treatment of acute mania

Several medications have proven effective for the acute management of mania. These include lithium, valproate, carbamazepine, atypical antipsychotics such as risperidone, olanzapine, quetiapine, ziprasidone, aripiprazole, and other medications that have not been approved as yet, such as paliperidone and asenapine (see Bond *et al.*, 2008, and Yatham *et al.*, 2009 for details). All these medications have Level I evidence for their efficacy in the treatment of acute mania. There is no evidence to suggest that there is any difference in efficacy between these agents when given in monotherapy, as each medication appears to provide about 20 to 25% incremental response rate over placebo (Yatham, 2005). The decision as to which medication or combination of medications would be used in the treatment of mania should be based on three criteria: efficacy, onset of action and adverse events.

Efficacy

A number of studies have also examined the efficacy of a combination of lithium or divalproex with an atypical antipsychotic in comparison to lithium or divalproex plus placebo. These studies have shown that risperidone or olanzapine or quetiapine or ziprasidone or aripiprazole, when given in conjunction with lithium or valproate, is more effective than lithium or valproate plus placebo. On average, approximately 20% more patients will improve with a combination treatment compared with lithium or valproate plus placebo over the same period of time, and there do not appear to be any significant differences in efficacy between various atypical antipsychotic and mood-stabiliser combinations. However, no study to date has examined lithium or divalproex plus an atypical antipsychotic combination versus atypical antipsychotic plus placebo. Therefore, in summary, combination treatment is more effective than monotherapy, but there is no difference between the various monotherapies or various combination therapies in terms of response rates for acute mania.

Onset of action

How fast a medication works can be determined by examining at what point an active medication separated from placebo in double-blind, placebo-controlled trials that assessed the efficacy of various medications for acute mania. These studies suggest that in general most medications separate from placebo within the first 7 d, the only exceptions being lithium and valproate. However, studies that have used valproate oral loading have shown that valproate also separates from placebo if it is dosed aggressively from the beginning. Therefore, there does not appear to be any significant difference in the onset of action between various atypical antipsychotic medications and divalproex. It is likely that lithium works a bit slower than other antimanic medications, but by 3 weeks there does not appear to be any difference in response rates between various antimanic agents.

Regarding combination treatments, there is evidence that they separate from monotherapy within the first week, suggesting that combination treatments work faster than monotherapy. However, there does not

Table 5.3 Recommendations for pharmacological treatment of acute mania

First line	Lithium, divalproex, olanzapine, risperidone, quetiapine, aripiprazole, ziprasidone, lithium or divalproex + risperidone, lithium or divalproex + quetiapine, lithium or divalproex + olanzapine, ***lithium or divalproex + aripiprazole***[a]
Second line	Carbamazepine, oxcarbazepine, ECT, lithium + divalproex, ***asenapine***[a]***, lithium or divalproex + asenapine***[a]***, paliperidone monotherapy***[a]
Third line	Haloperidol, chlorpromazine, lithium or divalproex + haloperidol, lithium + carbamazepine, clozapine, ***tamoxifen***[a]
Not recommended	Monotherapy with gabapentin, topiramate, lamotrigine, verapamil, tiagabine, risperidone + carbamazepine, ***olanzapine + carbamazepine***[a]***, lithium or divalproex + paliperidone***[a]

ECT, electroconvulsive therapy.
[a] New.

appear to be any difference in onset of action between various combinations of treatments.

Adverse events

There are clear differences in adverse-event profiles between various atypical antipsychotic medications as well as mood stabilisers such as lithium and divalproex. These are familiar to all psychiatrists and are covered in other chapters in this book.

Clinical management of mania

Given that there is no difference in efficacy or onset of action between various antimanic agents when used in monotherapy or in combination therapy, the decision as to which medication a given patient should be prescribed should be based on the adverse-event profile of the medications as well as the patient's previous history of response or non-response and preference with regard to the side effects.

The first step in the management of mania would include assessing risk for aggressive behaviour and violence towards others and using pharmacological strategies and restraints to manage this. Patients who are taking antidepressants should be taken off them, and any co-morbid substance abuse or complications of co-morbid substance abuse or co-morbid medical problems should be addressed. If a patient has been taking one of the first-line treatments for acute mania, this should be optimised before further changes are considered. A list of first-, second- and third-line treatments, as well as the treatments that are not recommended for mania, are listed in Table 5.3.

If a patient has not been on any medication, they should be commenced on one of the first-line treatments. The decision as to which first-line treatment the patient would be treated with should be based on previous history of response or non-response to a given treatment, family history of response, any clinical features that might predict a response to one medication or another and patient or family preference. A treatment algorithm for the management of patients with acute mania is illustrated in Figure 5.1.

For patients who are commenced on monotherapy with one of the first-line agents, it should be dosed adequately, and if the improvement is less than 20% by the end of the second week then it would be appropriate to either add, or switch the patient to, an alternative first-line treatment. For example, if the patient has been taking lithium or valproate in monotherapy and has not responded by at least 20% by the end of the second week, the patient may either have an atypical antipsychotic added or be switched to an atypical antipsychotic. Similarly, if a patient was taking an atypical antipsychotic and did not show a response they could either have lithium or valproate added, be switched to lithium or valproate or be switched to a different atypical antipsychotic.

In those patients taking a combination and showing a poor response, either the mood stabiliser or the atypical antipsychotic could be switched. If several first-line treatments, either in the monotherapy or in the combination therapy, failed, second-line treatments should be considered, and these may include medications such as carbamazepine or oxcarbazepine; if there is no response, it might be appropriate to try third-line treatment, such as conventional antipsychotics or clozapine. There are also several newer novel options, which have shown some promise, such as tamoxifen, but these medications should only be tried after the first-, second- and third-line treatment options have been exhausted. The CANMAT guidelines suggest that gabapentin, topiramate and lamotrigine should not be used in monotherapy for treatment of acute mania. The CANMAT guidelines also suggest that carbamazepine should not be combined with risperidone or olanzapine, as carbamazepine induces the metabolism of these medications, and the efficacy of the combination does

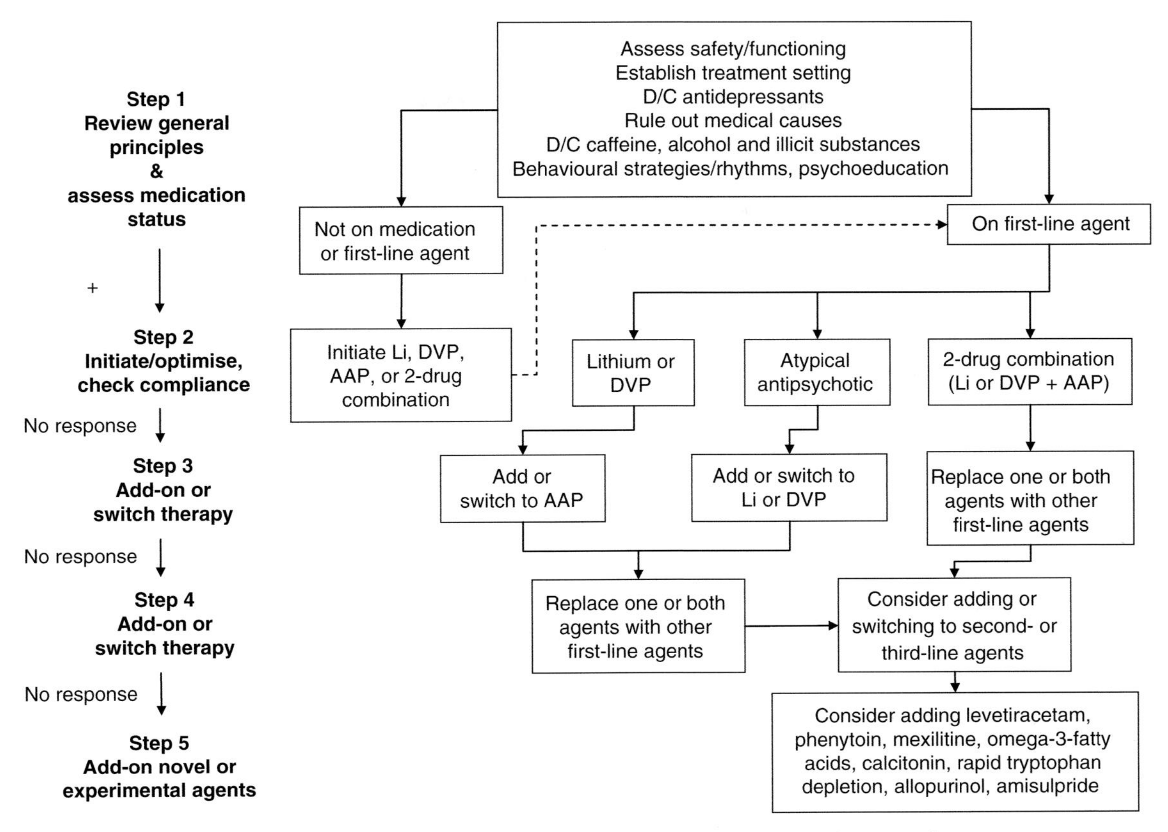

Figure 5.1 Treatment algorithm for acute mania. AAP, atypical antipsychotic; DVP, divalproex; Li, lithium; D/C, discontinue.

not appear to be superior to the efficacy of carbamazepine plus placebo in such patients.

Mania with psychotic features

A significant proportion of bipolar patients with mania have psychotic symptoms. However, there does not appear to be any evidence to suggest that the psychotic symptoms predict poor outcome in BD type I patients. Studies suggest that the response rates and improvement in manic symptoms are similar in patients with or without psychotic symptoms with various atypical antipsychotics such as olanzapine, risperidone or quetiapine. The prognosis, however, does appear to be somewhat poorer in BD type I patients with mood incongruent psychotic symptoms. However, whether this has an impact on acute antimanic treatments is currently unknown, as studies have not specifically looked at this particular outcome measure.

Mixed episodes

There is some evidence that patients with mixed episodes do not respond to lithium as well as those who have classical mania. By contrast, divalproex appears to be equally effective for classical mania and those with mixed episodes. Studies indicate that atypical antipsychotics such as olanzapine and other mood stabilisers such as carbamazepine are equally effective in those with mixed episodes. Therefore, the CANMAT guidelines suggest that for patients with mixed episodes, divalproex sodium, olanzapine or carbamazepine might be preferable to lithium. The efficacy of other atypical antipsychotics such as risperidone has not been specifically examined in patients with mixed episodes.

Management of bipolar depression

In contrast to the treatment of mania, for which many evidence-based pharmacological treatment options are available, few treatments have been investigated in studies with sound methodological designs for the treatment of acute bipolar depression. Therefore, treatment of bipolar depression is one of the most controversial areas in the management of BD, as there is wide

Table 5.4 Recommendations for pharmacological treatment of acute bipolar I depression[a]

First line	Lithium, lamotrigine, quetiapine, lithium or divalproex + SSRI, olanzapine + SSRI, lithium + divalproex, lithium or divalproex + bupropion
Second line	Quetiapine + SSRI, ***divalproex***[b], lithium or divalproex + lamotrigine, ***adjunctive modafinil***[b]
Third line	Carbamazepine, olanzapine, lithium + carbamazepine, lithium + pramipexole, lithium or divalproex + venlafaxine, lithium + MAOI, ECT, lithium or divalproex or AAP + TCA, lithium or divalproex or carbamazepine + SSRI + lamotrigine, adjunctive EPA, adjunctive riluzole, adjunctive topiramate
Not recommended	Gabapentin monotherapy, ***aripiprazole monotherapy***[b]

AAP, atypical antipsychotic; ECT, electroconvulsive therapy; EPA, ethyl-eicosapentaenoic acid; MAOI, monoamine oxidase inhibitor; SSRI, selective serotonin reuptake inhibitor; TCA, tricyclic antidepressant.
[a] The management of a bipolar depressive episode with antidepressants remains complex. The clinician must balance the desired effect of remission with the undesired effect of switching.
[b] New.

disagreement about the role of antidepressants and other medications in managing acute bipolar depressive symptoms.

Pharmacological treatment of acute bipolar depression

The CANMAT group took into consideration the evidence for efficacy from double-blind placebo-controlled trials and adverse events of medications, including their propensity to destabilise the course of BD, as well as clinical experience and support for the intervention, in formulating the first-, second- and third-line treatments for bipolar depression; these are listed in Table 5.4.

There was a general consensus with regard to recommending quetiapine as one of the first-line treatments for acute bipolar depression. This was based on the evidence for efficacy from five double-blind placebo-controlled trials of acute bipolar depression and two double-blind extension studies showing that the patients who responded to quetiapine monotherapy were less likely to relapse into a depressive episode compared with those who were switched to placebo. Olanzapine plus fluoxetine combination was also included as one of the first-line treatments, given its superiority over placebo as well as olanzapine monotherapy and its comparable efficacy to lamotrigine monotherapy in treating acute bipolar depressive symptoms.

There was a debate about whether lamotrigine monotherapy, lithium monotherapy or lithium/valproate plus a selective serotonin reuptake inhibitor (SSRI) or bupropion combination should be recommended as first-line treatments or not. Although lithium was shown to be superior to placebo in treating acute bipolar depressive symptoms in small crossover placebo-controlled trials conducted in the 1970s and 1980s, a more recent trial comparing lithium vs placebo vs quetiapine showed that lithium was not superior to placebo (Young *et al.*, 2008). However, in this study, the mean serum lithium levels were 0.6 mmol/l, and the group felt this may not have been a true test of the efficacy of lithium, as a previous study by Nemeroff *et al.* (2001) showed a qualitative distinction in the treatment efficacy of lithium at serum levels of 0.8 mmol/l or more compared with levels of 0.6 mmol/l or less: namely, that serum levels of 0.8 mmol/l or more were as effective as lithium plus antidepressant combination, but that this was not true of serum levels of 0.6 mmol/l or less. Given this, the group agreed that lithium should continue to be recommended as one of the first-line treatments until further evidence confirms or refutes its efficacy.

Similarly, although lamotrigine failed to beat placebo in four recent GlaxoSmithKline-sponsored double-blind placebo-controlled trials, a meta-analysis combining these four trials and a previous positive study showed that lamotrigine was significantly superior to placebo in treating acute bipolar depressive symptoms (Geddes *et al.*, 2007). Furthermore, a recent double-blind study showed that lamotrigine add-on to lithium was more effective than placebo add-on in treating bipolar depressive symptoms (van der Loos *et al.*, 2007). This, combined with a wealth of clinical experience, convinced the CANMAT group that lamotrigine should also continue to be recommended as one of the first-line treatments.

Antidepressants are widely used in combination with an atypical antipsychotic or lithium or valproate, and a meta-analysis confirmed their efficacy in treating acute bipolar depression. However, a recent study using more practical outcome measures showed that antidepressant therapy was no more effective than placebo adjunctive therapy on the primary outcome

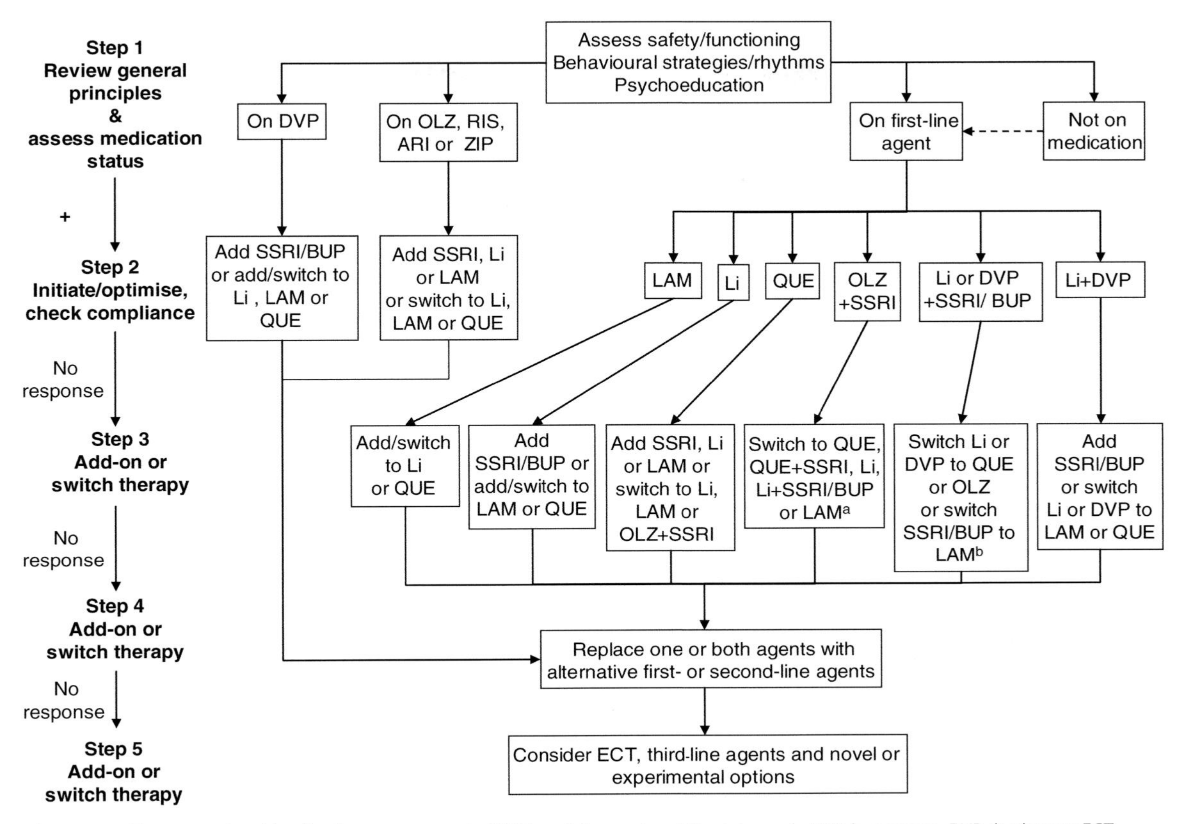

Figure 5.2 Treatment algorithm for the management of BD type I depression. ARI, aripiprazole; BUP, bupropion; DVP, divalproex; ECT, electroconvulsive therapy; LAM, lamotrigine; Li, lithium; OLZ, olanzapine; QUE, quetiapine; RIS, risperidone; SSRI, selective serotonin reuptake inhibitor; ZIP, ziprasidone.
[a] Or switch the SSRI to another SSRI.
[b] Or switch the SSRI or BUP to another SSRI or BUP.

measure of durable recovery, which was defined as a patient experiencing two or fewer depressive symptoms for eight consecutive weeks during the six-month study (Sachs *et al.*, 2007). However, given the fact that this study used an unusual outcome measure, and given the fact that there was a positive meta-analysis and a wealth of experience on the utility of antidepressants as adjunctive therapy for bipolar depression, the CANMAT group continued to recommend this as one of the first-line agents. This would of course be subject to change if further studies become available that refute the efficacy of adjunctive antidepressant treatment.

Divalproex has been recommended as one of the second-line options, based on three small positive control trials. Adjunctive modafinil and adjunctive lamotrigine to lithium or divalproex have been recommended as second-line treatment, based on the recent double-blind controlled trials. The SSRI and quetiapine combination has also been included as a second-line option, given that quetiapine monotherapy has been effective, and other studies showing that the atypical antipsychotics augment the efficacy of SSRIs in treating depressive symptoms.

Clinical management of acute bipolar depression

The CANMAT treatment algorithm for acute bipolar depression is represented in Figure 5.2.

Patients presenting with bipolar depressive symptoms should be assessed for risk of suicide, and if the risk is high the patient should be managed on an inpatient basis rather than an outpatient basis, unless significant psychosocial support is available. All patients with acute bipolar depression should receive psychosocial interventions, including psychoeducation, when possible, along with pharmacotherapy. Those who are not on any medications should be started on one of the first-line agents, which include lamotrigine monotherapy, lithium monotherapy, quetiapine monotherapy, olanzapine plus SSRI combination or lithium or valproate plus an SSRI or bupropion. If a patient already has been taking one of the mood stabilisers such as lithium, and if this was ineffective, combining with a second mood stabiliser would also be appropriate.

If a patient does not respond to an adequate trial of one of the first-line treatments, the next step might include switching the patient to a different medication, or combining two different agents or augmenting the first agent. The decision as to which approach would be used for a particular patient depends on whether a patient has shown any response to the first treatment. A switch to an alternative first-line agent would be appropriate if the first-line agent that the patient is currently taking has not led to any improvement in depressive symptoms. An example might be that if a patient has been taking lamotrigine monotherapy and has not shown any response, they could be switched to quetiapine or lithium or olanzapine/fluoxetine combination (OFC) or lithium or divalproex plus an SSRI or bupropion. Given that the evidence for the efficacy of first-line agents is of better quality, these options should be tried first before moving on to the next step.

If a patient has shown a partial response to one of the first-line options, the next step might be a combination of two agents. For instance a patient who is taking lithium and is showing a partial response could have either lamotrigine or quetiapine or an antidepressant such as an SSRI, bupropion or divalproex added. If these strategies are ineffective, the next step is to add agents that might augment the efficacy of some of the first-line treatments; these include agents such as modafinil and pramipexole. Patients who have significant severe psychotic symptoms and intense suicidal ideation might be suitable for electroconvulsive therapy (ECT), as this is very effective in treating acute bipolar depressive symptoms.

If lithium is used for treating acute bipolar depression, serum lithium levels should be at least 0.8 mmol/l. In the case of lamotrigine, in order to avoid the risk of skin rash this should be started at a low dose of 12.5 to 25 mg daily and increased by 12.5 to 25 mg increments every 1 to 2 weeks while the patient is being monitored for skin rash. In addition, the risk of Stevens–Johnson syndrome with lamotrigine should be clearly explained to all patients. An adequate trial for treating acute bipolar depression is at least 6 to 8 weeks, and if a patient does not show any response at all during this period, switching to a newer strategy or combining with another agent would be appropriate.

Maintenance treatment of bipolar disorder

Why and when prophylaxis?

There is evidence that 80% of bipolar patients will have a relapse of a mood episode within 2.5 years (Prien *et al.*, 1973a, 1973b) and 95% within 5 years of a previous episode. Furthermore, recent double-blind, placebo-controlled trials of maintenance pharmacotherapy reveal that the relapse rates are at least 50% in those assigned to placebo (Calabrese *et al.*, 2003; Tohen *et al.*, 2004). Physicians are well aware of the risks and consequences associated with an acute manic or a depressive episode for bipolar patients. In addition, the risk of treatment resistance increases with more episodes. This may be related to increased oxidative stress, worsening cognitive symptoms and changes in brain volume with successive mood episodes. Given this, it is prudent that prophylaxis be offered to all bipolar patients right from the first manic episode.

However, many patients, particularly those who had their first manic episode at a young age, will have significant difficulty in accepting the bipolar diagnosis. Psychosocial interventions, including psychoeducation, should be offered to these patients to help them understand the nature of BD, the consequences of a mood episode, the role of medication and side effects of medications, etc. In other words, patients need to be empowered to take responsibility for managing their BD through better understanding of the illness and treatment options available.

Role of psychosocial treatments

Treatment adherence is a significant problem in patients with BD. Therefore, psychosocial strategies should be used in conjunction with pharmacotherapy to prevent relapse of mood episodes in bipolar patients. There is evidence that psychoeducation, interpersonal and social rhythm therapy (IPSRT), family-focused treatment and intensive psychotherapies, including IPSRT and cognitive behavioural therapy given weekly and bi-weekly for up to 30 sessions over a 9-month period are more effective than three sessions of individual psychoeducation in preventing relapse of mood episodes in bipolar patients (Miklowitz *et al.*, 2007).

Table 5.5 Recommendations for maintenance pharmacotherapy of bipolar disorder

First line	Lithium, lamotrigine monotherapy (mainly for those with mild manias), divalproex, olanzapine, ***quetiapine, lithium or divalproex + quetiapine***[a], ***risperidone LAI***[a], ***adjunctive risperidone LAI***[a], ***aripiprazole (mainly for preventing mania)***[a], ***adjunctive ziprasidone***[a]
Second line	Carbamazepine, lithium + divalproex, lithium + carbamazepine, lithium or divalproex + olanzapine, lithium + risperidone, lithium + lamotrigine, olanzapine + fluoxetine
Third line	Adjunctive phenytoin, adjunctive clozapine, adjunctive ECT, adjunctive topiramate, adjunctive omega-3-fatty acids, adjunctive oxcarbazepine, or adjunctive gabapentin
Not recommended	Adjunctive flupenthixol, monotherapy with gabapentin, topiramate or antidepressants

ECT, electroconvulsive therapy; LAI, long-acting injection; SSRI, selective serotonin reuptake inhibitor.
[a] New

Pharmacological prophylaxis treatment for bipolar disorder

CANMAT recommendations for maintenance treatment of BD are listed in Table 5.5.

Double-blind, controlled trials suggest that monotherapy with lithium, lamotrigine, olanzapine, quetiapine, aripiprazole or risperidone is more effective than placebo in preventing relapse/recurrence of mood episode in bipolar patients (see Bond *et al.*, 2008 and Yatham *et al.*, 2009 for evidence). Although a valproate study was negative on the primary efficacy measure, fewer patients who had been treated for acute mania with divalproex and subsequently randomised to receive divalproex relapsed than those who were switched to placebo. Further, there is evidence from other randomised open studies and a wealth of clinical experience supporting the efficacy of divalproex, and therefore divalproex is also recommended as one of the first-line treatments for the maintenance treatment of BD by the CANMAT group. In addition, recent studies suggest that adjunctive quetiapine, adjunctive ziprasidone and adjunctive risperidone are more effective than placebo adjunctive therapy, and therefore these also have been included as first-line treatments.

Aripiprazole monotherapy is effective mainly in preventing manic relapses, and therefore clinicians should use aripiprazole primarily in situations in which mania prevention is the goal. Similarly, lamotrigine monotherapy has limited efficacy in preventing mania but has significant efficacy in preventing depressive episodes. Therefore, lamotrigine is most appropriate in situations in which prevention of depressive episodes is the main goal, either because manic relapse is being controlled by another agent or because the patient has a history of very infrequent and mild manic episodes.

Carbamazepine has been recommended as a second-line agent, primarily because of the lack of large double-blind placebo-controlled trials as well as the propensity of carbamazepine to induce hepatic microsomal enzymes, which makes it difficult to combine carbamazepine with other medications. A combination of two first-line agents, such as lithium or divalproex or lithium plus lamotrigine, might be appropriate in patients who are not responding to prophylaxis with one of the first-line agents alone. Adjunctive clozapine as well as adjunctive omega fatty acids have been recommended as third-line options for prophylaxis of BD, but adjunctive flupentixol or monotherapy with gabapentin, topiramate or antidepressants is not recommended, as they do not appear to have any efficacy in preventing mood episodes in bipolar patients.

Clinical management and practical considerations/monitoring

There is evidence that the subsyndromal symptoms predict a relapse of mood episode, and therefore patients should be treated aggressively to euthymia to reduce the risk of relapse. In addition, there is evidence that the polarity of the index episode predicts the polarity of relapse into a subsequent episode and therefore should be a consideration for determining the most appropriate pharmacological treatment options for prophylaxis. For instance, a patient who has had a recent index depressive episode is much more likely to relapse into a depressive episode, and therefore a suitable prophylaxis for that patient might be an agent that has been shown to have significant efficacy in preventing relapse of depressive episodes, including lamotrigine or quetiapine but not aripiprazole. Similarly, a patient who has had a recent manic episode might require a stronger antimanic prophylactic agent, such as lithium, risperidone, olanzapine, or divalproex rather than lamotrigine to prevent further relapse of manic episodes. The patient's serum levels should be checked periodically to ensure that he/she is adherent to the treatment. Further, at each follow-up visit, a few minutes should be spent discussing treatment adherence issues, as this may help to improve treatment adherence in bipolar patients. In addition, all side effects should be addressed, as these may have a significant negative impact on treatment adherence if not addressed.

In general, controlled trials suggest that patients who responded to an acute episode with a particular treatment have a better chance of staying well if continued on the same treatment, and therefore that should be the goal for all patients with BD. For example, if a patient was treated for an acute manic episode with lithium monotherapy and it was effective, then lithium should be offered as the treatment for prophylaxis. Similarly, if a patient has been treated with risperidone or quetiapine plus a mood stabiliser, that combination should be offered for prophylaxis, as the combination therapy appears to be more beneficial than monotherapy with a mood stabiliser in prophylaxis. However, the efficacy of continuation of combination therapy must be weighed against the risk of side effects, including weight gain, hypercholesterolaemia and other metabolic issues, and these should be addressed to ensure that treatment adherence is not compromised.

At each visit the patient should be monitored for any subsyndromal symptoms and should be asked about side effects. All patients should have baseline laboratory investigations, including complete blood count, fasting glucose, fasting lipid levels, electrolytes, liver enzymes, urine toxicology and TSH before maintenance therapy, and should have their weight and BMI assessed at baseline and at regular intervals thereafter. Serum levels of lithium or divalproex should be checked at least once every 6 to 12 months and as clinically indicated.

References

APA (1994). Practice guideline for the treatment of patients with bipolar disorder. *American Journal of Psychiatry*, **151**, 1–36.

Bond, D., Vieta, E., Tohen, M.F., *et al.* (2008). Antipsychotic medications in bipolar disorder: a critical review of randomized controlled trials. In: Yatham, L.N & Kusumakar, V. (Eds). *Bipolar Disorder: A Clinician's Guide to Treatment Management*. New York: Taylor and Francis, pp. 295–364.

Calabrese, J.R., Bowden, C.L., Sachs, G., *et al.* (2003). A placebo-controlled 18-month trial of lamotrigine and lithium maintenance treatment in recently depressed patients with bipolar I disorder. *Archives of General Psychiatry*, **64**, 1013–24.

Colom, F. & Lam, D. (2005). Psychoeducation: improving outcomes in bipolar disorder. *European Psychiatry*, **20**, 359–64.

Geddes, J., Huffman, R., Paska, W., *et al.* (2007). Lamotrigine for acute treatment of bipolar depression: individual patient data meta-analysis of 5 randomized, placebo-controlled trials. *Bipolar Disorders*, **9**, 42–3.

Hirschfeld, R.M., Williams, J.B., Spitzer, R.L., *et al.* (2000). Development and validation of a screening instrument for bipolar spectrum disorder: the Mood Disorder Questionnaire. *American Journal of Psychiatry*, **157**, 1873–5.

Judd, L.L., Schettler, P.J., Akiskal, H.S., *et al.* (2008). Residual symptom recovery from major affective episodes in bipolar disorders and rapid episode relapse/recurrence. *Archives of General Psychiatry*, **65**, 386–94.

Miklowitz, D.J. (2008). Adjunctive psychotherapy for bipolar disorder: state of the evidence. *American Journal of Psychiatry*, **165**, 1408–19.

Miklowitz, D.J., Otto, M.W., Frank, E., *et al.* (2007). Intensive psychosocial intervention enhances functioning in patients with bipolar depression: results from a 9-month randomized controlled trial. *The American Journal of Psychiatry*, **164**, 1340–7.

Nemeroff, C.B., Evans, D.L., Gyulai, L., *et al.* (2001). Double-blind, placebo-controlled comparison of imipramine and paroxetine in the treatment of bipolar depression. *The American Journal of Psychiatry*, **158**, 906–12.

Prien, R.F., Caffey Jr, E.M., Klett, C.J. (1973a). Prophylactic efficacy of lithium carbonate in manic-depressive illness. Report of the Veterans Administration and National Institute of Mental Health collaborative study group. *Archives of General Psychiatry*, **28**, 337–41.

Prien, R.F., Klett, C.J., Caffey Jr, E.M. (1973b). Lithium carbonate and imipramine in prevention of affective episodes. A comparison in recurrent affective illness. *Archives of General Psychiatry*, **29**, 420–5.

Sachs, G.S., Nierenberg, A.A., Calabrese, J.R., *et al.* (2007). Effectiveness of adjunctive antidepressant treatment for bipolar depression. *New England Journal of Medicine*, **356**, 1711–22.

Tohen, M., Bowden, C., Calabrese, J., *et al.* (2004). Olanzapine's efficacy for relapse prevention in bipolar disorder: a randomized double-blind placebo-controlled 12-month clinical trial. *Bipolar Disorders*, **6**, 307.

van der Loos, M., Nolen, W.A., Vieta, E., *et al.* (2007). Lamotrigine as add-on to lithium in bipolar depression. *Bipolar Disorders*, **9** (Suppl. 1), 107.

Yatham, L.N. (2005). Atypical antipsychotics for bipolar disorder. *The Psychiatric Clinics of North America*, **28**, 325–47.

Yatham, L.N., Kennedy, S.H., O'Donovan, C., *et al.* (2005). Canadian Network for Mood and Anxiety Treatments (CANMAT) guidelines for the management of patients with bipolar disorder: consensus and controversies. *Bipolar Disorders*, **7** (Suppl. 3), 5–69.

Yatham, L.N., Kennedy, S.H., O'Donovan, C., *et al.* (2006). Canadian Network for Mood and Anxiety Treatments (CANMAT) guidelines for the management of patients with bipolar disorder: update 2007. *Bipolar Disorders*, **8**, 721–39.

Yatham, L.N., Kennedy, S.H., Schaffer, A., *et al.* (2009). Canadian Network for Mood and Anxiety Treatments (CANMAT) and International Society for Bipolar Disorders (ISBD) collaborative update of CANMAT guidelines for the management of patients with bipolar disorder: update 2009. *Bipolar Disorders*, **11**, 225–255.

Young, A.H., Carlsson, A., Olausson, B., *et al.* (2008). A double-blind, placebo-controlled study with acute and continuation phase of quetiapine and lithium in adults with bipolar depression (EMBOLDEN I). *Bipolar Disorders*, **10**, 451.

Chapter 6

Psychosocial interventions in bipolar disorder: theories, mechanisms and key clinical trials

Sagar V. Parikh and Vytas Velyvis

Bipolar disorder (BD) is the disorder with 'the most' – the most lethality, the most recurrences and the most co-morbidity – of any of the major psychiatric disorders. Given its complexity, it is not surprising that medication alone is inadequate to control the disorder or allow for recovery from its sequelae. After a long period of relatively few psychosocial interventions as a result of the early excitement about the efficacy of lithium and the hope that other medications would also be as effective, the beginning of the twenty-first century has seen a burst of activity. Major randomised controlled trials have been conducted using a variety of individual and group interventions, encompassing several theoretical approaches. This chapter begins with a section that reviews the theoretical underpinnings of bipolar psychosocial interventions and summarises the key features of the dominant models. The second part of the chapter reviews the key trials that have been published over the past decade, and the final part explores the practical implications of the research, taking into account effectiveness and training issues that may guide recommendations about the utility of a specific intervention. Of note, the interventions reviewed are treatments for the BD itself, not for coexisting conditions such as substance abuse or anxiety disorders co-morbid with BD; these are reviewed elsewhere in this book.

Theoretical basis for models

Treatment interventions in medicine follow one of two paradigms: one derived from an understanding of the disorder (a theoretical model), and one derived from empirical findings based on expert opinion and open clinical experience. Within psychological disorders, major models exist to explain the aetiology and symptoms of some anxiety disorders and depression. The negative cognitive triad model of Beck for depression, for instance, has not only been tested for validity but provides a rationale for treatment manoeuvres such as automatic thought records (Beck *et al.*, 1979; Beck, 2008). Such theoretical models for BD are relatively scarce, based in part on the lack of basic research that identifies plausible psychosocial causes or moderators. Furthermore, existing models show considerable overlap, which in turn leads to considerable overlap in the design of psychosocial interventions designed on the basis of these models.

A further challenge to developing a theoretical model is the inherent complexity of the disorder in terms of phases of the illness, and the specific targets for intervention based on phase, symptoms and sequelae. Although bipolar disorder is currently understood as a single illness, no unitary theory explains all the different types of symptom. As a consequence, different models may be necessary to explain the illness in acute depression, in acute mania and in maintenance. Even within the maintenance phase, any bipolar intervention would have to be designed with possibly two different agendas: either to prevent relapse or to treat the sequelae of acute episodes, be they a damaged self-concept, a broken relationship or a lost job. It might be that many of these concerns lead to a unitary model – for instance, treating the sequelae might be directly linked to relapse prevention – but it is also possible that several separate models may need to be retained.

In an attempt to clarify the role of psychosocial interventions in BD, we begin by broadly reviewing the literature in search of putative models. Only four types of model emerged, underlying the four types of intervention that have been tested: psychoeducation (PE), cognitive behavioural therapy (CBT), family interventions (mostly family-focused therapy, FFT) and interpersonal and social rhythm therapy (IPSRT). These models are reviewed below, with attention to the core elements of the ensuing therapy.

Psychoeducation

Psychoeducation is broadly defined as the provision of information about an illness that assists a patient,

Practical Management of Bipolar Disorder, eds. Allan H. Young, I. Nicol Ferrier and Erin E. Michalak. Published by Cambridge University Press.

Table 6.1 Comparison of the Barcelona Bipolar Disorders Program and the Life Goals Program

Barcelona Bipolar Disorders Program, 21-session model
• 14 sessions – education regarding: • BD diagnosis and symptoms • identification of triggers for relapse • medications • alternative therapies • risks of treatment withdrawal and substance use
• 2 sessions – early detection of depressive and manic/hypomanic episodes
• 1 session – coping strategies for relapse prevention
• 1 session – introducing lifestyle regularity
• 1 session – stress management and problem-solving
Life Goals Program, Phase 1: 6 sessions
• 1 session – education re: review of bipolar disorders, symptoms and stigma
• 2 sessions – identify personal profile of symptoms of mania and depression and personal triggers for each mood state
• 2 sessions – identify past coping responses and role of substance abuse and medication adherence and creation of more effective coping strategies for mania and depression (personal action plan)
• 1 session – education re: regulation of circadian rhythms, medication compliance and alternative self-help programmes and psychosocial treatments

and potentially significant other facts about the nature of the illness, its treatments and key coping strategies (Sperry, 1995). Such a model postulates that patients armed with such knowledge can alter their experience of an illness. Psychoeducation has been widely tested across a host of medical disorders, with empirical studies confirming its efficacy in most cases. In the case of BD, PE interventions have followed this approach. Small early studies demonstrated an impact of increased understanding, improved medication compliance and, in some cases, improved outcomes (reviewed in Huxley *et al.*, 2000). A major goal of early psychoeducation was enhanced lithium compliance, as it was hoped that proper lithium use would fully control the disorder. Current psychoeducational models now include skill development in detecting prodromes of depression and mania, stress management, problem-solving, education to diminish the effects of stigma and denial of the illness, information regarding lifestyle regularity, and the creation of personalised coping strategies to prevent relapse of mood episodes. Specific studies on coping strategies are rare; Parikh *et al.* (2007) noted that the coping style may also differ with the type of bipolar disorder (BD type II patients use a narrower range of strategies than do BD type I patients), and this will necessitate tailoring of the intervention. Psychoeducation methods also vary and may include not only didactic presentation of information about BD and its treatment, but also elements of motivational interviewing as well as cognitive and behavioural strategies, further blurring distinctions between models.

Based on these broad constructs of psychoeducation, only two programmes have been developed to the level of a published manual. These programmes, the Barcelona Bipolar Disorders Program (Colom *et al.*, 2006) and the Life Goals Program (Bauer & McBride, 2003) Phase I, are compared in Table 6.1 concerning content and number of sessions. Both interventions are group treatments, but they differ in dosing (number of sessions) and in type of treatment provider. The Barcelona programme was designed and delivered by expert psychologists, whereas the Life Goals programme was explicitly designed to be delivered by mental health nurses using a highly specific manual that is both easy to learn (and train) and easy to adhere to.

Cognitive behavioural therapy

The most developed psychosocial model for bipolar disorder currently is CBT. As empirical support is robust for the CBT model of depression (Clark *et al.*, 1999), a corresponding model for bipolar disorder has been pursued. Curiously, though, all bipolar CBT manuals incorporate an extended section of psychoeducational techniques and interventions, sometimes before broaching any material that is explicitly related to CBT theory. Apart from this PE component of a CBT treatment, a model emerges postulating that bipolar patients possess idiosyncratic beliefs and assumptions that contribute to long-term vulnerability to depressed and manic or hypomanic mood states (Leahy & Beck, 1988; Basco & Rush, 1996; Leahy, 1999; Lam *et al.*, 2000; Newman *et al.*, 2002; Leahy, 2003). The bipolar CBT model further develops on a premise that the physiological predisposition for BD interacts with life events and coping abilities that are moderated by cognitive 'styles' that confer vulnerability. Thus genetic predisposition for bipolar illness may be catalysed by negative or positive events that are filtered through the individual's cognitive 'schemas'. These schemas are reflected in the content and structure of the automatic thoughts and maladaptive assumptions that characterise either the depressed or manic phase (Leahy & Beck, 1988; Lam *et al.*, 2000).

Jan Scott (2001) extended this further, describing Beck's cognitive model of mania as the mirror image of his model of depression, thus reflecting a 'positive cognitive triad', in which the self was seen as lovable and powerful and the future was overly optimistic. Such hyper-positive thinking was buttressed by cognitive distortions that provide biased confirmation of this hyper-positive cognitive triad. Early cognitive formulations of BD thus seem to reflect mood-state-specific schemas – one for depressive mood states, and another for manic mood states. These mood specific underlying beliefs and schemata seem to be represented as polar opposite concepts (e.g. manic self = 'I'm terrific'; depressed self = 'I'm a failure') that prove to be difficult, if not impossible, for the bipolar patient to reconcile when euthymic. As such, CBT treatment for BD may appear also to take on the quality of addressing two sets of mood-specific opposing schemas and underlying beliefs, leaving little attention to periods of normal mood states (and thus suggesting a need for a CBT model of euthymic BD).

One alternative area within CBT that has been explored by Lam and colleagues (2004) is that high levels of trait perfectionism and excessive striving for achievement may confer increased vulnerability to experience mood relapse. Relevant underlying beliefs consistent with this model may include beliefs that one is invulnerable, cannot lose or can take risks without any consequences – which may predispose patients to mania. These ambitious traits may interact with elevations in self-esteem and increased confidence during early prodromes of mania to elicit a full-blown manic or hypomanic episode. In addition, the extant research evidence suggests that many bipolar patients experience negative thoughts and beliefs, and some of these patients continue to think negatively even after their depression remits. Johnson & Tran (2007) review the evidence for both negative and hyperpositive schemata relevant to BD and suggest that this apparent conflict may be reconciled in that it is possible that bipolar patients may be overly influenced by environmental feedback.

Since 1990, several CBT manuals have been developed and tested in randomised controlled trials. Although the manuals may differ in some respects, all begin with a number of 'psychoeducation' sessions and then move on to address the unique CBT focus of restructuring dysfunctional thoughts and beliefs that may otherwise predispose patients to depressive or manic relapse. Whereas the specific dysfunctional beliefs related to depression tend to mimic those of depression (i.e. cognitive triad of negative bias towards self, future and the world), dysfunctional beliefs related to mania may include unrealistic beliefs of 'high goal attainment' (i.e. beliefs that one is invulnerable, can take risks without consequences). Finally, behavioural activation strategies and problem-solving skills for depression (e.g. activity scheduling) as well as time-delay and stimulus control strategies for hypomania/mania are also often discussed.

Table 6.2 Lam's cognitive behavioural therapy programme

Stage 1 (sessions 1–4): introduction to diathesis-stress model, review symptom history, self-monitoring and generate list of goals
Stage 2 (sessions 5–16): introduction to activity-scheduling, thought-monitoring, thought-challenging, challenging dysfunctional assumptions, behavioural experiments, identification of early warning signs and medication adherence (cost–benefit analysis)
Stage 3 (sessions 17–20): review of cognitive behavioural techniques, self-management issues and other issues related to interpersonal relationships and stigma

As the manual by Lam *et al.* (1999) has been tested most clearly in a randomised controlled trial, its session guidelines are presented in Table 6.2.

Family-focused therapy

Family-focused therapy (Miklowitz & Goldstein, 1997) for BD was inspired by successful family-based interventions for schizophrenia, which reduced communication breakdowns and levels of expressed emotion (EE) in the course of preventing relapse (Falloon *et al.*, 1985). Family-focused therapy presumes that managing BD can be optimised through the support and cooperation of family or significant others. Families characterised by high levels of EE have been found to have an increased risk for relapse or recurrence. Subsequently, this encouraged treatment studies that have shown that family therapy can reduce relapse rates in BD patients with high-EE families (Miklowitz *et al.*, 1988; Honig *et al.*, 1997; Butzlaff & Hooley, 1998). Family-focused therapy focuses on communication styles between patients and their families or marital relationships with the goal of improving relationship functioning. Unlike PE or CBT, which are usually undertaken after recovery from an acute episode, FFT can be initiated even during an acute episode. Family-focused therapy consists of a 21-session, 9-month-long treatment programme that uses a module on PE, a module on training in communication and a final module on problem-solving. Details of Milkowitz's family treatment programme for BD are shown in the Table 6.3.

Table 6.3 Family-focused therapy model in 3 modules (21 sessions/9 months)

Psychoeducation module: addresses symptoms, course, treatment and self-management of BD (including creation of relapse drill, which helps identify early warning signs of recurrences and rehearsal of early preventative intervention plan); it also addresses denial as a reaction to the disorder and importance of medication adherence, usually involving 7 or 8 sessions
Communication enhancement module: consists of behavioural rehearsal of effective speaking and listening strategies, including reduction of uncontrolled expressions of negative affect, active listening, giving positive feedback, making positive requests for changes in others' behaviours and giving constructive criticism; 7–10 sessions are suggested
Problem-solving module: involves patients and family members in identifying specific problems related to family life, the resumption of social and occupational roles following an episode, intimacy, boundaries, intrafamilial communication and treatment adherence; they are then taught problem-solving skills that assist them in developing plans to reinforce adaptive, desirable behaviours within the family; 4 or 5 sessions are suggested

Interpersonal and social rhythm therapy

Interpersonal therapy (IPT) was originally developed for the treatment of unipolar depression by Klerman and colleagues (1984). Depression was postulated to be a biomedical disorder with one or several of key psychosocial deficits – unresolved grief, interpersonal deficits, role conflict and difficulties in role transitions. Based on these roots, Frank and colleagues (1997) developed interpersonal and social rhythm therapy (IPSRT), in which patients are encouraged to explore feelings about having BD, are assisted in grieving the loss of their formerly healthy self and learn to accept how the disorder has affected their lives. But IPSRT further encompassed evidence that mood episode recurrences may be precipitated by disruptions in biological or social rhythms. The IPSRT model explored how the timing of daily events and related fluctuations in stimulation may negatively affect the affected person's daily routines or 'social rhythms'. Interpersonal and social rhythm therapy attempts to improve the functioning of the patient by attempting to stabilise interpersonal relationships and reduce interpersonal stress, as well as to stabilise circadian (primarily sleep) rhythms. This is accomplished by providing PE about the illness and the importance of medication adherence, by teaching social skills and interpersonal insight, by improving coping with interpersonal problems and fostering maintenance of routine and structure in daily activities. All patients are provided with a social rhythm metric, which is a daily self-report log to record sleep/wake times, level of social stimulation, timing of daily routines and daily mood. Frank (2005) summarised refinements of IPSRT and has published a manual (see Table 6.4).

Methods for summary of treatment trials

A systematic review of the literature published on psychosocial interventions in BD was conducted from 2000 until the time of writing this book (July 2008). Randomised controlled trials were emphasised. The search was conducted primarily through the use of electronic databases such as MEDLINE, PubMed and PsycINFO, as well as through observational review of psychiatric literature and references. Only publications with interventions for adults were chosen. This review considers the evidence for four empirically supported psychosocial interventions, including PE, FFT, IPSRT and CBT. Although other types of possible intervention were investigated, no other empirically supported therapeutic models were found.

The keywords employed for the electronic database search included: MeSH terms of Bipolar Disorder, Psychotherapy, Cognitive Therapy, Family Therapy, Psychotherapy, Group; also included were keyword searches using manic-depression, psychosocial, psychotherapy, psychoeducation, family-focused therapy, interpersonal therapy, interpersonal and social rhythm therapy, CBT, FFT, PE and IPSRT.

Note that psychoeducation delivered as a part of multi-tiered therapeutic intervention is discussed as part of our review of PE, although it is difficult to explicitly identify its unique contribution. Furthermore, PE trials are discussed together, regardless of whether they were group or individual interventions. It should also be noted that all psychosocial interventions were given as adjuncts to pharmacological treatment of BD.

Outcomes regarding the efficacy of psychosocial interventions have also been examined according to specific therapeutic targets as well as the different phases of bipolar illness. When statistical analyses were lacking, the studies were still mentioned in our evidence table and indicated with an asterisk, but they were not taken into account in our critical review.

Review of psychoeducation studies

Since 2000 there have been a host of publications examining psychoeducation for BD, but these include multiple articles flowing from a relatively small number

Table 6.4 IPSRT model with 4 phases – 24 sessions/36 weeks

Initial – explores how disruptions in social routines and interpersonal problems are associated with mood episodes, using instruments such as social rhythm metric and an interpersonal inventory; basic PE is provided, and a focus for IPT therapy is chosen, all over 3–5 sessions
Intermediate – regularise social rhythms through emphasis on schedules and intervene in chosen IPT therapy problem areas, over 5–10 sessions
Continuation or maintenance – enhance patient confidence in abilities learned earlier, and begin tapering session frequency, over 5–10 sessions
Termination – prepare patient for termination by reviewing and emphasising skills, reducing session frequency, over 3–5 sessions

of studies. Studies could be classified as one of two types: those in which the PE was a principal intervention (three published studies) and those in which the PE was offered as part of an integrated disease-management programme (three studies). Finally, we comment on our own published and in-press work in this area.

The pivotal study of Colom *et al.* (2003a) examined the effects of 21 sessions of group PE versus 21 non-structured group meetings in 120 fully remitted BD type I and type II patients. Follow-up was monthly for 2 years. Targets measured were number of bipolar recurrences, time to recurrence and hospitalisations. Those who participated in the PE group had fewer total recurrences – manic or hypomanic recurrences – fewer mixed episodes and fewer depressive recurrences compared with the control group. Time to recurrence was longer for the PE group, and they also evidenced fewer hospitalisations at 12 and 18 months and overall fewer days in hospital.

From the original landmark study, Colom conducted a key subgroup analysis to examine whether personality disorders (PDs) affected the impact of PE (Colom *et al.*, 2004). Out of the original sample of 120 patients, 37 had co-morbid PDs; through the original randomisation process, 15 had been assigned to PE and 22 to control. Axis II disorders were assessed using the structural clinical interview for DSM-IV-TR (SCID-II), while outcomes were recurrences and number of hospitalisations over a 2-year follow-up. Results show that patients receiving PE had significantly longer times to relapse over the 24-month period compared with controls and spent far fewer days in hospital. The strongest effects were found in reductions in depressive relapses at 12–24 months. Manic relapses were significantly reduced only at the 18-month and 24-month follow-up visits. This subanalysis suggests that PE may be beneficial in reducing relapse and hospitalisations, even for patients who may be more difficult to treat because of personality disorders.

Colom *et al.* (2005) reported another subanalysis of their landmark study, looking at the impact of PE on blood lithium levels. Out of the original sample of 120, 93 were on lithium. Of these 93 subjects, 49 had been randomised to PE, and 44 to the control group. Lithium was measured every 6 months. Both groups had lithium levels in the therapeutic range throughout the study, but the PE group had slightly higher lithium levels and fewer fluctuations at three of the four measurement points (and by inference from the original study, better clinical outcomes, although this was not specified in the paper). Although it is impossible to ascertain what portion of the benefit of PE is a result of higher lithium levels, the fact that both groups had therapeutic lithium levels suggests that small differences in lithium levels were unlikely to be the main cause of improved outcome in PE.

Colom *et al.* also conducted a study with identical design to the pivotal 2003 study, but looking specifically at patients with high medication compliance (Colom *et al.*, 2003b). In this study (n=50), the benefits of the same 21 sessions of group PE for 25 fully remitted and treatment-compliant BD type I patients were compared with 25 BD type I patients who received standard care only, with a control condition of an equal number of patient meetings in a group. Follow-up was monthly for up to 2 years. Outcomes were number of bipolar episodes, hospitalisations and time to relapse. Those receiving PE had fewer recurrences of mood episodes during both the treatment and 2-year follow-up phase, fewer depressions and longer times to relapse in spite of equal pharmacotherapy treatment and blood lithium levels.

Suppes *et al.* (2003) provided PE for both inpatients and outpatients as part of the Texas Medication Algorithm Project (TMAP) for bipolar disorder. This project enrolled 459 bipolar patients across multiple sites for a complex intervention that included opportunities for patient education but focused primarily on a provider intervention that included systematic monitoring of symptoms by care coordinators (who also were the primary sources of patient education) and medication advice to the physicians. The targets were symptom severity and social functioning. The PE included providing handouts and opportunities for individual encounters and group sessions that were offered 'as desired' to the participants by a trained clinical care coordinator. Out of the 141 bipolar patients in the intervention group, 96% got at least one session, with a median exposure of 124 min of education; less than 25% participated in any group PE (Toprac *et al.*, 2006). This exposure to PE was thus brief and not a specific uniform intervention. Overall results of TMAP showed benefit in reduction of manic and psychotic symptoms, but not depression. It is doubtful if such a low dose of education would have had much impact on clinical outcome, but it is useful to see how little uptake there was when the education intervention was left to patient choice. Such results suggest that the provision of a specific PE course or group as a necessary part of treatment may be important to ensure delivery of the intervention.

Simon *et al.* (2006) provided structured group PE (Life Goals Phase 1 & 2) (Bauer & McBride, 2003) in conjunction with a 'systematic care management program', which included assessment and care planning, monthly telephone (symptom and medication) monitoring and feedback to the treatment team, all done by trained nurse case managers. Using 441 adult subjects drawn from a group pre-paid health plan, 212 were randomised to the intervention, and 229 were randomised to usual care (control). Follow-up assessments were quarterly using longitudinal interval follow-up evaluation (LIFE), which provides separate weekly ratings for manic symptoms and depressive symptoms, across a 6-point scale from 0 through subsyndromal to severe syndromal symptoms. This instrument, by capturing all levels of symptoms, provides more data on mood states than simple episode counting, and is thus a better measure of outcome. The intervention group had significantly lower mania scores through 2 years of follow-up. No differences were found in depressive symptoms. As is typical with an effectiveness study, patients were offered the PE intervention, but it was not mandatory; 59% of patients completed the five sessions of group PE that constituted Phase 1 of the Life Goals Program, with about 50% going on to receive more sessions (Phase 2). Of course, the specific benefits of PE for reduction in manic symptoms cannot be disentangled from the impact of the multifocal interventions, but it is reasonable to surmise that PE was an important component.

Similarly, Bauer *et al.* (2006a, 2006b) conducted a trial with a 'systematic care management program' using the same Life Goals Phase 1 and 2 manualised group PE programme versus usual care. A sample of 306 individuals (91% male) recruited from US veterans' hospitals were randomised to receive either usual care (control, n=149) or a complex intervention (n=157) incorporating care coordination by a nurse, group PE and treatment guideline support to treating physicians. Primary outcomes included LIFE ratings to assess targets of depressive and manic symptom severity and a modified social adjustment scale over a 3-year follow-up period. No differences were found between intervention and control groups in the LIFE ratings, but there were a number of modestly positive secondary outcomes. Here, as the PE intervention (five group sessions as Phase 1) was received by more than 78% of the experimental sample, it is reasonable to surmise that PE contributed to the positive outcomes.

Summary of psychoeducation studies

Overall, there have been many good-quality randomised controlled studies investigating PE. Several studies have employed long-term multi-year follow-up periods. The majority of the evidence points to the benefits of PE compared with treatment as usual in reducing recurrences and relapses, and increasing time to recurrence. For those studies that measured subsyndromal symptoms longitudinally (e.g. Simon *et al.*, 2006; Bauer *et al.*, 2006a, 2006b), differences between treatment and control groups were found in manic symptoms but were less compelling when it came to improving depressive symptoms. Many of the studies cited in this review employed PE programmes that lasted 9 months or longer and probably represent the most elaborate PE programmes currently available. Thus, it is somewhat disappointing that despite the comprehensiveness and duration of group PE treatment, better results were not found in several studies, particularly as it pertained to improving depressive symptoms and psychosocial functioning. Although Colom and colleagues generally reported excellent outcomes with respect to depressive recurrences (Colom *et al.*, 2003a), time spent in full episode is not likely to be the best indicator of improvement, given that most bipolar patients suffer with prominent subclinical symptoms that would not be included as a recurrence per se.

Review of cognitive behavioural therapy trials

Since 2000, one important pilot study and three major published randomised controlled trials of CBT (one of which generated separate outcome papers at 12 and 30 months) have been published. Studies involving CBT as one arm of multiple psychosocial interventions are considered separately, later in this chapter. Here, 'CBT' is used, although some prefer to use 'CT' to signify cognitive therapy; in fact, all the bipolar CBT or CT manuals include behavioural techniques, and so 'CBT' and 'CT' are interchangeable.

Scott and colleagues conducted a pilot study of cognitive therapy comparing 21 bipolar outpatients who received up to 25 individual sessions of CBT over a 6-month period to 21 waiting-list controls (Scott *et al.*, 2001). Outcomes targeted were relapse rates, depressive and manic symptoms, psychosocial functioning and general psychopathology. At 6 months, the CBT intervention subjects showed improvements compared with controls. At 18 months after baseline (i.e. 12 months after terminating treatment), those who received CBT showed a 60% reduction in relapse rates and fewer hospitalisations compared with their own reported relapses and hospitalisations before commencing treatment. This pilot study provided further support for conducting a full randomised controlled trial.

Lam *et al.* (2003) randomised 103 patients with BD type I to either individual CBT (n=51) or 'treatment as usual' (n=52). The CBT intervention consisted of 18 sessions of CBT over 6 months followed by two booster sessions over the following 6 months. Primary outcomes were number of episodes, number of days in episodes, social functioning, coping strategies and dysfunctional attitudes. Results of this study showed important effects at the end of year one that diminished subsequently. During the 12-month initial study period (which includes the treatment time), the CBT group had fewer bipolar episodes, days in episodes and admissions. The intervention group also had higer social functioning and better coping skills at the 12-month mark.

Lam *et al.* (2005a, 2005b) reported 30-month data (reported as a 6-month intervention period and a 24-month follow-up period) of the original 2003 study. Overall, the CBT intervention group still had better outcomes in terms of time to relapse, but the effect over the 30 months was almost entirely due to the benefits conferred over the first year; from months 12 to 30, there were almost no differences between CBT and controls. Furthermore, the improvements in social functioning and coping skills at month 12 were not seen at later assessments. Given the initial effects of reducing relapse and time in episode, CBT was recommended, with a suggestion that ongoing maintenance CBT might be effective in maintaining the benefits seen in year one.

Given the early promise of CBT, Scott and colleagues (2006) sought to extend the treatment to a broader cohort of bipolar patients in an 'effectiveness' study. They recruited 253 subjects who were almost twice as likely to have serious co-morbidity or higher bipolar severity than the seven major preceding psychosocial studies in BD. In their randomised controlled trial, 126 participants were assigned to treatment as usual (TAU), while 127 were assigned to receive 20 sessions of individual CBT. Over one-quarter of the intervention group attended fewer than 13 sessions, and 40% did not achieve the stated goals of CBT, reflecting the challenge of treatment compliance in a more ill sample. Primary outcomes included time to recurrence of an episode, with an additional outcome a comparison of the weekly mood symptom ratings from the LIFE instrument. Results of the 18-month study showed no differences in recurrences or average mood ratings between CBT and TAU, except for a small subsample with fewer than 12 previous episodes who benefited from CBT. The authors concluded that their findings did not support the use of CBT to prevent recurrence, except possibly in those with fewer episodes.

Ball *et al.* (2006) conducted a randomised controlled trial comparing 20 weekly sessions of schema therapy (CBT with emotive techniques; n=25) versus a control group who received usual treatment (n=27). Outcomes included relapse rates, dysfunctional attitudes, psychosocial functioning, hopelessness, self-control and medication adherence. At the end of the 6-month treatment period, the CBT group had modestly lower depressive symptoms. All subjects were assessed every 3 months for an additional year, with the CBT group experiencing non-significant trends in a greater time to relapse, lower mania scores (Young Mania Rating Scale) and improved behavioural self-control. Given the relatively small number of participants as well as its modest findings, the study suggests the value of CBT as an acute intervention but echoed the Lam study in showing the loss of impact of CBT after acute treatment ceased. These results are in contrast to the results of CBT trials for unipolar depression and anxiety disorders, which have shown long-term and enduring effects from CBT.

Summary of cognitive behavioural therapy trials

Although the results of initial studies in CBT for BD were generally positive, later results have often been modest; some did not show maintenance of gains over longer follow-up intervals (e.g. Lam *et al.*, 2005a, 2005b), while others did not show any gains over control groups (e.g. Scott *et al.*, 2006). In addition, there have not always been associated changes in the central mechanism of action underlying this therapeutic strategy – namely changes in underlying dysfunction beliefs (Zaretsky *et al.*, 2008). Given that the central and distinguishing focus of CBT is its emphasis on ameliorating dysfunctional thoughts, the very modestly positive clinical outcomes and modest effects on dysfunctional attitudes suggest that CBT is of limited value to the average bipolar patient.

Review of family-focused therapy trials

There have been six randomised controlled trials in FFT since 2000. Miklowitz *et al.* (2000) compared the relative efficacy of FFT for 31 BD patients and two family education sessions and follow-up crisis management as a control group (*n*=70). Both treatments were delivered over 9 months. Follow-up was for 1 year, conducted every 3 months. Targets were relapse rates, symptom severity and medication compliance. Patients who received FFT had fewer relapses over the course of the 1-year follow-up, and showed greater improvements in depressive symptoms compared with the crisis management group. No differences in manic symptoms were found. The greatest improvements were evidenced in the patients whose families were high in EE.

Miklowitz *et al.* (2003a) is a 2-year continuation of follow-up from the previous (2000) study of 101 BD patients, 31 of whom received FFT whereas 70 received crisis management (CM). The FFT patients continued to have fewer relapses and longer survival intervals over 2 years of follow-up than those who received crisis management. Using the schedule for affective disorders and schizophrenia (SADS-C) total affective symptom scores (composite score for depression and mania), these authors found that FFT participants had lower total affective symptoms over the 24 months compared with the CM group (this was largely due to the depressive symptoms, not the manic symptoms, which were not significantly different).

Miklowitz *et al.* (2003b) used the same control group (CM=70) as the above study, but the treatment group this time was a combined treatment using family PE and some elements of individual IPSRT (integrated family and individual therapy (IFIT)) for up to 50 sessions in total. The average number of sessions received was 29.4. Follow-up was 1 year (i.e. during the active phase of treatment). Targets were time to relapse and symptom severity. Patients in IFIT evidenced longer time to relapse and greater reductions in depressive symptoms, but not mania symptoms, compared with the CM group. Note: this was not a randomised controlled trial but an open treatment trial compared with a matched historical comparison group.

Rea *et al.* (2003) evaluated the effect of 9 months of FFT for 28 recently hospitalised bipolar, manic patients versus an individually focused patient treatment (*n*=25). Follow-up was 1 year at 3-month intervals. Outcomes evaluated were hospitalisations and relapses. The FFT patients were less likely to be re-hospitalised, and had fewer relapses over 2 years compared with those in individual treatment. There was no difference in time to first relapse, however, between the two treatment groups.

Miklowitz *et al.* (2004) examined the relative effects of 21 sessions of FFT for adolescents who had had an exacerbation of bipolar mood symptoms within 3 months before study entry. This was a 1-year open treatment trial involving 20 bipolar adolescents. This pilot study found that FFT for adolescents was associated with improvements in depressed, manic symptoms and behavioural problems over a 1-year period; a randomised trial of this approach is currently under way.

Miller *et al.* (2004) examined the relative benefits of psychosocial treatment for 92 patients meeting criteria for a current BD type I mood episode who were randomly assigned to receive either 12 sessions of family therapy, 6 sessions of multi-family PE group therapy, or control (just pharmacotherapy). Target was time to recovery only. Follow-up was up to 28 months. Results suggest that neither adjunctive family therapy nor multi-family group PE improved recovery from BD type I mood episodes compared with pharmacotherapy alone.

Summary of family-focused therapy trials

Several randomised controlled trials employing FFT demonstrate that FFT improves relapse rates and time to relapse as well as overall mood symptoms. There is less evidence to support the idea that FFT improves symptoms of mania or hypomania over TAU. Unlike other psychosocial interventions, which are largely geared to being delivered to patients during times of relative mood stability for the purposes of reducing relapse or recurrence, FFT is well suited to be delivered to the family and patient during a mood episode to promote a more effective recovery. In fact, the value of FFT as well as other interventions in acute bipolar depression is reviewed later in this chapter. One drawback of the FFT studies is the relatively small number of participants who have actually received the treatment, given the small size of the studies and the use of larger control groups. Although it is encouraging to witness the benefits of FFT in coping with depressive symptoms and recurrences, FFT treatment may be more limited when it comes to preventing hypomanic or manic symptoms and relapse in general. As with other individual intervention research, there is limited information on persistense of gains at follow-up as well as applicability of the treatment to patients who have little or no contact with their families.

Review of interpersonal and social rhythm therapy trials

Before 2000, there were only two reports of IPSRT in the treatment of bipolar disorder, both of which are from Frank *et al.* (1997, 1999). One of these two reported no differences in symptoms, or time to relapse between patients receiving IPSRT versus clinical management. Since 2000, there have been three publications all stemming from the same study – a complex one also from Frank and colleagues (2005).

In this landmark study, 175 BD type I subjects participated in a two-stage (acute and maintenance phases) study, with randomisation at the start of the study to one of four possible treatment conditions. This was a full factorial 2×2 design with phase (acute vs maintenance) and treatment (IPSRT vs intensive clinical management (ICM)). In the acute phase, subjects could receive either acute IPSRT or acute ICM. If they responded well enough, they then proceeded to the maintenance stage, where once again they would receive either IPSRT or ICM, as determined at the original study entry. Thus one could receive either one treatment acutely and then the other treatment in the maintenance phase, or stay in the same treatment for both stages. The group receiving acute and maintenance intervention (IPSRT/IPSRT) included 39 subjects, of whom 27 remitted acutely and proceeded to the maintenance. The IPSRT/ICM group had 48 subjects initially, of whom 34 remitted with acute IPSRT and proceeded to maintenance treatment with ICM. The ICM/ICM group started with 43 subjects, of whom 32 remitted and proceeded to continue with ICM in the maintenance phase. Finally, 45 subjects were randomised to ICM/IPSRT, of whom 32 remitted with acute ICM and proceeded to maintenance IPSRT. Prinicipal outcomes were time to stabilisation in the acute phase and time to recurrence in the maintenance phase.

Initial results from the study (Frank *et al.*, 1999, 2000) pointed to the importance of continuity of treatment; individuals who were relapsing were more likely to have switched treatment modalities from the acute to the maintenance phase. Somewhat surprisingly, the first primary outcome of time to stabilisation was no different for acute ICM versus acute IPSRT. Similarly, individuals receiving maintenance ICM did not differ in terms of time to recurrence compared with those receiving maintenance IPSRT. However, individuals treated with IPSRT in the acute treatment phase experienced longer survival time without a new affective episode and were more likely to remain well for the full 2 years of the preventative maintenance phase (regardless of the treatment received in the preventative maintenance phase). Similarly, recipients of acute phase IPSRT had greater regularity of social rhythms at the end of the acute phase, and the ability to increase regularity of social rhythms during acute treatment was associated with reduced likelihood of recurrence, suggesting a mechanism of action that was consistent with the theory of IPSRT.

An additional outcome on suicidality was reported on this sample by Rucci *et al.* (2002). All 175 subjects were interviewed with rigorous life-charting techniques to establish data on previous suicide attempts. During the acute treatment phase, 53% were being treated for depression, 23% for mania and 25% for mixed episodes. Suicide attempts were tracked during their participation during both the acute and maintenance phases of the study, and were reported in terms of a rate per 100 person-months (five occurred during the study). Compared with pre-treatment, the rate of suicide was reduced by threefold in the acute phase, and nearly 18-fold in the maintenance phase. These results emphasised the value of intensive attention – whether ICM or IPSRT – in reducing suicidality.

Summary of interpersonal and social rhythm therapy

A single, well-designed and well-executed randomised controlled trial has shown acute IPSRT to be helpful in reducing relapse during the maintenance phase of BD. Importantly, it did not show any advantage over acute ICM during an acute episode of either depression, mania or mixed episode. One measure linked to the mechanism of action, namely the increased regularity of social rhythms, is a particularly promising finding. However, these results need to be replicated, particularly in a setting other than the superior academic centre that created the therapy.

Summary of comparative studies

Studies reviewed so far have compared a single major treatment modality to a control condition. Three key studies have now examined the comparative benefits of different psychosocial interventions.

In the largest study of psychosocial intervention in the acute phase of illness, Miklowitz and colleagues (2007) involved multiple sites from the Sytematic Treatment Enhancement Program for Bipolar Disorder (STEP-BD) (Sachs *et al.*, 2003). This programme enrolled over 4361 bipolar subjects in a variety of studies, some open and some randomised controlled trials. From this large sample, 15 sites contributed 293 participants to a study of acute psychosocial intervention. Adding to the complexity was that 236 of these individuals were also enrolled simultaneously in a randomised controlled trial of mood stabiliser plus either antidepressant or placebo for bipolar depression. The remaining 57 subjects (293 minus 236) were not eligible for the antidepressant study but were treated openly with pharmacotherapy according to common treatment guidelines.

Psychosocial treatment options included IPSRT, FFT and CBT. However, assignment of the psychosocial treatment was further complicated by the fact that some members did not have family members available (*n*=134 – almost half of the 293 subjects). Furthermore, for practical reasons, few sites could offer treatment in all three modalities. Therefore, the study was designed so that at any one site, only two intensive psychosocial treatments would be offered. Thus, 10 sites offered CBT, 9 provided FFT and 11 offered IPSRT. All 15 sites offered the control condition, which was three individual sessions of 'collaborative care' (CC). Collaborative care was described as containing elements of providing 'a brief version of the most common psychosocial strategies shown to offer benefit for bipolar disorder' (Miklowitz *et al.*, 2007). Thus, it is worth noting that CC is not a specific brief intervention for bipolar depression, but a more general brief psychosocial manoeuvre. As such, it would be reasonable to call the CC control condition a type of enhanced TAU rather than a depression-specific intervention. The CC was contrasted to up to 30 sessions of the intensive psychosocial interventions given over a 9-month period, with a total study period of 12 months (including the 9-month treatment time). Primary outcomes were time to recovery and the proportion of patients classified as well during each of the 12 study months. Major findings of the study were that all three intensive interventions, versus the CC control, were able to achieve higher recovery rates by year end and also shorten time to recovery. No differences were seen between the three intensive treatments. The authors concluded that intensive psychosocial intervention was more beneficial than brief intervention for bipolar depression, but acknowledged that cost-effectiveness merited further study.

Finally, our group conducted two randomised controlled trials that involved PE in comparison with CBT in the maintenance phase. In our first study (Zaretsky *et al.*, 2008), 79 subjects were randomised to receive individual interventions, either PE (*n*=39) or PE plus additional CBT (*n*=40). The same manual (Basco & Rush, 1996) was used to deliver both the PE (the first seven PE sessions) and the PE plus CBT treatment (the same first seven PE sessions plus 13 additional CBT sessions). There were no differences in the primary outcome of relapses over a 1-year period, but on the measure of number of days of depressed mood (of any severity, not necessarily in a full depressive episode), there was an advantage to CBT.

Our second comparative study (Parikh *et al.* – currently under review) was also a study in the maintenance phase. However, it was designed more explicitly as an effectiveness study, with broad entry criteria, multiple sites and sensitivity to cost and efficiency of treatment delivery. As we have seen, various psychosocial interventions have demonstrated efficacy, but differ in intensity, cost and ease of dissemination. Such differences may provide more direction for public health recommendations of which treatments to recommend. A total of 204 participants (ages 18–64) with either type I or type II BD participated from four Canadian academic centres. Participants were assigned to receive either 20 individual sessions of CBT (*n*=95) or six sessions of group PE (*n*=109).

Primary outcome of symptom course and morbidity was assessed prospectively over 78 weeks using LIFE, which yields depression and mania symptom burden scores for each week. Additional outcomes included time to relapse. Both treatments had similar outcomes with respect to reduction of symptom burden and the likelihood of relapse. Approximately 8%

of subjects dropped out before receiving PE, while 64% were treatment completers; rates were similar for CBT (6 and 66%, respectively). Psychoeducation cost $160 per subject compared with CBT, at $1200 per subject. Despite longer treatment duration and individualised treatment, CBT did not show a significantly greater clinical benefit compared with group PE. Similar efficacy but lower cost and potential ease of dissemination suggest that the group PE intervention might be preferable from a public health perspective as an adjunct to medication for BD.

Conclusions and future directions

We began this chapter by highlighting the lethality and complexity of BD, and by acknowledging the limitations of pharmacotherapy. Such a compelling need for better treatments has led to excellent scientific investigation of psychosocial interventions as adjuncts to medication. Significant progress has been made in identifying a number of key psychosocial elements that are important. This chapter reviewed four theoretical models of psychosocial treatments for BD: namely, PE, CBT, IPSRT and FFT (see Table 6.5). Although there are different targets of change within these models, all address multiple concerns presumed to either cause or exacerbate bipolar symptoms and poor functioning. These shared targets include: disruptions in circadian rhythms; early detection of prodromal mood symptoms; medication adherence; communication styles and interpersonal stress; coping with stigma; reducing EE; managing stress; enhancing problem-solving; correcting maladaptive thoughts and beliefs; education about BD and risks; and addressing substance use.

However, we must also acknowledge that there are major problems with psychosocial interventions in BD. These begin with the relative paucity of information on mechanisms of action, and the remarkable overlap of manoeuvres and techniques among the psychosocial interventions. It is clear that PE is a major component of every intervention, and as our recent two comparative studies have shown, it appears that CBT adds little to an effective PE treatment. The lack of demonstration of enduring effects of even intensive psychosocial intervention is also troubling. Overall, these modest and sometimes disappointing outcomes from psychosocial clinical trials in BD invite reflection on what are reasonable expectations for psychosocial interventions. Perhaps the complexities of BD are too great for *any one* psychosocial theory or intervention to accommodate. As we lack robust psychological or psychosocial models in BD, particularly ones that can suffice in capturing mood oscillations as well as extremes of depression and mania, future research must focus efforts on expanding current models and identify additional mechanisms of action. In addition, it is important that our models be capable of accommodating new putative mechanisms of action and newly discovered moderating or mediating influences on bipolar outcomes within their theoretical structures. Both PE and CBT appear to be open models capable of adaptation. For example, PE has expanded its scope of treatment considerably to target a much wider range of outcomes compared with earlier goals of treatment.

Similarly, CBT has been able to increase its range of both cognitive and behavioural targets to cope with both bipolar manic and depressive relapse as well as new schemas or underlying beliefs. In order for future research in CBT to be fruitful, efforts must continue to identify bipolar-related schemata that may not be fully understood or articulated at this time. Some current research points to perfectionistic or high-achievement needs as a potential destabilising personality factor in BD. However, it is unclear to what extent most bipolar individuals have such unusually high needs for achievement. Other possible areas to explore in future CBT research might be how meta-cognitive beliefs about one's emotions and/or (mis)perceived emotional lability may undermine a stable and healthy schema or self-concept. Conversely, in the language of CBT, a fragile or 'damaged' core belief, may, in turn, result in hopelessness and an internalised stigma of being chronically ill. Beyond such psychological models must then come the mapping of such contructs onto a neuroscience-based framework; where do these constructs lie neuroanatomically, how are they biologically mediated, and how would biological as well as psychological manoeuvres serve to alter symptom expression and ultimately patient functioning?

Outside of these four main psychosocial models discussed, it should be noted that there are other psychosocial treatments that have not been explored or modified to address BD. Emotion-focused treatment, mindfulness-based treatment and narrative therapy are all examples of treatments that have shown some promise in other populations but have not yet been rigorously explored in the context of BD.

Finally, although it has begun to be mentioned, the issue of cost and ease of dissemination of intensive psychosocial treatments will play a major role in

Table 6.5 Key psychosocial clinical trials in bipolar disorder, 2000–2008

PE studies (including chronic disease management intervention trials utilising PE)							
Year	**Authors**	**Study treatments**	**Study design**	**Follow-up**	**Outcomes**	**Participants**	**Key results/implications**
2003	Suppes *et al.*	PE – various (part of patient care in Texas Medication Algorithm Project)	RCT – complex intervention involving treatment guidelines, extensive physician education and supervision and opportunities for patient education (voluntary)	1 year	Symptom measures of mania, depression and psychosis	459 bipolar patients across multiple sites; sites, not patients were randomised; 141 participants received intervention; 318 controls	Patients were invited to participate in individual and/or group education but mostly received a single session and some written materials; overall trial results showed benefits in reducing manic symptoms, not depression
2003b	Colom *et al.*	Group PE (Barcelona manual)	RCT but *only* with subjects with high medication compliance	2 years	Recurrences; time to recurrence	25 PE; 25 controls	PE had fewer overall recurrences, longer time to recurrence
2003a	Colom *et al.*	Group PE (Barcelona manual)	RCT with control group receiving non-structured group therapy	2 years	Recurrences; time to recurrence	60 PE; 60 controls	PE had fewer overall recurrences, both depressive and manic; longer time to recurrence; fewer hospitalisations
2004	Colom *et al.*	Group PE	Subanalysis of 2003 study looking at impact of PE on those with co-morbid personality disorders	2 years	Recurrences; time to recurrence	15 PE; 22 controls	PE had fewer overall recurrences; longer time to recurrence; all patients had co-morbid personality disorders; and 100% in control group relapsed
2005	Colom *et al.*	Group PE	Subanalysis of 2003 study looking at effect of PE on plasma lithium levels	2 years	Recurrences; time to recurrence ; lithium levels obtained at 0, 6, 12, 18, 24 months	49 PE; 44 controls	PE had slightly higher lithium levels and fewer recurrences; as controls still had therapeutic levels, PE superiority not just due to higher lithium
2006	Simon *et al.*	Group PE (Life Goals)	RCT; multifaceted chronic disease management intervention, including PE group treatment, medication monitoring, feedback to MD, etc.	2 years	Symptoms as measured by LIFE mania and depression ratings	212 intervention; 229 TAU control group	Intervention had lower mean mania ratings over 12 months and one-third less time spent in hypomania/mania, but no impact on depression; complex intervention moderately helpful
2006a	Bauer *et al.*	Group PE (Life Goals)	RCT; multifaceted chronic disease management intervention, including PE group treatment, treatment guidelines, etc.	3 years	Symptoms (LIFE), recurrences, no. of weeks in episode, quality of life, functioning, treatment satisfaction	166 intervention; 164 TAU control group; participants 72% male, and military veterans, so more challenging population	PE showed fewer weeks in affective episode, improved role functioning, quality of life, treatment satisfaction; no difference in mean manic and depressive symptoms; complex intervention modestly helpful

Table 6.5 *(cont.)*

PE studies (including chronic disease management intervention trials utilising PE)							
Year	**Authors**	**Study treatments**	**Study design**	**Follow-up**	**Outcomes**	**Participants**	**Key results/implications**
Cognitive behavioural therapy (CBT)							
2001	Scott *et al.*	CBT (up to 25 sessions over 6 months)	Pilot study comparing CBT with waitlist control group	18 months	Recurrences, symptoms, functioning, hospitalisations	21 CBT, 21 waitlist controls receiving treatment as usual	Modest improvement at 6 months CBT vs controls; at 18 months, fewer recurrences and hospitalisations in a pre-post comparison in CBT participants
2003	Lam *et al.*	CBT (Lam manual)	RCT; individual CBT (18 sessions in first 6 months plus 2 boosters in next 6 months) versus treatment as usual	1 year	Recurrences, no. of days in episode, functioning, coping, attitudes	51 CBT, 52 treatment as usual controls	CBT had significantly fewer bipolar episodes, days in bipolar episode, number of admissions, higher psychosocial functioning, and fewer symptoms; coped better with manic prodromes
2005a	Lam *et al.*	CBT (Lam manual)	Extended follow-up of 2003 study	30 months	Recurrences, no. of days in episode, functioning, coping, attitudes	51 CBT, 52 treatment as usual controls	CBT still superior in time to relapse, but due to effect in first year; no final differences in functioning or coping skills; suggests need for CBT booster sessions beyond year one; first-year benefits may justify cost of CBT intervention
2006	Ball *et al.*	CBT (Schema therapy – CBT with emotive techniques)	RCT; individual CBT (20 weekly sessions) vs treatment as usual	12 months/6 months follow-up	Recurrences, dysfunctional attitudes, functioning, hopelessness, self-control, compliance	25 CBT, 27 treatment as usual controls	CBT no difference in recurrence, but showed fewer depressive symptoms, fewer dysfunctional attitudes and increased self-control at 6 months
2006	Scott *et al.*	CBT (derived from Beck model)	RCT; individual CBT (20 regular sessions over 6 months, plus 2 boosters) vs treatment as usual	18 months	Time to recurrence, weekly symptom ratings (LIFE), costs	127 CBT, 126 treatment as usual; effectiveness study with broader inclusions, including patients in acute episodes	CBT no differences in any outcomes, except for small subgroup with fewer lifetime episodes who did benefit from CBT; more ill group than other studies, with fewer able to complete CBT; nonetheless, probably shows true CBT performance in real world
Family-focused therapy (FFT) trials							
2000	Miklowitz *et al.*	FFT vs crisis management (CM)	RCT; both treatments occurred over 9 months (FFT = 21 sessions; CM = 2 family education sessions plus prn crisis sessions)	1 year	Recurrences, time to recurrence, symptom severity and medication compliance	31 FFT, 70 CM	FFT had fewer relapses and longer delays before relapses during the study year and showed greater improvements in depressive but not manic symptoms

2003a	Miklowitz *et al.*	FFT vs CM	RCT; further results from study with 1-year data published in 2000	2 years	Recurrences, time to recurrence, symptom severity and medication compliance	31 FFT, 70 CM	FFT had fewer relapses and longer delays before relapses during 2-year period and showed greater improvements in depressive but not manic symptoms; improved medication compliance in FFT
2003b	Miklowitz *et al.*	Integrated family and individual therapy (IFIT) vs CM	Cohort study, using open treatment with IFIT (up to 50 sessions) compared with previous study's CM group	1 year	Recurrences, time to recurrence	30 IFIT, 70 CM	IFIT had longer survival intervals; greater reductions in depressive symptoms
2003	Rea *et al.*	FFT vs Individual treatment, both given over 9 months	RCT; FFT = 21 sessions of 1 h; individual therapy involved 21 sessions of 30 min, essentially psychoeducational	2 years	Recurrences, time to recurrence	28 FFT, 25 individual therapy	FFT subjects less likely to to be rehospitalised, experienced fewer mood episodes; FFT did not differ from individual therapy in likelihood for first relapse
2004	Miller *et al.*	Multifamily psychoeducational group therapy (MPE) or family therapy (FT) vs pharmacotherapy (PT)	RCT; MPE = 6 sessions, FT = 12 sessions, control was PT treatment as usual	Up to 28 months	Time to recovery (survival analysis)	30 MPE, 33 FT, 29 PT	No differences between interventions in time to recovery
IPSRT studies							
1999 and 2000	Frank *et al.*	IPSRT vs intensive clinical management (ICM) with 2 parts, acute and maintenance	RCT; IPSRT approximately 20 sessions, ICM essentially treatment as usual at a very thorough academic medical centre	Initial report on acute phase results over 1 year	Time to stabilisation from acute mood episode	41 IPSRT, 41 ICM	No difference between groups in time to stabilisation; IPSRT group had significantly more regular routines; after acute phase, participants randomised again to maintenance IPSRT or ICM; changing modalities was associated with relapse
2005	Frank *et al.*	IPSRT vs ICM	RCT; 2×2 design with initial randomisation to acute IPSRT or ICM, then maintenance phase randomisation to IPSRT or ICM	2 years	Time to recurrence in maintenance phase	175 participants in 4 arms: 39 IPSRT acute/27 IPSRT maintenance (main); 48 IPSRT acute/34 ICM main; 43 ICM acute/32 ICM main; 45 ICM acute/32 IPSRT main	IPSRT patients in acute phase survived longer without relapse regardless of type of maintenance phase treatment

Table 6.5 *(cont.)*

PE studies (including chronic disease management intervention trials utilising PE)							
Year	**Authors**	**Study treatments**	**Study design**	**Follow-up**	**Outcomes**	**Participants**	**Key results/implications**
Comparative studies							
2007	Miklowitz *et al.*	Intensive psychotherapy (STEP-BD) vs collaborative care (CC)	RCT of BD; intensive psychotherapy included either 30 sessions over 9 months of IPSRT, CBT or FFT; CC involved 3 sessions	9 months	Function outcomes included LIFE-RIFT relationship, work, satisfaction with activities, recreational activities	84 intensive psychotherapy; 68 CC in acute bipolar depression	Intensive psychotherapy had better total functioning, relationship functioning and life satisfaction scores over 9 months than those in CC; no differences in vocational or recreational functioning
2007	Miklowitz *et al.*	Intensive psychotherapy (STEP-BD) vs CC	RCT of BD; extended sample and duration of above study, with different outcome	1 year	Time to recovery, proportion of patients classified as well during each of the 12 study months	163 intensive psychotherapy, 130 CC in acute bipolar depression	Intensive psychotherapy had higher recovery rates, shorter times to recovery and greater likelihood of being well during any study month compared with CC; no differences seen between the 3 intensive psychotherapies
2008	Zaretsky *et al.*	PE vs CBT, both drawn from Basco & Rush manual	RCT; PE (7 individual sessions) vs CBT (same 7 PE sessions plus 13 CBT sessions)	1 year	Recurrences, weekly mood symptoms	39 PE, 40 CBT	No differences between PE and CBT in number of recurrences but reduction in days with any degree of depressed mood by CBT
2008	Parikh *et al.*	Group PE (Life Goals) vs individual CBT (Lam manual)	RCT; PE (6 group sessions) vs CBT (20 individual sessions)	18 months	Weekly LIFE mania and depression symptom scores, time to recurrence	109 PE, 95 CBT	No differences between PE and CBT in symptom scores or time to recurrence; PE intervention vastly cheaper and easier to disseminate, recommended from public health perspective

the adoption of these interventions. Further research should clarify which mechanisms of action may be most potent, and should also explicitly examine the question of cost and efficiency of delivery of treatment. For instance, we noted that the excellent Barcelona Bipolar Disorders Program was designed and delivered originally by psychologists, while the Life Goals Program was designed to be taught by psychiatric nurses. In addition to the obvious higher cost of psychologists versus nurses, the vastly larger number of nurses and the concomitant presence of nurses in many different treatment settings, in rural and urban areas, in academic and community settings, argues for the use of such well-placed providers. Furthermore, given the training traditions of each profession, it is worth noting our experience in training primarily nurses, as well as some other professionals, in the Life Goals Program for our four-site study of group PE versus individual CBT. After carefully inviting only experienced psychiatric nurses and some similar Masters level trained psychotherapists, we used a single day training event followed by accompanying the trainee through a single cycle of six PE group sessions. Additional supervision was provided as needed but typically totalled to no more than 1 h over the six sessions. By contrast, providing training in FFT or CBT takes much longer, and involves much more supervision, as both of the pivotal trials in FFT and CBT mention in the key papers. In our PE versus CBT study, we noted that the cost of delivery was $160 per participant for PE, as opposed to $1200 for the individual CBT. While the true cost of each therapy in real practice would necessitate a properly powered trial for financial outcomes (which would require thousands of participants), it is still reasonable to conclude that group PE is going to be much less expensive than individual CBT. By respecting such financial parameters, it would then be possible to follow a model of 'stepped care' (Parikh & Kennedy, 2004), in which all patients would be offered group PE, and the more expensive individual treatments would be reserved for those with persistent needs after finishing PE. Ideally, individuals might be evaluated after completion of PE for the presence of either persistent dysfunctional attitudes, disrupted sleep cycles or highly negative family situations, and then respectively referred to CBT, IPSRT or FFT. For those with enduring deficits about sense of self, or co-morbid problems such as anxiety, personality or substance, additional psychodynamic therapy or targeted therapy for the co-morbidity could be invoked. In this book, other chapters address psychosocial interventions for co-morbid disorders, and a companion chapter explores the use of rating scales and methods of obtaining training in BD that will assist the individual in the development of expertise in the delivery of psychosocial interventions.

References

Ball, J.R., Mitchell, P.B., Corry, J.C., Skillecorn, A., Smith, M., Malhi, G.S. (2006). A randomized controlled trial of cognitive therapy for bipolar disorder: focus on long-term change. *Journal of Clinical Psychiatry*, **67**, 277–86.

Basco, M.R. & Rush, A.J. (1996). *Cognitive-Behavioural Therapy for Bipolar Disorder*. New York: Guilford Press.

Bauer, M.S. & McBride, L. (2003). *Structured Group Psychotherapy for Bipolar Disorder: The Life Goals Program*, 2nd edition. New York: Springer.

Bauer, M.S., McBride, L., Williford, W.O., *et al.* (2006a). Collaborative care for bipolar disorder: Part I. Intervention and implementation in a randomized effectiveness trial. *Psychiatric Services*, **57**, 927–36.

Bauer, M.S., McBride, L., Williford, W.O., *et al.* (2006b). Collaborative care for bipolar disorder: Part II. Impact on clinical outcomes, function, and cost. *Psychiatric Services*, **57**, 937–45.

Beck, A.T. (2008). The evolution of the cognitive model of depression and its neurobiological correlates. *American Journal of Psychiatry*, **165**, 969–77.

Beck, A.T., Rush, A.J., Shaw, B.F., Emergy, G. (1979). *Cognitive Therapy of Depression*. New York: Guilford Press.

Butzlaff, R. L. & Hooley, J. M. (1998). Expressed emotion and psychiatric relapse. *Archives of General Psychiatry*, **55**, 547–52.

Clark, D.A., Beck, A.T., Alford, B.A. (1999). *Scientific Foundations of Cognitive Theory and Therapy of Depression*. Hoboken, NJ: John Wiley.

Colom, F., Vieta, E., Martinez-Aran, A., *et al.* (2003a). A randomized trial on the efficacy of group psychoeducation in the prophylaxis of recurrences in bipolar patients whose disease is in remission. *Archives of General Psychiatry*, **60**, 402–7.

Colom, F., Vieta, E., Reinares, M., *et al.* (2003b). Psychoeducation efficacy in bipolar disorders: beyond compliance enhancement. *Journal of Clinical Psychiatry*, **64**, 1101–5.

Colom, F., Vieta, E., Sanchez-Moreno, J., *et al.* (2005). Stabilizing the stabilizer: group psychoeducation enhances the stability of serum lithium levels. *Bipolar Disorders*, **7** (Suppl. 5), 32–6.

Colom, F., Vieta, E., Sanchez-Moreno, J., *et al.* (2004). Psychoeducation in bipolar patients with comorbid personality disorders. *Bipolar Disorders*, **6**, 294–8.

Colom, F., Vieta, E., Scott, J. (2006). *Psychoeducation Manual for Bipolar Disorder*. Cambridge University Press.

Falloon, I., Boyd, J.L., McGill, C.W., *et al.* (1985). Family management in the prevention of morbidity of schizophrenia: clinical outcome of a two year longitudinal study. *Archives of General Psychiatry*, **42**, 887–96.

Frank, E. (2005). *Treating Bipolar Disorder: a Clinician's Guide to Interpersonal and Social Rhythm Therapy*. New York: Guilford Press.

Frank, E., Hlastala, S., Ritenour, A., *et al.* (1997). Inducing lifestyle regularity in recovering bipolar disorder patients: Results from the maintenance therapies in bipolar disorder protocol. *Biological Psychiatry*, **41**, 1165–73.

Frank, E., Kupfer, D.J., Thase, M.E., *et al.* (2005). Two-year outcomes for interpersonal and social rhythm therapy in individuals with bipolar I disorder. *Archives of General Psychiatry*, **62**, 996–1004.

Frank, E., Swartz, H.A., Kupfer, D.J. (2000). Interpersonal and social rhythm therapy: managing the chaos of bipolar disorder. *Biological Psychiatry*, **48**, 593–604.

Frank, E., Swartz, H.A., Mallinger, A.G., Thase, M.E., Weaver, E.V., Kupfer, D.J. (1999). Adjunctive psychotherapy for bipolar disorder: effects of changing treatment modality. *Journal of Abnormal Psychology*, **14**, 1014–18.

Honig, A., Hofman, A., Rozendaal, N., Dingemans, P. (1997). Psycho-education in bipolar disorder: effect on expressed emotion. *Psychiatry Research*, **72**, 17–22.

Huxley, N.A., Parikh, S.V., Baldessarini, R.J. (2000). Effectiveness of psychosocial treatments in bipolar disorder: state of the evidence. *Harvard Review of Psychiatry*, **8,** 126–40.

Johnson, S. & Tran, T. (2007). Bipolar disorder: what can psychotherapists learn from the cognitive research? *Journal of Clinical Psychology*, **63**, 425–32.

Klerman, G., Weissman, M., Rousanville, B., Cheveron, E. (1984). *Interpersonal Psychotherapy of Depression*. New York: Basic Books.

Lam, D.H., Bright, J., Jones, S., *et al.* (2000). Cognitive therapy for bipolar disorder – a pilot study of relapse prevention. *Cognitive Therapy and Research*, **24**, 503–20.

Lam, D.H., Hayward, P., Watkins, E.R., Wright, K., Sham, P. (2005a). Relapse prevention in patients with bipolar disorder: cognitive therapy outcome after 2 years. *American Journal of Psychiatry*, **162**, 324–9.

Lam, D.H., Jones, S.H., Hayward, P., Bright, J.A. (1999). *Cognitive Therapy for Bipolar Disorder*. Chichester, UK: Wiley.

Lam, D.H., McCrone, P., Wright, K., Kerr, N. (2005b). Cost-effectiveness of relapse-prevention cognitive therapy for bipolar disorder: 30-month study. *The British Journal of Psychiatry*, **186**, 500–6.

Lam, D.H., Watkins, E.R., Hayward, P., *et al.* (2003). A randomized controlled study of cognitive therapy for relapse prevention for bipolar affective disorder: outcome for the first year. *Archives of General Psychiatry*, **60**, 145–52.

Lam, D.H., Wright, K., Smith, N. (2004). Dysfunctional assumptions in bipolar disorder. *Journal of Affective Disorders*, **79**, 193–9.

Leahy, R.L. (1999). Decision making and mania. *Journal of Cognitive Psychotherapy: An International Quarterly*, **13**, 83–105.

Leahy, R.L. (2003). *Cognitive Therapy Techniques: A Practitioner's Guide*. New York: Guilford Press.

Leahy, R.L. & Beck, A.T. (1988). Cognitive therapy of depression and mania. In: Cancro, R. & Georgotas, R. (Eds). *Depression and Mania*. New York: Elsevier, pp. 517–37.

Miklowitz, D.J. & Goldstein, M.J. (1997). *Bipolar Disorder: A Family-focused Treatment Approach*. New York, NY: Guilford Press.

Miklowitz, D.J., George, E.L., Axelson, D.A., *et al.* (2004). Family-focused treatment for adolescents with bipolar disorder. *Journal of Affective Disorders*, **82** (Suppl. 1), S113–S128.

Miklowitz, D.J., George, E.L., Richards, J.A., Simoneau, T.L., Suddath, R.L. (2003a). A randomized study of family-focused psychoeducation and pharmacotherapy in the outpatient management of bipolar disorder. *Archives of General Psychiatry*, **60**, 904–12.

Miklowitz, D.J., Goldstein, M.J., Nuechterlein, K.H., Snyder, K.S., Mintz, J. (1988). Family factors and the course of bipolar affective disorder. *Archives of General Psychiatry*, **45**, 225–31.

Miklowitz, D.J., Otto, M.W., Frank, E., *et al.* (2007). Psychosocial treatments for bipolar depression: a 1-year randomized trial from the Systematic Treatment Enhancement Program. *Archives of General Psychiatry*, **64**, 419–26.

Miklowitz, D.J., Richards, J.A., George, E.L., *et al.* (2003b). Integrated family and individual therapy for bipolar disorder: results of a treatment development study. *Journal of Clinical Psychiatry*, **64**, 182–91.

Miklowitz, D.J., Simoneau, T.L., George, E.L., *et al.* (2000). Family-focused treatment of bipolar disorder: 1-year effects of a psychoeducational program in conjunction with pharmacotherapy. *Biologial Psychiatry*, **48**, 582–92.

Miller, I.W., Solomon, D.A., Ryan, C.E., Keitner, G.I. (2004). Does adjunctive family therapy enhance recovery from bipolar I mood episodes? *Journal of Affective Disorders*, **82**, 431–6.

Newman, C.F., Leahy, R.L., Beck, A.T., Reilly-Harrington, N.A., Gyulai, L. (2002). *Bipolar Disorder: A Cognitive Therapy Approach*. Washington DC: American Psychological Association.

Parikh, S.V. & Kennedy, S.H. (2004). Integration of patient, provider, and systems treatment approaches in bipolar disorder. In: Power, M . (Ed.). *Mood Disorders: A Handbook of Science and Practice*. London: Wiley.

Parikh, S.V., Velyvis, V., Yatham, L.N., *et al.* (2007). Coping styles in the prodromes of mania. *Bipolar Disorders*, **9**, 589–95.

Rea, M., Tompson, M., Miklowitz, D., Goldstein, M., Hwang, S., Mintz, J. (2003). Family-focused treatment versus individual treatment for bipolar disorder: results of a randomized clinical trial. *Journal of Consulting and Clinical Psychology*, **71**, 482–92.

Rucci, P., Frank, E., Kostelnik, B., *et al.* (2002). Suicide attempts in patients with bipolar I disorder during acute and maintenance phases of intensive treatment with pharmacotherapy and adjunctive psychotherapy. *American Journal of Psychiatry*, **159**, 1160–4.

Sachs G.S., Thase, M.E., Otto, M.W., *et al.* (2003). Rationale, design, and methods of the systematic treatment enhancement program for bipolar disorder (STEP-BD). *Biological Psychiatry*, **53**, 1028–42.

Scott, J. (2001). Cognitive therapy as an adjunct to medication in bipolar disorder. *British Journal of Psychiatry*, **41**, S164–8.

Scott, J., Garland, A., Moorhead, S. (2001). A pilot study of cognitive therapy in bipolar disorder. *Psychological Medicine*, **31**, 459–67.

Scott, J., Paykel, E., Morriss, R., *et al.* (2006). Cognitive behavioural therapy for severe and recurrent bipolar disorders. *British Journal of Psychiatry*, **188**, 313–20.

Simon, G.E., Ludman, E.J., Bauer, M.S., Unutzer, J., Operskalski, B. (2006). Long-term effectiveness and cost of a systematic care program for bipolar disorder. *Archives of General Psychiatry*, **63**, 500–8.

Sperry, L. (1995). *Psychopharmacology and Psychotherapy: strategies for maximizing treatment outcomes.* New York: Brunner/Mazel.

Suppes, T., Rush, A.J., Dennehy, E.B., *et al.* (2003). Texas Medication Algorithm Project, Phase 3 (TMAP-3): clinical results for patients with a history of mania. *Journal of Clinical Psychiatry*, **64**, 370–82.

Toprac, M.G., Ennehy, E.B., Carmody, T.J., *et al.* (2006). Implementation of the Texas Medication Algorithm Project Patient and Family Education Program. *Journal of Clinical Psychiatry*, **67**, 1362–72.

Zaretsky, A., Lancee, W., Miller, C., Harris, A., Parikh, S.V. (2008). Is cognitive-behavioural therapy more effective than psychoeducation in bipolar disorder? *Canadian Journal of Psychiatry*, **53**, 441–8.

Chapter 7

Physical treatments in bipolar disorder

Marisa Le Masurier, Lucie L. Herrmann, Louisa K. Coulson and Klaus P. Ebmeier

Although the mainstay of treatment in bipolar affective disorder is pharmacological, a tradition of physical treatments has continued to survive, often in the face of bad press. Side effects and the controversial image of electroconvulsive therapy (ECT) have spawned a multitude of alternative treatments in an attempt to emulate its superior efficacy in the treatment of depression. After an initial account of the use and efficacy of ECT in bipolar depression and mania, we will give a description and preliminary evaluation of some of these alternative techniques.

Electroconvulsive therapy in bipolar disorder

In 1776 the Viennese physician Leopold Auenbrugger described the use of camphor-induced seizures in the treatment of mania (Lesky, 1959). Electroconvulsive therapy was first employed in a manic patient in 1938, and by the 1950s ECT was widely used to treat a range of psychiatric disorders (Fink, 2001). It was superseded in the 1960s with the advent of psychotropic drugs, but experienced a later resurgence in treatment-resistant cases. Electroconvulsive therapy has been successfully used to treat the varying presentations of bipolar disorder (BD), including bipolar depression, mania, catatonia and mixed affective states.

Electroconvulsive therapy in bipolar depression

A systematic review and meta-analysis conducted by The UK ECT Review Group identified 73 randomised trials of ECT efficacy and safety in depressive illness (UK ECT Review Group, 2003). The authors concluded that ECT is an effective treatment for depressive disorders, and is probably more effective than drug therapy. They also concluded that bilateral ECT is moderately more effective than unilateral ECT and high dose more effective than low dose. Of the trials in the review, some included depressed patients with a diagnosis of BD, but in most of the trials numbers were small, and no subgroup analysis was done to answer the question of whether ECT has different effects in bipolar depression compared with unipolar depression. More recent randomised controlled trials (RCTs) comparing ECT with transcranial magnetic stimulation, and contrasting bilateral and unilateral ECT in depression, included some bipolar patients, but these were not analysed separately (Eschweiler *et al.*, 2007; McLoughlin *et al.*, 2007).

The effectiveness of ECT in bipolar depression has been examined directly, although most of these studies are old and small, only two were RCTs (Greenblatt *et al.*, 1964; Stromgren, 1988) and they differed in their methodology and administration of treatment. Although results vary, ECT was as effective in bipolar depression as in unipolar depression in the majority of studies (Perris & d'Elia, 1966; Abrams & Taylor, 1974; Avery & Winokur, 1977; Avery & Lubrano, 1979; Homan *et al.*, 1982; Black *et al.*, 1986; Stromgren, 1988). All studies of ECT in bipolar depression showed that ECT was superior or equal to antidepressant drugs (Greenblatt *et al.*, 1964; Bratfos & Haug, 1965; Perris & d'Elia, 1966; Avery & Winokur, 1977; Avery & Lubrano, 1979; Homan *et al.*, 1982; Black *et al.*, 1986). Two recent naturalistic studies of ECT in bipolar depression represent extremes in response rates: 76% of patients responded to ECT in a US sample (Devanand *et al.*, 2000), but only 26% in an Italian study of medication-resistant patients (Ciapparelli *et al.*, 2001).

More recently, differences in response to ECT between bipolar and unipolar depression have been addressed by retrospective analyses of three randomised trials. The authors found an equivalent outcome in unipolar and bipolar depression (approximately 50% response rate, defined by a 60% reduction in the Hamilton depression rating scale 1 week after final ECT), but a greater speed of response in bipolar patients (Daly *et al.*, 2001; Dolberg *et al.*, 2001). The reason for

Practical Management of Bipolar Disorder, eds. Allan H. Young, I. Nicol Ferrier and Erin E. Michalak. Published by Cambridge University Press.

this is uncertain, but it has been suggested that it may be related to a greater and more rapid increase in seizure threshold in response to ECT in bipolar patients compared with that in unipolar patients. It is estimated that about 50% of antidepressant-resistant bipolar patients respond to ECT (Daly *et al.*, 2001).

One disadvantage of ECT in bipolar depression is the risk of a switch to hypomania, although this is thought to be less of a risk than with antidepressants (Zornberg & Pope, 1993). Other disadvantages of using ECT alone are its relatively high relapse rate, and the lack of information it provides about the best prophylactic medication for ECT responders (Compton & Nemeroff, 2000).

Although early experience suggested lithium should be stopped during ECT because of increased risk of neurotoxicity (Fink, 1994), recent reviews have not found an increased number of adverse events associated with lithium and ECT (Jha *et al.*, 1996; Dolenc & Rasmussen, 2005), although there continue to be case reports of neurotoxicity (Sartorius *et al.*, 2005).

Autobiographical memory loss is associated with ECT. This can be reduced by using brief pulse rather than sine wave ECT, by unilateral positioning of electrodes, by using the least number of treatments possible and by titrating electrical current relative to the patient's seizure threshold (UK ECT Review Group, 2003; Fraser *et al.*, 2008).

In summary, ECT is as effective in bipolar as in unipolar depression, and may be especially useful in antidepressant-resistant cases. Electroconvulsive therapy has a rapid onset of action, which makes it an important potential choice of treatment for severe illness with suicidal risk.

Electroconvulsive therapy in mania

There have been very few RCTs of ECT in mania, probably because of the practical and ethical difficulties inherent in conducting such studies. A review of ECT for the treatment of acute manic episodes summarises a large number of retrospective and prospective studies carried out since the 1940s (Mukherjee *et al.*, 1994). The authors conclude that remission or marked clinical improvement was achieved in 80% of 589 manic patients treated with ECT. To date, three randomised trials (with 69 acutely manic patients), each differing in their primary objectives and experimental design, have been reported (Small *et al.*, 1988; Mukherjee, 1989; Sikdar *et al.*, 1994). In the first, designed to compare ECT with lithium, the group treated with a combination of ECT, lithium and neuroleptic drugs entered remission more quickly over the 8-week trial period than the group receiving lithium and neuroleptics alone (Small *et al.*, 1988). However, the eventual outcomes did not differ between the two groups. The second study examined whether ECT was effective in manic patients resistant to pharmacotherapy, and randomly assigned patients to one of four treatment groups (right unilateral, left unilateral, bilateral ECT or lithium–haloperidol combination pharmacotherapy). Thirteen of 22 patients remitted fully with ECT; the remainder either showed no response or deteriorated. None of those treated with pharmacotherapy improved (Mukherjee, 1989). The third study sought to compare ECT with sham-ECT, both in combination with chlorpromazine. The ECT group showed a significantly greater and faster improvement than the sham group, with 12 out of 15 patients achieving complete remission after eight sessions of ECT.

Retrospective controlled studies have found ECT to be either superior or comparable to antipsychotic drugs (McCabe, 1976; McCabe & Norris, 1977; Thomas & Reddy, 1982) and lithium (Thomas & Reddy, 1982; Black *et al.*, 1987). Naturalistic studies have shown ECT to be effective in achieving remission in approximately 60% of pharmacotherapy-resistant manic patients (Alexander *et al.*, 1988; Stromgren, 1988), 78% of routinely treated hospitalised manic patients (Black *et al.*, 1987) and 100% of manic patients treated with unmodified ECT in clinics in India (Mukherjee & Debsikdar, 1992) and Nigeria (Ikeji *et al.*, 1999). The naturalistic study by Black and colleagues showed ECT to be superior to lithium (Black *et al.*, 1987) in treating mania. In a recent naturalistic study in Brazil, the use of ECT to treat manic patients did not decrease the length of inpatient stay, but reduced re-admission. ECT use was associated with psychotic symptoms, a high number of previous hospitalisations, female gender and having a psychiatrist who had treated more than 20 manic inpatients in the previous 5 years (Volpe & Tavares, 2004).

In summary, although well-controlled prospective data are lacking, clinical experience and the available evidence suggest that ECT has robust antimanic effects. Recent case reports do testify to its continuing use in treating pharmacotherapy-resistant mania (Ditmore *et al.*, 1992; Hanin *et al.*, 1993; Finnerty *et al.*, 1996; Chanpattana, 2000; Macedo-Soares *et al.*, 2005).

Electroconvulsive therapy in mixed affective states

No RCTs have been carried out in patients with a mixed affective state, although two naturalistic retrospective studies and two case series have been published;

response rates varied from 56% in medication-resistant patients to 100% in a small case series (Tundo *et al.*, 1991; Devanand *et al.*, 2000; Gruber *et al.*, 2000; Ciapparelli *et al.*, 2001). Electroconvulsive therapy may thus be effective in treating mixed affective states, but conclusive evidence is lacking.

Electroconvulsive therapy for catatonia

There are no RCTs of ECT in catatonic patients, although retrospective chart reviews and case reports have consistently shown ECT to be effective (Pataki *et al.*, 1992; Rohland *et al.*, 1993; Taylor & Fink, 2003), particularly in patients with affective disorder (Fink, 1990; Pataki *et al.*, 1992; Rohland *et al.*, 1993), and in patients who fail to respond to first-line treatment with benzodiazepines (Bush *et al.*, 1996; Ungvari *et al.*, 2001).

Maintenance electroconvulsive therapy

A recent review of continuation and maintenance ECT in treatment-resistant BD identified a number of studies, mostly case reports, with only a few prospective studies of heterogeneous groups of patients (Vaidya *et al.*, 2003). This included a 16-month prospective study of patients with affective disorder, with a reduction in hospitalisation time from 44 to 7% of the year (Vanelle *et al.*, 1994). Other authors also report successful outcomes using maintenance ECT in bipolar patients who relapsed on pharmacotherapy alone (Petrides *et al.*, 1994), or who had a rapid-cycling illness (Fink, 1991).

Transcranial stimulation in bipolar disorder

Transcranial magnetic stimulation (TMS) uses fast discharges of electricity through electrically insulated coils to generate a magnetic field, which in turn penetrates the skull without impediment and can induce a current in conducting structures, such as neurones in superficial (2 cm) cortex. The induced current triggers action potentials in certain populations of brain cells. Depending on the stimulation frequency, repetitive TMS appears to have different effects on cortex: low frequencies at about 1 Hz or lower have a quenching effect; higher frequencies, e.g. 10 and 20 Hz, increase cortical excitability (Ziemann & Hallett, 2007). Thus left dorsolateral hypoperfusion in depression is to be remedied either by stimulating left dorsolateral TMS at 10 or 20 Hz (George *et al.*, 1997), or by quenching stimulation at 1Hz over the reciprocally connected right dorsolateral cortex (Klein *et al.*, 1999).

Transcranial magnetic stimulation has been licensed as a treatment for depression in a number of countries and is privately available in others. A recent review on rapid (r)TMS in depression by NICE in the UK concluded that the potential impact of the procedure on the National Health Service may be moderate to large (NICE, 2007). Yet, there is uncertainty regarding the efficacy of the procedure, reflected by inconsistent results of individual RCTs. The majority of published RCTs are underpowered, so that only a meta-analysis could generate results that inspired any confidence. A recent comprehensive meta-analysis indicated that rTMS is superior to sham treatment, with a moderate to large effect size (details of studies available from the authors). Moreover, owing to the lack of significant predictors of efficacy, together with the poor quality of available sham treatments, the possibility that TMS has a non-specific effect on depression cannot be excluded (Herrmann & Ebmeier, 2006).

Table 7.1 and Figure 7.1 indicate that none of the pooled effect sizes is statistically homogeneous, i.e. the magnitude of effects varies more than expected by chance. This can be caused by differences in method, from sampling strategies to diagnostic criteria, assessment scales, frequency of treatment, strength and frequency of stimulation, train length, total number of stimuli or position of the stimulation coil. Consequently, none of the single studies included can serve as a paradigm of treatment, i.e. a treatment protocol that is likely to generate the representative polled effect size. In a larger, previously published analysis, we found only strength of the stimulation relative to motor threshold to be a significant predictor of treatment outcome (Herrmann & Ebmeier, 2006). There is no evidence that bipolar patients may show differential responsivity during their depressed phase.

The few studies focusing on manic symptoms have taken a complementary approach by stimulating right prefrontal cortex with high frequencies, as stimulation of left prefrontal cortex has been said to be associated with emergent mania. However, a systematic review of the literature does not seem to confirm this concern (Xia *et al.*, 2008). In addition, although initial open studies appeared to suggest efficacy of real TMS in mania, the only sham-controlled study published so far has not been able to confirm this. For obvious practical and ethical reasons, it is more difficult to study manic than mild or moderately depressed patients, so at this

Table 7.1 Sensitivity analysis – efficacy of TMS in depressed patients with unipolar and bipolar course

Measure	Unipolar patients only	Bipolar and mixed patient group	Diagnosis unclear
No. of studies included	12	10	7
Random effects (DerSimonian-Laird) pooled *d*+	*d*+=0.84 (95% CI 0.47–1.21) $P<0.0001$	*d*+=0.83 (95% CI 0.34–1.33) $P<0.0009$	*d*+=0.85 (95% CI 0.36–1.33) $P<0.0006$
Non-combinability of studies – Cochrane's Q	$Q_{(df=11)}=48.5$, $P<0.0001$	$Q_{(df=9)}=48.1$, $P<0.0001$	$Q_{(df=6)}=15.4$, $P<0.02$
Bias indicator – Kendall's τ	τ=0.35, p=0.12	τ=0.20, p=0.48	τ=-0.14, p=0.56

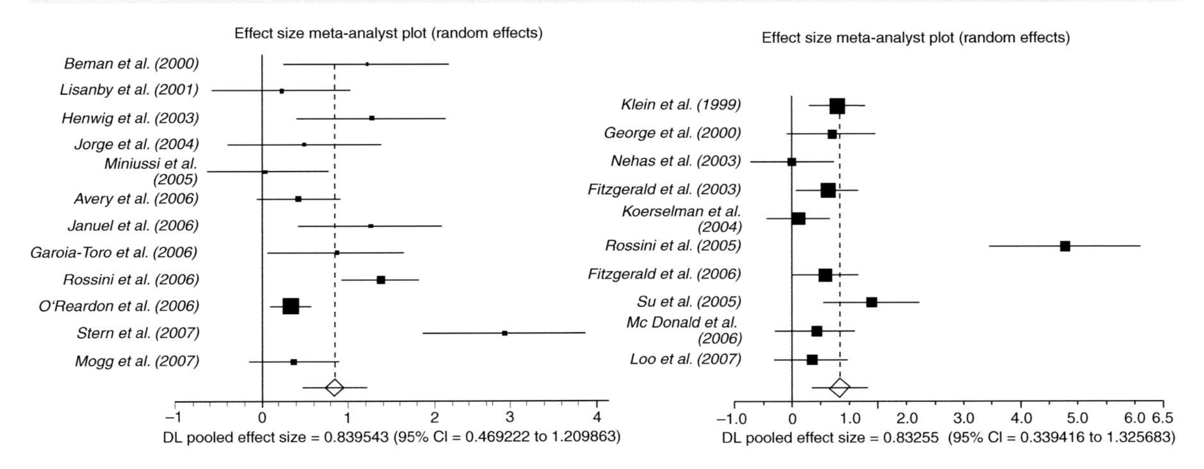

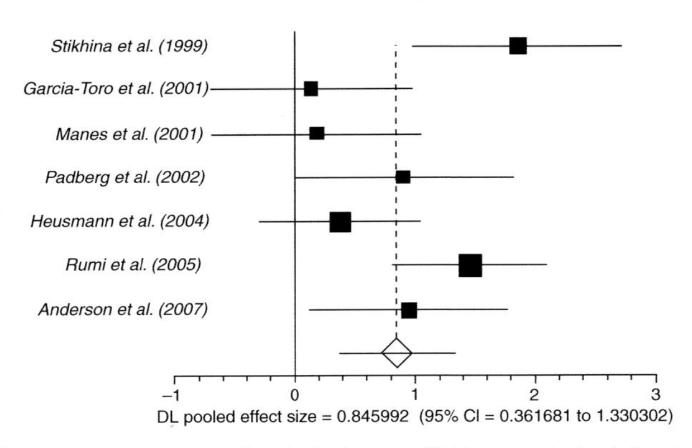

Figure 7.1 Forest plots of studies examining patients with unipolar (upper left), bipolar and mixed disorder (upper right), and 'diagnosis unclear' (lower).

stage the evidence on TMS efficacy in mania is too slim to call. In summary, TMS is user-friendly, safe if used within established parameters of stimulation strength and frequency, requires no anaesthesia and generates little in terms of side effects. It is unclear to what extent it can modify symptoms of mania, but in depression the active treatment on average results in a large treatment effect ($d>0.8$).

Vagus nerve stimulation

Stimulation of the left vagus nerve increases regional cerebral blood flow in bilateral orbito-frontal, anterior cingulate and right superior and medial frontal cortex. Reductions in blood flow have been described in bilateral temporal cortex and right posterior cingulate gyrus (Sheline *et al.*, 2006). There is evidence of direct and secondary projections of the vagus nerve to noradrenergic neurones in the locus ceruleus and raphe serotonergic neurones that may be responsible for any antidepressant effects (Nemeroff *et al.*, 2006). First observations of the antidepressant effect of vagus nerve stimulation (VNS) were made in epileptic patients implanted to reduce their seizure frequency, although induction of psychiatric syndromes has been described, too. A number of efficacy studies have been conducted by Cyberonics (Houston, TX, USA; see Sackeim *et al.*, 2007), who also applied for and gained approval of the US Food and Drug Administration for this intervention in treatment-resistant depression (TRD). The limited success rate of VNS has to be appreciated in relation to the special selection of patients entering VNS trials, i.e. patients who have not responded at least to a number of antidepressant trials (Sackeim, 2001). Although implantation of the VNS device involves minor surgery, significant side effects of chronic stimulation are, among others: hoarseness, hypaesthesia, headache, infection, dyspnoea, insomnia, cough, laryngism, nausea, ataxia, pain, vomiting, paresthesia, dyspepsia and pharyngitis. At the time of writing, a number of new study results are in press or about to be published, so it is hoped that more convincing efficacy data will soon be available to establish this treatment modality as part of the standard repertoire of antidepressant treatments, with particular regard to bipolar, and also rapid-cycling bipolar, illness.

Deep brain stimulation

Modern neurosurgical treatments for psychiatric conditions, such as anterior cingulotomy, subcaudate tractotomy, limbic leucotomy and anterior capsulotomy, regulate brain function by the removal or destruction of neural pathways (Binder & Iskandar, 2000). Such 'psychosurgical' procedures are reserved for symptomatic relief from psychiatric conditions that are refractory to pharmacological, psychotherapeutic or electroconvulsive therapies. Similarly, the use of reversible neuromodulation techniques to treat otherwise treatment-resistant psychiatric conditions is based on the premise that modulation of brain regions that mediate emotional and/or cognitive functions associated with mental illness could potentially provide symptomatic relief from these conditions (Heller *et al.*, 2006).

Deep brain stimulation (DBS) induces reversible, adjustable modification of brain function by the placement of current-carrying electrodes each with four contact points to subcortical structures of the brain, known as target regions. Contact points are activated by a pulse generator, which is placed subcutaneously below the clavicle. Settings including pulse width, frequency and amplitude of stimulation can be adjusted after the operation to suit an individual patient's needs. It has been suggested that high-frequency (>100 Hz), chronic stimulation may have effects functionally equivalent to ablation procedures (Breit *et al.*, 2004).

The use of DBS is approved as a treatment for symptoms of otherwise medically intractable movement disorders. In psychiatric conditions such as major depression (MD), DBS is experimental. Experimental stimulation in mood disorders has shown efficacy with electrodes placed in the subgenual cingulate (Cg25), ventral striatum/ventral internal capsule and inferior thalamic peduncle.

The use of deep brain stimulation in psychiatric conditions: major depression

Deep brain stimulation in the treatment of psychiatric conditions such as MD targets those regions that in neurobiological models and neuroimaging studies are implicated in the anatomy and physiology of depressive symptoms, i.e. brain regions that have connections with anatomical networks mediating the core behavioural symptoms: a persistent negative mood state and anhedonia (APA, 2000). Major depression is a complex disorder characterised by abnormalities of affective, cognitive, somatic and autonomic function, and is widely accepted as a systems level disorder comprising multiple distributed neural circuits. The pathophysiology of MD may be understood in the context of dysfunctional brain activity in the anatomic substrates of cortico-striatal-pallido-thamalo-cortical circuits (Alexander *et al.*, 1986) and their connections with other subcortical and brainstem areas, as well as dysfunction of the coordinated interactions between networks of cortico-striatal and cortico-limbic pathways (Mayberg, 1997).

The cingulate gyrus as a target for deep brain stimulation in treatment-resistant depression

Neurobiological models of depression suggest that mood state and associated cognitive, somatic and autonomic behaviours are mediated by the coordinated interaction of a distributed network of pathways in dorsal cortical (neocortical and midline limbic elements) and ventral frontal-limbic (paralimbic cortical, subcortical and brainstem) regions (Mayberg *et al.*, 1997).

The subgenual prefrontal cortex (Brodman (BA) area 25) has been identified as a target region for DBS. This is based on evidence of both hypermetabolism in this area associated with TRD and of pre-treatment activity as a predictor of treatment response.

Neuroimaging studies have suggested that rostral cingulate gyrus (Cg24a) hypermetabolism *before* treatment favours antidepressant treatment response (Mayberg *et al.*, 1997; Pizzagalli *et al.*, 2001). Mayberg and colleagues suggest Cg24a as the substrate integrating the processing of affective, cognitive, somatic and autonomic behaviours and therefore a potential target for DBS (Sakas & Panourias, 2006).

A reciprocal pattern of changes in brain activity in ventral (subgenual cingulate (BA 25 and insula) and dorsal (dorsal prefrontal (BA 46/9), inferior parietal (BA 40) and dorsal anterior (BA 24b) and posterior (BA 23/31) cingulate) regions was found to be associated with transient induced states of sadness in healthy volunteers (increased ventral and decreased dorsal activity) and resolution of sustained negative mood state in clinically depressed patients treated with antidepressant medication (decreased ventral and increased dorsal activity) (Mayberg *et al.*, 1999). Mean prefrontal cortical activity was 5% greater in patients with familial pure depressive disease (FPDD) compared with healthy controls (Drevets *et al.*, 1992). Post-hoc analyses demonstrated a 6% reduction in prefrontal activity in remitted (asymptomatic) depressed patients compared with FPDD (symptomatic) patients. Prefrontal hyperactivity in the FPDD group thus represented a state-related abnormality.

Subgenual prefrontal cortex (BA 25) activity in mood-disordered patients and healthy controls demonstrated decreased cerebral blood flow (CBF) and ^{18}F-fluoro-deoxyglucose (FDG) uptake of 8 and 12% in bipolar and unipolar patients, respectively. However, MRI images acquired to correct for the partial volume effect of PET data demonstrated 39 and 48% reduced grey matter volume in the same area (Drevets *et al.*, 1997), therefore metabolic activity of the BA 25 was actually increased in the mood disordered individuals relative to controls. A functional MRI (fMRI) study provided cross-modal support for resting state increased network functional connectivity involving the subgenual prefrontal cortex (BA 25) (Greicius *et al.*, 2007).

A meta-analysis of studies using voxel-based techniques of neuroimaging in patients with MD (42 studies) and healthy individuals (44 studies) demonstrated that those regions found to be maximally abnormal in patients with MD (dorsal anterior cingulate (BA 32) extending to subgenual cingulate (BA 25)) corresponded to those regions that were active in healthy individuals undertaking emotional tasks (Steele *et al.*, 2007). This supports the notion that abnormal cingulate activity is associated with dysfunctional emotional regulation.

Clinical response following high-frequency bilateral subgenual stimulation of cingulate white matter (Cg25WM) was reported in four out of six patients with TRD, and remission or near remission in three of the treatment responders (Mayberg *et al.*, 2005). Patterns of pre-treatment subgenual cingulate hyperactivity and prefrontal hypoactivity were similar across both responders and non-responders; however, the magnitude of prefrontal hypoactivity was greater in responders. Comparisons of pre- and post-treatment metabolic activity using PET scans showed decreased subgenual cingulate and orbito-frontal cortex activity in short-term responders, and decreased hypothalamus, anterior insula and medial frontal cortex as well as increased dorsolateral prefrontal cortex and dorsal anterior and posterior cingulate cortical activity in long-term responders. This pattern of reduction of pre-operative cingulate hypermetabolism and normalisation of pre-operative cortical frontal hypometabolism observed in treatment responders was also reported in eight out of twelve TRD patients treated with bilateral Cg25 DBS in a different study (Lozano, 2006).

Diffusion MRI tractography was used to investigate the hypothesis that the anterior cingulate cortex (ACC) and its connectivity patterns with reciprocal cortical and subcortical regions is the anatomical substrate mediating the changes in brain activity following DBS for TRD (Johansen-Berg *et al.*, 2008). The electrode contact points for those patients who responded to DBS were located within (six out of nine patients) and just outside (three out of nine patients) the subgenual ACC. Connectivity-based parcellation suggested that the efficacy of DBS as a neurosurgical procedure in the treatment of MD is a function of the stimulation of the subgenual ACC and its reciprocal connections with the amygdala, hippocampus, nucleus accumbens, mid-cingulate cortex and orbito-frontal cortex.

Anatomical correlates of the brain reward system as targets for deep brain stimulation

Reward, the experience of neuronal responses to the positive or negative valence of a stimulus (object or event), induces subjective feelings of pleasure (Hêdonê) and a positive emotional state, which in turn provide the motivation for goal-directed approach and consummatory behaviour (Schultz, 2001). Anhedonia, a loss of pleasure or interest in previously enjoyed and rewarding activities, is a core behavioural symptom of MD.

Affect and motivation are psychological states mediated by the physiological processes of the limbic system and its reciprocal connections with cortical and subcortical regions (Morgane *et al.*, 2005). It has been suggested that neuro-anatomical (amygdala, anterior cingulate, left dorsolateral prefrontal cortex, orbito-frontal cortex and subgenual prefrontal cortex) and neuropharmacological (dopamine (DA), serotonin (5-HT) and acetylcholine (Ach)) substrates common to both the pathophysiology of MD and the neurobiology of the behavioural reward system (BRS), mediate the core depressive symptom of anhedonia (Naranjo *et al.*, 2001). Therefore, investigation of the anatomical structures comprising the neural pathways of the BRS may identify potential targets for the treatment of anhedonia and psychomotor slowing in MD with DBS (Schlaepfer & Lieb, 2005).

A dopaminergic probe, dextroamphetamine sulphate, was used with fMRI to investigate brain activity in structures associated with reward processing in MD and healthy controls (Tremblay *et al.*, 2005). A significant hypersensitive response to the rewarding effects of this probe (measured with the Addiction Research Centre Inventory) was observed in the ventrolateral prefrontal and orbito-frontal cortices, the caudate nucleus and putamen of patients with MD. The ventral striatum (VS; nucleus accumbens (NAcc) in rodents) is thought to be a key anatomical structure of the BRS, providing the interface between affective and motor circuits that mediate reward-related behaviour (Breiter & Rosen, 1999). An investigation of the anatomical substrates associated with reward in rats found that opioid neurotransmission in the NAcc and ventral pallidum (VP) contributes to reward motivation in general ('wanting') and amplifies affective reactions to reward (hedonic impact or 'liking') in a subregion of the NAcc (Pecina *et al.*, 2006). Furthermore, the VP mediates homeostatically induced changes in sensory pleasure. It is concluded that both the NAcc and VP are part of the widespread hedonic system (or BRS) mediating reward-related behaviour.

The efficacy of NAcc DBS in TRD was assessed by comparing clinical ratings of depression, measured with the Hamilton depression rating scale (HAM-D, or HDRS) and the Montgomery–Asberg depression rating scale (MADRS), with PET images taken at pre-treatment baseline and 1 week after stimulation onset (Schlaepfer *et al.*, 2008). There were decreases in HDRS (33.7 to 19.7; P=0.02) and MADRS (35.7 to 24.7; P=0.02) scores. Positron emission tomography images demonstrated increased metabolism in bilateral VS, bilateral dorsolateral and dorsomedial prefrontal cortex, cingulate cortex and bilateral amygdala and decreased metabolism in ventrolateral and ventromedial prefrontal cortex, dorsal caudate nucleus and thalamus. The authors conclude that DBS in those structures mediating reward and motivational processes (namely NAcc) has efficacy as a neurosurgical procedure in alleviating anhedonia as a symptom of depression.

The globus pallidus has been identified as a component of the BRS that mediates positive emotional states and subjective feelings of pleasure (hedonia) in response to positively reinforcing stimuli. Stimulation-induced significant improvement in pre-operative depressive symptoms (Rettig *et al.*, 2000) and clinically significant response to TRD, measured using HDRS and Beck depression inventory (BDI) scores (Kosel *et al.*, 2007), have been reported following DBS in internal globus pallidus to treat patients with co-morbid movement disorders.

A preliminary report of chronic bilateral DBS in the anterior limb of the internal capsule (a striatal region separating the caudate nucleus and thalamus from the lenticular nucleus (putamen and GP)) and NAcc for patients with TRD reported significant clinical response (P=0.003) in HDRS scores at between 3 and 8 months follow-up in four patients (Eskandar *et al.*, 2007). Similar results have been reported for three patients with TRD treated with bilateral DBS in the ventral part of the anterior limb of the internal capsule (Gabriels *et al.*, 2003). Clinical response (measured using MADRS score) was achieved by all three patients at 6 months follow-up, with two achieving remission.

Inferior thalamic peduncle

There has been a case report of bilateral DBS in the inferior thalamic peduncle for a patient with TRD with co-morbid borderline personality disorder (BPD) and bulimia (Jimenez *et al.*, 2005). The patient demonstrated a reduction in HDRS and BDI scores taken at pre-treatment baseline and bi-monthly follow-up and reached remission. Between 8 and 20 months after stimulation onset, the stimulator was turned off in a double-blind procedure. There was no immediate relapse of depression, and the patient remained in remission for two months; following that, the patient showed spontaneous fluctuations in depression score and a tendency towards abnormalities demonstrated in pre-treatment baseline. Fibres travelling in the inferior thalamic peduncle provide connections between the orbito-frontal cortex and ventral anterior and mediodorsal thalamic nuclei in both the direct, topographically organised thalamo-cortical loops and cortico-striatal-pallido-thamalo-cortical loops (Alexander *et al.*, 1986). The authors suggest that chronic bilateral stimulation may have disrupted overactive orbito-frontal cortex, which has been implicated in the pathophysiology of MD (Drevets, 2000).

Although these results are very preliminary, DBS has potential as both a clinical intervention for the symptomatic relief of treatment-resistant psychiatric conditions and to elucidate the fundamental mechanisms of human brain function (Kringelbach *et al.*, 2007). However, the heterogeneous nature of depressive symptoms with the frequent presence of co-morbid Axis I and II disorders is likely to create difficulties in identifying target regions for effective stimulation, thus impeding the development of a universally applicable neurosurgical treatment for severe depression.

Acknowledgements

We wish to acknowledge the financial support of the Gordon Small Charitable Trust for KPE and LLH and the Norman Collisson Foundation for KPE and LKC.

References

Abrams, R. & Taylor, M.A. (1974). Unipolar and bipolar depressive illness. Phenomenology and response to electroconvulsive therapy. *Archives of General Psychiatry*, **30**, 320–1.

Alexander, G.E., DeLong, M.R., Strick, P.L. (1986). Parallel organization of functionally segregated circuits linking basal ganglia and cortex. *Annual Review of Neuroscience*, **9**, 357–81.

Alexander, R.C., Salomon, M., Ionescu-Pioggia, M., Cole, J.O. (1988). Convulsive therapy in the treatment of mania: McLean Hospital 1973–1986. *Convulsive Therapy*, **4**, 115–25.

APA (2000). *The Diagnostic and Statistical Manual of Mental Disorders*, 4th edition. American Psychiatric Association, Arlington, VA, USA. www.psychiatryonline.com/resourceTOC.aspx?resourceID=1 [accessed 23 December 2009].

Avery, D. & Lubrano, A. (1979). Depression treated with imipramine and ECT: the DeCarolis study reconsidered. *American Journal of Psychiatry*, **136**, 559–62.

Avery, D. & Winokur, G. (1977). The efficacy of electroconvulsive therapy and antidepressants in depression. *Biological Psychiatry*, **12**, 507–23.

Binder, D.K. & Iskandar, B.J. (2000). Modern neurosurgery for psychiatric disorders. *Neurosurgery*, **47**, 9–21; discussion 21–3.

Black, D.W., Winokur, G., Nasrallah, A. (1986). ECT in unipolar and bipolar disorders: a naturalistic evaluation of 460 patients. *Convulsive Therapy*, **2**, 231–8.

Black, D.W., Winokur, G., Nasrallah, A. (1987). Treatment of mania: a naturalistic study of electroconvulsive therapy versus lithium in 438 patients. *Journal of Clinical Psychiatry*, **48**, 132–9.

Bratfos, O. & Haug, J.O. (1965). Electroconvulsive therapy and antidepressant drugs in manic-depressive disease. Treatment results at discharge and 3 months later. *Acta Psychiatrica Scandinavica*, **41**, 588–96.

Breit, S., Schulz, J.B., Benabid, A.L. (2004). Deep brain stimulation. *Cell and Tissue Research*, **318**, 275–88.

Breiter, H.C. & Rosen, B.R. (1999). Functional magnetic resonance imaging of brain reward circuitry in the human. *Annals of the New York Academy of Sciences*, **877**, 523–47.

Bush, G., Fink, M., Petrides, G., Dowling, F., Francis, A. (1996). Catatonia. II. Treatment with lorazepam and electroconvulsive therapy. *Acta Psychiatrica Scandinavica*, **93**, 137–43.

Chanpattana, W. (2000). Combined ECT and clozapine in treatment-resistant mania. *Journal of ECT*, **16**, 204–7.

Ciapparelli, A., Dell' Osso, L., Tundo, A., *et al.* (2001). Electroconvulsive therapy in medication-nonresponsive patients with mixed mania and bipolar depression. *Journal of Clinical Psychiatry*, **62**, 552–5.

Compton, M.T. & Nemeroff, C.B. (2000). The treatment of bipolar depression. *Journal of Clinical Psychiatry*, **61** (Suppl. 9), 57–67.

Daly, J.J., Prudic, J., Devanand, D.P., *et al.* (2001). ECT in bipolar and unipolar depression: differences in speed of response. *Bipolar Disorders*, **3**, 95–104.

Devanand, D.P., Polanco, P., Cruz, R., *et al.* (2000). The efficacy of ECT in mixed affective states. *Journal of ECT*, **16**, 32–7.

Ditmore, B.G., Malek-Ahmadi, P., Mills, D.M., Weddige, R.L. (1992). Manic psychosis and catatonia stemming from systemic lupus erythematosus: response to ECT. *Convulsive Therapy*, **8**, 33–7.

Dolberg, O.T., Schreiber, S., Grunhaus, L. (2001). Transcranial magnetic stimulation-induced switch into mania: a report of two cases. *Biological Psychiatry*, **49**, 468–70.

Dolenc, T.J. & Rasmussen, K.G. (2005). The safety of electroconvulsive therapy and lithium in combination: a case series and review of the literature. *Journal of ECT*, **21**, 165–70.

Drevets, W.C. (2000). Neuroimaging studies of mood disorders. *Biological Psychiatry*, **48**, 813–29.

Drevets, W.C., Price, J.L., Simpson, J.J.R., *et al.* (1997). Subgenual prefrontal cortex abnormalities in mood disorders. *Nature*, **386**, 824–7.

Drevets, W.C., Videen, T.O., Price, J.L., Preskorn, S.H., Carmichael, S.T., Raichle, M.E. (1992). A functional anatomical study of unipolar depression. *Journal of Neuroscience*, **12**, 3628–41.

Eschweiler, G.W., Vonthein, R., Bode, R., *et al.* (2007). Clinical efficacy and cognitive side effects of bifrontal versus right unilateral electroconvulsive therapy (ECT): a short-term randomised controlled trial in pharmaco-resistant major depression. *Journal of Affective Disorders*, **101**, 149–57.

Eskandar, E., Rauch, S., Jameson, M., Flaherty, A., Darin, D. (2007). Feasibility study of the safety, tolerability, and efficacy of deep brain stimulation in the anterior internal capsule for intractable major depression. *Neurosurgery*, **61**, 214–15.

Fink, M. (1990). Is catatonia a primary indication for ECT? *Convulsive Therapy*, **6**, 1–4.

Fink, M. (1991). ECT in the long-term follow-up of BPD. *Biological Psychiatry*, **30**, 1172.

Fink, M. (1994). Optimizing ECT. *L'Encéphale*, **20**, 297–302.

Fink, M. (2001). Convulsive therapy: a review of the first 55 years. *Journal of Affective Disorders*, **63**, 1–15.

Finnerty, M., Levin, Z., Miller, L.J. (1996). Acute manic episodes in pregnancy. *American Journal of Psychiatry*, **153**, 261–3.

Fraser, L.M., O ' Carroll, R.E., Ebmeier, K.P. (2008). The effect of ECT on autobiographical memory: a systematic review. *Journal of ECT*, **24**, 10–17.

Gabriels, L., Cosyns, P., Nuttin, B., Demeulemeester, H., Gybels, J. (2003). Deep brain stimulation for treatment-refractory obsessive-compulsive disorder: psychopathological and neuropsychological outcome in three cases. *Acta Psychiatrica Scandinavica*, **107**, 275–82.

George, M.S., Wassermann, E.M., Kimbrell, T.A., *et al.* (1997). Mood improvement following daily left prefrontal repetitive transcranial magnetic stimulation in patients with depression: a placebo-controlled crossover trial. *American Journal of Psychiatry*, **154**, 1752–6.

Greenblatt, M., Grosser, G.H., Wechsler, H. (1964). Differential response of hospitalized depressed patients to somatic therapy. *American Journal of Psychiatry*, **120**, 935–43.

Greicius, M.D., Flores, B.H., Menon, V., Glover, G.H., Solvason, H.B., Kenna, H. (2007). Resting-state functional connectivity in major depression: abnormally increased contributions from subgenual cingulate cortex and thalamus. *Biological Psychiatry*, **62**, 429–37.

Gruber, N.P., Dilsaver, S.C., Shoaib, A.M., Swann, A.C. (2000). ECT in mixed affective states: a case series. *Journal of ECT*, **16**, 183–8.

Hanin, B., Srour, N., Margolin, J., Braun, P., Levitin, A.L., Ritsner, M. (1993). Electroconvulsive therapy in mania: successful outcome despite short duration of convulsions. *Convulsive Therapy*, **9**, 50–3.

Heller, A.C., Amar, A.P., Liu, C.Y., Apuzzo, M.L. (2006). Surgery of the mind and mood: a mosaic of issues in time and evolution. *Neurosurgery*, **59**, 720–33; discussion 733–9.

Herrmann, L.L. & Ebmeier, K.P. (2006). Factors modifying the efficacy of transcranial magnetic stimulation in the treatment of depression: a review. *Journal of Clinical Psychiatry*, **67**, 1870–6.

Homan, S., Lachenbruch, P.A., Winokur, G., Clayton, P. (1982). An efficacy study of electroconvulsive therapy and antidepressants in the treatment of primary depression. *Psychological Medicine*, **12**, 615–24.

Ikeji, O.C., Ohaeri, J.U., Osahon, R.O., Agidee, R.O. (1999). Naturalistic comparative study of outcome and cognitive effects of unmodified electro-convulsive therapy in schizophrenia, mania and severe depression in Nigeria. *East African Medical Journal*, **76**, 644–50.

Jha, A.K., Stein, G.S., Fenwick, P. (1996). Negative interaction between lithium and electroconvulsive therapy – a case-control study.[see comment]. *British Journal of Psychiatry*, **168**, 241–3.

Jimenez, F., Velasco, F., Salin-Pascual, R., *et al.* (2005). A patient with a resistant major depression disorder treated with deep brain stimulation in the inferior thalamic peduncle. *Neurosurgery*, **57**, 585–93; discussion 585–93.

Johansen-Berg, H., Gutman, D.A., Behrens, T.E., *et al.* (2008). Anatomical connectivity of the subgenual cingulate region targeted with deep brain stimulation for treatment-resistant depression. *Cerebral Cortex*, **18**, 1374–83.

Klein, E., Kolsky, Y., Puyerovsky, M., Koren, D., Chistyakov, A., Feinsod, M. (1999). Right prefrontal slow repetitive transcranial magnetic stimulation in schizophrenia: a double-blind sham-controlled pilot study. *Biological Psychiatry*, **46**, 1451–4.

Kosel, M., Sturm, V., Frick, C., *et al.* (2007). Mood improvement after deep brain stimulation of the internal globus pallidus for tardive dyskinesia in a patient suffering from major depression. *Journal of Psychiatric Research*, **41**, 801–3.

Kringelbach, M.L., Owen, S.L.F., Aziz, T.Z. (2007). Deep-brain stimulation. *Future Neurology*, **2**, 633–46.

Lesky, E. (1959). Auenbrugger's camphor therapy and the convulsive therapy of psychoses: on the occasion of the 150th anniversary of Auenbrugger's death on 18th May 1959. *Wiener Klinische Wochenschrift*, **71**, 289–93.

Lozano, A.M. (2006). Deep brain stimulation: the new neurosurgery for depression? *European Neuropsychopharmacology*, **16**, S191.

Macedo-Soares, M.B., Moreno, R.A., Rigonatti, S.P., Lafer, B. (2005). Efficacy of electroconvulsive therapy in treatment-resistant bipolar disorder: a case series. *Journal of ECT*, **21**, 31–4.

Mayberg, H S. (1997). Limbic-cortical dysregulation: a proposed model of depression. *Journal of Neuropsychiatry and Clinical Neurosciences*, **9**, 471–81.

Mayberg, H.S., Brannan, S.K., Mahurin, R.K., *et al.* (1997). Cingulate function in depression: a potential predictor of treatment response. *Neuroreport*, **8**, 1057–61.

Mayberg, H.S., Liotti, M., Brannan, S.K., *et al.* (1999). Reciprocal limbic-cortical function and negative mood: converging PET findings in depression and normal sadness. *American Journal of Psychiatry*, **156**, 675–82.

Mayberg, H.S., Lozano, A.M., Voon, V., *et al.* (2005). Deep brain stimulation for treatment-resistant depression. *Neuron*, **45**, 651–60.

McCabe, M.S. (1976). ECT in the treatment of mania: a controlled study. *American Journal of Psychiatry*, **133**, 688–91.

McCabe, M.S. & Norris, B. (1977). ECT versus chlorpromazine in mania. *Biological Psychiatry*, **12**, 245–54.

McLoughlin, D.M., Mogg, A., Eranti, S., *et al.* (2007). The clinical effectiveness and cost of repetitive transcranial magnetic stimulation versus electroconvulsive therapy in severe depression: a multicentre pragmatic randomised controlled trial and economic analysis. *Health Technology Assessment (Winchester, England)*, **11**, 1–54.

Morgane, P.J., Galler, J.R., Mokler, D.J. (2005). A review of systems and networks of the limbic forebrain/limbic midbrain. *Progress in Neurobiology*, **75**, 143–60.

Mukherjee, S. (1989). Mechanisms of the antimanic effect of electroconvulsive therapy. *Convulsive Therapy*, **5**, 227–43.

Mukherjee, S. & Debsikdar, V. (1992). Unmodified electroconvulsive therapy of acute mania: a retrospective naturalistic study. *Convulsive Therapy*, **8**, 115–25.

Mukherjee, S., Sackeim, H.A., Schnur, D.B. (1994). Electroconvulsive therapy of acute manic episodes: a review of 50 years' experience.[see comment]. *American Journal of Psychiatry*, **151**, 169–76.

Naranjo, C.A., Tremblay, L.K., Busto, U.E. (2001). The role of the brain reward system in depression. *Progress in Neuro-Psychopharmacology & Biological Psychiatry*, **25**, 781–823.

Nemeroff, C.B., Mayberg, H.S., Krahl, S.E. *et al.* (2006). VNS therapy in treatment-resistant depression: clinical evidence and putative neurobiological mechanisms. *Neuropsychopharmacology*, **31**, 1345–55.

NICE (2007). *Interventional Procedures Programme: Interventional procedure overview of transcranial magnetic stimulation for severe depression.* www.nice.org.uk/ip346overview [accessed 6 January 2010].

Pataki, J., Zervas, I.M., Jandorf, L. (1992). Catatonia in a university inpatient service (1985–1990). *Convulsive Therapy*, **8**, 163–73.

Pecina, S., Smith, K.S., Berridge, K.C. (2006). Hedonic hot spots in the brain. *The Neuroscientist*, **12**, 500–11.

Perris, C. & d ' Elia, G. (1966). A study of bipolar (manic-depressive) and unipolar recurrent depressive psychoses. X. Mortality, suicide and life-cycles. *Acta Psychiatrica Scandinavica, Supplementum*, **194**, 172–89.

Petrides, G., Dhossche, D., Fink, M., Francis, A. (1994). Continuation ECT: relapse prevention in affective disorders. *Convulsive Therapy*, **10**, 189–94.

Pizzagalli, D., Pascual-Marqui, R.D., Nitschke, J.B., *et al.* (2001). Anterior cingulate activity as a predictor of degree of treatment response in major depression: evidence from brain electrical tomography analysis. *American Journal of Psychiatry*, **158**, 405–15.

Rettig, G.M., York, M.K., Lai, E.C., *et al.* (2000). Neuropsychological outcome after unilateral pallidotomy for the treatment of Parkinson's disease. *Journal of Neurology, Neurosurgery, and Psychiatry*, **69**, 326–36.

Rohland, B.M., Carroll, B.T., Jacoby, R.G. (1993). ECT in the treatment of the catatonic syndrome. *Journal of Affective Disorders*, **29**, 255–61.

Sackeim, H.A. (2001). The definition and meaning of treatment-resistant depression. *Journal of Clinical Psychiatry*, **62** (Suppl. 16), 10–17.

Sackeim, H.A., Brannan, S.K., Rush, A.J. George, M.S., Marangell, L.B., Allen, J. (2007). Durability of antidepressant response to vagus nerve stimulation (VNS). *International Journal of Neuropsychopharmacology*, **10**, 817–26.

Sakas, D.E. & Panourias, I.G. (2006). Rostral cingulate gyrus: a putative target for deep brain stimulation in treatment-refractory depression. *Medical Hypotheses*, **66**, 491–4.

Sartorius, A., Wolf, J., Henn, F.A. (2005). Lithium and ECT – concurrent use still demands attention: three case reports. *World Journal of Biological Psychiatry*, **6**, 121–4.

Schlaepfer, T.E., Cohen, M.X., Frick, C., *et al.* (2008). Deep brain stimulation to reward circuitry alleviates anhedonia in refractory major depression. *Neuropsychopharmacology*, **33**, 368–77.

Schlaepfer, T.E. & Lieb, K. (2005). Deep brain stimulation for treatment of refractory depression. *Lancet*, **366**, 1420–2.

Schultz, W. (2001). Reward signaling by dopamine neurons. *The Neuroscientist*, **7**, 293–302.

Sheline, Y.I., Chibnall, J.T., George, M.S., Fletcher, J.W., Mintun, M.A. (2006). Cerebral blood flow changes during vagus nerve stimulation for depression. *Psychiatry Research*, **146**, 179–84.

Sikdar, S., Kulhara, P., Avasthi, A., Singh, H. (1994). Combined chlorpromazine and electroconvulsive therapy in mania. *British Journal of Psychiatry*, **164**, 806–10.

Small, J.G., Klapper, M.H., Kellams, J.J., *et al.* (1988). Electroconvulsive treatment compared with lithium in the management of manic states. *Archives of General Psychiatry*, **45**, 727–32.

Steele, J.D., Currie, J., Lawrie, S.M., Reid, I. (2007). Prefrontal cortical functional abnormality in major depressive disorder: a stereotactic meta-analysis. *Journal of Affective Disorders*, **101**, 1–11.

Stromgren, L.S. (1988). Electroconvulsive therapy in Aarhus, Denmark, in 1984: its application in nondepressive disorders. *Convulsive Therapy*, **4**, 306–13.

Taylor, M.A. & Fink, M. (2003). Catatonia in psychiatric classification: a home of its own.[see comment]. *American Journal of Psychiatry*, **160**, 1233–41.

Thomas, J. & Reddy, B. (1982). The treatment of mania. A retrospective evaluation of the effects of ECT, chlorpromazine, and lithium. *Journal of Affective Disorders*, **4**, 85–92.

Tremblay, L.K., Naranjo, C.A., Graham, S.J., *et al.* (2005). Functional neuroanatomical substrates of altered reward processing in major depressive disorder revealed by a dopaminergic probe. *Archives of General Psychiatry*, **62**, 1228–36.

Tundo, A., Decina, P., Toro, L. (1991). Mixed depressive syndrome and response to ECT [abstract]. *Biological Psychiatry*, **29**, 212.

UK ECT Review Group. (2003). Efficacy and safety of electroconvulsive therapy in depressive disorders: a systematic review and meta-analysis [see comment]. *Lancet*, **361**, 799–808.

Ungvari, G.S., Kau, L.S., Wai-Kwong, T., Shing, N.F. (2001). The pharmacological treatment of catatonia: an overview. *European Archives of Psychiatry & Clinical Neuroscience*, **251** (Suppl. 1), 131–4.

Vaidya, N.A., Mahableshwarkar, A.R., Shahid, R. (2003). Continuation and maintenance ECT in treatment-resistant bipolar disorder. *Journal of ECT*, **19**, 10–16.

Vanelle, J.M., Loo, H., Galinowski, A., *et al.* (1994). Maintenance ECT in intractable manic-depressive disorders. *Convulsive Therapy*, **10**, 195–205.

Volpe, F.M. & Tavares, A. (2004). Manic patients receiving ECT in a Brazilian sample. *Journal of Affective Disorders*, **79**, 201–8.

Xia, G., Gajwani, P., Muzina, D.J., *et al.* (2008). Treatment-emergent mania in unipolar and bipolar depression: focus on repetitive transcranial magnetic stimulation. *International Journal of Neuropsychopharmacology*, **11**, 119–30.

Ziemann, U. & Hallett, M. (2007). Basic neurophysiologial studies with transcranial magnetic stimulation. In: George, M.S. & Belmaker, R.H. (Eds). *Transcranial Magnetic Stimulation in Clinical Psychiatry*. Washington, DC: American Psychiatric Publishing, Inc., pp. 59–84.

Zornberg, G.L. & Pope Jr, H.G. (1993). Treatment of depression in bipolar disorder: new directions for research. *Journal of Clinical Psychopharmacology*, **13**, 397–408.

Chapter 8

Treating bipolar disorder in the early stages of illness

E. Jane Garland and Anne Duffy

Introduction

At least one-quarter of patients with bipolar disorder (BD) experience the onset of their first major mood episode during adolescence (Leboyer *et al.*, 2005). Recently, there has been an increase in the diagnosis of BD in young people in North America, as evidenced by a 40-fold increase over an 8-year period in office visits (Moreno *et al.*, 2007), and a three-fold increase in paediatric hospital discharge diagnoses (Blader & Carlson, 2007). This trend has, in part, been associated with the broadening of the interpretation of diagnostic criteria.

Despite an apparent increase in the prevalence of BD in children and adolescents and in publications addressing this, there is limited research on acute treatment effectiveness, and even less on long-term treatment and outcome. Many recommendations are derived from open studies and extrapolations from adult research. Therefore, this review integrates both research evidence and clinical experience in order to guide the clinician in rational and practical management.

Diagnostic issues

Diagnostic issues: adolescent-onset bipolar disorder

Effective treatment is predicated on an accurate diagnosis. We are now beginning to understand and recognise the early stages of BD, which often appear during adolescence. The earliest mood manifestations may be recurrent brief episodes of depression (Duffy *et al.*, 2007a), and the first major mood episode is typically depressive in polarity. Clinical indicators suggestive of underlying or latent bipolarity in a depressed adolescent include psychomotor retardation, psychosis and a positive family history of bipolar disorder (Strober & Carlson, 1982). Hypomanic and manic episodes typically occur several years following the index episode of depression, although subthreshold activation may manifest earlier than full-blown episodes. Although there is a concern that earlier-onset BD may be associated with a worse outcome, the limited information available suggests that the long-term outcome is similar to that of adult-onset BD. Early-onset disorder has also been associated with higher frequency of psychotic features, but this might be a manifestation of a particular subtype or developmental vulnerability.

During adolescence, evaluation of potential hypomanic symptoms can be complicated by developmental parent–child dynamics, characterised by rebellion and the need for autonomy, recreational drug experimentation or peer-influenced experimental behaviour (Carlson, 2005). Although overt mania is evident through the profound and persistent changes in mood, sleep, energy, activity, thinking and judgement, hypomania is more difficult to clinically differentiate from normal elation. This challenge is illustrated by the observation that hypomanic symptoms have been reported at an implausibly high rate among high-school students (Lewinsohn *et al.*, 1995). Follow-up study has demonstrated that these symptomatic subjects do not develop BD in adulthood (Lewinsohn *et al.*, 2000). There is a need for more systematic research on the frequency and the meaning of hypomanic-like symptoms in adolescents, specifically to evaluate whether they are a normal variant or a non-specific indicator of risk. Confirmation of the diagnosis will require careful delineation of a clear onset of illness, recurrent episodes and a distinct quality of euphoric or depressed mood, along with associated changes in neurovegetative symptoms and functioning. Accurate history, longitudinal observation and good collateral information are essential to distinguish true BD from other primary problems such as anxiety disorders or substance-use disorders with cocaine and crystal methamphetamine, which can produce episodic, cyclical mood and behavioural changes.

The most robust risk factor predicting BD is a positive family history. The heritability of bipolar disorder

Practical Management of Bipolar Disorder, eds. Allan H. Young, I. Nicol Ferrier and Erin E. Michalak. Published by Cambridge University Press.

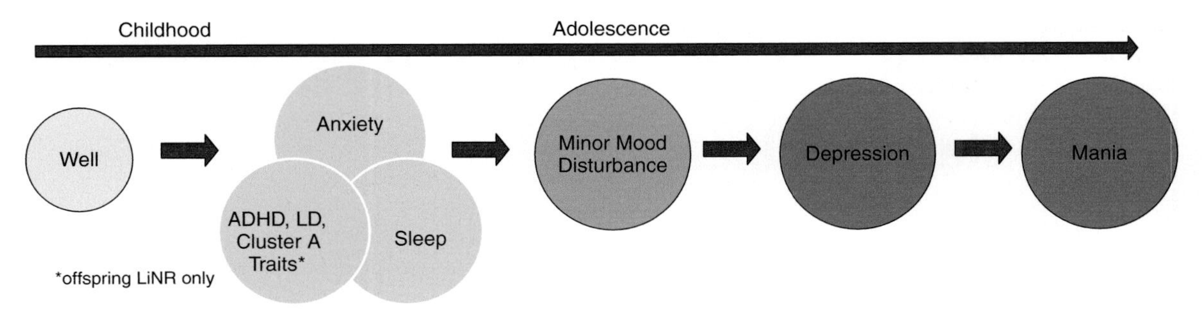

Figure 8.1

is the highest of all psychiatric disorders, including schizophrenia (Farmer *et al.*, 2007). Because of potential overreporting of BD, one should ideally interview any reportedly affected adult family member to verify that the symptoms are consistent with BD rather than another psychiatric or psychological disturbance; if the affected adult family member is not available, detailed questioning of informants regarding symptomatology should be undertaken to try to clarify the basis of the presumed BD. Research has shown that family history is sensitive enough to identify illness, but not specific enough to determine with certainty the nature of the disorder (Andreasen *et al.*, 1997).

There are numerous reports of high rates of 'co-morbid' disorders associated with early-onset BD. In high-risk studies, cognitive symptoms, sleep disturbances and/or anxiety disorders have been described as antecedent early stages in the emerging mood disorder (Duffy *et al.*, 2007a).

In clinical samples of early-onset BD, symptoms of attention deficit hyperactivity disorder (ADHD), other externalising disorders and anxiety disorders, among others, have been reported and considered to be unrelated disturbances. However, rather than multiple co-morbidities, the likelihood should be considered that there are antecedent conditions that represent the earliest stages of the development of BD; that is, there is one core illness, rather than several illnesses manifesting in the same individual. The fact that the *Diagnostic and Statistical Manual of Mental Disorders* (APA, 2000) (DSM-IV-TR) enables the coding of several separate diagnoses in the same individual reinforces the concept of several concurrent problems. A recent study showed significant differences in child psychiatric diagnostic practices between clinicians in the UK and the USA, the former opting for a single diagnosis (ADHD) and the latter for concurrent co-morbid diagnoses (i.e. ADHD and mania) for the same case presentations (Dubicka *et al.*, 2007). Diagnosing co-morbid disorders increases the probability of polypharmacy, potentially increasing side effects, or the use of medications that may adversely affect the course of illness, potentially leading to earlier onset or rapid cycling (Reichart & Nolen, 2004).

Diagnostic issues: manic-like symptoms in pre-adolescent children

Although classical BD is rare in children, over the past 10 years there have been reports of children manifesting chronic symptoms of severe irritability and behavioural dysregulation, hypothesised by some as representing pre-adolescent mania (Carlson, 2005). Modifications or re-interpretations of the DSM-IV-TR criteria have been proposed to allow for the diagnosis of BD in the presence of a non-episodic illness or a chronically irritable, emotionally reactive state, which is inconsistent with the formal diagnostic criteria and the description of the disorder in adolescents and adults. Conceptualisation of these presentations as 'paediatric bipolar disorder' may be an assumptive leap (Leibenluft *et al.*, 2003; Harrington & Myatt, 2003); therefore, an alternative category of 'severe mood and behavioural dysregulation' (SMD) has been proposed to prevent premature labelling and to provide a basis for systematic study of children manifesting this syndrome.

Studies in both clinical and community-based samples have shown that children meeting criteria for SMD do not develop BD in adolescence, and they differ from adolescent bipolar subjects in important ways (Brotman *et al.*, 2006). Specifically SMD children do not have an elevated family risk of BD (Brotman *et al.*, 2007) and have different cognitive and clinical correlates compared with bipolar youth (Dickstein *et al.*, 2005). Rather than a form of very early-onset or pre-adolescent mania, the emerging data suggest that this syndrome may represent a general indicator of risk for a broad range of psychiatric, psychological and behavioural problems. In order to safeguard the integrity of research strategies and clinical practice guidelines, it is essential to clearly separate 'narrow phenotype' classical BD from the broader and more common group of children with mixed psychopathology.

Although the syndrome of SMD in children does not appear to be consistent with BD, high-risk studies of children at genetic risk have mapped out the early childhood psychopathological manifestations of the illness. Specifically, children at specific familial risk have been reported to manifest episodic or 'mini-clusters' of depressive and manic-like symptoms (Shaw *et al.*, 2005), which may meet criteria for cyclothymia. In addition, some of these children manifest non-mood antecedents, including sleep and anxiety disorders, followed by subsyndromal depressive symptoms, before manifesting syndromal mood disorder in early adolescence (Duffy *et al.*, 2007a).

Diagnostic issues: summary

More research is needed to enable us to reliably describe the early stages of BD. However, it appears at this time that adolescent-onset BD represents the early stages of an adult-type BD, whereas pre-pubertal onset of SMD represents a more general risk of mixed psychopathology. There is evidence that the early stages of BD evolve through antecedents such as anxiety, sleep disturbances and subthreshold mood disturbances, to full threshold mood episodes, usually initially depressive.

Pharmacotherapy

Overview of evidence-based treatment

Compared with the wealth of research investigating effective treatments for various phases of BD in adults (mania, depression and prophylaxis), there is a relative lack of data regarding treatment during the early stages of illness in affected youth. However, it is clear that severe, persistent mood disorders onset in adolescence, and that effective treatment of these acute episodes and prophylaxis against recurrence are appropriate to protect against burden of illness effects. This review summarises the available mood stabilisers, current level of evidence for efficacy in young people and guiding symptoms for medication choice.

Pharmacotherapy of acute mania

There are few double-blind placebo-controlled (DBPC) studies of mood-stabilising pharmacotherapy for children and adolescents. However, emerging evidence indicates the efficacy of mood stabilisers in adolescent mania, similar to that seen in adults. Treatment recommendations for the early stages of BD are therefore extrapolated from a small number of DBPC trials, open studies, adult research and clinical experience.

The strongest evidence for antimanic effectiveness in adolescent patients is for lithium, including a DBPC trial in substance-abusing adolescents (Geller *et al.*, 1998); lithium was also efficacious in several small earlier DBPC trials (see Duffy, 2006, for review). There is a surprising lack of placebo-controlled trials of valproate in adolescents, although it is presumed to be effective based on adult data and several prospective open trials (see Smarty & Findling, 2007, for review). A double-blind controlled trial in 50 hospitalised adolescents with BD type I mania or mixed state compared valproate to quetiapine (DelBello *et al.*, 2006) without placebo control over a 28 d trial; both were effective, although quetiapine at a dose of 400–600 mg daily was more rapidly effective than valproate titrated to a serum level of 80–120 μg/ml. In a post-hoc analysis from this trial, in a subset of these bipolar adolescents who also had disruptive behaviour disorders, both valproate and quetiapine were equally effective in specifically reducing symptoms of impulsivity and reactive aggression (Barzman *et al.*, 2006). An often-referenced open-label comparison of lithium, valproic acid and carbamazepine (Kowatch *et al.*, 2000) adds little value to our understanding, as the patients were heterogeneous and only three of 42 patients who entered actually completed the 8-week trial. Regarding other anticonvulsants, data are limited. There is a moderately sized (n=116, manic or mixed-state outpatients) failed placebo-controlled trial of oxcarbazine (Wagner *et al.*, 2006) and a failed placebo-controlled topiramate pilot study in 50 adolescents (DelBello *et al.*, 2005). Retrospective trials of carbamazpezine are reviewed elsewhere (Smarty & Findling, 2007). Although recent research in adults is supporting the effectiveness of lamotrigine for mania, depression and maintenance therapy in both DBPC and double-blind lithium-controlled trials, in adolescents there is only an open-label trial of lamotrigine in bipolar depression (Chang *et al.*, 2006b) and some retrospective data.

Promising new data support the efficacy of atypical neuroleptics in adolescent mania, similar to emerging data in adult BD. Quetiapine was at least as effective as valproate in reducing acute manic symptoms in adolescents, as noted above (DelBello *et al.*, 2006), and as add-on therapy to valproate, quetiapine at an average dose of 450 mg daily was more effective than placebo in reducing manic symptoms in hospitalised adolescents (DelBello *et al.*, 2002). In a large (n=161) multisite DBPC

Table 8.1 Evidence and guiding symptoms for pharmacotherapy of early-onset bipolar disorder

Medication	Guiding symptoms	Evidence in youth[a] (see text)	Other target symptoms	Dosing comments: initial/ target	Adverse effects of special relevance to youth
Lithium	First line for classical BD type I or II (mania or depression) acute and long-term treatment; family history of response	I	Aggression; psychosis; substance abuse	300 mg daily, then 300 mg bid, titrate to serum level (0.8–1.0 mmol/l acute mania) 0.6–0.8 mmol/l maintenance	Rare renal toxicity risk; thyroid dysfunction; polyuria; weight gain; acne
Anticonvulsants	Mania, mixed state, co-morbid anxiety or seizures, rapid cycling				
Valproic acid	First line for mixed manic episodes	II–III (open; one DB with quetiapine)	Rages; anxiety	Hospitalised mania, may use loading dose; 250–500 mg daily titrate to therapeutic blood level	Weight gain; ovarian dysfunction; hair loss; sedation
Carbamazepine	Third line	IV	Aggression	100 mg daily increase every few days to 400–600 mg (serum level guides)	Weight gain; sedation; agranulocytosis (rare)
Lamotrigine	BPII, depression; hypomania & maintenance in adults	III	Anxiety, cluster B personality, obsessive–compulsive symptoms	Titrate VERY SLOWLY (12.5 mg bid weekly increase by 12.5 mg bid to range of 100–150 mg)/150–400 mg daily	Risk of serious hypersensitivity rash 1/100 for children vs 1/300 for adults
Topiramate	Mania, weight gain; 3rd line	IV		25 mg bid to 200–300 mg daily	Cognitive dysfunction; metabolic acidosis
Atypical neuroleptics	Psychosis, anxiety, agitation; in combination with lithium or valproate for stabilisation or maintenance				Weight gain; insulin resistance; dyslipidaemia; prolactin elevation; neuroleptic malignant syndrome; small risk of tardive dyskinesia
Quetiapine	First line for psychotic mania; second line for nonpsychotic mania, depression	II (add on; DB with valproate)		Acute mania: from 100 mg, titrate to 400–600 mg daily	Weight gain and prolactin elevation least likely
Olanzapine	First line for psychotic mania, second line for nonpsychotic mania	II		Acute mania: initiate with 5–10 mg daily, titrate to15–20 mg daily	Most pronounced weight gain; prolactin elevation may occur
Risperidone	Second line for mania, psychotic mania	II	Disruptive behaviour disorders; autism spectrum	<3 mg/day as effective as 3–6 mg	Prolactin elevation; EPS
Aripiprazole	Third line for mania, psychotic mania	II	—		Low risk of weight gain

[a]Evidence criteria: level I, meta-analysis or replicated double-blind (DB), randomised controlled trial (RCT) with placebo; level II, at least one DB-RCT with placebo or active comparison condition; level III, prospective uncontrolled trial with 10 or more subjects; level IV, anecdotal reports or expert opinion, retrospective chart reviews.

trial in acute mania, olanzapine was superior to placebo in achieving both response and remission, at a mean daily dose of about 10 mg (Tohen *et al.*, 2007). Risperidone was effective in an as-yet unpublished DBPC trial in manic 10- to 17-year-olds, with doses up to 2.5 mg as effective as higher doses (Mathis, 2007). Similarly, results of a large multicentre DBPC trial of aripiprazole (n=296), as yet unpublished, reportedly demonstrate significant superiority over placebo (Chang *et al.*, 2007); there are case reports but no DBPC trials of ziprasidone to date. Previous open trials and retrospective chart reviews have also supported effectiveness of atypical neuroleptics (see Smarty & Findling, 2007, for review).

Choice of initial agent

Although guidelines tell us what agents are effective across groups of patients, they do not indicate which agent might work best as the initial mood stabiliser in a particular patient. Evidence suggests that there is differential response between identifiable subgroups (Alda *et al.*, 2004) (see Table 8.1). For example, lithium appears to be most effective for classical or typical mania, whereas in mania with prominent psychotic features, atypical antipsychotics may be more appropriate monotherapy (Kafantaris *et al.*, 2001; Duffy, 2006). In acutely agitated mania, combination therapy may more rapidly reduce agitation and limit the risk to self and others.

Guiding features supporting lithium initiation include discrete episodes, clear manic episodes, family history of response to lithium and familial loading for recurrent mood disorders, not schizophrenia (Duffy, 2006). Lithium was also effective for bipolar adolescents with co-morbid substance abuse (Geller *et al.*, 1998).

Indications for primary mood stabilisation with an anticonvulsant agent, particularly valproate, include mixed, irritable and rapid-cycling mania, and co-morbid anxiety. Alternative anticonvulsants include lamotrigine, especially for prominent depressive symptoms and anxiety. Topiramate may be considered a third-line agent, especially with marked weight-gain issues and failure of other agents for mania. Carbamazepine may be an appropriate third-line alternative in the presence of aggression and self-injurious behaviour, especially in developmentally delayed individuals. Individuals with neurodevelopmental disorders accompanied by seizures may be treated with anticonvulsants as a first line.

Guiding features suggesting initiation of treatment with atypical neuroleptics, especially quetiapine and olanzapine, include the presence of psychotic features, agitation and anxiety. For co-morbid disruptive behaviour disorders, risperidone is also an appropriate initial medication. Benzodiazepines are another class of medication commonly used adjunctively with mood stabilisers for the specific purpose of short-term stabilisation of acute agitated mania.

Treatment of bipolar depression

Although most research addresses acute mania, the depressive polarity is often more recurrent and persistent over the course of the illness. Lithium has been proven effective as a treatment for both acute mania and acute depression in BD and recurrent depressive illness in adults (Schou, 1999). Lamotrigine is effective in the treatment of adult bipolar depression and in prolonging the time before recurrence; however, there is only one open-label trial of lamotrigine in adolescents with bipolar depression (Chang *et al.*, 2006b). New data affirm the antidepressant efficacy of some atypical neuroleptics (quetiapine, olanzapine) in adults with bipolar depression (reviewed in Chapter 4 of this book), but adolescent data are lacking.

Generally, prescription of antidepressants without a mood stabiliser is not recommended. Adjunctive antidepressants should be short term, as they risk converting an episodic illness into a chronic fluctuating illness. However, for patients who are not accepting treatment with mood stabilisers, a short course of antidepressant medication at the lowest effective dose may be indicated. The patient must be closely monitored for paradoxical response, switching, rapid cycling, agitation and/or suicidal ideation.

With regard to 'breakthrough depressions' in patients taking mood stabilisers, the most frequent cause of relapse is non-adherence. Thyroid dysfunction should be ruled out during depressive recurrences, especially in lithium-treated patients. The primary treatment for bipolar depression is to continue and optimise the primary mood-stabilising agent, and to augment with the psychotherapeutic and supportive coping skills, as discussed below.

Maintenance therapy

Bipolar disorder is by nature episodic and recurrent, and therefore the use of a maintenance mood stabiliser to prevent recurrences is recommended. Whereas long-term data in adolescents are limited, evidence from case series and open trials suggests that mood stabilisers are well tolerated and effective as maintenance therapy, as in adults.

Treatment guidelines currently recommend several potentially effective agents, including lithium, anticonvulsants and atypical antipsychotics. However, we have good evidence that BD is a heterogeneous diagnostic category capturing different subtypes of illness (Alda *et al.*, 2004), and studies in adult patients have shown that there is a differential response to individual mood stabilisers (Grof, 2003; Passmore *et al.*, 2003), which has a familial association (Grof *et al.*, 2002). By selecting a mood stabiliser for individual patients based on their clinical characteristics and family history of response, the response rate increases to a range of 70–80% (Schou, 1999; Grof, 2003), whereas arbitrary selection of a mood stabiliser drops long-term stabilisation rates to only 20–30% (Garnham *et al.*, 2007). Poor response may then lead to combination therapy, which might have been unnecessary had the effective monotherapy been selected in the first place. While there is a suggestion from discontinuation studies (Kafantaris *et al.*, 2004; Smarty & Findling, 2007) that early-onset BD will require combination therapy, other research evidence suggests that affected adolescents tolerate and respond to selected monotherapy (Duffy *et al.*, 2007b).

It is not clear from the literature if intervening successfully early in the course of BD changes outcome in terms of burden of illness, although logically this seems plausible. From adult studies, we know that lithium-treated bipolar patients have a normalised suicidal risk and a substantially reduced morbidity risk. There is mounting evidence supporting a neuroprotective effective of both lithium and valproate, operating through several mechanisms involving reduction of oxidative damage, and these effects may even be synergistic (Leng *et al.*, 2008). Neurocognitive and neuroimaging studies suggest a positive correlation between the number of mood episodes and the cognitive deficit and/or anatomical changes (Sheline *et al.*, 2003). These data suggest that prevention of recurrent illness is of paramount importance, particularly at a crucial time for academic, vocational and social development.

There are no data as yet with regard to the relative risks and benefits of very early pharmacological intervention even before the first threshold mood episode in adolescents at genetic risk. Although there has been some preliminary work in this area (Chang *et al.*, 2006a), the risks and benefits of preventative intervention are not clear. This is an important direction for future research.

Medical and mood monitoring

Standard medical and laboratory initial workup and ongoing monitoring for mood stabilisation medication is indicated as in adults, including baseline weight and height, thyroid function, electrocardiogram, and specific metabolic monitoring (including fasting blood sugar, cholesterol, triglycerides) and monitoring for potential movement disorders when neuroleptics are prescribed. Blood levels of mood stabilisers such as lithium and valproate will guide dosing.

During medication trials, systematic charting of mood and key target symptoms in response to each change is essential for evaluating effectiveness in the individual patient. Data from parent and teacher observations are also helpful. 'Mood chart' self-monitoring should include hours of sleep, direction and intensity of mood changes, while making note of potential de-stabilisers such as menstrual cycles, sleep deprivation, significant stressors and travel. Standardised clinician rating scales are recommended for evaluating severity and response during acute episodes (e.g. Young mania rating scale and Hamilton depression rating scale, reviewed in Chapter 18 of this book). Lifeline graphing of polarity, severity and duration of recurrent mood episodes is recommended for evaluating effectiveness of prophylaxis, and potentially predicting times of risk.

Adverse effects of concern in young people

Potential adverse effects need to be discussed in detail with young people and their guardians. Besides the narrow therapeutic index, which requires responsible dosing, long-term risks of lithium include the relatively rare occurrence of nephrotoxicity (see Duffy, 2006, for review). Common concerns that may affect patient acceptance of lithium treatment include polydipsia, weight gain and exacerbation of acne and other skin conditions. Valproate can be associated with weight gain, ovarian dysfunction associated with menstrual irregularities and hair loss, which can be particularly upsetting to adolescents. Education must be provided regarding those anticonvulsants that might interfere with oral contraceptive efficacy (lamotrigine and carbamazepine). The risk of teratogenicity with all mood stabilisers must be discussed, including anticonvulsants in particular. Serious hypersensitivity rash with lamotrigine is more prevalent in individuals under the age of 16 than in adults, at a rate of 1/100 versus 1/300 in adults; the risk may be reduced with slow titration, and is increased by concurrent valproate therapy.

Although atypical neuroleptics are associated with lower rates of prolactin elevation, extrapyramidal side effects (EPS) and risk of tardive dyskinesia than typical neuroleptics, risperidone at doses higher than 2–3 mg is more typical in its dopamine-blocking effects. Aripiprazole is associated with EPS. In young people, prolactin elevation may be persistent with both risperidone and olanzapine (Alfaro *et al.*, 2002). Quetiapine is least associated with this effect, although galactorrhoea has been reported with quetiapine as well as other atypicals. All of the atypical neuroleptics have been associated with case reports of neuroleptic malignant syndrome.

Weight gain is one of the most distressing side effects of atypical neuroleptics, significantly higher than seen with lithium or valproate (see Correll, 2007, for review). These gains can be dramatic, with an average reported increase of 3.8 kg on olanzapine versus 0.3 kg on placebo in a 3-week adolescent mania trial (Tohen *et al.*, 2007). This weight gain is often associated with other indicators of metabolic syndrome, including glucose intolerance and hyperlipidaemia, which have potential long-term adverse effects on cardiovascular health. Among the atypical neuroleptics indicated for BD, quetiapine is least strongly associated with weight gain. Aripiprazole, which appears to be effective in mania (Chang, 2007), and ziprasidone are not associated with weight gain. Among the anticonvulsants, topiramate and lamotrigine are anticonvulsant mood stabilisers that are not associated with weight gain, although the efficacy of topiramate as a long-term mood stabiliser is questionable.

Because weight gain associated with mood stabilising agents is not only a health concern, but also a factor that may interfere with medication adherence, this specific risk needs to be addressed proactively with educational and preventative strategies. For example, lithium-associated weight gain is partially correlated with polydipsia, and ensuring that non-calorie containing beverages are consumed may reduce the risk considerably. For atypical neuroleptics in particular, lack of normal satiety signals can contribute to intractable persistent overeating. With this knowledge, the patient can learn to estimate visually what is the right amount of food and to distract or reassure themselves even though they may still feel 'hungry' after eating a sufficient amount. Bulkier food and eating by the clock may assist with this problem. Addressing the potential weight gain associated with mood stabilisers provides an opportunity to enhance other lifestyle factors such as nutrition, including ensuring that there are adequate essential fatty acids, which have been reported to augment antidepressant treatment in depression, and regular daily exercise, which may also enhance mood and reduce irritability as well as improve cardiovascular health.

Enhancing medication adherence

Medication adherence was a key predictor of outcome in a 12-month follow-up of bipolar adolescents after index hospitalisation (DelBello *et al.*, 2007), and full medication adherence was found to be low, at 35%. In order to improve adherence, the clinician, patient and family should be in agreement with the treatment plan. This requires a clear discussion of diagnosis and treatment approach, as well as detailed and frequent monitoring of acceptance, tolerability and other concerns that may interfere with adherence.

Vulnerable times for non-adherence include stable periods and at the onset of mania, when insight deteriorates. Factors enhancing adherence may include a positive therapeutic alliance, regular monitoring even when well, and continued psychoeducation about the effectiveness of medication and the benefits of staying well. Trusted support people can be enlisted, with the agreement of the patient, to remind them of the importance of regular appointments, and of checking in with the doctor at any sign of destabilisation. Acute medical responsiveness to the family's or patient's concerns is then essential to abort preventable deterioration. Routine follow-up reviews can be scheduled at identified times in the yearly cycle when the individual has relapsed in the past.

Non-adherence may also result from a patient's perceptions or experiences of adverse effects that affect their wellbeing and self-concept. In the acute phase, sedation and gastrointestinal effects may be concerns. Over time, specific issues such as hair loss with valproate or exacerbation of skin problems with lithium may become problematic. Most commonly, weight gain – common to many mood stabilisers, as discussed above – interferes with the adolescent's willingness to take medication or parents' willingness to continue medication in younger children. Proactively addressing this issue in collaboration with family members can potentially reduce the risk of non-adherence.

Evidence-based psychotherapy

Psychoeducation regarding lifestyle stabilisation and medication adherence, as well as supportive monitoring are essential components of bipolar treatment in young people. Psychoeducation appears to be helpful as adjunctive therapy in adolescent bipolar disorder (Fristad *et al.*, 2003), although controlled data are lacking.

Specific psychotherapies have also been researched as adjunctive therapy for BD. Controlled trials in adult BD indicate that family-focused therapy (FFT), targeting communication, emotional reactivity and problem-solving, is more beneficial in preventing relapse than usual care. In adults, several therapeutic modalities, including FFT, interpersonal therapy (IPT) and social rhythm therapy (SRT), are more effective than usual care in preventing relapse. Of these, FFT is more effective than individual psychotherapy in reducing relapses in 2-year follow-up (reviewed in Chapter 6 of this book). The primary benefit of adjunctive psychotherapy appears to be reduction of depressive symptoms and relapse reduction, which may relate to more appropriate lifestyle management, enhanced medication adherence or early intervention with recurrence.

Effectiveness data on psychotherapeutic modalities for early-onset BD are lacking. An open trial of FFT modified for bipolar adolescents is promising (Miklowitz *et al.*, 2004). Interpersonal therapy and SRT have also been adapted to the developmental needs of bipolar adolescents (Hlastala & Frank, 2006).

Resource issues

Locus of treatment

Young people with BD often come to medical attention with acute mania or a suicide attempt associated with depression. Acute hospitalisation may lead to diagnosis and acute stabilisation, but the long-term outcome depends on effective and appropriate aftercare, which will allow optimisation of treatment, enhance treatment adherence and assist in prevention of complications such as substance abuse. Comprehensive support for bipolar youth requires expert psychiatric consultation within a specialised multidisciplinary team with assertive case management, in conjunction with family and educational options, including day treatment or brief hospitalisations when indicated.

Family and school supports

Bipolar disorder places a burden of care and stress upon the family. Affordable medical care and medications must be provided, in addition to parent and family psychoeducation and practical supports.

Adaptation of the school programme will usually be required as a result of the episodic disruptions caused by illness during this crucial educational phase, and the impairment of concentration and function associated with both mania and depression. Supportive alternatives include part-time attendance supplemented by tutors and online courses, or smaller educational settings. A day programme provides comprehensive educational options, specific psychotherapeutic components, medication monitoring and systematic ongoing evaluation of mental state for optimisation of treatment. Supported school programmes or day programmes can assist young people to accomplish adolescent developmental tasks despite the disruptive effects of BD.

Continuity of care during transition to young adulthood

Older adolescents, especially 17- to 24-year-olds, are often lost in the transition from a family or school-focused child psychiatry system to an individualised adult system, at a time of very high risk for onset and recurrence of mood episodes. The clinician in adult-oriented services lacks access to collateral information on symptoms and functioning. The young adult with BD loses a whole network of advocating adults to contact the physician at the first signs of deterioration, or to advocate for access to services and resources. These young adults are at risk of disrupted schooling, unemployment and living in poverty, as well as unstable living situations. All these factors may further destabilise their illness or lead to problems paying for medication. Increased substance abuse in this age range also may result in the young person having difficulty accessing care or being excluded from some resources.

As BD is a lifelong illness, transition planning and continuity of care must be priorities in service design, as well as flexibility in age limits to ensure that a young person makes the transition smoothly when they are stable enough to adapt to the new care system, rather than at an arbitrary time determined by their birth date. A bipolar subspecialty team spanning adolescence and adulthood, or a community mental health team specialised in chronic mental illness, with active case management and outreach, can prevent disengagement from care at this vulnerable stage of life. Patient support groups and public advocacy for disability services can also help bridge the gap. At all stages of life, continued involvement of the family and significant support people should be maintained if possible, but this needs to be negotiated with the patient in a respectful and constructive way that ultimately aims to enhance their healthy independence.

Management of bipolar disorder in youth: summary

Bipolar disorder often onsets with subthreshold and then full-threshold mood episodes in adolescence, with the polarity of the first episode often being depressive. Acute and maintenance mood-stabilising treatments appear to be as efficacious in the early stages of illness as they are in adulthood. Non-adherence and low acceptability of pharmacological treatment are major problems to be addressed, as is loss of identified or at-risk young people to follow-up.

The treatment approach should emphasise a systematic medication protocol combined with psychoeducation and mood-management skills. Medication adherence and avoiding substance abuse can be encouraged with active case management and supportive resources. Stabilisation may be facilitated by family interventions and an appropriate educational setting adapted for an individual with a potentially relapsing mood disorder.

References

Alda, M., Grof, P., Rouleau, G., Turecki, G., Young, L.T. (2004). Neurobiology of treatment response in affective disorders. *European Neuropsychopharmacology*, **13**, S155.

Alfaro, C.L., Wudarsky, M., Nicolson, R., *et al.* (2002). Correlation of antipsychotic and prolactin concentrations in children and adolescents acutely treated with haloperidol, clozapine, or olanzapine. *Journal of Child and Adolescent Psychopharmacology*, **12**, 83–91.

Andreasen, N.C., Endicott, J., Spitzer, R.L., Winokur, G. (1997). The family history method using diagnostic criteria: reliability and validity. *Archives of General Psychiatry*, **34**, 1229–35.

APA (2000). *The Diagnostic and Statistical Manual of Mental Disorders*, 4th edition. American Psychiatric Association, Arlington, VA, USA. www.psychiatryonline.com/resourceTOC.aspx?resourceID=1 [accessed 23 December 2009].

Barzman, D.H., DelBello, M.P., Adler, C.M., Stanford, K.E., Strakowski, S.M. (2006). The efficacy and tolerability of quetiapine versus divalproex for the treatment of impulsivity and reactive aggression in adolescents with co-occurring bipolar disorder and disruptive behavior disorders. *Journal of Child & Adolescent Psychopharmacology*, **16**, 665–70.

Blader, J.C. & Carlson, G.A. (2007). Increased rates of bipolar disorder diagnoses among U.S. child, adolescent and adult inpatients, 1996–2004. *Biological Psychiatry*, **62**, 107–14.

Brotman, M.A., Schmajuk, M., Rich, B.A., *et al.* (2006). Prevalence, clinical correlates, and longitudinal course of severe mood dysregulation in children. *Biological Psychiatry*, **60**, 991–7.

Brotman, M.A., Kassem, L., Reising, M.M., *et al.* (2007). Parental diagnoses in youth with narrow phenotype bipolar disorder or severe mood dysregulation. *American Journal of Psychiatry*, **164**, 1238–41.

Carlson, G. (2005). Early onset bipolar disorder: clinical and research considerations. *Journal of Clinical Child and Adolescent Psychology*, **34**, 333–43.

Chang, K., Howe, M., Gallellii, K., Miklowitz, D. (2006a). Prevention of pediatric bipolar disorder: integration of neurobiological and psychosocial processes. *Annals of the New York Academy of Sciences*, **1094**, 235–47.

Chang, K.D., Nyilas, M., Aurang, C., *et al.* (2007). *Efficacy of aripiprazole in children (10–17 years old) with mania.* Abstract presented at the 54th annual meeting of the Academy of Child and Adolescent Psychiatry, Boston, MA, 23–28 October 2007.

Chang, K., Saxena, K., Howe, M. (2006b). An open-label study of lamotrigine adjunct or monotherapy for the treatment of adolescents with bipolar depression. *Journal of the American Academy of Child and Adolescent Psychiatry*, **45**, 298–304.

Correll, C.U. (2007). Weight gain and metabolic effect of mood stabilizers and antipsychotics in pediatric bipolar disorder: a systematic review and pooled analysis of short-term trials. *Journal of the American Academy of Child and Adolescent Psychiatry*, **46**, 687–700.

DelBello, M.P., Findling, R.L., Kushner, S., *et al.* (2005). A pilot controlled trial of topiramate for mania in children and adolescents with bipolar disorder. *Journal of the American Academy of Child and Adolescent Psychiatry*, **44**, 539–47.

DelBello, M.P., Hanseman, D., Adler, C.M., Fleck, D.E., Strakowski, S.M. (2007). Twelve-month outcome of adolescents with bipolar disorder following first hospitalization for a manic or mixed episode. *American Journal of Psychiatry*, **164**, 582–90.

DelBello, M.P., Kowatch, R.A., Adler, C.M., *et al.* (2006). A double-blind randomized pilot study comparing quetiapine and divalproex for adolescent mania. *Journal of the American Academy of Child and Adolescent Psychiatry*, **45**, 305–13.

DelBello, M.P., Schwiers, M.L., Rosenberg, H.L., Strakowski, S.M. (2002). A double-blind, randomized, placebo-controlled study of quetiapine as adjunctive treatment for adolescent mania. *Journal of the American Academy of Child and Adolescent Psychiatry*, **41**, 1216–23.

Dickstein, D.P., Milham, M.P., Nugent, A.C., *et al.* (2005). Frontotemporal alterations in pediatric bipolar disorder: results of a voxel-based morphometry study. *Archives of General Psychiatry*, **62**, 734–41.

Dubicka, B., Carlson, G., Vail, A., Harrington, R. (2007). Prepubertal mania: diagnostic differences between US and UK clinicians. *European Child & Adolescent Psychiatry*, **17**, 153–61.

Duffy, A. (2006). Lithium treatment in children and adolescents: a selected review and integration of research findings. In: Bauer, M., Grof, P., Muller-Oerlinghausen, B. (Eds). *Lithium in Neuropsychiatry: the Comprehensive Guide*. London: Taylor & Francis Books Ltd., pp. 193–206.

Duffy, A., Alda, M., Crawford, L., Milin, R., Grof, P. (2007a). The early manifestations of bipolar disorder: a longitudinal prospective study of the offspring of bipolar parents. *Bipolar Disorders*, **9**, 828–38.

Duffy, A., Alda, M., Milin, R., Grof, P. (2007b). A consecutive series of treated affected offspring of parents with bipolar disorder: is response associated with the clinical profile? *Canadian Journal of Psychiatry*, **52**, 369–75.

Farmer, A., Elkin, A., McGuffin, P. (2007). The genetics of bipolar affective disorder. *Current Opinion in Psychiatry*, **20**, 8–12.

Fristad, M.A., Gavazzi, S.M., Mackinaw-Koons, B. (2003). Family psychoeducation: an adjunctive intervention for children with bipolar disorder. *Biological Psychiatry*, **53**, 1000–1008.

Garnham, J., Munro, A., Slaney, C., *et al.* (2007). Prophylactic treatment response in bipolar disorder: results of a naturalistic observation study. *Journal of Affective Disorders*, **104**, 185–90.

Geller, B., Cooper, T.B., Sun, K., *et al.* (1998). Double-blind and placebo-controlled study of lithium for adolescent bipolar disorders with secondary substance dependency. *Journal of the American Academy of Child and Adolescent Psychiatry*, **37**, 171–8.

Grof, P. (2003). Selecting effective long-term treatment for bipolar patients: monotherapy and combinations. *Journal of Clinical Psychiatry*, **64** (Suppl. 5), 53–61.

Grof, P., Duffy, A., Cavazzoni, P., *et al.* (2002). Is response to prophylactic lithium a familial trait? *Journal of Clinical Psychiatry*, **63**, 942–7.

Harrington, R. & Myatt, T. (2003). Is pre-adolescent mania the same condition as adult mania? A British perspective. *Biological Psychiatry*, **53**, 961–9.

Hlastala, S.A. & Frank, E. (2006). Adapting interpersonal and social rhythm therapy to the developmental needs of adolescents with bipolar disorder. *Development and Psychopathology*, **18**, 1267–99.

Kafantaris, V., Coletti, D.J., Dicker, R., Padula, G., Kane, J.M. (2001). Adjunctive antipsychotic treatment of adolescents with bipolar psychosis. *Journal of the American Academy of Child and Adolescent Psychiatry*, **40**, 1448–56.

Kafantaris, V., Coletti, D.J., Dicker, R., Padula, G., Pleak, R.R., Alvir, J.M. (2004). Lithium treatment of acute mania in adolescents: a placebo-controlled discontinuation study. *Journal of the American Academy of Child and Adolescent Psychiatry*, **43**, 984–93.

Kowatch, R.A., Suppes, T., Carmody, T.J., *et al.* (2000). Effect size of lithium, divalproex sodium, and carbamazepine in children and adolescents with bipolar disorder. *Journal of the American Academy of Child and Adolescent Psychiatry*, **39**, 713–20.

Leboyer, M., Henry, C., Paillere-Martinot, M.L., Bellivier, F. (2005). Age at onset in bipolar affective disorders: a review. *Bipolar Disorders*, **7**, 111–18.

Leibenluft, E., Charney, D.S., Towbin, K.E., Bhangoo, R.K., Pine, D.S. (2003). Defining clinical phenotypes of juvenile mania. *American Journal of Psychiatry*, **160**, 430–7.

Leng, Y., Liang, M.H., Ren, M., Marinova, Z., Leeds, P., Chuang, D.M. (2008). Synergistic neuroprotective effects of lithium and valproic acid or other histone deacetylase inhibitors in neurons: roles of glycogen synthase kinase-3 inhibition. *Journal of Neuroscience*, **28**, 2576–88.

Lewinsohn, P.M., Klein, D.N., Seeley, J.R. (1995). Bipolar disorders in a community sample of older adolescents: prevalence, phenomenology, comorbidity, and course. *Journal of the American Academy of Child and Adolescent Psychiatry*, **34**, 454–63.

Lewinsohn, P.M., Klein, D.N., Seeley, J.R. (2000). Bipolar disorder during adolescence and young adulthood in a community sample. *Bipolar Disorders*, **2**, 281–93.

Mathis, M.V. (2007). *Recommendation of approvable action for risperidone (Risperdal®) for the treatment of schizophrenia and bipolar I disorder in pediatric patients*. www.psychrights.org/Drugs/risperidone_clinical_BPCA070618.pdf [accessed 10 January 2010].

Miklowitz, D.J., George, E.L., Axelson, D.A., *et al.* (2004). Family-focused treatment for adolescents with bipolar disorder. *Journal of Affective Disorders*, **82S**, S113–S128.

Moreno, C., Laje, G., Blanco, C., Jiang, H., Schmidt, A.B., Olfson, M. (2007). National trends in the outpatient diagnosis and treatment of bipolar disorder in youth. *Archives of General Psychiatry*, **64**, 1032–9.

Passmore, M., Garnham, J., Duffy, A., *et al.* (2003). Phenotypic spectra of bipolar disorder in responders to lithium versus lamotrigine. *Bipolar Disorders*, **5**, 110–14.

Reichart, C.G. & Nolen, W.A. (2004). Earlier onset of bipolar disorder in children by antidepressants or stimulants? A hypothesis. *Journal of Affective Disorders*, **78**, 81–4.

Rybakowski, J., Chlopocka-Wozniak, M., Suwalska, A. (2001). The prophylactic effect of long-term lithium administration in bipolar patients entering treatment in the 1970s and 1980s. *Bipolar Disorders*, **3**, 63–7.

Schou, M. (1999). Perspectives on lithium treatment: action, efficacy, effect on suicidal behavior. *Bipolar Disorders*, **1**, 5–10.

Shaw, J.A., Egeland, J.A., Endicott, J., Allen, C.R., Hostetter, A.M. (2005). A 10-year prospective study of prodromal patterns for bipolar disorder among Amish youth. *Journal of the American Academy of Child and Adolescent Psychiatry*, **44**, 1104–11.

Sheline, Y.I., Gado, M.H., Kraemer, H.C. (2003). Untreated depression and hippocampal volume loss. *American Journal of Psychiatry*, **160**, 1–3.

Smarty, S. & Findling, R.L. (2007). Psychopharmacology of pediatric bipolar disorder: a review. *Psychopharmacology*, **191**, 39–54.

Strober, M. & Carlson, G. (1982). Bipolar illness in adolescents with major depression. *Archives of General Psychiatry*, **39**, 549–55.

Tohen, M., Kryzhanovskaya, L., Carlson, G., *et al.* (2007). Olanzapine versus placebo in the treatment of adolescents with bipolar mania. *American Journal of Psychiatry*, **164**, 1547–56.

Wagner, K.D., Kowatch, R.A., Emslie, G.J., *et al.* (2006). A double-blind, randomized, placebo-controlled trial of oxcarbazepine in the treatment of bipolar disorder in children and adolescents. *American Journal of Psychiatry*, **163**, 1179–86.

Chapter 9

Special populations: the elderly

Alan J. Thomas

The management of bipolar disorder (BD) in older people is plagued by the same kind of difficulties as for the management of schizophrenia and depression for this age group. These include two assumptions: first, that the clinical features remain essentially the same into old age and that therefore the diagnostic criteria in DSM-IV-TR (APA, 2000) and ICD-10 (WHO, 2007) can be applied without modification; and second, that the treatments used in younger adults also remain the appropriate treatments for older adults with BD. The paucity of good clinical research into late-life BD makes entertaining these assumptions unavoidable but the clinician should be aware that both assumptions may be wrong. Such difficulties are in turn related to the likely aetiological differences in late-life BD, with clinicians continuing to find attractive the concept of secondary mania, which has some recognition in DSM-IV-TR in the category 'mood disorder due to a general medical condition' (APA, 2000: 293.83). This reflects the reality that a manic presentation in an older person is much more likely to be related to physical illness or organic brain disease.

The first part of this chapter will review the (limited) evidence for such differences in late-life BD and the second part will consider the management of BD in older people, taking into account the evidence for differences in clinical presentation.

Clinical features of late-life bipolar disorder

Age of onset and clinical features

The age of onset of BD has long been regarded as an important factor in distinguishing different subpopulations of older people with BD, with those with a late age of onset having more 'organic' features. Although it may be possible to distinguish three BD subgroups by age of onset (early, intermediate and late onset) (Leboyer *et al.*, 2005), all the studies of older adults with BD have compared late-onset with early-onset groups. Although studies suggest such a split is valid, the age at which this split occurs remains unclear. Analysing the Danish case register for BD, Kessing reported a bimodal distribution of age of illness onset with a high intermode age of 65 (Kessing, 2006). Cassidy and Carroll also found a bimodal age of onset of BD among 366 bipolar patients but with a much younger intermode, of 47 (Cassidy & Carroll, 2002). Comparing studies is thus made more difficult because different age splits are used, with cut-offs varying from 45 to 65.

Although higher figures have been produced for inpatients (Tohen *et al.*, 1990), community studies report the mean age of onset of BD to be about 22 (Kessler *et al.*, 1997; Schaffer *et al.*, 2006). By contrast, studies of older bipolar patients report a relatively late age of onset of the illness of about 50 years (Broadhead & Jacoby, 1990; Snowdon, 1991; Shulman *et al.*, 1992) and that among older inpatients with BD there were few subjects with a first episode of mania occurring before 40 (Snowdon, 1991; Shulman *et al.*, 1992). Although these findings might reflect selection bias, they show that psychiatrists managing older people are looking after a different group of patients and not merely a group of older patients.

A population-based study in Denmark (Kessing, 2006) reported that those with a late-onset BD (first episode after 50 years of age) had fewer psychotic symptoms due to fewer manic episodes but more often presented with psychotic symptoms during their depressive episodes. Although this may appear unusual, this is consistent with an earlier study (Broadhead & Jacoby, 1990), which found older manic patients to have attenuated manic symptoms than younger patients and with literature in unipolar depression reporting older people with major depression to have more severe psychotic symptoms (Brodaty *et al.*, 1997).

Practical Management of Bipolar Disorder, eds. Allan H. Young, I. Nicol Ferrier and Erin E. Michalak. Published by Cambridge University Press.

Earlier studies reported evidence of cognitive impairment in older people with mania (Stone, 1989; Broadhead & Jacoby, 1990; Dhingra & Rabins, 1991), and a study of euthymic bipolar patients finding that over half of the elderly subjects scored more than a standard deviation below age-matched controls on the mini-mental state examination (MMSE) and the Mattis dementia rating scale (Gildengers *et al.*, 2004) showed that this impairment appears to be a trait feature of the illness itself. Further analysis of these patients has reported impairment in information processing speed and executive function, and such deficits are associated with significantly impaired functioning (Gildengers *et al.*, 2007). This has previously been reported in younger adults with euthymic BD (Ferrier *et al.*, 1999), and although neurocognitive impairment is also a feature of BD in earlier life it seems the degree of impairment is likely to be more severe in older bipolar patients. However, a direct comparison with younger bipolar patients would strengthen this widely held view (Young *et al.*, 2006). It would also be important to demonstrate that such findings are not confounded by the inclusion of people with early dementia, especially fronto-temporal dementia (FTD) (Neary *et al.*, 1998). The potential for misdiagnosing BD in people presenting with early dementia, especially FTD, has been subject to little comment or research, unlike the parallel situation with depression and dementia, but the NICE bipolar guidelines do refer to its importance (National Collaborating Centre for Mental Health, 2006). It is well recognised that dementia in older people may present as depression (a prodromal presentation) and a disinhibited, chaotic and overactive clinical presentation could be due to an analogous prodromal presentation of FTD (or a frontal presentation of other dementia). Differentiating dementia from BD has important implications for treatment, because studies have failed to show a benefit from using anticonvulsants in behavioural disturbance in dementia (Konovalov *et al.*, 2008), whereas they have a clear place in bipolar management.

It seems old-age psychiatry services see people with an older onset of BD who have attenuated mania, more depression and related psychosis and possibly more severe neurocognitive deficits. It may be that people whose BD begins in early adult life disappear by old age through a combination of high mortality, 'burn out' of manic symptoms with a change to depressive symptoms, and perhaps loss of contact with specialist services and effective treatment of mania. At the same time perhaps a new group of later-onset patients emerges who present to older people's services, who have more cognitive impairment and who account for those in such services with relatively late onset of their mania.

Illness course and outcome

Although most bipolar subjects experience an illness course with both manic and depressive episodes, this is not always the case, and the delay to the first manic episode can be many years. One important finding from studying older bipolar patients is that over half of those whose first episode is a depression go on to have at least three depressive episodes before developing their first manic episode (Shulman & Post, 1980; Stone, 1989; Snowdon, 1991). Furthermore, the latency from this first episode of depression to the first manic episode is a long one – a mean of 15 years with almost a quarter having a latency of over 25 years (Shulman & Post, 1980; Shulman *et al.*, 1992). Such findings emphasise the importance of maintaining a high suspicion about the potential presence of BD, even among older, apparently unipolar, patients. A second group of late-life bipolar patients are those discussed above with a late onset, however this is defined. One study reported that about a quarter of 92 patients over 65 with BD had no previous history of any affective illness (Stone, 1989). Another important group of patients are those with unipolar mania, with up to 12% of older bipolar patients in one study meeting the criteria of three clear manic episodes and no depressive episode over an illness period of at least 10 years (Shulman & Tohen, 1994). As might be expected, this subgroup had a younger age of illness onset than their peers.

Mortality rates are increased in late-life BD patients compared with the general elderly population (Dhingra & Rabins, 1991). In a study comparing age- and sex-matched elderly unipolar and bipolar subjects, Shulman and colleagues found that over 6 years more than half of the bipolar patients had died compared with about one-fifth of those with unipolar disorder (Shulman *et al.*, 1992). Suicide has often been reported as increased in this patient group (Tsuang, 1978; Tsuang *et al.*, 1980; Weeke & Vaeth, 1986) and in a 40-year follow-up of patients with severe mental illness, suicide was significantly increased, at 8.5%, compared with controls and to a comparable extent with unipolar disorder (10.6%) (Winokur & Tsuang, 1975). However, the increased mortality in BD is not only due to suicide but also due to infective diseases and vascular diseases (Tsuang *et al.*, 1980).

Secondary mania and neurological co-morbidity

'Secondary mania' is a term used by clinicians to indicate that some individuals with a late-life BD appear to have developed manic presentation due to organic brain disease or other physical illness. Whether the term 'secondary mania' is used or not, it is important in older patients to give careful consideration to the possible role of 'organic conditions' in the development of late-life mania (National Collaborating Centre for Mental Health, 2006). In recent years much interest has focused more specifically on a possible role for vascular disease in the development of BD in older people (see below). However, secondary mania has been reported as resulting from a wide range of conditions, including cerebral infarctions (Jampala & Abrams, 1983), white matter lesions (Fujikawa *et al.*, 1995), following open heart surgery (Isles & Orrell, 1991), traumatic brain injury (Jorge *et al.*, 1993), thyrotoxicosis (Lee *et al.*, 1991), HIV infection (Lyketsos *et al.*, 1993) and dementia due to Alzheimer's disease (Lyketsos *et al.*, 1995).

Closely related to the secondary mania concept is the reported high prevalence of neurological co-morbidity in older people diagnosed with BD (Shulman, 1997b). Although a higher co-morbidity is bound to be present compared with younger adults, this phenomenon may also be related to age of onset, because a comparison of older inpatients with BD found that those with a late onset of the illness (mean age of onset, 75) had a much higher prevalence of comorbid neurological illness (71 vs 28%) than older patients who had had multiple previous episodes and thus an early age of onset (Tohen *et al.*, 1994). Within neurological literature, the term 'disinhibition syndrome' is sometimes used for clinical presentations similar to secondary mania and in which lesions in the orbito-frontal cortex and lateralisation to the right hemisphere are frequently reported (Shulman, 1997a). However, the evidence base consists of case reports and small cases, and although plausible, more robust studies are needed to validate such a neuroanatomical relationship with manic syndromes.

Neuroimaging studies and vascular disease

An MRI study of the caudate nucleus in 36 subjects with late-life BD reported a significant reduction on only the right side compared with matched controls, but late-onset patients (first episode after 45) had a decrease in total brain volume compared with early-onset patients and in total caudate volume compared with the controls (Beyer *et al.*, 2004a). The same group also reported the left hippocampus to be larger in patients but no differences were found by age of onset (Beyer *et al.*, 2004b). Volumetric studies of these same anatomical structures in late-life depression have been inconsistent, and further and larger studies are needed to determine whether consistent volume changes are present and how any such abnormalities might relate to age of illness onset.

Paralleling the literature on unipolar disorder in later life, in a cross-sectional study of 74 older people with late-onset BD, Hays and colleagues found them to have a weaker family history of BD but stronger for cerebrovascular disease than those with an early onset of BD (Hays *et al.*, 1998). However, a difficulty in interpreting these findings is that this and other studies examining family history (Broadhead & Jacoby, 1990; Snowdon, 1991; Shulman *et al.*, 1992; Hays *et al.*, 1998) have produced a wide range, from 24 to 88%, for a positive family history of BD, meaning that such an apparent difference might simply be an artefact, and larger studies with more robust case finding for family history are needed to clarify this issue. By contrast, several studies have confirmed an increase in vascular disease in older people with BD.

In a cross-sectional study of 62 people with late-life BD (all were over 60), those with a first episode after 50 were found to have more cerebrovascular risk factors than the early-onset group (Wylie *et al.*, 1999). The above case-note review of 366 bipolar patients found age of onset to be bimodal, and dividing the group by the intermodal age – 47 – it was found that those with a late onset had higher cholesterol levels and more vascular risk factors (Cassidy & Carroll, 2002). Subramaniam and colleagues compared older people with BD (all over 65) who had early onset (n=30, first episode before 60) with those with a late onset (n=20) (Subramaniam *et al.*, 2007). None of the individuals had dementia, and nor did they differ in cognition. The late-onset group had a higher Framingham stroke risk score (Wolf *et al.*, 1991).

In a related parallel to the unipolar literature, white matter and subcortical grey matter hyperintense lesions on MRI appear to be increased in those with late-life BD. An early report comparing 12 late-life bipolar patients (all over 50) with age- and sex-matched controls found an increase in large white matter hyperintensities (WMHs) in the deep white matter (Mcdonald *et al.*, 1991). The same group examined 70 consecutive admissions to their affective disorders unit with BD (mean age 50) and again found that compared with 70 age- and sex-matched controls these bipolar patients had more WMHs in the deep white matter and also an increase in hyperintense lesions in the subcortical grey matter (Mcdonald *et al.*, 1999). De Asis and colleagues compared WMHs and subcortical grey matter hyperintensities in 40 late-life bipolar patients (>60 years) with 15 age-matched controls and found a highly significant excess of these lesions in the frontal white matter in the bipolar group and that age of illness onset was associated with this excess of WMH in the right frontal lobe (De Asis *et al.*, 2006). However, there was no difference in subcortical hyperintensities.

As these WMHs have been shown to be due to cerebral ischaemia in late-life depression (Thomas *et al.*, 2002), it is likely that such lesions in BD are also a result of cerebrovascular disease, and the increase in late-life BD therefore supports a role for cerebrovascular disease in the development or maintenance of this illness in older people. However, an increase in these lesions has also been reported in younger people with BD (Altshuler *et al.*, 1995; Dupont *et al.*, 1995; Ferrier *et al.*, 1999), and it is not clear whether older bipolar patients have an increase in these lesions compared with younger patients. Such evidence has led to the proposal that there is a subtype of BD related to vascular disease, termed 'vascular mania' (Steffens & Krishnan, 1998), which again parallels the 'vascular depression' hypothesis (Alexopoulos *et al.*, 1997).

Management of late-life bipolar disorder

Management of older people with BD follows the same principles as for younger adults, but with two key differences. First, there is a more pressing need for a thorough physical assessment, including a detailed neurological examination and a structural brain scan because of the issues of neurological co-morbidity and cerebrovascular disease discussed above. Such an assessment may identify specific additional treatment needs, e.g. the presence of 'silent' cerebral infarcts, meriting vascular secondary prophylaxis for vascular disease, or age-associated impairments that affect bipolar management, e.g. significantly impaired renal function. Second, and relatedly, pharmacodynamic and pharmacokinetic changes (as discussed below) due to gerontological and geriatric factors alter the risk:benefit ratio in older people, reducing the use of medication. Finally, in considering bipolar management, we will examine only the evidence for the use of medication, because there is almost no evidence specific to older people to inform the use of psychoeducation, psychosocial interventions or electroconvulsive therapy (ECT). Although the presence of neurocognitive impairments may make the use of psychotherapies more challenging in older people with BD, therapeutic nihilism should not prevent their judicious use. Similarly, cognitive impairment probably makes older people more prone to confusion post-ECT, but the evidence for the efficacy of ECT in unipolar disorder (Tew *et al.*, 1999) should encourage its use in BD where it would be used in younger adults.

Altered pharmacology

Although the mode of action of psychotropic agents remains the same for older people, changes in neuronal structure and possibly number seem to make neurones more sensitive to such medication, and this might explain the increase in adverse effects, e.g. tardive dyskinesia and sedation (Jeste, 2000), in older people even on lower doses of medication. Absorption of psychotropic medication when given orally, sublingually or by intramuscular injection is not altered in older people, and although first-pass metabolism is reduced, this by itself does not lead to clinically important increases in active drug in the systemic circulation. Age-related reduction in muscle bulk leads to an increase in the relative amount of fat and thus an increase in the volume of distribution. There is some loss of albumin, but of more clinical importance is competition with warfarin for albumin binding, and care needs to be taken with the use of fluoxetine in particular in patients on regular warfarin. As mentioned, hepatic changes reduce biotransformation, but the impact is probably less than that from genetic variation (Ritchie, 2008). The major gerontological change is the reduction in glomerular filtration rate, typically by about 35%, but there is marked individual variation. This, together with the reduced volume of distribution, frequently makes the safe use of lithium difficult or impossible in older people. One study found that lithium in older people was excreted at half the rate of younger patients (Hardy *et al.*, 1987). For most drugs, the combination of these changes in volume of distribution and biotransformation, and perhaps the increased pharmacodynamic sensitivity, means that lower doses of psychotropic agents are needed compared with those for younger adults. Such changes in the volume of distribution and in biotransformation, for example, strongly affect diazepam, which can build up, leading to prolonged and unhelpful effects, and which is therefore best avoided in older people. By contrast, the biotransformation of lorazepam is unaltered, as it is inactivated by phase II conjugation, and this together with its shorter half-life makes it the preferred benzodiazepine in older people (Ritchie, 2008).

Although not pharmacological matters in a narrow sense, two other issues are important to consider when prescribing for older people. First, polypharmacy continues to be a common feature in older people in general and in those with psychiatric illness in particular, with the attendant increase in drug–drug interactions and adverse effects. For example, a recent study found the mean number of prescriptions among older people in the community to be eight (Cannon *et al.*, 2006) (and it is of course higher in secondary care and among psychiatry patients). Second, it is also important to remember that frailty with ageing and/or the effects of co-morbid diseases make adverse effects in older people more likely to have serious consequences, e.g. oversedation with diazepam is more likely to lead to falls, and these are more likely to lead to fractures.

Lithium

There are no randomised controlled trials on the use of lithium in late-life BD, either for use in acute mania or for maintenance, but it continues to be widely used (Head & Dening, 1998; Shulman *et al.*, 2003). This is changing, however. In part this is due to a cohort effect, with younger bipolar patients growing old on newer maintenance treatments, especially valproate, and in part this reflects a change in prescribing patterns among psychiatrists managing older people, who are tending to initiate these other treatments in preference to lithium, even though there is little specific evidence to guide such a shift in prescribing patterns (Shulman *et al.*, 2003). Although there are no randomised trials, there are a few retrospective studies of lithium for acute mania that suggest that lithium is effective and can be safely used in older people with mania (Van Der Velde, 1970; Himmelhoch *et al.*, 1980; Chen *et al.*, 1999). For example, one study compared 30 'elderly' people (over 55) presenting with mania and treated using lithium with 29 people treated with valproate and found 67% to be rated as improved on lithium compared with 38% on valproate; however, after looking at serum levels of these medications, the authors concluded that this difference was probably due to suboptimal use of valproate (Chen *et al.*, 1999). Evidence for maintenance use of lithium in BD in the elderly is also sparse, with studies again having several limitations, including a mixed age range, retrospective analysis and unblinded ratings (Hewick *et al.*, 1977; Murray *et al.*, 1983; Stone, 1989). These studies suggest that lithium continues to be effective as a maintenance agent but that, as expected, lower levels are associated with poorer outcome. The best maintenance dose for older people remains unclear. Although 0.5 mmol/l has been recommended (Shulman *et al.*, 1987), others have reported better outcome using higher serum levels (Chen *et al.*, 1999). However, higher levels lead to an increase in adverse effects. The reduction in renal clearance referred to above suggests that on average about half the dose for younger adults should be used, but given the wide variability in alterations in renal function, it is probably better to be guided by someone's glomerular filtration rate rather than by their chronological age and to aim for higher maintenance levels if possible.

A further difficulty with using lithium in older people is that the adverse effects of lithium seem to occur more often in older people and to be more severe. Lithium-induced tremor is frequently disabling and a common cause of switching to an alternative maintenance treatment, and neurocognitive effects occur in 31% of elderly people on lithium (Young *et al.*, 2004). A cross-sectional study of older bipolar patients reported 58% to have ECG abnormalities (Roose *et al.*, 1979) and polyuria, polydipsia, weight gain and oedema occurred in 46% of older lithium users in one study (Hewick *et al.*, 1977). A large study of lithium users over the age of 65 found 6% to be on treatment for hypothyroidism – twice the expected prevalence in the general population (Shulman *et al.*, 2005b). Delirium is another adverse effect and frequently leads to hospitalisation (Shulman *et al.*, 2005a), and such problems develop in spite of the use of lower doses and serum levels (Roose *et al.*, 1979; Murray *et al.*, 1983). The problem of drug interactions due to polypharmacy was mentioned earlier, and an observational study of over 10 000 elderly lithium users found 3.9% had been admitted to hospital for lithium toxicity (Juurlink *et al.*, 2004). This study also reported that the initiation of a loop diuretic or an ACE inhibitor was associated with a significant increase (by fivefold and sevenfold, respectively) in lithium toxicity, highlighting the dangers of polypharmacy in older people.

Anticonvulsants

Valproate has been shown to be an effective maintenance treatment in BD, and its use is probably gradually replacing that of lithium (Shulman *et al.*, 2003). Once again, although there are no randomised trials in elderly bipolar patients, several reports have been published. All these studies have been uncontrolled, retrospective assessments, with mixed diagnoses (including unipolar disorder, and dementia in one case (Niedermier & Nasrallah, 1998)) and modest subject numbers (range 13–39). However, they consistently report improvements on valproate and that better outcomes appear to occur in those on higher maintenance doses. Side effects were predictable (sedation, confusion, dysarthria, ataxia and tremor) and dose related, but overall valproate was well tolerated, and adverse effects can be modulated by dose reduction (Shulman & Herrmann, 1999). If efficacy of valproate is equivalent to lithium, for acute and maintenance treatment, in younger adults and as there is no clear evidence of superiority in this age group, then clinically the decision about which drug to use in late-life BD should be guided by tolerability. Concerns about lithium use in older people, justified by the evidence above, have probably helped prompt the reported shift towards preferential valproate use (Shulman *et al.*, 2003). Neurocognitive side effects have been reported less frequently on valproate (13%) than on lithium (31%) (Young *et al.*, 2004). One report, presumably not including those with overt sedation, found no impairment on neuropsychological testing in elderly people taking valproate (Craig & Tallis, 1994). Clinical experience and expert opinion suggest that valproate is reasonably well tolerated (Young *et al.*, 2004; Shulman & Herrmann, 2008), although admission rates for delirium in elderly valpraote users were no different from those of people on lithium (Shulman *et al.*, 2005a).

Carbamazepine

Evidence on the use of carbamazepine in late-life BD is sparse. One double-blind comparison of lithium and carbamazepine included a few older people with BD and claimed equivalent efficacy, although lithium levels were low (Okuma *et al.*, 1990), and a few case reports have made similar claims. As a potent inducer of cytochrome P450, 2D6 carbamazepine is recognised to reduce the serum levels of other medication, e.g. warfarin, valproate and haloperidol, and its quinidine-like properties make it a cause of cardiac rhythm abnormalities, even at therapeutic levels (Kasarskis *et al.*, 1992). Such factors, together with the other recognised adverse effects of carbamazepine, including significant gastrointestinal disturbances, hyponatraemia, rashes and leucopenia, suggest that it is not an appropriate agent to use in older people before other options have been explored.

Other anticonvulsants

There is very little evidence to inform the use of other anticonvulsants in late-life BD. A case series of five patients reported benefits of treating depression with lamotrigine in late-life BD as augmentation for lithium, with three patients achieving and maintaining remission (Robillard & Conn, 2002), and lamotrigine has been studied in a subanalysis of 33 'older' (mean age only 62) bipolar patients on maintenance therapy, and was reported to reduce time to intervention for affective episodes compared with placebo (Sajatovic *et al.*, 2005). Case series and case reports have also reported benefits from using gabapentin (Shulman & Herrmann, 2008).

Antipsychotics

Again, there are no good quality clinical trials examining the use of antipsychotics for BD in older patients, and no reports. Antipsychotics are associated with a range of well-known adverse effects, including weight gain, hyperlipidaemia and diabetes mellitus for atypical agents and parkinsonism, tardive dyskinesia and dystonias for typical drugs, and elderly people appear to tolerate the atypical agents better (Jeste *et al.*, 1999). However, the use of antipsychotics in older people has been complicated by evidence that in dementia they appear to worsen cognitive decline (Ballard *et al.*, 2005) and increase cerebrovascular adverse events and mortality (Schneider *et al.*, 2005). Although there is no evidence that this applies in BD, caution needs to be used in older people with risk factors for cerebrovascular disease.

Antidepressants

There appear to be no data on the use of antidepressants for the management of depression in late-life BD. As with unipolar disorder, selective serotonin reuptake inhibitors are the preferred class of antidepressant, but extrapolation of the data from younger adults suggests that antidepressants should not be used without a concomitant 'mood stabiliser', such as lithium or valproate, because of the risk of precipitating hypomania or mania (National Collaborating Centre for Mental

Health, 2006). However, it may be that this concern is less relevant in older people, because a large retrospective cohort study of the use of antidepressants in late-life BD found that the use of antidepressants was associated with a 50% reduction in the rate of hospitalisation for manic/mixed episodes (Schaffer *et al.*, 2006).

Summary

Distinguishing BD from dementia has important implications for management because of the well-publicised risks associated with using antipsychotics in dementia (Schneider *et al.*, 2005) and the lack of evidence for efficacy of anticonvulsants and lithium for behavioural disturbances in dementia (Konovalov *et al.*, 2008). The high rates of adverse effects from lithium and the increased problems these cause, together with the management difficulties caused by impaired renal function, now make lithium an unattractive option compared with valproate for the maintenance phase of BD in older people. In the acute phase, clinician confidence in the ease of using olanzapine or other antipsychotics and experience with these treatments in other illnesses tend to favour these over lithium and valproate.

References

Alexopoulos, G.S., Meyers, B.S., Young, R.C., Campbell, S., Silbersweig, D., Charlson, M. (1997). 'Vascular depression' hypothesis. *Archives of General Psychiatry*, **54**, 915–22.

Altshuler, L.L., Curran, J.G., Hauser, P., Mintz, J., Denicoff, K., Post, R. (1995). T2 hyperintensities in bipolar disorder: magnetic resonance imaging comparison and literature meta-analysis. *American Journal of Psychiatry*, **152**, 1139–44.

APA (2000). *The Diagnostic and Statistical Manual of Mental Disorders*, 4th edition. American Psychiatric Association, Arlington, VA, USA. www.psychiatryonline.com/resourceTOC.aspx?resourceID=1 [accessed 23 December 2009].

Ballard, C., Margallo-Lana, M., Juszczak, E., *et al.* (2005). Quetiapine and rivastigmine and cognitive decline in Alzheimer's disease: randomised double blind placebo controlled trial. *BMJ*, **330**, 74.

Beyer, J.L., Kuchibhatla, M., Payne, M., *et al.* (2004a). Caudate volume measurement in older adults with bipolar disorder. *International Journal of Geriatric Psychiatry*, **19**, 109–14.

Beyer, J.L., Kuchibhatla, M., Payne, M.E., *et al.* (2004b). Hippocampal volume measurement in older adults with bipolar disorder. *American Journal of Geriatric Psychiatry*, **12**, 613–20.

Broadhead, J. & Jacoby, R. (1990). Mania in old age: a first prospective study. *International Journal of Geriatric Psychiatry*, **5**, 215.

Brodaty, H., Luscombe, G., Parker, G., *et al.* (1997). Increased rate of psychosis and psychomotor change in depression with age. *Psychological Medicine*, **27**, 1205–13.

Cannon, K.T., Choi, M.M., Zuniga, M.A. (2006). Potentially inappropriate medication use in elderly patients receiving home health care: a retrospective data analysis. *American Journal of Geriatric Pharmacotherapy*, **4**, 134–43.

Cassidy, F. & Carroll, B.J. (2002). Vascular risk factors in late onset mania. *Psychological Medicine*, **32**, 359–62.

Chen, S.T., Altshuler, L.L., Melnyk, K.A., Erhart, S.M., Miller, E., Mintz, J. (1999). Efficacy of lithium vs. valproate in the treatment of mania in the elderly: a retrospective study. *Journal of Clinical Psychiatry*, **60**, 181–6.

Craig, I. & Tallis, R. (1994). Impact of valproate and phenytoin on cognitive function in elderly patients: results of a single-blind randomized comparative study. *Epilepsia*, **35**, 381–90.

De Asis, J.M., Greenwald, B.S., Alexopoulos, G.S., *et al.* (2006). Frontal signal hyperintensities in mania in old age. *American Journal of Geriatric Psychiatry*, **14**, 598–604.

Dhingra, U. & Rabins, P.V. (1991). Mania in the elderly: a 5–7 year follow-up. *Journal of the American Geriatrics Society*, **39**, 581–3.

Dupont, R.M., Jernigan, T.L., Heindel, W., *et al.* (1995). Magnetic resonance imaging and mood disorders. Localization of white matter and other subcortical abnormalities. *Archives of General Psychiatry*, **52**, 747–55.

Ferrier, I.N., Stanton, B.R., Kelly, T.P., Scott, J. (1999). Neuropsychological function in euthymic patients with bipolar disorder. *British Journal of Psychiatry*, **175**, 246–51.

Fujikawa, T., Yamawaki, S., Touhouda, Y. (1995). Silent cerebral infarctions in patients with late-onset mania. *Stroke*, **26**, 946–9.

Gildengers, A.G., Butters, M.A., Chisholm, D., *et al.* (2007). Cognitive functioning and instrumental activities of daily living in late-life bipolar disorder. *American Journal of Geriatric Psychiatry*, **15**, 174–9.

Gildengers, A.G., Butters, M.A., Seligman, K., *et al.* (2004). Cognitive functioning in late-life bipolar disorder. *American Journal of Psychiatry*, **161**, 736–8.

Hardy, B.G., Shulman, K.I., Mackenzie, S.E., Kutcher, S.P., Silverberg, J.D. (1987). Pharmacokinetics of lithium in the elderly. *Journal of Clinical Psychopharmacology*, **7**, 153–8.

Hays, J.C., Krishnan, K.R., George, L.K., Blazer, D.G. (1998). Age of first onset of bipolar disorder: demographic, family history, and psychosocial correlates. *Depression & Anxiety*, **7**, 76–82.

Head, L. & Dening, T. (1998). Lithium in the over-65s: who is taking it and who is monitoring it? A survey of older adults on lithium in the Cambridge Mental Health Services catchment area.[see comment]. *International Journal of Geriatric Psychiatry*, **13**, 164–71.

Hewick, D.S., Newbury, P., Hopwood, S., Naylor, G., Moody, J. (1977). Age as a factor affecting lithium therapy. *British Journal of Clinical Pharmacology*, **4**, 201–5.

Himmelhoch, J.M., Neil, J.F., May, S.J., Fuchs, C.Z., Licata, S.M. (1980). Age, dementia, dyskinesias, and lithium response. *American Journal of Psychiatry*, **137**, 941–5.

Isles, L.J. & Orrell, M.W. (1991). Secondary mania after open-heart surgery. *British Journal of Psychiatry*, **159**, 280–2.

Jampala, V.C. & Abrams, R. (1983). Mania secondary to left and right hemisphere damage. *American Journal of Psychiatry*, **140**, 1197–9.

Jeste, D.V. (2000). Tardive dyskinesia in older patients. *Journal of Clinical Psychiatry*, **61**, 27–32.

Jeste, D.V., Lacro, J.P., Bailey, A., Rockwell, E., Harris, M.J., Caligiuri, M.P. (1999). Lower incidence of tardive dyskinesia with risperidone compared with haloperidol in older patients. *Journal of the American Geriatrics Society*, **47**, 716–19.

Jorge, R.E., Robinson, R.G., Starkstein, S.E., Arndt, S.V., Forrester, A.W., Geisler, F.H. (1993). Secondary mania following traumatic brain injury. *American Journal of Psychiatry*, **150**, 916–21.

Juurlink, D.N., Mamdani, M.M., Kopp, A., Rochon, P., Shulman, K.I., Redelmeier, D.A. (2004). Drug-induced lithium toxicity in the elderly: a population-based study. *Journal of the American Geriatrics Society*, **52**, 794–8.

Kasarskis, E.J., Kuo, C.S., Berger, R., Nelson, K.R. (1992). Carbamazepine-induced cardiac dysfunction. Characterization of two distinct clinical syndromes. *Archives of Internal Medicine*, **152**, 186–91.

Kessing, L.V. (2006). Diagnostic subtypes of bipolar disorder in older versus younger adults. *Bipolar Disorders*, **8**, 56–64.

Kessler, R.C., Rubinow, D.R., Holmes, C., Abelson, J.M., Zhao, S. (1997). The epidemiology of DSM-III-R bipolar I disorder in a general population survey. *Psychological Medicine*, **27**, 1079–89.

Konovalov, S., Muralee, S., Tampi, R.R. (2008). Anticonvulsants for the treatment of behavioural and psychological symptoms of dementia: a literature review. *International Psychogeriatrics*, **20**, 293–308.

Leboyer, M., Henry, C., Paillere-Martinot, M.L., Bellivier, F. (2005). Age at onset in bipolar affective disorders: a review. *Bipolar Disorders*, **7**, 111–18.

Lee, S., Chow, C.C., Wing, Y.K., Leung, C.M., Chiu, H., Chen, C.N. (1991). Mania secondary to thyrotoxicosis. *British Journal of Psychiatry*, **159**, 712–13.

Lyketsos, C.G., Corazzini, K., Steele, C. (1995). Mania in Alzheimer's disease. *Journal of Neuropsychiatry & Clinical Neurosciences*, **7**, 350–2.

Lyketsos, C.G., Hanson, A.L., Fishman, M., Rosenblatt, A., Mchugh, P.R., Treisman, G.J. (1993). Manic syndrome early and late in the course of HIV. *American Journal of Psychiatry*, **150**, 326–7.

Mcdonald, W.M., Krishnan, K.R., Doraiswamy, P.M., Blazer, D.G. (1991). Occurrence of subcortical hyperintensities in elderly subjects with mania. *Psychiatry Research*, **40**, 211–20.

Mcdonald, W.M., Tupler, L.A., Marsteller, F.A., *et al.* (1999). Hyperintense lesions on magnetic resonance images in bipolar disorder. *Biological Psychiatry*, **45**, 965–71.

Murray, N., Hopwood, S., Balfour, D.J., Ogston, S., Hewick, D.S. (1983). The influence of age on lithium efficacy and side-effects in out-patients. *Psychological Medicine*, **13**, 53–60.

National Collaborating Centre for Mental Health (2006). *Bipolar Disorder, the Management of Bipolar Disorder in Adults, Children and Adolescents, in Primary and Secondary Care*. London: NICE. http://guidance.nice.org.uk/CG38 [accessed 3 January 2010].

Neary, D., Snowden, J.S., Gustafson, L., *et al.* (1998). Frontotemporal lobar degeneration: a consensus on clinical diagnostic criteria. *Neurology*, **51**, 1546–54.

Niedermier, J.A. & Nasrallah, H.A. (1998). Clinical correlates of response to valproate in geriatric inpatients. *Annals of Clinical Psychiatry*, **10**, 165–8.

Okuma, T., Yamashita, I., Takahashi, R., *et al.* (1990). Comparison of the antimanic efficacy of carbamazepine and lithium carbonate by double-blind controlled study. *Pharmacopsychiatry*, **23**, 143–50.

Ritchie, C.W. (2008). Psychopharmacology in the elderly. In: Jacoby, R., Oppenheimer, C., Dening, T., Thomas, A. (Eds). *The Oxford Textbook of Old Age Psychiatry*. Oxford University Press.

Robillard, M. & Conn, D.K. (2002). Lamotrigine use in geriatric patients with bipolar depression. *Canadian Journal of Psychiatry*, **47**, 767–70.

Roose, S.P., Bone, S., Haidorfer, C., Dunner, D.L., Fieve, R.R. (1979). Lithium treatment in older patients. *American Journal of Psychiatry*, **136**, 843–4.

Sajatovic, M., Gyulai, L., Calabrese, J.R., *et al.* (2005). Maintenance treatment outcomes in older patients with bipolar I disorder. *American Journal of Geriatric Psychiatry*, **13**, 305–11.

Schaffer, A., Cairney, J., Cheung, A., Veldhuizen, S., Levitt, A. (2006). Community survey of bipolar disorder in

Canada: lifetime prevalence and illness characteristics. *Canadian Journal of Psychiatry*, **51**, 9–16.

Schneider, L.S., Dagerman, K.S., Insel, P. (2005). Risk of death with atypical antipsychotic drug treatment for dementia: meta-analysis of randomized placebo-controlled trials.[see comment]. *JAMA*, **294**, 1934–43.

Shulman, K. & Post, F. (1980). Bipolar affective disorder in old age. *British Journal of Psychiatry*, **136**, 26–32.

Shulman, K.I. (1997a). Disinhibition syndromes, secondary mania and bipolar disorder in old age. *Journal of Affective Disorders*, **46**, 175–82.

Shulman, K.I. (1997b). Neurologic comorbidity and mania in old age. *Clinical Neuroscience*, **4**, 37–40.

Shulman, K.I. & Herrmann, N. (1999). Bipolar disorder in old age. *Canadian Family Physician*, **45**, 1229–37.

Shulman, K.I. & Herrmann, N. (2008). Manic syndromes in old age. In: Jacoby, R., Oppenheimer, C., Dening, T., Thomas, A. (Eds). *The Oxford Textbook of Old Age Psychiatry*. Oxford University Press.

Shulman, K.I., Mackenzie, S., Hardy, B. (1987). The clinical use of lithium carbonate in old age: a review. *Progress in Neuro-Psychopharmacology & Biological Psychiatry*, **11**, 159–64.

Shulman, K.I., Rochon, P., Sykora, K., *et al.* (2003). Changing prescription patterns for lithium and valproic acid in old age: shifting practice without evidence. *BMJ (Clinical research ed.)*, **326**, 960–1.

Shulman, K.I., Sykora, K., Gill, S., *et al.* (2005a). Incidence of delirium in older adults newly prescribed lithium or valproate: a population-based cohort study.[see comment]. *Journal of Clinical Psychiatry*, **66**, 424–7.

Shulman, K.I., Sykora, K., Gill, S.S., *et al.* (2005b). New thyroxine treatment in older adults beginning lithium therapy: implications for clinical practice. *American Journal of Geriatric Psychiatry*, **13**, 299–304.

Shulman, K.I. & Tohen, M. (1994). Unipolar mania reconsidered: evidence from an elderly cohort. *British Journal of Psychiatry*, **164**, 547–9.

Shulman, K.I., Tohen, M., Satlin, A., Mallya, G., Kalunian, D. (1992). Mania compared with unipolar depression in old age. *American Journal of Psychiatry*, **149**, 341–5.

Snowdon, J. (1991). A retrospective case-note study of bipolar disorder in old age. *British Journal of Psychiatry*, **158**, 485–90.

Steffens, D.C. & Krishnan, K.R. (1998). Structural neuroimaging and mood disorders: recent findings, implications for classification, and future directions. *Biological Psychiatry*, **43**, 705–12.

Stone, K. (1989). Mania in the elderly. *British Journal of Psychiatry*, **155**, 220–4.

Subramaniam, H., Dennis, M.S., Byrne, E.J. (2007). The role of vascular risk factors in late onset bipolar disorder. *International Journal of Geriatric Psychiatry*, **22**, 733–7.

Tew Jr, J.D., Mulsant, B.H., Haskett, R.F., *et al.* (1999). Acute efficacy of ECT in the treatment of major depression in the old-old. *American Journal of Psychiatry*, **156**, 1865–70.

Thomas, A.J., O ' brien, J.T., Davis, S., *et al.* (2002). Ischemic basis for deep white matter hyperintensities in major depression: a neuropathological study. *Archives of General Psychiatry*, **59**, 785–92.

Tohen, M., Shulman, K.I., Satlin, A. (1994). First-episode mania in late life. *American Journal of Psychiatry*, **151**, 130–2.

Tohen, M., Waternaux, C.M., Tsuang, M.T. (1990). Outcome in mania. A 4-year prospective follow-up of 75 patients utilizing survival analysis. *Archives of General Psychiatry*, **47**, 1106–11.

Tsuang, M.T. (1978). Suicide in schizophrenics, manics, depressives, and surgical controls. A comparison with general population suicide mortality. *Archives of General Psychiatry*, **35**, 153–5.

Tsuang, M.T., Woolson, R.F., Fleming, J.A. (1980). Premature deaths in schizophrenia and affective disorders. An analysis of survival curves and variables affecting the shortened survival. *Archives of General Psychiatry*, **37**, 979–83.

Van der Velde, C.D. (1970). Effectiveness of lithium carbonate in the treatment of manic-depressive illness. *American Journal of Psychiatry*, **127**, 345–51.

Weeke, A. & Vaeth, M. (1986). Excess mortality of bipolar and unipolar manic-depressive patients. *Journal of Affective Disorders*, **11**, 227–34.

WHO (2007). *International Classification of Diseases*, 10th edition. World Health Organization, Geneva. http://apps.who.int/classifications/apps/icd/icd10online/ [accessed 14 December 2009].

Winokur, G. & Tsuang, M. (1975). The Iowa 500: suicide in mania, depression, and schizophrenia. *American Journal of Psychiatry*, **132**, 650–1.

Wolf, P.A., D 'agostino, R.B., Belanger, A.J., Kannel, W.B. (1991). Probability of stroke: a risk profile from the Framingham Study. *Stroke*, **22**, 312–18.

Wylie, M.E., Mulsant, B.H., Pollock, B.G., *et al.* (1999). Age at onset in geriatric bipolar disorder. Effects on clinical presentation and treatment outcomes in an inpatient sample. *American Journal of Geriatric Psychiatry*, **7**, 77–83.

Young, R.C., Gyulai, L., Mulsant, B.H., *et al.* (2004). Pharmacotherapy of bipolar disorder in old age: review and recommendations. *American Journal of Geriatric Psychiatry*, **12**, 342–57.

Young, R.C., Murphy, C.F., Heo, M., *et al.* (2006). Cognitive impairment in bipolar disorder in old age: literature review and findings in manic patients. *Journal of Affective Disorders*, **92**, 125–31.

Chapter 10

Special populations: women and reproductive issues

Karine A.N. Macritchie and Carol Henshaw

Introduction

Bipolar women experience their illness differently to men. They face reproductive health issues that healthy women do not, and in general they carry heavy psychosocial burdens. This chapter begins with an exploration of the course of bipolar illness in women. Then the nature and management of reproductive health issues in bipolar women, including the management of pregnancy and the puerperium, are described. Finally, the influence of the illness on a woman's self-identity and on her chosen roles is considered.

Bipolar illness in women

Gender differences exist in the nature and course of bipolar disorder, in the occurrence of co-morbidities and in the consequences of treatment.

The course of the illness

The prevalence of bipolar disorder (BD) type I is similar between the sexes. Although studies are not entirely consistent, BD type II appears to occur more frequently in women (Barnes & Mitchell, 2005). This difference in prevalence has been observed as early as adolescence and young adulthood (Wittchen *et al.*, 1998). In paediatric BD, girls present with higher rates of depressed mood and boys present with higher rates of manic mood (Duax *et al.*, 2007). Over the course of their illness, women are at greater risk of depressive episodes, mixed states and subsyndromal symptoms than of pure mania (Angst, 1978; Suppes *et al.*, 2005). Depressive episodes are of longer duration in women and are more likely to be treatment refractory (Goodwin & Jamison, 1990).

Rapid-cycling disorder is found to be more common in women than in men (Tondo & Baldessarini, 1998). One reason for this might be the association between rapid cycling and BD type II (Baldessarini *et al.*, 2000). Another might be its association with hypothyroidism, also more common in women (Bauer *et al.*, 1990). The role of sex steroids is unclear, but in one study, a third of women with 'continuous circular' BD developed this pattern in the perimenopausal period (Kukopulos *et al.*, 1980). On the other hand, oestrogen therapy has been reported to induce mania and rapid cycling in post-menopausal women (Young *et al.*, 1997). Psychotropic factors such as greater exposure to antidepressants and inadequate mood stabilisation might also play a role.

The premenstrual and menstrual phases of the menstrual cycle are associated with an increase in rates of hospitalisation, suicide rates and in the severity of suicidal intent in bipolar women. The majority of bipolar women report regular premenstrual exacerbations of mood symptoms, but the effect of the menstrual phase on mood in women with rapid-cycling BD is unclear (see Barnes & Mitchell, 2005, for brief review). Interestingly, one recent placebo-controlled pilot study on tamoxifen reported a significant antimanic effect (Kulkarni *et al.*, 2006).

The perimenopausal period may be associated with a change in the course of their illness: Soares & Taylor (2007), reviewing the effects of 'menopausal transition' on the course of bipolar illness, reported a deterioration in mood symptoms, particularly in those patients who were not receiving hormone replacement therapy. Marsh *et al.* (2008) reported that depressive episodes increased in frequency during the perimenopausal years. Robertson *et al.* (2008) reported that five women with a history of bipolar affective postpartum psychosis experienced major mood episodes at the time of surgical or natural menopause. In three cases, the women had remained entirely well in the intervening period. The fourth had a history of severe premenstrual mood symptoms, and the fifth had suffered two depressive episodes in the interim.

Little is yet known regarding the effects of hormone replacement on the mood symptoms of bipolar women. In women suffering non-bipolar depression

Practical Management of Bipolar Disorder, eds. Allan H. Young, I. Nicol Ferrier and Erin E. Michalak. Published by Cambridge University Press.

in the perimenopause, studies support the efficacy of unopposed oestrogen, but it appears to be ineffective in post-menopausal women. In the studies that were reviewed by Soares and Taylor, oestrogen monotherapy was administered for short periods only (Soares & Taylor, 2007). Longer-term use of unopposed oestrogen carries a risk of endometrial carcinoma.

The most dramatic impact on the course of bipolar illness in women occurs following childbirth. Bipolar women are at high risk of illness in the puerperium: puerperal psychosis occurs after 25–50% of deliveries in bipolar mothers (see Jones & Craddock, 2005, for review), compared with an underlying rate of approximately 0.1% deliveries in the general population.

Co-morbidity

Co-morbid anxiety disorder is common in bipolar patients, but women have higher rates than men. Panic disorder and social phobia are especially problematic (Barnes & Mitchell, 2005). Frye *et al.* (2003) found that BD more than doubled the risk of alcoholism in men compared with those of their gender in the general population. In women, the risk was seven times greater. Alcoholism was strongly associated with polysubstance use in bipolar women. For bipolar women, substance abuse greatly increases the risk of arrest and detention in police custody (McDermott *et al.*, 2007).

Childhood abuse

A study of 650 bipolar subjects found that 65% of the women and 35% of the men gave a history of physical, sexual or verbal abuse in childhood or adolescence. Abused groups had a longer duration of illness and longer periods of untreated illness (Leverich *et al.*, 2002). In adulthood, compared with the general female population, bipolar women are at increased risk of both sexual and physical abuse (Coverdale & Turbott 2000).

Diagnostic delay

In general, women with BD suffer longer periods of illness before the diagnosis is recognised and they receive treatment. McElroy highlighted the diagnostic difficulties presented by bipolar women (McElroy, 2004): the 'presentation of female bipolar disorder may resemble depressive disorders, co-morbid Axis I disorders, Axis II personality disorders, behavioral dysregulation, or general medical disorders; thus, it is critically important for clinicians to assess for a history of hypomania or mania when determining diagnosis in any woman presenting with psychological symptoms'.

Suicidality

A recent population-based study of suicidality in those receiving treatment for BD found that women had significantly higher suicide attempt rates, but lower rates of completed suicide than men (Simon *et al.*, 2007). However, pregnancy and the puerperium are times of high vulnerability for women with severe mental illness. In the UK, the Confidential Enquiries into Maternal Deaths (1997–1999) (CEMD, 2001) found that suicide was one of the leading causes of maternal death, at a rate of 1–2 per 100 000 maternities. The majority of suicides occurred in women with a severe mental illness. The number who suffered with BD was not detailed in the report, but in 17 of the 28 cases of suicide, the women were suffering from depression, depressive psychosis or puerperal psychosis. Many of the suicides were violent in nature. Older age and advantageous social circumstances were not protective.

In summary, the bipolar woman's illness may go unrecognised for a substantial part of her life. She may suffer more depressive, mixed and treatment-resistant episodes than her male counterpart. She is at risk of rapid-cycling disorder and puerperal psychosis, major challenges for the clinician. Pregnancy is not protective against episodes of mental illness, and the puerperal period is a time of high risk for puerperal psychosis and suicide.

Management of reproductive health issues in bipolar women

Reproductive health is a major consideration in the management of bipolar women, and an area of sensitivity for patients. It appears that BD is associated with menstrual dysfunction, even before the onset of the illness. Psychotropic medications prescribed in BD have direct consequences on the wellbeing of the bipolar woman with childbearing potential and may have long-term health consequences for her. Particular care is required in the matters of contraception and planning for pregnancy and the puerperium.

Underlying hypothalamo–pituitary–gonadal axis dysfunction in bipolar disorder

Preliminary evidence suggests that BD is associated with hypothalamo–pituitary–gonadal axis dysfunction in women. Baron *et al.* (1982) reported decreased fertility rates before illness onset. In a large study of menstrual-cycle dysfunction in affective disorders, 34% of bipolar women reported abnormalities before treatment – a significantly higher proportion than those with unipolar depression (24%) and healthy subjects (22%) (see Joffe, 2007, for a description of the trial and review).

Hyperprolactinaemia

The conventional serum level of the pituitary hormone prolactin is below 500 mU/l for men and women. Typical, and some atypical, antipsychotics (especially risperidone) cause hyperprolactinaemia. Prolactin secretion is inhibited by dopaminergic (D_2) hypothalamic neurones acting on the pituitary cells. Antipsychotic agents with a strong affinity for the D_2 receptors block these inhibitory receptors and enhance prolactin release. Haddad & Wieck (2004) provide a detailed review of hyperprolactinaemia and guidelines on its management. These are summarised here, but the reader is encouraged to refer to the original article.

A great deal of individual variation exists in the effects of antipsychotic medications on prolactin levels and on the degree to which symptoms are experienced. The onset of symptoms usually occurs within the first few months of treatment. Compared with men, women exhibit significantly greater elevations in prolactin for the same dose of antipsychotic.

Hyperprolactinaemia manifests in anovulation, menstrual irregularities, galactorrhoea and sexual dysfunction. Oligo- or amenorrhoea is a sign of infrequent ovulation or anovulation, which may lead to subfertility and osteoporosis (see Haddad & Wieck, 2004; Joffe, 2007). Children and adolescents may be especially sensitive to the prolactin-raising effects of antipsychotics (Haddad & Wieck, 2004). Few data exist on the long-term effects of antipsychotic-induced hyperprolactinaemia in this group. However, the potential inhibition of the pubertal growth spurt through the induction of anovulatory states and the effect on sexual development are obvious sources of concern.

Physiological hyperprolactinaemia may arise during lactation, pregnancy, stress and sexual activity. Other agents, such as some antihypertensive medications, histamine receptor antagonists and amphetamines may induce hyperprolactinaemia. Medical causes of elevated prolactin include epileptic seizures, hypothalamic, pituitary and end-organ endocrine abnormalities, chronic renal failure, hepatic failure and its ectopic production from small cell bronchial carcinoma.

An important differential diagnosis is a prolactinoma: a benign adenoma of the anterior pituitary. Visual field defects and the presence of headaches suggest a lesion occupying the sellar space, but their absence does not rule it out. A prolactin level of more than 4500 mU/l is highly suggestive of a prolactinoma. Concentrations of more than 2500 mU/l are consistent with a micro-prolactinoma or a non-functioning adenoma (Haddad & Wieck, 2004). Where no temporal relationship can be established between the commencement of treatment and the onset of hyperprolactinaemia, or where hyperprolactinaemia is severe, a referral to an endocrinologist and MRI neuroimaging should be considered. Management guidelines after Haddad & Wieck (2004) are outlined in Table 10.1.

A small but significant increase in the risk of breast cancer was reported in a large retrospective study of women who were prescribed typical antipsychotics or antiemetic dopamine antagonists (Wang *et al.*, 2002). A 16% increased risk of breast cancer (an adjusted hazard ratio, 1.16; 95% CI 1.07–1.26) was found in this group, and a dose–risk relationship was observed. This was described as a statistically significant preliminary result, reflecting a small added risk in absolute terms. The need for further adequately powered and rigorous studies was highlighted. However, as a small number of breast cancers are prolactin sensitive, prolactin-raising agents are better avoided in patients with this history (Haddad & Wieck, 2004).

Polycystic ovary syndrome

Valproate is a commonly used agent in BD, but its use is problematic in women because of its association with polycystic ovary syndrome (PCOS), a syndrome with serious reproductive, dermatological and metabolic manifestations. The clinical picture of PCOS is that of menstrual irregularity, hirsuitism, alopecia, acne and obesity. Hormone assays show elevated serum testosterone, increased luteinising hormone with a reduced or normal follicle-stimulating hormone, insulin resistance and hyperlipidaemia. There are important associated risks and long-term sequelae: infertility, endometrial hyperplasia (a precursor of endometrial cancer) and the risk of type 2 diabetes mellitus. There is an associated risk of cardiovascular disease due to type 2 diabetes mellitus, dyslipidaemia and systolic hypertension (Lane, 2006).

In the general population, PCOS occurs in up to 7% of women of reproductive age (Lane, 2006). Joffe (2007) reviewed his recent work in valproate-treated bipolar women. In one study, his team compared 86 premenopausal bipolar women taking valproate with 144 bipolar women taking another anticonvulsant or mood stabiliser. Of the valproate-treated group, 10.5% developed new-onset menstrual abnormalities

Table 10.1 Management of women on prolactin-elevating medication

Record the patient's pattern of menstrual cycle: duration, intervals and amount of bleeding
Record any history of previous lactation/sexual dysfunction and assess risk factors for osteoporosis
Obtain a pre-treatment serum prolactin level
Warn patient of, and monitor for, symptoms of hyperprolactinaemia
If the patient develops symptoms:
Check serum morning prolactin level (and further blood tests below)
For mild-to-moderate elevations in prolactin, repeat to exclude physiological surges
Consider the differential diagnosis described in the text
History/physical examination/pregnancy test/thyroid function test/blood urea and creatinine
Consider an MRI to exclude a sellar-space-occupying lesion
Consider bone mineral density measurements if the patient is hyperprolactinaemic + amenorrhoeic for 12 months
Seek the opinion of an endocrinologist, if in doubt about the causal link with the antipsychotic treatment
Features that suggest antipsychotic-induced hyperprolactinaemia, but do not conclusively exclude other causes:
The agent is well documented to cause hyperprolactinaemia
Onset of symptoms occurred shortly after antipsychotic commenced/dose increased
There is an absence of signs and symptoms of a sellar-space-occupying lesion
A prolactin level of <2000 mU/l is obtained
Other laboratory tests are normal
A space-occupying lesion is excluded on MRI
Other factors to consider:
The degree of distress caused by the patient's symptoms
The benefit of antipsychotic treatment and the risk of relapse on withdrawal of the antipsychotic
The anticipated duration of secondary amenorrhoea and the associated risk of decreased bone mineral density
Possible treatment options:
Decrease the dose of antipsychotic
Switch to a prolactin-sparing antipsychotic
A dopamine receptor agonist for symptoms of hyperprolactinaemia
A combined oral contraceptive for symptoms of oestrogen deficiency

Adapted from Haddad & Wieck, 2004
NB. If stopping an oral antipsychotic is not feasible, but clarification of its role in precipitating hyperprolactinaemia is required, a short diagnostic cessation of medication over 3 d should result in a fall in prolactin levels. However, prolactin levels can remain elevated for 6 months after stopping a depot antipsychotic.

and hyperandrogenism compared with 1.4% in the non-valproate-treated group. Menstrual abnormalities on commencing valproate were apparent within the first year. There is preliminary evidence that valproate-induced PCOS remits on withdrawal, although menstrual cycles may not normalise for up to a year.

The differential diagnoses of valproate-induced PCOS should be considered, especially when the temporal relationship with treatment is unclear. Several other endocrine disorders can be mistaken for PCOS. Hypogonadotrophic hypogonadism may occur due to nutritional and metabolic disturbances, including anorexia nervosa and bulimia. Tumours of the central nervous system such as craniopharyngiomas and germinomas may disrupt follicle-stimulating hormone secretion. Hyperprolactinaemia, hypothyroidism, hyperadrenalism and androgen-producing tumours should also be considered (see Lane, 2006, for review).

Although other mechanisms have been proposed, valproate probably induces PCOS through increased androgen production. This may occur by a direct stimulatory effect on theca cell androgen synthesis or through the inhibition of oestradiol production from testosterone (see Joffe, 2007).

These consequences of valproate treatment, in addition to its teratogenic potential, have led to the recommendation that it not be prescribed in bipolar women of childbearing potential, and adolescents in particular (NICE, 2007). For the individual cases in which valproate treatment is judged to be clinically essential, guidelines on the management of valproate-induced polycystic ovary syndrome are summarised in Table 10.2.

Table 10.2 Management of premenopausal women on valproate

Before beginning treatment:
Discuss the risk of developing PCOS on valproate, the teratogenic risk and the need for contraception
Assess pre-treatment menstrual cycle patterns/gynaecological history (e.g. endometrial disease)
Assess for hirsuitism, acne and hair loss
Record the patient's weight and blood pressure
Consider history of hyperlipidaemia/diabetes and measurement of serum lipid and glucose levels
After valproate treatment is commenced:
Monitor for the onset of symptoms closely, especially in the first year
In the presence of menstrual dysfunction, consider the differential diagnosis in the text
In particular, check for pregnancy, hyperprolactinaemia and the perimenopause
Consider referral to a gynaecologist, who may request hormonal assays and a pelvic ultrasound
Abnormalities in the hormone profile of women with PCOS[a]:
LH:FSH ratio in early follicular phase (between D3 and D6 of the cycle) (>2)
'Mid-luteal' progesterone low (<6 nmol/l)
Testosterone early follicular phase (between D3 and D6 of the cycle) (>4 nmol/l)
Fasting glucose >7.8 mmol/l indicative of diabetes mellitus
Glucose:insulin ratio >4 suggestive of reduced insulin sensitivity
Treatment options:
Change of the mood stabiliser
Continue treatment while seeking an endocrinology and/or gynaecology opinion

Adapted from Joffe, 2007 and Bauer *et al.*, 2002.
[a] Values may vary with local laboratory reference ranges.

General and sexual health care

Few studies have been conducted into the sexual health-care needs of bipolar women. Coverdale *et al.* (1997) interviewed women with major psychiatric disorders (including BD) regarding their contraceptive needs and their sexual behaviour. A control group of women with no psychiatric history, individually matched for age and ethnic background, were also interviewed. The group with chronic mental illness had undergone significantly more induced abortions. Heterosexual women with chronic mental illness reported a significantly greater likelihood of having more than one partner, to have been pressured into unwanted sexual intercourse in the previous year and to have had other forms of 'high-risk' sexual experience. These results emphasise the need for women with chronic mental illness to have access to contraception and protection from sexually transmitted disease. It is estimated that two-thirds of all sexually transmitted infections occur in teenagers and in those in their early twenties: young bipolar patients are therefore a vulnerable group who require help to address their risk (Dehne & Riedner, 2001). The need for effective contraceptive treatment is especially important in bipolar women on potentially teratogenic medication and those under 18 years old who are at increased risk of unplanned pregnancy.

Patients with long-term mental illness appear less likely to engage in screening programmes and to seek medical help (Wang *et al.*, 2002). In a study of patients with schizophrenia, women with this enduring mental illness were less likely to receive mammograms, pelvic examinations and cervical screening and to be prescribed hormone replacement therapy (Lindamer *et al.*, 2003).

Contraception

Most mood stabilisers are teratogenic (see Table 10.3). It is essential to ensure that the patient has been made aware of the risks and is using reliable contraception. Folic acid 5 mg daily is recommended to reduce (but does not eliminate) the risk of neural tube defects resulting from first-trimester anticonvulsant exposure (Kjaer *et al.*, 2008).

Table 10.3 Teratogenic potential of mood stabilisers

	Risk	Study
Lithium		
Risk ratios for:		
1. All malformations	1.5–3.0	
2. Cardiac malformations	1.2–7.7	
		Cohen *et al.*, 1994
Sodium valproate		
Relative risk of any malformation, compared with:		
Other anticonvulsants	2.59	
Untreated patients with epilepsy	3.16	
General population	3.77	
Relative risk increases if prescribed with other anticonvulsants, the dose is above 600 mg/d and is greatest at doses >1000 mg/d		
CMs noted:		
Neural tube defects	1–5%	
Cranial facial defects		
Fetal valproate syndrome		
brachycephaly	small nose and mouth	
high forehead	low-set posteriorly rotated ears	
shallow orbits	long overlapping fingers and toes	
ocular hypertelorism	hyperconvex fingernails	
Fingernail hypoplasia		
Cardiac defects		
		Koren *et al.*, 2002
Carbamazepine		
Rates of:		
Major malformations	6.7%	
Neural tube defects	0.5–1%	
Cardiovascular	1.5–2%	
		Ornoy, 2006
Lamotrigine		
Relative risk of oral clefts	11.1	Viguera *et al.*, 2007b
Risk of all CMs	3.2%	
Risk of all CMs if dose >200 mg daily	5.4%	
Risk of all CMs if combined with valproate	9.6%	Morrow *et al.*, 2006

Carbamazepine and topiramate reduce the efficacy of oral contraception via induction of cytochrome P450 (CYP) 3A4 isoenzymes, which metabolise the exogenous oestrogen and may result in an unplanned pregnancy. A preparation containing at least 50 μg ethinylestradiol or another contraceptive method is advised. Oral contraceptives (OCs) reduce serum levels of lamotrigine by approximately 50% by increasing clearance, so a higher dose of lamotrigine may be required if an OC is introduced. Lamotrigine can reduce serum levels of synthetic progestogens, which may reduce the efficacy of progestogen-only contraception.

Pre-pregnancy planning

Freeman *et al.* (2002) observed that 67% of a cohort of bipolar women experienced recurrences after childbirth. Those who have had a previous puerperal psychotic episode have a risk of recurrence after subsequent pregnancies of between 50 and 100% (Freeman *et al.*, 2002; Robertson *et al.*, 2005). Women with a family history of puerperal relapse in a first-degree relative are more likely to relapse postpartum than women without such a history (73 vs 30%) (Jones & Craddock, 2001).

It is therefore important to discuss possible pregnancy and the risks involved with all women of reproductive potential, whether they are currently planning a pregnancy or not. Specific pre-conceptual counselling involving an individual risk–benefit analysis regarding their risk of recurrence after delivery, medication during pregnancy, potential risks to the fetus and breastfeeding for those planning a pregnancy should be undertaken for those who wish to become pregnant. However, half of all pregnancies are unplanned, and most women do not realise that they are pregnant until they are at least at 6 weeks' gestation, by which time drug exposure has already happened. Women may discontinue medication abruptly themselves or be advised to do so by others, including health professionals.

Pregnancy and the puerperium

Psychotropics cross the placenta, and antidepressants and their metabolites have been found in amniotic fluid (AF). The fetus inhales AF and swallows it (the amount increasing as pregnancy progresses), and there is also possible transcutaneous absorption.

Before prescribing during pregnancy it is essential to consider *all* the potential risks to the fetus:

1. intrauterine death
2. risk of teratogenesis and organ malformation
3. growth impairment
4. risk of neonatal toxicity or withdrawal syndromes after delivery
5. risk of long-term neurobehavioural sequelae.

It is also crucial to take into account other potential teratogens that might act in an additive or synergistic way, including non-prescribed drugs, herbal and homeopathic remedies, smoking, alcohol and environmental toxins. Any woman with first-trimester exposure to lithium or anticonvulsants requires a second-trimester high-resolution detailed ultrasound examination for fetal anomaly.

Balanced against the risks of medication is the risk to the mother and fetus if medication is discontinued and relapse occurs. Depression in pregnancy is associated with reduced appetite, substance misuse, smoking and poor engagement with antenatal care. Manic relapses may involve high-risk behaviours that compromise both maternal and fetal safety. Treating the relapse may involve higher doses and a greater number of medications than maintenance medication.

Pregnancy is not protective against recurrence (Viguera *et al.*, 2000). Pregnant women who discontinue mood stabilisers are twice as likely to experience a recurrence during pregnancy, have a shorter time to recurrence and are ill for longer than those who remain on medication (Viguera *et al.*, 2007a). Most recurrences are depressive or mixed episodes. The time to recurrence is shortened with a rapid (within 2 weeks) rather than a slower (over 2–4 weeks) discontinuation.

A pregnant woman who has previously remained well when medication has been withdrawn may be able to discontinue at least for the first trimester. Others, who have relapsed on previous discontinuations, whose relapses have involved self-harm, or who have rapid-cycling, more severe or psychotic disorders, may need to remain on medication.

The largest body of safety data for antipsychotics is for older, typical drugs, which do not appear to increase the rate of congenital malformations (CMs). There are limited data available regarding the atypical antipsychotics and the risk of CMs, but there are concerns about potential weight gain associated with olanzapine, which might increase the risk of gestational diabetes and a large-for-gestation infant. Tricyclic antidepressants do not appear to increase rates of CMs but are associated with neonatal withdrawal syndrome. There does appear to be a small increased risk of CM (particularly cardiac defects with paroxetine) with first-trimester exposure to selective serotonin reuptake inhibitors. There is also increasing evidence that there is an association with pregnancy exposure and withdrawal symptoms (see Nordeng *et al.*, 2001, for review) and possibly a small reduction in gestational age.

If a woman continues to take lithium, the dose may need to be increased as renal clearance increases during pregnancy. Levels should be monitored more frequently than usual: every 2–4 weeks in the first and second trimesters and weekly in the third. Lithium should be withheld before labour, and fluid balance must be closely monitored to avoid toxicity. Renal clearance rapidly returns to normal after delivery, and therefore the dose needs adjusting back to the prepregnant dose.

Women on prophylactic medication after delivery are less likely to have a recurrence than those who are not (Cohen *et al.*, 1995). Those who do relapse will do so soon after delivery. Heron *et al.* (2008) reported that 73% of women who developed a postpartum relapse had developed hypomanic symptoms by the third postpartum day. The majority of episodes had begun within 4 weeks. It is therefore crucial that any medication that has been withdrawn during pregnancy is reinstated. If a woman chooses not to restart medication, she must be extremely closely monitored.

Uncontrolled studies suggest that postpartum lithium prophylaxis is effective (Stewart *et al.*, 1991; Austin, 1992), but a small non-randomised study failed to show a benefit of divalproex over placebo (Wisner *et al.*, 2004).

Breastfeeding

Psychotropics pass into breast milk. The infant plasma level and amount reaching the infant's central nervous system are dependent upon several variables, including whether or not the infant is exclusively breastfed, the proportions of fore or hind milk ingested and timing of the feed since the last maternal dose. Neonates have a reduced capacity to metabolise drugs in the first 2 weeks of life, and their kidneys are less efficient, the glomerular filtration rate not reaching that of an adult until 2–5 months of age. The infant blood–brain barrier is also immature.

General guidelines to consider when deciding whether to allow breastfeeding when a mother is taking psychotropics are as follows.

1. Sick or preterm infants are less able than healthy term babies to metabolise drugs.
2. Using drugs with a short half-life may make it possible to time feeds to the period when plasma levels are lowest.
3. Once-daily medication could be taken just after feeding and before the infant's longest sleep time. However, this may not be possible with a hungry, demand-fed baby.
4. Expressing milk for feeding by others may help if it is important for the mother to sleep.
5. Monitor the feeding, activity level, sleep and consciousness level of any breastfed infant whose mother is taking psychotropics, and if there are any concerns, suspend breastfeeding and seek a paediatric opinion.

Case studies have observed infant serum levels of lithium ranging from 30 to 200% of maternal levels, and as the infant kidney is immature, lithium is best avoided when breastfeeding. Carbamazepine and valproate are considered by the American Academy of Pediatrics as safe to take if breastfeeding, whereas NICE (2007) advises against lamotrigine because of the risk of Stevens–Johnson syndrome in the infant.

A systematic review (Weissman *et al.*, 2004) concluded that nortriptyline, sertraline and paroxetine are the preferred antidepressant choices for prescribing during lactation, as they tend to produce low or undetectable levels in infant serum and are unlikely to develop elevated levels, whereas fluoxetine and citalopram appear likely to produce elevated levels, particularly if there has been exposure during pregnancy.

There are more safety data available for typical antipsychotics than for atypical antipsychotics. Clozapine is generally best avoided due to the risk of neutropenia in the infant, but if it is deemed necessary, the infant's full blood count must be monitored regularly.

Notes on pharmacokinetics in women

Female patients have a 1.5- to 1.7-fold greater risk of developing an adverse drug reaction compared with men (Rademaker, 2001). Pharmacokinetic differences may play a part in this difference. In general, women have lower gastrointestinal drug absorption rates than men, slower renal clearance and lower hepatic metabolic rates. Differences in the activity of CYP enzymes exist with female gender: CYP3A4 activity is increased, and CYP2D6, CYP2C19 and CYP1A2 activities are decreased (Rademaker, 2001). In addition, rates of hepatic metabolism appear to fluctuate in relation to the menstrual cycle, with the peak occurring in mid-cycle, and with slower rates occurring in the follicular phase (Barnes & Mitchell, 2005). Thus higher serum levels may occur in the follicular phase, potentially increasing the risk of adverse effects. Changes in renal clearance during pregnancy, and their implications for treatment, are described in the section on pregnancy and the puerperium above.

Psychological and psychosocial issues: the influence of bipolar disorder on the development of self-identity

'One is not born, but rather becomes, a woman', Simone de Beauvoir (1949; English translation, 1989) famously observed. What challenges does the woman with BD face on this journey? Bipolar disorder may impact on her developing gender identity and her personal and work roles, key factors in her sense of self.

The development of self-identity in the bipolar adolescent girl

For the bipolar adolescent, the onset of her illness may occur just as she begins to explore her future identity as an adult woman. The transition from childhood to adolescence is an important stage in the development of gender identity (Bancroft, 1989). Puberty, with its hormonal changes, the development of secondary sexual characteristics and the onset of menstruation, coincides with emotional instability experienced by many healthy adolescents.

Because of its private nature, little information exists regarding the development of self-identity in bipolar adolescent girls. The adolescent faces an uncertain world of dating, sexual encounters and social and academic competition. The bipolar adolescent may face these challenges while experiencing severe affective episodes. Her illness may be dismissed as the mood disturbances expected at that age. She may suffer from co-morbid illnesses such as anxiety, attention deficit hyperactivity disorder (ADHD), eating disorders or substance misuse. She is more likely than her healthy colleagues to have suffered abuse. Key figures in her family of origin may themselves be affected by psychiatric illness. She may be exposed to the psychodynamic effects of a childhood disrupted by parental illness and by the role of a child carer. The bipolar adolescent girl faces a higher risk of delayed menarche and early menstrual dysfunction compared with her peers (Joffe *et al.*, 2006).

Given these potential burdens, it is unsurprising that Rucklidge (2006), examining psychosocial functioning in adolescents with BD, reported more traumatic events and negative life experiences, lower self-esteem, more hopelessness and more difficulties in controlling emotion in this group than in his control group. Wozniak *et al.* (2004) found that young adolescents with bipolar depression and ADHD had greater problems with leisure activities, shared activities with peers and close family relationships in comparison with healthy control subjects.

Self-identity and relationship issues in the adult bipolar woman

For the young adult woman, the pursuit of further education or a career may be disrupted by her illness. The establishment and maintenance of relationships with a partner and friends may prove difficult. A systematic review of bipolar patients' perception of the psychosocial burdens of their illness found that relationships with family members and partners were severely affected by the illness (Elgie & Morselli, 2007). Further, problems in the relationships persisted long after affective symptoms had abated. Kessler *et al.* (1998) reported that bipolar patients had more than three times the population risk of divorce. To the young adult bipolar woman, perhaps struggling with a new diagnosis and with medication, independence, marriage and parenthood may seem problematic or unachievable.

The findings of Coverdale *et al.* (1997) regarding the risk of unwanted intercourse and the rates of termination of pregnancy in women with chronic mental illness are concerning. These women were also significantly more likely to have given up their own children for others to raise. As mentioned above, divorce rates are higher in this group. What do these stark data signify for each woman as an individual? Unwanted sexual experiences may occur during episodes of mental illness or at times of alcohol or substance misuse, or they may be manifestations of a lack of self-care. A variety of other psychodynamic factors might play a role. Bipolar women may struggle with grief resulting from the loss of key relationships and the custody of their children – personal issues that may never be discussed with health-care professionals. These life events and their consequences may impact heavily on their inner lives and their happiness.

Mothers with bipolar disorder

Bipolar illness may influence a woman's decision on whether to start a family. However, many bipolar women wish to be, and do become, successful mothers. Although there are some personal accounts in the literature (e.g. Fox, 1999), few psychosocial studies exist regarding the experience of bipolar mothers. However, more general studies of women with severe mental illness, in which women with BD form a subgroup, are available.

Two recent papers provide vivid descriptions of motherhood in the face of chronic severe mental illness (Diaz-Caneja & Johnson, 2004; Montgomery *et al.*, 2006). At interview, women expressed a need to appear 'normal', to be 'a good mother' who placed her child's needs before her own, creating security for them. Fulfilling these duties brought a sense of self-worth. However, illness often undermined these mothers' views of themselves in this role and brought others to question their ability to be responsible for their children. When they were ill, mothers described 'masking' their symptoms from their children and others, censoring their speech and performing childcare tasks in a mechanical, automatic way (Montgomery *et al.*, 2006).

Mothers with chronic mental illness reported ambivalence regarding mental health services (Diaz-Caneja & Johnson, 2004). They were fearful that disclosure of mental ill health or parenting difficulties would result in the loss of custody or access to their children. Inpatient admission provided the opportunity to recover from acute illness and meant that children were protected from seeing their mother ill. However, the separation meant that they could not continue to mother their children for that time, and might increase the risk of losing their custody. During periods of hospitalisation, children might be placed with unknown foster parents, a further anxiety for their mothers. On discharge, the women reported that they were expected to move from the role of hospital inpatient to that of full-time, often single-handed, parent with little support (Diaz-Caneja & Johnson, 2004).

The stigma of mental illness and the fear that their children may become mentally ill were prominent concerns. 'If other mothers knew I had a mental illness, they might not allow their children to play with mine' (Diaz-Caneja & Johnson, 2004). There is a strong link between familial bipolar illness and childhood psychiatric morbidity. The lifetime prevalence of affective disorders is high in this group (Sadovnick *et al.*, 1994). Anxiety and substance abuse disorders are also common, with rates of 24 and 12%, respectively (Edmonds *et al.*, 1998).

Despite these difficulties, children and motherhood were generally experienced as 'rewarding and central to (the) lives' of these women, an aid and motivation to recover and to remain well (Montgomery *et al.*, 2006). Family-based treatments may be used to address parenting, communication and problem-solving skills, but further studies are required regarding their effectiveness. Practical measures, such as the provision of respite childcare and planning for hospitalisation, may help. Advocating these approaches as part of a treatment plan for every mother with mental illness, Fox (1999) said 'Professionals need to be aware that parenting and mental illness are not mutually exclusive. We must listen to what people want and need and must support them in working toward their goals.'

In summary, little research exists regarding the interplay between a woman's bipolar illness and her confidence in her gender and self-identity and in the roles she progresses through in life. Clearly, these are highly personal issues. The adolescent or adult woman with BD often carries a heavy psychosocial burden, not shared by her healthy peers. She may face challenges in her roles as partner, mother, friend and homemaker, and in her professional career, which she finds difficult to discuss with mental health professionals. Many of these women show considerable courage in coping with their illness and the demands of their lives. In general, there is a need for more rigorous study of psychosocial interventions in BD (Beynon *et al.*, 2008): work on the support of bipolar mothers would be timely and welcome.

Conclusions

Many bipolar women face the difficult treatment challenges of long depressive episodes, rapid-cycling disorder and severe puerperal illness. Reproductive health issues require careful attention, both before the commencement of pharmacological treatment and over its course. The issues of contraception, prepregnancy planning, pregnancy and the puerperium demand detailed planning and vigilance. The psychological and psychosocial burdens carried by bipolar women are heavy, potentially beginning in adolescence or earlier. The successful management of BD and its sequelae in women requires sensitive and individually tailored planning to address the psychiatric, psychological and practical aspects of their care.

References

Angst, J. (1978). The course of affective disorders II. Typology of bipolar manic-depressive illness. *Archiv für Psychiatrie und Nervenkrankheiten*, **226**, 65–73.

Austin, M. (1992). Puerperal affective psychosis: is there a case for lithium prophylaxis? *British Journal of Psychiatry*, **161**, 692–4.

Baldessarini, R.J., Tondo, L., Floris, G., Hennen, J. (2000). Effects of rapid cycling on response to lithium maintenance treatment in 360 bipolar I and II disorder patients. *Journal of Affective Disorders*, **61**, 13–22.

Bancroft, J.H.J. (1989). Sexual development: childhood to adolescence. In: Bancroft, J. (Ed.). *Human Sexuality and Its Problems*, 2nd editon. London: Churchill Livingstone, pp. 163–5.

Barnes, C. & Mitchell, P. (2005). Considerations in the management of bipolar disorder in women. *Australian and New Zealand Journal of Psychiatry*, **39**, 662–73.

Baron, M., Risch, N., Mendlewicz, J. (1982). Differential fertility in bipolar affective illness. *Journal of Affective Disorders*, **4**, 103–12.

Bauer, M.S., Whybrow, P.C., Winokur, A. (1990). Rapid cycling bipolar affective disorder I: association with grade I hypothyroidism. *Archives of General Psychiatry*, **47**, 427–32.

Bauer, J., Isojärvi, J.I.T., Herzog, A.G., *et al.* (2002). Reproductive dysfunction in women with epilepsy: recommendations for evaluation and management. *Journal of Neurology, Neurosurgery and Psychiatry*, **73**, 121–5.

Beauvoir, S de. (1989). The Second Sex, translated by H. M. Parshley . New York: Vintage Books, *English translation of Le Deuxième Sexe*. Paris: Gallimard (1949).

Beynon, S., Soares-Weiser, K., Woolacott, N., Duffy, S., Geddes, J.R. (2008). Psychosocial interventions for the prevention of relapse in bipolar disorder: systematic review of controlled trials. *British Journal of Psychiatry*, **192**, 5–11.

CEMD (2001). *Why Mothers Die 1997–1999. The Confidential Enquiries into Maternal Deaths in the United Kingdom*. London: RCOG Press.

Cohen, L.S., Friedman, J.M., Jefferson, J.W., Johnson, E.M., Weiner, M.L. (1994). A re-evaluation of risk of in utero exposure to lithium. *Journal of the American Medical Association*, **271**, 146–50.

Cohen, L.S., Sichel, D.A., Robertson, L.M., Heckscher, E., Rosenbaum, J.F. (1995). Postpartum prophylaxis for women with bipolar disorder. *American Journal of Psychiatry*, **152**, 1641–5.

Coverdale, J.H. & Turbott, S.H. (2000). Sexual and physical abuse of chronically ill psychiatric outpatients compared with a matched sample of medical outpatients. *Journal of Nervous and Mental Disease*, **188**, 440–5.

Coverdale, J.H., Turbott, S.H., Roberts, H. (1997). Family planning needs and STD risk behaviours of female psychiatric out-patients. *British Journal of Psychiatry*, **171**, 69–72.

Dehne, K.L. & Riedner, F. (2001). Sexually transmitted infections among adolescents: the need for adequate health services. *Reproductive Health Matters*, **9**, 170–83.

Diaz-Caneja, A. & Johnson, S. (2004). The views and experiences of severely mentally ill mothers. A qualitative study. *Social Psychiatry and Psychiatric Epidemiology*, **39**, 472–82.

Duax, J.M., Youngstrom, E.A., Calabrese, J.R., Findling, R.L. (2007). Sex differences in paediatric bipolar disorder. *Journal of Clinical Psychiatry*, **68**, 1565–73.

Edmonds, L.K., Mosley, B.J., Admiraal, A.J., Olds, R.J., Romans, S.E., Silverstone, T., *et al.* (1998). Familial bipolar disorder: preliminary results from the Otago Familial Bipolar Genetic Study. *Australian and New Zealand Journal of Psychiatry*, **32**, 823–9.

Elgie, R. & Morselli, P.L. (2007). Social functioning in bipolar patients: the perception and perspective of patients, relatives and advocacy organizations –a review. *Bipolar Disorders*, **9**, 144–57.

Fox, L. (1999). Missing out on motherhood. *Psychiatric Services*, **50**, 193–4.

Freeman, M.P., Smith, K.W., Freeman, S.A., *et al.* (2002). The impact of reproductive events on the course of bipolar disorder in women. *Journal of Clinical Psychiatry*, **63**, 284–7.

Frye, M.A., Altshuler, L.L., McElroy, S.L., *et al.* (2003). Gender differences in prevalence, risk and clinical correlates of alcoholism comorbidity in bipolar disorder. *American Journal of Psychiatry*, **60**, 883–9.

Goodwin, F.K. & Jamison, K.R. (Eds). (1990). *Manic-depressive Illness*. New York: Oxford University Press.

Haddad, P.M. & Wieck, A. (2004). Antipsychotic-induced hyperprolactinaemia. Mechanisms, clinical features and management. *Drugs*, **20**, 2291–314.

Heron, J., McGuiness, M., Blackmore, E.R., Craddock, N., Jones, I. (2008). Early symptoms in puerperal psychosis. *British Journal of Obstetrics and Gynaecology*, **115**, 348–53.

Joffe, H., Kim, D.R., Foris, J.M., *et al.* (2006). Menstrual dysfunction prior to onset of psychiatric illness is reported more commonly by women with bipolar disorder than by women with unipolar depression and healthy controls. *Journal of Clinical Psychiatry*, **67**, 297–304.

Joffe, H. (2007). Reproductive biology and psychotropic treatments in premenopausal women with bipolar disorder. *Journal of Clinical Psychiatry*, **68** (Suppl. 9), 10–15.

Jones, I. & Craddock, N. (2001). Familiality of the puerperal trigger in bipolar disorder: results of a family study. *American Journal of Psychiatry*, **158**, 913–17.

Jones, I. & Craddock, N. (2005). Bipolar disorder and childbirth: the importance of recognising risk. *British Journal of Psychiatry*, **186**, 453–4.

Kessler, R.C., Walters, E.E., Forthofer, M.S. (1998). The social consequences of psychiatric disorders. III. Probability of marital stability. *American Journal of Psychiatry*, **155**, 1092–6.

Kjaer, D., Horvath-Puho, E., Christensen, J., *et al.* (2008). Antiepileptic drug use, folic acid supplementation, and congenital abnormalities: a population-based case-control study. *British Journal of Obstetrics and Gynaecology*, **115**, 98–103.

Koren, G., Cohn, T., Chitayat, D., *et al.* (2002). Use of atypical antipsychotics during pregnancy and the risk of neural tube defects in infants. *American Journal of Psychiatry*, **159**, 136–7.

Kukopulos, A., Reginaldi, D., Laddomada, P., Floris, G., Serra, G., Tondo, L. (1980). Course of the manic-depressive cycle and changes caused by treatment. *Pharmakopsychiatrie, Neuro-Psychopharmakologie*, **13**, 156–67.

Kulkarni, J., Garland, K.A., Scaffidi, A., *et al.* (2006). A pilot study of hormone modulation as a new treatment for mania in women with bipolar affective disorder. *Psychoneuroendocrinology*, **31**, 543–7.

Lane, D.E. (2006). Polycystic ovary syndrome and its differential diagnosis. *Obstetrical & Gynecological Survey*, **61**, 125–35.

Leverich, G.S., McElroy, S.L., Suppes, T., *et al.* (2002). Early physical and sexual abuse associated with an adverse course of bipolar illness. *Biological Psychiatry*, **51**, 288–97.

Lindamer, L.A., Buse, D.C., Auslander, L., Unutzer, J., Bartels, S.J., Jeste, D. (2003). A comparison of gynaecological variables and service use among older women with and without schizophrenia. *Psychiatric Services*, **54**, 902–4.

Marsh, W.K., Templeton, A., Ketter, T.A., Rasgon, N.L. (2008). Increased frequency of depressive episodes during the menopausal transition in women with bipolar disorder: preliminary report. *Journal of Psychiatric Research*, **42**, 247–51.

McDermott, B.E., Quanbeck, C.D., Frye, M.A. (2007). Comorbid substance use disorder in women with bipolar disorder associated with criminal arrest. *Bipolar Disorder*, **9**, 536–40.

McElroy, S.L. (2004). Bipolar disorders: special diagnostic and treatment considerations in women. *CNS Spectrums*, **9** (Suppl. 7), 5–18.

Montgomery, P., Tomkins, C., Forchuk, C., French, S. (2006). Keeping close: mothering with serious mental illness. *Issues and Innovations in Nursing Practice*, **54**, 20–8.

Morrow, J., Russell, A., Guthrie, E., *et al.* (2006). Malformation risks of antiepileptic drugs in pregnancy: a prospective study from the UK Epilepsy and Pregnancy Register. *Journal of Neurology, Neurosurgery and Psychiatry*, **77**, 193–8.

NICE (2007). *Antenatal and Postnatal Mental Health: Clinical Management and Service Guidance.* London: National Institute for Health and Clinical Excellence. http://guidance.nice.org.uk/CG45/niceguidance/pdf/English [accessed 9 January 2010].

Nordeng, H., Lindemann, R., Perminov, K.V., Reikvam, A. (2001). Neonatal withdrawal syndrome after in utero exposure to selective serotonin reuptake inhibitors. *Acta Paediatrica*, **90**, 288–91.

Ornoy, A. (2006). Neuroteratogens in man: an overview with special emphasis on the teratogenicity of antiepileptic drugs in pregnancy. *Reproductive Toxicology*, **22**, 214–26.

Rademaker, M. (2001). Do women have more adverse drug reactions? *American Journal of Clinical Dermatology*, **2**, 349–51.

Robertson, E., Jones, I., Haque, S., Holder, R., Craddock, N. (2005). Risk of puerperal and non-puerperal recurrences following bipolar affective puerperal (postpartum) psychosis. *British Journal of Psychiatry*, **186**, 258–9.

Robertson Blackmore, E., Craddock, N., Walters, J., Jones, I. (2008). Is the perimenopause a time of increased risk of recurrence in women with a history of bipolar affective post-partum psychosis? A case series. *Archives of Women's Mental Health*, **11**, 75–8.

Rucklidge, J.J. (2006). Psychosocial functioning of adolescents with and without paediatric bipolar disorder. *Journal of Affective Disorders*, **91**, 181–8.

Sadovnick, A.D., Remick, R.A., Lam, R., *et al.* (1994). Mood Disorder Service Genetic Database: morbidity risks for mood disorders in 3,942 first-degree relatives of 671 index cases with single depression, recurrent depression, bipolar I, or bipolar II. *American Journal of Medical Genetics*, **15**, 132–40.

Simon, G.E., Hunkeler, E., Fireman, B., Lee, J.Y., Savarino, J. (2007). Risk of suicide attempt and suicide death in patients treated for bipolar disorder. *Bipolar Disorder*, **9**, 526–30.

Soares, C.N. & Taylor, V. (2007). Effects and management of the menopausal transition in women with depression and bipolar disorder. *Journal of Clinical Psychiatry*, **68**, 16–21.

Stewart, D.E., Klompenhouwer, J.L., Kendell, R.E., van Hulkst, A.M. (1991). Prophylactic lithium in puerperal psychosis. The experience of three centres. *British Journal of Psychiatry*, **158**, 393–7.

Suppes, T., Mintz, J., McElroy, S.L., *et al.* (2005). Mixed hypomania in 908 patients with bipolar disorder evaluated prospectively in the Stanley Foundation Bipolar Treatment Network: a sex-specific phenomenon. *Archives of General Psychiatry*, **62**, 1089–96.

Tondo, L. & Baldessarini, R.J. (1998). Rapid cycling in women and men with bipolar manic-depressive disorders. *American Journal of Psychiatry*, **155**, 1434–6.

Viguera, A.C., Nonacs, R., Cohen, L.S., Tondo, L., Murray, A., Baldessarini, R.J. (2000). Risk of recurrence of bipolar disorder in pregnant and nonpregnant women after discontinuing lithium maintenance. *American Journal of Psychiatry*, **157**, 179–84.

Viguera, A.C., Whitfield, T., Baldessarini, R.J., *et al.* (2007a). Risk of recurrence in women with bipolar disorder during pregnancy: prospective study of mood stabilizer discontinuation. *American Journal of Psychiatry*, **164**, 1817–24.

Viguera, A.C., Koukopoulos, A., Muzina, D.J., Baldessarini, R.J. (2007b). Teratogenicity and anticonvulsants: lessons from neurology to psychiatry. *Journal of Clinical Psychiatry*, **68**, 29–33.

Wang, P.S., Walker, A.M., Tsuang, M.T., *et al.* (2002). Dopamine antagonists and the development of breast cancer. *Archives of General Psychiatry*, **59**, 1145–54.

Weissman, A.M., Levy, B.T., Hartz, A.J., *et al.* (2004). Pooled analysis of antidepressant levels in lactating mothers, breast milk, and nursing infants. *American Journal of Psychiatry*, **161**, 1066–78.

Wisner, K.L., Hanusa, B.H., Peindl, K.S., Perel, J.M. (2004). Prevention of postpartum episodes in women with bipolar disorder. *Biological Psychiatry*, **56**, 592–6.

Wittchen, H.U., Nelson, C.B., Lachner, G. (1998). Prevalence of mental disorders and psychosocial impairments in adolescents and young adults. *Psychological Medicine*, **28**, 109–26.

Wozniak, J., Spencer, T., Biederman, J., *et al.* (2004). The clinical characteristics of unipolar versus bipolar major depression in ADHD youth. *Journal of Affective Disorders*, **82**, S59–S69.

Young, R.C., Moline, M., Kleyman, F. (1997). Hormone replacement therapy and late-life mania. *American Journal of Geriatric Psychiatry*, **5**, 179–81.

Chapter 11

Physical health issues

Chennattucherry John Joseph and Yee Ming Mok

Cardiovascular diseases (CVDs) and metabolic disorders are growing public health problems in most countries. The situation is worse among individuals with serious mental health problems such as bipolar disorder (BD) and schizophrenia, which in turn leads to substantial economic burden to the overstretched health-care resources. The undesirable effects of the various pharmacological treatments for BD and lifestyle factors add more to the medical co-morbidity associated with BD.

The data collected from public mental health agencies in the USA as part of the sixteen-state study on mental health performance measures showed that CVD is the primary cause of death in people with mental illness (Colton & Manderscheid, 2006). It has already been established that individuals with BD are at higher risk of premature death than the general population and those with major depressive disorder (Harris & Barraclough, 1998). The results from a national data analysis in the UK (Hippisley-Cox *et al.*, 2007) that examined the health inequalities experienced by people with diagnoses of schizophrenia or BD in comparison to those without those diagnoses showed that individuals with those diagnoses had higher prevalence of CVDs and were less likely to receive appropriate attention and treatment for raised cholesterol levels despite their higher level of risk factors. The mortality due to CVDs has declined in the general population in recent years as a result of improved diagnosis and availability of better treatments. However, the evidence so far indicates that cardiovascular mortality remains high in people with severe mental health problems without any appreciable decline (Newcomer & Hennekens, 2007). The standardised mortality ratios for all natural causes of death were calculated as 1.9 for males and 2.1 for females with diagnosis of BD, whereas the figures were 1.5 and 1.6, respectively, for individuals with diagnosis of unipolar depression (Osby *et al.*, 2001). Most excess deaths in BD were from natural causes, whereas unnatural causes such as suicide, violence and accidents accounted for most premature deaths in unipolar depression.

Metabolic state in bipolar disorder

There is substantial evidence indicating that the presence of BD itself is an independent risk factor for weight gain, diabetes mellitus, hypertension, dyslipidaemia, CVD and metabolic syndrome. Although lifestyle factors such as lack of exercise, unhealthy diet, smoking, psychoactive substance misuse, poor health awareness and access to health care intensify the medical burden in individuals with BD, there is evidence that the fundamental biological processes involved in the pathophysiology of BD and metabolic syndrome share many common characteristics. Metabolic syndrome is a complex multifactorial condition that is poorly defined and understood. There is also lack of consensus among experts about the usefulness of having such a diagnostic category. There are many different definitions of metabolic syndrome. The diagnostic criteria described in the third report of the National Cholesterol Education Program Adult Treatment Panel (NCEP ATP III, 2002) are given below (Table 11.1). While it is evident that individuals with BD have increased rates of obesity, it has been difficult to evaluate the independent contribution of BD to obesity because of the presence of multiple confounding factors, including the effects of medication and lifestyle.

Bipolar disorder and metabolic syndrome share many physical symptoms in common, and both conditions have common risk factors such as obesity, alterations in glucose metabolism and insulin regulation, dysregulation of the hypothalamic–pituitary–adrenal (HPA) and the hypothalamic–pituitary–thyroid (HPT) axes, deranged regulation of haemostasis and the sympathetic nervous system. See Fagiolini *et al.* (2008) and Taylor & MacQueen (2006) for further details.

Practical Management of Bipolar Disorder, eds. Allan H. Young, I. Nicol Ferrier and Erin E. Michalak. Published by Cambridge University Press.

Table 11.1 Metabolic syndrome NCEP ATP III criteria

Metabolic syndrome is present when three or more of the following five criteria are met. Risk factor defining level
Abdominal obesity, waist circumference Men >102 cm (40″), Women >88 cm (35″)
Serum triglycerides ≥150 mg/dl (≥1.7 mmol/l)
HDL cholesterol Men <40 mg/dl (<1.03 mmol/l) Women <50 mg/dl (<1.29 mmol/l)
Blood pressure ≥130/≥85 mmHg
Fasting blood glucose ≥110 mg/dl (≥6.1 mmol/l)

Derived from NCEP ATP III, 2002.

According to the World Health Organization (WHO), a BMI of ≥25 kg/m^2 is defined as overweight and a BMI of ≥30 kg/m^2 is defined as obesity. Although obesity in general is linked with higher risk of CVDs, the evidence indicates that body fat distribution is far more important in predicting the risk of CVD and diabetes. The BMI gives an overall impression of the level of fat, whereas measures such as the waist-to-hip ratio and waist circumference, which evaluate the degree of central obesity, help to predict the risk for hypertension, coronary artery diseases and diabetes better than BMI. There is evidence that the excess fat in individuals with BD tends to be more centrally distributed, as opposed to the pattern of distribution in the general population. In a study of bipolar patients (Fagiolini *et al.*, 2005), 30% of patients had metabolic syndrome, 74% were either obese (45%) or overweight (29%), 48% met criteria for either hypertriglyceridaemia or were receiving cholesterol-lowering medication, 39% had hypertension and 8% had high fasting glucose or were taking antidiabetic medications.

Both depressed and manic phases of BD are associated with elevated levels of cortisol (Cassidy *et al.*, 1998). The chronically elevated glucocorticoids can cause insulin resistance (Zakrzewska *et al.*, 1997) and increased visceral fat deposition. It is proposed that the activity of lipoprotein lipase regulates the fat influx into adipocytes through the binding of cortisol–glucocorticoid receptor complex of the lipase gene (Bjorntorp, 1996). As the density of the glucocorticoid receptors is higher in visceral adipose tissue than elsewhere, more cortisol is bound here and more triglycerides are assimilated into adipose tissue (Holmang & Bjorntorp, 1992). It is also suggested that dysregulation of the HPA axis might be associated with obesity through inefficient signalling of leptin, a satiety hormone (Rosmond & Bjorntorp, 1998). Studies in rats have shown that glucocorticoids diminish leptin signals (Zakrzewska *et al.*, 1997). There is also evidence linking immune function and metabolic syndrome in BD. In depression and mania, elevated levels of interleukin-6 and C-reactive protein have been observed. Interleukin-6 is a potent stimulator of corticotropin-releasing hormone production, which in turn stimulates the HPA axis, ultimately resulting in an increase in cortisol levels (Dentino *et al.*, 1999). Taylor & MacQueen (2006) conclude that the dysregulation of cortisol in BD might therefore contribute to many components of the metabolic syndrome including insulin resistance, abdominal obesity and dyslipidaemia.

A number of studies have shown that BD and schizophrenia are associated with an increased risk of diabetes. There is also evidence indicating that BMI, but not medication use, is positively correlated with new-onset type 2 diabetes mellitus (Regenold *et al.*, 2002). High rates of insulin resistance and impaired glucose tolerance in patients with serious mental illness were documented as early as the first half of the twentieth century, long before the introduction of antipsychotic drugs. Increased platelet activity (up to 40% more than controls) and hypercoagulability of blood have also been demonstrated in individuals with depression, which in turn increases the risk of CVDs.

Taylor & MacQueen (2006) suggest that abnormalities in the regulation of the sympathetic nervous system are involved in the pathophysiology of metabolic syndrome. Elevated activity of the sympathetic nervous system has been demonstrated in BD, obesity and hypertension. It is proposed that the hypothalamic centres controlling the HPA axis and sympathetic response are closely linked, to the extent that the HPA axis and sympathetic system are activated in parallel (Chrousos & Gold, 1992). This link may partially explain the expression of some of the physical symptoms commonly observed in BD and metabolic syndrome.

It is widely known that calorie intake in excess of amount of energy expenditure can lead to weight gain and obesity in adults. Another factor that is proposed to explain the increased propensity of bipolar patients to develop obesity is that the resting energy expenditure (REE) is lower in individuals with BD. Soreca *et al.* (2007) measured the REE in bipolar patients on maintenance treatment and in healthy controls using indirect calorimetry. They observed that the measured REE in bipolar patients on long-term psychopharmacologic treatment was lower than that of healthy controls, indicating reduced energy expenditure in bipolar patients and therefore increased vulnerability to weight gain. There is also evidence that binge-eating disorder and excessive calorie intake are common in individuals with BD. Therefore it is sensible to conclude that a combination of reduced energy expenditure and excessive calorie intake puts individuals with BD at risk of obesity.

Health issues associated with medications in bipolar disorder

There is an increasing body of evidence pointing to BD itself being associated with health problems. However, the medications used to treat BD are also associated with side effects on physical health.

The treatment of BD can be divided into the acute phase and the maintenance phase. The acute treatment of BD is usually a short period during which treatment is aimed at treating the depressive or manic episode. Once treated, medications may be used to help the patient remain in remission. While there is some overlap, the medications used in the maintenance phase will be discussed here, as they are used in the longer term and hence are more likely to produce adverse effects.

The main medications used to treat BD in the maintenance phase can be broadly divided into two categories: mood stabilisers and atypical antipsychotics.

Mood stabilisers

Lithium

Up to three-quarters of patients treated with lithium will experience some side effects related to various organ systems. Dose-related side effects of lithium include renal (e.g. diabetes insipidus, tubulointerstitial renal disease), gastrointestinal (e.g. nausea, vomiting, abdominal pain, loss of appetite, diarrhoea), neurological (e.g. tremors, raised intracranial pressure), haematological (e.g. benign leucocytosis), endocrine (e.g. thyroid and parathyroid dysfunction), cardiac (e.g. benign electrocardiogram changes, conduction abnormalities), dermatological (e.g. acne, psoriasis, hair loss) and weight gain.

Lithium has been associated with chronic renal disease. Approximately 10 to 20% of patients receiving long-term lithium treatment will have some morphological changes in the kidneys, and usually these changes are interstitial fibrosis, tubular atrophy and sometimes glomerular sclerosis. A retrospective study has shown that a small number of patients on lithium developed renal insufficiency (Lepkifker *et al*., 2004). Additionally, some studies have shown that a small percentage of patients treated with lithium may develop rising serum creatinine levels after 10 years or more of lithium treatment (Gitlin, 1993; Markowitz *et al*., 2000). Hypothyroidism occurs in patients treated with lithium and seems to occur more frequently in women (Johnston & Eagles, 1999). Hyperparathyroidism has also been noted with lithium treatment (Szalat *et al*., 2009), but the prevalence of this complication has not been established. Lithium has also been known to cause benign ECG changes. A small number of case reports have described psoriasis associated with lithium treatment. Some people treated with lithium may experience severe pustular acne.

Lithium has a narrow therapeutic index and can be lethal in overdose. Signs and symptoms of early intoxication include tremor, nausea and diarrhoea, blurred vision, vertigo, confusion and increased reflexes. At higher serum levels, patients may experience more severe neurological complications and eventually experience seizures, coma and cardiac arrhythmias. Some drugs may increase lithium levels and therefore increase the risk of lithium toxicity. Lithium levels may increase with the concomitant use of NSAIDs. Thiazide diuretics, ACE inhibitors and some antibiotics (e.g. metronidazole) may also increase lithium levels.

Anticonvulsants

Sodium valproate

Although minor side effects of valproate, such as sedation or gastrointestinal problems, are common initially, they typically resolve with continued treatment or dose adjustment. Common side effects of valproate include benign hepatic transaminase elevations, osteoporosis, tremors and sedation. Mild asymptomatic leucopenia and thrombocytopenia are less frequent and are reversible upon stopping valproate. However, more severe cases of thrombocytopenia have been reported. Bone marrow suppression and aplastic anaemia have been reported in association with valproate, but are rare. Other side effects include hair loss, increased appetite and weight gain.

The relationship between polycystic ovary syndrome and valproate treatment is unclear, but there is a suggestion that valproate is associated with features of polycystic ovary syndrome (Joffe *et al.*, 2006). Rare, idiosyncratic, but potentially fatal adverse events with valproate include irreversible hepatic failure, hyperammonaemic encephalopathy, haemorrhagic pancreatitis and agranulocytosis.

Carbamazepine

Up to half of patients receiving carbamazepine experience side effects, and the drug is associated with potentially serious adverse reactions. The most common dose-related side effects of carbamazepine are diplopia, blurred vision, fatigue, nausea and ataxia. Carbamazepine is known to cause elevations in serum liver enzymes. Less frequent side effects include skin rashes, mild leucopenia, mild thrombocytopenia, hyponatraemia and hypo-osmolality. Rare, idiosyncratic, but serious and potentially fatal, adverse effects of carbamazepine include agranulocytosis, aplastic anaemia, thrombocytopenia, hepatic failure, exfoliative dermatitis (e.g. Stevens–Johnson syndrome) and pancreatitis. Other rare side effects include systemic hypersensitivity reactions, cardiac arrhythmias, and, very rarely, renal effects (including renal failure, oliguria, haematuria and proteinuria).

Lamotrigine

The most common side effects of lamotrigine in the treatment of depression are headache, nausea and dry mouth. While the risk of a benign rash is about 8.3%, the risk of serious rash, including Stevens–Johnson syndrome and toxic epidermal necrolysis, was initially reported as 0.3% in adults (Guberman *et al.*, 1999). However, with a slow titration schedule, the risk of serious rash fell to 0.01% in adults (Calabrese *et al.*, 1999). Rash may occur at any time but is more likely to occur early in treatment. It is possibly more likely if lamotrigine and valproate are administered concomitantly.

Of all drug interactions involving carbamazepine, valproate and lamotrigine, the interaction between lamotrigine and valproate has the most potentially serious consequences. Valproate can increase lamotrigine levels, thereby increasing the risk of Stevens–Johnson syndrome. It should be noted that valproate reduces lamotrigine clearance and increases its half-life even in the presence of hepatic enzyme inducers such as carbamazepine.

Atypical antipsychotics

The atypical antipsychotics aripiprazole, olanzapine and quetiapine have demonstrated efficacy in randomised controlled trials for BD (Scherk *et al.*, 2007; Jarema, 2007). There is an increased risk of metabolic complications with these atypical antipsychotics. Other major concerns are cardiac effects and hyperprolactinaemia. There is as yet little evidence for paliperidone and amisulpride in the treatment of BD. Clozapine is not commonly used in the treatment of BD.

The aforementioned atypical antipsychotics have known associations with weight gain, dyslipidaemias and diabetes mellitus. Hypertension is not uncommon and can develop with the weight gain. Although dyslipidaemias and diabetes mellitus can occur independently of weight gain, the relative risk profiles for individual atypical agents are somewhat similar for these metabolic conditions. Besides metabolic effects, atypical antipsychotics can cause cardiac effects. Autonomic side effects such as postural hypotension and tachycardia/bradycardia may occur. Myocarditis and cardiomyopathy are infrequent, but potentially fatal complications, and have been most strongly associated with clozapine. Other than sertindole, ziprasidone is considered the most likely of the atypicals to cause QT prolongation on ECG.

Among the atypical antipsychotics mentioned, risperidone has the greatest propensity to cause hyperprolactinaemia, which may be comparable with, or greater than, that associated with typical antipsychotics. Olanzapine generally does not increase prolactin, but may cause non-significant sustained hyperprolactinaemia at higher doses. Ziprasidone may increase prolactin, but generally only transiently, and neither aripiprazole nor quetiapine has been associated with prolactin elevation. There have been isolated reports of neutropenia and thrombocytopenia with olanzapine, risperidone and quetiapine, and neutropenia with ziprasidone. However, apart from clozapine, the other atypical antipsychotics are generally considered to have a benign haematological profile.

Weight gain

Pooled data from two double-blind placebo controlled maintenance studies in BD type I examined weight changes in patients on lithium, lamotrigine and placebo

(Bowden *et al.*, 2006; Sachs *et al.*, 2006). Comparing those who were and were not obese, weight gain was only significant for the obese group treated with lithium. This is consistent with clinical studies showing greater weight gain among those already overweight.

Valproate and, to a lesser extent, carbamazepine are associated with weight gain, whereas lamotrigine is considered to be weight neutral (Ben-Menachem, 2007). Other trials have shown higher rates and degree of weight gain in association with valproate relative to carbamazepine and to lamotrigine. In a randomised, controlled 12-month trial, weight gain was more frequently associated with valproate compared with lithium and placebo, but lamotrigine has not been associated with weight gain (Bowden *et al.*, 2000). Valproate is associated with less weight gain than with olanzapine (Zajecka *et al.*, 2002). Evidence suggests that the risk of weight gain is present with all atypical antipsychotics. With the exception of clozapine, the likelihood of weight gain is highest with olanzapine, intermediate with risperidone and quetiapine, and lowest with aripiprazole and ziprasidone (see Table 11.2).

Table 11.2 Degree of weight gain

High	Intermediate	Low
Clozapine	Risperidone	Lamotrigine
Olanzapine	Quetiapine	Carbamazepine
	Lithium	Ziprasidone
	Valproate	

Antipsychotic drugs and extrapyramidal side effects

A discussion of the physical health complications of BD is not complete until the extrapyramidal side effects (EPSs) associated with the use of antipsychotic drugs are also considered. Although it is reassuring that the incidence and prevalence of EPSs have been significantly reduced with the introduction of atypical antipsychotics, it is important to remember that the newer antipsychotics have not eliminated them completely. As the longer-acting injectable preparations among the new-generation antipsychotics are currently limited to risperidone, which has only limited published data, old antipsychotic drugs are still at times necessary in the maintenance phase of BD patients with a history of limited concordance.

The EPSs can manifest in various forms, including tremors, rigidity, dystonia, akathisia, bradykinesia and tardive dyskinesia. The propensity to cause extrapyramidal symptoms varies with the dopamine D_2 receptor binding affinity among the antipsychotic drugs. Among the atypical antipsychotics, risperidone, ziprasidone and aripiprazole tend to produce more EPSs, owing to their higher affinity for D_2 receptors in comparison with the low D_2-binding drugs such as clozapine and quetiapine. Olanzapine occupies an intermediate position in terms of its D_2 receptor affinity and therefore its propensity to cause the extrapyramidal symptoms. Whereas the acute EPSs improve on withdrawing the antipsychotic drugs or can be controlled by medications to a large extent, tardive dyskinesia is a potentially irreversible syndrome of abnormal movements that can develop late in the course of treatment with antipsychotic drugs. After reviewing the epidemiology of tardive dyskinesia, Tarsy & Baldessarini (2006) commented that the second-generation antipsychotic drugs including aripiprazole, clozapine, olanzapine, risperidone, quetiapine and ziprasidone have relatively low risk for acute extrapyramidal syndromes characteristic of older antipsychotic drugs, but that the anticipated reduction in the risk of tardive dyskinesia was less well documented. Correll & Schenk (2008) published a systematic review that included 12 studies (*n*=28 051, age 39.7 years, 59.7% male, 70.9% white, followed for 463 925 person-years). They estimated the annualised tardive dyskinesia incidence as 3.9% for second-generation antipsychotic drugs and 5.5% for first-generation antipsychotic drugs. Therefore, tardive dyskinesia continues to pose significant health concerns for patients taking antipsychotic drugs, despite the arrival of second-generation antipsychotic drugs.

Management of physical health issues in bipolar disorder

Basic monitoring and general measures

Having reviewed the metabolic state in BD on its own and also in relation to the pharmacological agents used in its treatment and prophylaxis, it is now appropriate to examine the monitoring arrangements that need to be in place in order to facilitate the early detection and timely interventions to avoid or minimise the extent of physical health complications. Before examining the measures to control or limit physical health issues, it might be helpful to differentiate the modifiable and non-modifiable risk factors (Table 11.3). Although it is important to pay attention to the non-modifiable

Table 11.3 Risk factors

Non-modifiable risk factors	Modifiable risk factors
Gender	Obesity
Family history	Smoking
Personal history	Hyperglycaemia
Age	Hypertension
Ethnicity	Dyslipidaemia

factors as part of the assessment in order to identify the high-risk group that needs closer monitoring, only the modifiable factors can be manipulated to patients' benefit.

Different pharmacological treatments have different adverse-effect profiles, and therefore the presence of cardiovascular risk factors and established CVD are to be taken into account before making treatment decisions. The management and minimisation of metabolic risk are important when providing optimal care to patients receiving psychotropic drugs. A comprehensive approach in newly diagnosed patients, those switching medication, or currently receiving treatment will involve evaluation, individual tailoring of therapy and ongoing monitoring, together with patient education (Barnett *et al.*, 2007). During the evaluation of each patient attention should be paid to personal medical history and family history of CVD, diabetes, obesity and stroke. Recording of body weight, blood pressure, lipid profile and blood glucose level at the outset of treatment and the ongoing monitoring of these parameters form the foundation for good practice in reducing the risk of cardiovascular and metabolic adverse effects. Although BMI gives an estimate of weight gain, monitoring the waist circumference would make it more meaningful, as central obesity is more strongly linked to the risk of metabolic syndrome than the measure of general obesity. Cardiovascular risks can be reduced by selecting treatments with tolerable adverse-effect profiles tailored to the needs of individual patients and by using targeted interventions to manage the modifiable risk factors.

Lifestyle needs to be explored as smoking, drug and alcohol misuse, lack of exercise and unhealthy diet habits tend to be more common among people with BD compared with the general population. Patients with BD tend to have lower success rates in changing their lifestyle and behaviour. De Backer *et al.* (2003) noted that changes in patterns of individual behaviour are necessary in the vast majority of patients with established CVD, or those at high risk of CVD.

Table 11.4 Stop tobacco smoking – 5 A's approach

A	Ask – systematically identify all smokers at every opportunity
A	Assess – determine the patient's degree of addiction and readiness to stop smoking
A	Advise – urge strongly all smokers to quit
A	Assist – agree on a smoking cessation strategy, including behavioural counselling, nicotine replacement therapy and/or pharmacological intervention
A	Arrange – a schedule of follow-up

Reproduced with permission from De Backer *et al.*, 2003.

Studies in this area have demonstrated that physicians have not been able to give adequate emphasis on promoting behavioural change in patients in routine clinical practice. The management of behavioural risk factors such as unhealthy diet, smoking and sedentary lifestyle requires a multidisciplinary approach. The first step towards achieving this is to develop a therapeutic alliance with the patient. Creating awareness via education is pivotal in inducing motivation in patients for behavioural change. Once a commitment to changing the behaviour is ensured with the patient, a lifestyle modification plan needs to be designed and agreed. Patients will need regular follow-up and it will be necessary to involve other health-care staff in reinforcing the message and behavioural change. Specialist dietary advice is often needed to help patients make healthy food choices. Eating healthily will reduce the risk by helping to reduce calorie intake, lose weight, lower blood pressure and control blood glucose level. Total fat intake should account for no more than 30% of energy intake. Eating fruits, vegetables, wholegrain cereals, low-fat dairy products, fish and lean meat should be encouraged. Physical activity should also be encouraged in all age groups. The goal is to have at least 30 min of physical activity most days of the week, but more activity is associated with more health benefits. Patients who are reasonably healthy should be encouraged to exercise for 30 to 45 min four to five times a week at 60 to 75% of the average maximum heart rate. All patients who smoke and are at high risk for CVD should be encouraged to stop smoking. The 5 A's approach (De Backer *et al.*, 2003) is useful in helping patients in this area (Table 11.4).

Table 11.5 summarises the European guidelines for the prevention of CVDs in clinical practice (Graham *et al.*, 2007), and it shows different targets for the general population and patients at high risk for CVDs. In addition, specific treatment using cardio-protective

Table 11.5 European guidelines for the prevention of CVDs in clinical practice

Risk factor	Units	Target	Target in high-risk patients
Blood pressure	mmHg	<140/90	<130/80 if feasible
Total cholesterol	mmol/l (mg/dl)	<5 (190)	<4.5 (175) and <4 (155) if feasible
LDL cholesterol	mmol/l (mg/dl)	<3 (115)	<2.5 (100) and <2 (80) if feasible
BMI	kg/m²	<25 and avoid central obesity	
Blood glucose	mmol/l (mg/dl)	<6 (110)	<6 (110) and HbA_{1c} <6.5% if feasible
Smoking		No smoking	
Diet		Healthy food choices recommended	
Physical activity		Moderate activity 30 min/d	

Table 11.6 Basic monitoring and general measures in bipolar patients

Evaluation	**Examination**
Non-modifiable risk factors • Family history • Individual medical history Modifiable risk factors • Excessive body weight/ central obesity • Blood pressure/pulse • Lipid levels • Glucose levels • Smoking • Alcohol/drug use • Diet • Physical activity • Antipsychotic therapy	BMI ± waist circumference Blood pressure/pulse Full blood count Electrolytes, urea and creatinine Liver function tests Blood glucose Lipid profile Pregnancy test (if clinically indicated)
Education	**Further monitoring**
Lifestyle advice: Increased physical activity Reduced calorie intake Stop smoking – 5 A's approach Potential treatment effects Patient treatment preference/ education	Personal/family history Weight (BMI) ± waist circumference Blood pressure/pulse Glucose/lipid profiles

drugs might be warranted in high-risk patients, especially those with established atherosclerotic CVDs. This is an area in which psychiatrists, general practitioners and general physicians need to work collaboratively in order to achieve better outcomes for bipolar patients with significant cardiovascular co-morbidity. Although it is desirable to have fasting blood samples to check the lipid profiles and blood glucose level, it might not be practically feasible in some bipolar patients, and therefore non-fasting blood samples will need to be used as a compromise in trying to achieve a balance between what needs to happen ideally and what can happen in reality.

There is evidence that a small reduction in the independent risk factors can lead to significant reduction in CVDs (Hennekens, 1998; Li *et al.*, 2006). A 10% reduction in blood cholesterol can lead to 30% reduction in the risk of coronary heart disease. A reduction in the diastolic blood pressure by 6 mmHg can reduce the risk of coronary heart disease by 16% and the risk of stroke by 42%. Cessation of smoking will lead to 50% reduction in the risk of coronary heart disease and losing 4–10 kg of weight in obese people can help to reduce the risk of coronary heart disease by 27%. It shows how a small reduction in risk factors can have a big impact on the outcome. These findings should encourage all health professionals to work with their patients to help them change their unhealthy lifestyle and improve the relevant blood parameters, as this can lead to a huge positive impact on their outcomes and quality of life.

It is also important to enquire about history of haematological problems, liver disease and renal diseases during the evaluation of patients' personal and family histories. In addition to doing the screening tests for the cardiovascular risk factors, tests should also be conducted to check serum electrolytes, urea and creatinine, liver function tests and full blood count (FBC) including haemoglobin, white cell count and platelets, as part of the basic screening in bipolar patients. Many of the pharmacological agents used in the treatment of BD have the potential to cause blood dyscrasias (e.g. leucopenia with carbamazepine and clozapine, agranulocytosis and aplastic anaemia with carbamazepine, and thrombocytopenia with valproate), changes in renal function (e.g. lithium), electrolytes (e.g. hyponatraemia with carbamazepine) and liver function (e.g. carbamazepine and valproate). In women of childbearing age, the possible implications of medications for pregnancy and contraception should be discussed. If clinically indicated, a pregnancy test needs to be performed. The detailed evaluation of these issues follows later on under the treatment-specific measures.

Guidelines for safety monitoring in bipolar patients vary considerably across different countries depending on the local resources and clinical and political priorities. Some advocate thyroid function, screening for drug misuse, coagulation studies and routine ECG as part of baseline assessment. The divergence across different guidelines reflects the complexity of the issues involved and shows how difficult it is to reach a consensus on handling this complex area of clinical practice. The protocols need to be agreed locally, taking the principles on good practice, availability of resources, affordability and patient safety into account. The main points are summarised in Table 11.6.

Treatment-specific monitoring

The discussion of management of physical health issues associated with lithium and anticonvulsants will concentrate on the additional measures required with patients on these medications. General health measures are presumed to have been put in place. The following recommendations are based on the American Psychiatric Association's guidelines published in 2002 (APA, 2002) as well as the National Institute for Health and Clinical Excellence (NICE) guidelines (National Collaborating Centre for Mental Health, 2006).

Lithium

Before beginning lithium treatment, the patient's general medical history should be reviewed, with special reference to those systems that might affect or be affected by lithium therapy. In addition, the possibility of pregnancy or the presence of any dermatological disorder should be considered.

Besides the general health measures such as taking a history and physical examination, FBC, urea and electrolytes, pregnancy test (if appropriate), thyroid function evaluation and, for patients over age 40, ECG monitoring – preferably with a rhythm strip – should be performed before lithium initiation. Lithium levels should be checked before and after every dose increment. Steady state levels are likely to be reached approximately 5 to 7 days after dose adjustment. In general, renal function should be tested every 2 to 3 months during the first 6 months of treatment and thyroid function should be evaluated once or twice during the first 6 months of lithium treatment. Following that, renal and thyroid function may be checked every 6 months to 1 year, or whenever clinically indicated.

Anticonvulsants

Before initiating treatment, a general medical history with attention to hepatic, haematologic and bleeding abnormalities should be taken. A baseline evaluation would include FBC with differential and platelet count, liver function test (LFT), and renal function/electrolytes tests (for carbamazepine). Clinical assessments, including FBC and LFT, should be done at a minimum of every 6 months for stable patients who are taking valproate.

Valproate displaces highly protein-bound drugs from their protein binding sites. In addition, valproate inhibits lamotrigine metabolism and more than doubles its elimination half-life by competing for glucuronidation enzyme sites in the liver. Consequently, in patients treated with valproate, lamotrigine must be initiated at a dose that is lower than the normal dose. Increments should be slower, and the final dose lower than usual as well.

Full blood count, platelets and LFTs should be performed every few weeks during the first 2 months of carbamazepine treatment. Thereafter, if normal and no symptoms of bone marrow suppression or hepatitis appear, blood counts and LFTs should be performed at least every 3 months. If possible, it is preferable that serum levels should be determined 5 to 7 days after a dose change, or sooner if toxicity or noncompliance is suspected. Additionally, as carbamazepine can induce its own metabolism, a repeat level should be done a month later.

As mentioned, carbamazepine is able to induce drug metabolism, including its own, through cytochrome P450 oxidation and conjugation. This may decrease levels of concomitantly administered medications such as valproate, lamotrigine, oral contraceptives, protease inhibitors, benzodiazepines and many antipsychotic and antidepressant medications. In addition, carbamazepine is metabolised primarily through a single enzyme, cytochrome P450 isoenzyme 3A3/4, making drug–drug interactions even more likely. Consequently, carbamazepine levels may be increased by medications that inhibit the cytochrome P450 isoenzyme 3A3/4, such as fluoxetine, fluvoxamine and some antibiotics and calcium-channel blockers.

For patients on valproate, an enquiry into the menstrual pattern of female patients should be conducted at 3-monthly intervals for the first year and then yearly subsequently.

Should a rash occur while patients are on carbamazepine or lamotrigine, the attending psychiatrist should ascertain if the rash is clearly drug related. If so, the medications should be stopped. If the patient is unable to obtain a medical review immediately, the medications should be stopped as a precautionary measure.

Antipsychotics

Antipsychotics, including both first- and second-generation drugs, have been traditionally used in the treatment of BD, especially in the manic phase of the illness. Among the second-generation antipsychotics, there is sufficient evidence demonstrated through randomised controlled trials that aripiprazole, olanzapine, quetiapine, risperidone and ziprasidone are effective treatments for BD, and the practice of prescribing these drugs in the treatment of BD is on the rise globally. Clozapine has also at times been used in the treatment of difficult-to-treat BD. The main concerns in terms of adverse effects surrounding the prescription of antipsychotic drugs include weight gain, alteration in lipid profiles, increased risk of diabetes, metabolic syndrome, hormonal side effects from hyperprolactinaemia, haematological adverse reactions and cardiac adverse effects, such as myocarditis and increased risk of ventricular arrhythmias due to prolonged QTc interval, and also EPSs. Antipsychotic drugs vary considerably in their potential to cause these adverse effects. Among the potential symptoms, the cardiovascular and metabolic adverse effects are shared by most of the antipsychotics.

The American Diabetes Association (ADA), American Psychiatric Association (APA), American Association of Clinical Endocrinologists, and North American Association for the Study of Obesity (2004) jointly produced a consensus monitoring protocol for patients receiving second-generation antipsychotic drugs. The panel recommended that personal and family history of medical conditions should be assessed at the beginning of antipsychotic treatment and yearly thereafter. The baseline screening measures should be obtained before, or as soon as clinically feasible after, initiation of any antipsychotic medication. If abnormalities are identified, then appropriate treatment should be initiated. The patient's weight should also be re-evaluated at 4, 8 and 12 weeks after initiating or switching to second-generation antipsychotics, and quarterly thereafter. Fasting plasma glucose and blood pressure should also be assessed at the start, 3 months after initiation of treatment and then annually. Fasting lipid levels should be assessed at the start and at 3 months, and repeat testing in those with a normal lipid profile can be performed at 5-year intervals or more frequently if indicated. However, the consensus panel recognised that more frequent monitoring might be necessary, depending on the clinical status of the patient.

Guidelines vary from one another about the items to be monitored and the frequency for the monitoring of these parameters. A British expert consensus meeting on schizophrenia and diabetes in Dublin in 2003 produced a consensus summary for monitoring of patients receiving antipsychotic treatment (Expert group, 2004). They recommended that checking of random blood sugar and glycosylated haemoglobin (HbA_{1c}) should be done at the start or on switching antipsychotic therapy and that these measures should be repeated after 4 months. If the results are normal, further monitoring may be done yearly. Considering how difficult it might be to obtain fasting blood samples from mentally ill patients whose ability to cooperate with these tests may be limited by virtue of their illness, the idea of checking random blood sugar and HbA_{1c} instead of fasting blood sugar sounds more pragmatic. The Australian consensus working group on diabetes, psychotic disorders and antipsychotic therapy (Lambert & Chapman, 2004) recommended the monitoring of BMI and waist:hip ratio instead of waist circumference every visit or 3 monthly. The Australian group recommended that monitoring of random or fasting blood glucose should ideally take place monthly for the first 6 months followed by 3- to 6-monthly monitoring afterwards. Their recommended frequency for monitoring of lipid profiles and blood pressure were 6 monthly as opposed to the 12-monthly and 5-yearly monitoring of these parameters proposed by the ADA consensus group. The Australian group recommended HbA_{1c} monitoring only for those with established diagnosis of diabetes in contrast to the recommendation by the British expert group, who proposed a combined monitoring of random blood sugar and HbA_{1c} for all patients receiving antipsychotic drugs, irrespective of the diagnosis of diabetes. This diversity in the guidelines across different countries demonstrates the difficulty in reaching a consensus among experts in the field, and therefore local practice needs to be adapted in line with the local health priorities and resource availability.

Among the cardiac adverse effects, prolongation of QTc interval can at times be significant, especially with sertindole and ziprasidone. However, it is important to acknowledge that there is considerable variability in QTc interval among normal healthy adults who are not on any medication, and there is as yet a lack of consensus about the acceptable upper safe limit of QTc. The acceptable upper safe limit might be around 440–460 ms for men and 440–470 ms for women (Gupta *et al.*, 2007).

Monitoring of ECG and liaising with cardiologists may become necessary in managing patients who are at risk. As the incidence of dangerously prolonged QTc and arrhythmias is rare, ECG monitoring is better reserved for patients at risk based on the choice of antipsychotic drug and their pre-existing medical history. Myocarditis and cardiomyopathy are rare, but potentially fatal cardiac adversities associated with the use of clozapine and therefore clinical vigilance is necessary to facilitate early detection and intervention. Persistent tachycardia should prompt the search for other indicators for myocarditis or cardiomyopathy. The hormonal side effects as a result of hyperprolactinaemia can manifest in the form of galactorrhoea, breast tenderness and enlargement, abnormalities in menstrual cycle and sexual dysfunction. Among the second-generation antipsychotics, risperidone and amisulpride are relatively more likely to cause the hormonal adverse effects due to raised prolactin, and therefore monitoring of serum prolactin level may be appropriate when clinically indicated. If hyperprolactinaemia persists despite reducing or changing the antipsychotic, or if it is accompanied by headaches and neurological signs such as visual field defects, further investigations such as brain imaging will become necessary to rule out the possibility of pituitary tumours. At times it may become necessary to manage the hyperprolactinaemia with medication (after liaising with the endocrinologists) when withdrawing or reducing the antipsychotic, but this is not always a pragmatic option in view of the clinical presentation and its severity.

Haematological risks are usually associated with clozapine. The established guidelines for haematological monitoring in clozapine therapy need to be followed in patients receiving clozapine. Although haematological adverse effects have been reported with other antipsychotic drugs, the incidence is very rare, and therefore routine haematological monitoring cannot be justified except in the case of clozapine. Table 11.7 provides a compromise view of the different monitoring aspects discussed so far.

Special populations

The following recommendations are based on the APA guidelines published in 2002 (APA, 2002) as well as the NICE guidelines in 2006 (National Collaborating Centre for Mental Health, 2006).

Pregnancy

As pregnancy in BD is discussed in detail in Chapter 10, this section will only discuss the salient health-related aspects. Many medications used to treat BD are associated with a higher risk of birth defects, so psychiatrists should encourage contraceptive practices for all female bipolar patients of childbearing age. Carbamazepine is known to increase the metabolism of oral contraceptive pills (OCPs) and its use in women using OCPs should be approached with caution.

Pregnancy in a bipolar patient can give rise to a number of health issues; hence, whenever possible, women with BD planning to get pregnant should be encouraged to consult with their psychiatrist. In the discussion, the potential teratogenic risks of medications should be balanced against the risk of no treatment, with the resultant risk of a relapse. The antenatal period does not increase the incidence of relapses. However, in patients on maintenance lithium treatment, the rate of relapse is increased by lithium discontinuation, particularly when abrupt (Viguera *et al.*, 2000). The postpartum period is consistently associated with a markedly greater risk for relapse. It is therefore generally considered that prophylactic medications such as lithium or valproate may prevent such postpartum mood episodes, although there are few data to support this (Cohen *et al.*, 1995). During a manic episode patients are prone to risky behaviours such as alcohol and substance misuse, which confers further risks to the fetus.

Balanced against this, first-trimester exposure to lithium, valproate or carbamazepine is associated with a greater risk of birth defects (Holmes *et al.*, 2001). Lithium was initially considered to be highly teratogenic, especially for Ebstein's anomaly (Cohen *et al.*, 1994). However, the voluntary physician reporting basis of this register may have led to an overestimation of teratogenicity. Subsequent studies have suggested a lower risk (Yonkers *et al.*, 2004).

Both valproate and carbamazepine are associated with increased rates of major malformations, although the association is less strong for carbamazepine. The NICE guidelines advocate the avoidance of valproate use in women under the age of 18, owing to concerns about polycystic ovary syndrome and the heightened risk of unplanned pregnancies in this age group. These guidelines also discourage the routine prescription of valproate for women of childbearing potential, and emphasise the risks associated with its use during pregnancy and the need for adequate contraception. Although there initially were data suggesting that lamotrigine use in pregnancy results in a higher risk of oral clefts, recent data seem to dispute this (GlaxoSmithKline, 2009).

Table 11.7 Antipsychotic monitoring summary[a]

	Start	Ongoing monitoring
Personal and family history	To enquire about history of medical conditions, including cardiac problems	To repeat yearly
Weight	BMI ±	Monthly for the first 3 months followed by quarterly monitoring
	Waist circumference or Waist-to-hip ratio at the start, as agreed locally	Monitor yearly afterwards
Blood pressure	Measure at the start	Monitor quarterly for the first year and then yearly
Fasting glucose	Measure at the start	Monitor quarterly for the first year and then yearly
Fasting lipid profiles	Measure at the start	Repeat at 3 months and then yearly
ECG	Not required routinely, but to be done as clinically indicated	
Serum prolactin level	Not required routinely, but to be done as clinically indicated	
Haematological monitoring	Required for clozapine in line with the specific monitoring guidelines	
Extrapyramidal side effects	To enquire about symptoms and do physical examination as and when clinically appropriate	

[a] Patients with metabolic or cardiovascular risk factors may require more frequent and detailed investigations.

Antipsychotic agents may be needed to treat psychotic features of BD during pregnancy, but they may also represent an alternative for treating symptoms of mania. However, there are few systematic studies looking into the use of antipsychotics in pregnancy. High-potency antipsychotic medications are preferred during pregnancy, as they are less likely to have associated anticholinergic, antihistaminergic or hypotensive effects. For newer antipsychotic agents such as olanzapine, clozapine and quetiapine, little is known about the potential risks of teratogenicity or the potential effects in the neonate. Electroconvulsive therapy (ECT) is a possible treatment for severe mania or depression during pregnancy. In terms of teratogenicity, the short-term administration of anaesthetic agents with ECT may present less risk to the fetus than pharmacological treatment options (APA, 2001).

In 2000, the American Academy of Pediatrics recommended that women who choose to remain on regimens of lithium, valproate or carbamazepine during pregnancy should have maternal serum α-fetoprotein screening for neural tube defects before the 20th week of gestation, with amniocentesis as well as targeted ultrasonography performed for any elevated α-fetoprotein values (American Academy of Pediatrics, 2000). Women should also be encouraged to undergo high-resolution ultrasound examination at 16–18 weeks' gestation to detect cardiac abnormalities in the fetus.

As hepatic metabolism, renal excretion and fluid volume are all altered during pregnancy and the perinatal period, serum levels of medications should be monitored closely if possible and doses adjusted if indicated. The mother will especially undergo rapid fluid shifts during delivery and this may lead to drug serum level fluctuations. However, discontinuation of lithium during this period is, on the balance of risks and benefits, not indicated (Llewellyn *et al.*, 1998).

Breastfeeding

All medications used in the treatment of BD are secreted in breast milk in varying degrees, thereby exposing the neonate (Yoshida *et al.*, 1999). However, as with the use of medications during pregnancy, the risks of breastfeeding with psychotropic medications must be weighed against the benefits of breastfeeding. Because lithium is secreted in breast milk at 25% of maternal serum concentration, it is difficult to recommend for mothers who choose to breastfeed (Viguera *et al.*, 2007).

Fewer data on breastfeeding are available for carbamazepine and valproate (Chaudron & Jefferson, 2000). Although generally considered safe, potential risks should always be considered. Little is known about lamotrigine exposure in breastfed neonates; however, levels in the infant may reach 25% of maternal serum levels. Consequently, the potential for pharmacological effects, including a risk for life-threatening rash, should be taken into account. Apart from clozapine, there are few reports of specific adverse effects in breastfeeding infants associated with antipsychotics. Nonetheless, these drugs are found in breast milk and could affect the infant.

The elderly

As there is a chapter concentrating on elderly patients with BD, this section will focus on the general principles of management. Geriatric patients will usually require lower doses of medications, as ageing is associated with reductions in renal clearance and volume of distribution. Concomitant medications and medical co-morbidities may also interact with psychotropic medications. Older patients are also more sensitive to side effects. Older patients may be more likely to develop cognitive impairment with medications such as lithium or benzodiazepines (Van Gerpen *et al.*, 1999). They may have difficulty tolerating antipsychotic medications and are more likely to develop EPSs and tardive dyskinesia compared with younger individuals (Caligiuri *et al.*, 2000). With some antipsychotics, orthostatic hypotension might be particularly problematic and increase the risk of falls (Leipzig *et al.*, 1999). As a general rule of thumb, the adage of 'start low, go slow' for drug initiation and dose escalation, and clinical vigilance for adverse drug reactions and potential drug interactions, is difficult to argue with.

Children and adolescents

Where mood-stabilising medications are used in children, the developmental differences in pharmacokinetics and pharmacodynamics need to be considered, as these influence drug effects, dosages, toxicity and side effects (Kearns *et al.*, 2003). Safety data for psychotropic medications in children and adolescents are limited (Greenhill *et al.*, 2003). Hence, careful and monitored use of mood-stabilising medications in children and adolescents is recommended, with a strong emphasis on additional psychosocial interventions.

In addition to the baseline and first year monitoring recommendations as outlined for adults, it would be wise to monitor on a 6-monthly (rather than annual) basis in the long term for children treated with valproate and atypical antipsychotics. For patients on long-term valproate, carbamazepine or risperidone, advice should be given for bone density protection. If possible, medications that induce hyperprolactinaemia should be avoided, as this can have unknown effects on the developing adolescent.

References

American Academy of Pediatrics Committee on Drugs (2000). Use of psychoactive medication during pregnancy and possible effects on the fetus and newborn. *Pediatrics*, **105**, 880–7.

American Diabetes Association, American Psychiatric Association, American Association of Clinical Endocrinologists, North American Association for the Study of Obesity (2004). Consensus development conference on antipsychotic drugs and obesity and diabetes. *Diabetes Care*, **27**, 596–601.

APA (2001). *The Practice of Electroconvulsive Therapy: Recommendations for Treatment, Training, and Privileging: A Task Force Report of the American Psychiatric Association*, 2nd edition. Washington, DC: American Psychiatric Press.

APA (2002). Practice guideline for the treatment of patients with bipolar disorder (revision). *The American Journal of Psychiatry*, **159**, 1–50.

Barnett, A.H., Mackin, P., Chaudhry, I., *et al.* (2007). Minimising metabolic and cardiovascular risk in schizophrenia: diabetes, obesity and dyslipidaemia. *Journal of Psychopharmacology (Oxford, England)*, **21**, 357–73.

Ben-Menachem, E. (2007). Weight issues for people with epilepsy. *Epilepsia*, **48**, 42–5.

Bjorntorp, P. (1996). The regulation of adipose tissue distribution in humans. *International Journal of Obesity and Related Metabolic Disorders*, **20**, 291–302.

Bowden, C.L., Calabrese, J.R., Ketter, T.A., *et al.* (2006). Impact of lamotrigine and lithium on weight in obese and nonobese patients with bipolar 1 disorder. *The American Journal of Psychiatry*, **163**, 1199–201.

Bowden, C.L., Calabrese, J.R., McElroy, S.L., *et al.*; Divalproex Maintenance Study Group (2000). A randomised, placebo-controlled 12-month trial of divalproex and lithium in treatment of outpatients with bipolar disorder. *Archives of General Psychiatry*, **57**, 481–9.

Calabrese, J.R., Bowden, C.L., Sachs, G.S., *et al.*; Lamictal Study Group (1999). A double-blind placebo-controlled study of lamotrigine monotherapy in outpatients with bipolar 1 depression. *The Journal of Clinical Psychiatry*, **60**, 79–88.

Caligiuri, M.R., Jeste, D.V., Lacro, J.P. (2000). Antipsychotic-induced movement disorders in the elderly: epidemiology and treatment recommendations. *Drugs & Aging*, **17**, 363–84.

Cassidy, F., Ritchie, J.C., Carroll, B.J. (1998). Plasma dexamethasone concentration and cortisol response during manic episodes. *Biological Psychiatry*, **43**, 747–54.

Chaudron, L.H. & Jefferson, J.W. (2000). Mood stabilizers during breastfeeding: a review. *The Journal of Clinical Psychiatry*, **61**, 79–90.

Chrousos, G.P. & Gold, P.W. (1992). The concepts of stress and stress system disorders. Overview of physical and behavioural homeostasis. *JAMA*, **267**, 1244–52.

Cohen, L.S., Friedman, J.M., Jefferson, J.W., *et al.* (1994). A re-evaluation of risk of in utero exposure to lithium. *JAMA*, **271**, 146–50.

Cohen, L.S., Sichel, D.A., Robertson, L.M., *et al.* (1995). Postpartum prophylaxis for women with bipolar disorder. *The American Journal of Psychiatry*, **152**, 1641–5.

Colton, C.W. & Manderscheid, R.W. (2006). Congruencies in increased mortality rates, years of potential life lost, and causes of death among public mental health clients in eight states. *Preventing Chronic Disease*, **3**, A42.

Correll, C.U. & Schenk, E.M. (2008). Tardive dyskinesia and new antipsychotics. *Current Opinion in Psychiatry*, **21**, 151–6.

De Backer, G., Ambrosioni, E., Borch-Johnsen, K., *et al.* (2003). European guidelines on cardiovascular disease prevention in clinical practice. Third Joint Task Force of European and Other Societies on Cardiovascular Disease Prevention in Clinical Practice. *European Heart Journal*, **24**, 1601–10.

Dentino, A.N., Pieper, C.F., Rao, M.K., *et al.* (1999). Association of interleukin-6 and other biologic variables with depression in older people living in the community. *Journal of the American Geriatrics Society*, **47**, 6–11.

Expert group (2004). 'Schizophrenia and Diabetes 2003' Expert Consensus Meeting, Dublin, 3–4 October 2003: consensus summary. *The British Journal of Psychiatry Supplement*, **47**, S112–14.

Fagiolini, A., Frank, E., Scott, J.A., *et al.* (2005). Metabolic syndrome in bipolar disorder: findings from the Bipolar Disorder Center for Pennsylvanians. *Bipolar Disorder*, **7**, 424–30.

Fagiolini, A., Chengappa, K.N., Soreca, I., *et al.* (2008). Bipolar disorder and metabolic syndrome: causal factors, psychiatric outcomes and economic burden. *CNS Drugs*, **22**, 655–69.

Gitlin, M.J. (1993). Lithium-induced renal insufficiency. *Journal of Clinical Psychopharmacology*, **13**, 276–9.

GlaxoSmithKline (2009). *The Lamotrigine Pregnancy Registry. Interim report 1 September 1992 through 31 March 2009*. Issued July 2009. http://pregnancyregistry.gsk.com/documents/lam_report_spring2009.pdf [accessed 10 January 2010].

Graham, I., Atar, D., Borch- Johnsen, K., *et al.* (2007). European guidelines on cardiovascular disease prevention in clinical practice: executive summary. Fourth Joint Task Force of the European Society of Cardiology and Other Societies on Cardiovascular Disease Prevention in Clinical Practice (Constituted by representatives of nine societies and by invited experts). *European Heart Journal*, **28**, 2375–414.

Greenhill, L.L., Vitiello, B., Riddle, M.A., *et al.* (2003). Review of safety assessment methods used in pediatric psychopharmacology. *Journal of the American Academy of Child and Adolescent Psychiatry*, **42**, 627–33.

Guberman, A.H., Besag, F.M., Brodie, M.J., *et al.* (1999). Lamotrigine-associated rash: risk/benefit considerations in adults and children. *Epilepsia*, **40**, 985–91.

Gupta, A., Lawrence, A.T., Krishnan, K., *et al.* (2007). Current concepts in the mechanisms and management of drug-induced QT prolongation and torsade de pointes. *American Heart Journal*, **153**, 891–9.

Harris, E.C. & Barraclough, B. (1998). Excess mortality of mental disorder. *The British Journal of Psychiatry*, **173**, 11–53.

Hennekens, C.H. (1998). Increasing burden of cardiovascular disease: current knowledge and future directions for research on risk factors. *Circulation*, **97**, 1095–102.

Hippisley-Cox, J., Parker, C., Coupland, C., Vinogradova, Y. (2007). Inequalities in the primary care of patients with coronary heart disease and serious mental health problems: a cross-sectional study. *Heart*, **93**, 1256–62.

Holmang, A. & Bjorntorp, P. (1992). The effects of cortisol on insulin sensitivity in muscle. *Acta Physiologica Scandinavica*, **144**, 425–31.

Holmes, L.B., Harvey, E.A., Coull, B.A., *et al.* (2001). The teratogenicity of anticonvulsant drugs. *New England Journal of Medicine*, **344**, 1132–8.

Jarema, M. (2007). Atypical antipsychotics in the treatment of mood disorders. *Current Opinion in Psychiatry*, **20**, 23–9.

Joffe, H., Cohen, L.S., Suppes, T., *et al.* (2006). Valproate is associated with new-onset oligomenorrhea with hyperandrogenism in women with bipolar disorder. *Biological Psychiatry*, **59**, 1078–86.

Johnston, A.M. & Eagles, J.M. (1999). Lithium-associated clinical hypothyroidism: prevalence and risk factors. *The British Journal of Psychiatry*, **175**, 336–9.

Kearns, G.L., Abdel- Rahman, S.M., Alander, S.W., *et al.* (2003). Developmental pharmacology – drug disposition, action, and therapy in infants and children. *The New England Journal of Medicine*, **349**, 1157–67.

Lambert, T.J. & Chapman, L.H. (2004). Diabetes, psychotic disorders and antipsychotic therapy: a consensus statement. *The Medical Journal of Australia*, **181**, 544–8.

Leipzig, R.M., Cumming, R.G., Tinetti, M.E. (1999). Drugs and falls in older people: a systematic review and meta-analysis, I: psychotropic drugs. *Journal of the American Geriatrics Society*, **47**, 30–9.

Lepkifker, E., Sverflik, A., Iancu, I., *et al.* (2004). Renal insufficiency in long-term lithium treatment. *The Journal of Clinical Psychiatry*, **65**, 850–6.

Li, T.Y., Rana, J.S., Manson, J.E., *et al.* (2006). Obesity as compared with physical activity in predicting risk of coronary heart disease in women. *Circulation*, **113**, 499–506.

Llewellyn, A., Stowe, Z.N., Strader, J.R. (1998). The use of lithium and management of women with bipolar disorder during pregnancy and lactation. *The Journal of Clinical Psychiatry*, **59** (Suppl. 6), 57–64.

Markowitz, G.S., Radhakrishnan, J., Kambham, N., *et al.* (2000). Lithium nephrotoxicity: a progressive combined glomerular and tubulointerstitial nephropathy. *Journal of the American Society of Nephrology*, **11**, 1439–48.

National Collaborating Centre for Mental Health (2006). *Bipolar Disorder: the Management of Bipolar Disorder in Adults, Children and Adolescents, in Primary and Secondary Care.* London: NICE. www.nice.org.uk/CG38 [accessed 3 January 2010].

NCEP ATP III (2002). Third Report of the National Cholesterol Education Program (NCEP) Expert Panel on Detection, Evaluation and Treatment of High Blood Cholesterol in Adults (Adult Treatment Panel III) final report (2002). *Circulation*, **106**, 3143–421.

Newcomer, J.W. & Hennekens, C.H. (2007). Severe mental illness and risk of cardiovascular disease. *JAMA*, **298**, 1794–6.

Osby, U., Brandt, L., Correia, N., *et al.* (2001). Excess mortality in bipolar and unipolar disorder in Sweden. *Archives of General Psychiatry*, **58**, 844–50.

Regenold, W.T., Thapar, R.K., Marano, C., *et al.* (2002). Increased prevalence of type 2 diabetes mellitus among psychiatric inpatients with bipolar I affective and schizoaffective disorders independent of psychiatric drug use. *Journal of Affective Disorder*, **70**, 19–26.

Rosmond, R. & Bjorntorp, P. (1998). Blood pressure in relation to obesity, insulin and the hypothalamic-pituitary-adrenal axis in Swedish men. *Journal of Hypertension*, **16**, 1721–6.

Sachs, G., Bowden, C., Calabrese, J.R., *et al.* (2006). Effects of lamotrigine and lithium on body weight during maintenance treatment of bipolar 1 disorder. *Bipolar Disorders*, **8**, 175–81.

Scherk, H., Pajonk, F.G., Leucht, S. (2007). Second-generation antipsychotic agents in the treatment of acute mania: a systematic review and meta-analysis of randomized controlled trials. *Archives of General Psychiatry*, **64**, 442–5.

Soreca, I., Mauri, M., Castrogiovanni, S., *et al.* (2007). Measured and expected resting energy expenditure in patients with bipolar disorder on maintenance treatment. *Bipolar Disorders*, **9**, 784–8.

Szalat, A., Mazeh, H., Freund, H.R. (2009). Lithium-associated hyperparathyroidism: report of four cases and review of literature. *European Journal of Endocrinology*, **160**, 317–23.

Tarsy, D. & Baldessarini, R.J. (2006). Epidemiology of tardive dyskinesia: is risk declining with modern antipsychotics? *Movement Disorders*, **21**, 589–98.

Taylor, V. & MacQueen, G. (2006). Associations between bipolar disorder and metabolic syndrome: a review. *The Journal of Clinical Psychiatry*, **67**, 1034–41.

Van Gerpen, M.W., Johnson, J.E., Winstead, D.K. (1999). Mania in the geriatric patient population: a review of the literature. *The American Journal of Geriatric Psychiatry*, **7**, 188–202.

Viguera, A.C., Nonacs, R., Cohen, L.S., *et al.* (2000). Risk of recurrence of bipolar disorder in pregnant and nonpregnant women after discontinuing lithium maintenance. *The American Journal of Psychiatry*, **157**, 179–84.

Viguera, A.C., Whitfield, T., Baldessarini, R.J., *et al.* (2007). Risk of recurrence in women with bipolar disorder during pregnancy: prospective study of mood stabilizer discontinuation. *The American Journal of Psychiatry*, **164**, 1817–24; quiz 1923.

Yonkers, K.A., Wisner, K.L., Stowe, Z., *et al.* (2004). Management of bipolar disorder during pregnancy and the postpartum period. *The American Journal of Psychiatry*, **161**, 608–20.

Yoshida, K., Smith, B., Kumar, R. (1999). Psychotropic drugs in mothers' milk: a comprehensive review of assay methods, pharmacokinetics and of safety of breast-feeding. *Journal of Psychopharmacology (Oxford, England)*, **13**, 64–80.

Zajecka, J.M., Weisler, R., Sachs, G., *et al.* (2002). A comparison of the efficacy, safety, and tolerability of divalproex sodium and olanzapine in the treatment of bipolar disorder. *The Journal of Clinical Psychiatry*, **63**, 1148–55.

Zakrzewska, K.E., Cusin, I., Sainsbury, A., *et al.* (1997). Glucocorticoids as counter regulatory hormones of leptin: toward an understanding of leptin resistance. *Diabetes*, **46**, 717–19.

Chapter 12

Anxiety associated with bipolar disorder: clinical and pathophysiological significance

Pratap R. Chokka and Vikram K. Yeragani

Introduction

Bipolar disorder (BD) type I is characterised by the occurrence of one or more manic or mixed episodes, and although a depressive episode is not always required for a diagnosis, virtually all patients experience depressive episodes, often more commonly than manic episodes. Bipolar disorder type II is characterised by the occurrence of one or more major depressive episodes accompanied by at least one hypomanic episode. In order for a diagnosis to be made, mood symptoms must cause clinically significant distress or impairment in social, occupational or other important areas of functioning (APA, 2000). Sometimes, the affective symptoms can be subthreshold to qualify for a diagnosis of BD. In the US National Comorbidity Survey, the lifetime prevalence of BD types I and II and subthreshold BD was 4.5%, and the12-month prevalence was 2.8% in the general population (Merikangas *et al.*, 2007).

Bipolar disorder results in significant disability and negative impact on quality of life (see Table 12.1). General health and social functioning may be worse for patients with bipolar depression than those with unipolar depression (Kauer-Sant'Anna *et al.*, 2007). Bipolar disorder is associated with high rates of unemployment, even among college graduates, and a large percentage of individuals report job-related difficulties. Co-morbid disorders such as anxiety associated with BD carry a much worse prognosis (Simon *et al.*, 2004).

When compared to individuals with no diagnosed mental disorder, those with BD demonstrate significant increases in lifetime health-service utilisation, and the need for welfare and disability benefits. This is especially true when there are associated co-morbid conditions.

Anxiety disorders, such as panic disorder (PD), generalised anxiety disorder (GAD) and obsessive–compulsive disorder (OCD) are chronic and disabling. They are relatively common and cause a significant loss of productivity at work. The common symptom of anxiety disorders is mainly the fear of the unknown and is often accompanied by increased heart rate, difficulty in breathing, choking sensation, hyperventilation and other autonomic symptoms. One should also note that there is a strong association between anxiety and depression, and most often, patients who suffer from anxiety disorders also suffer from various degrees of depression.

Table 12.1 Impact of BD

• Patients with BD report significant difficulties in:
• work-related performance
• leisure activities
• social and family interactions
• marital functioning
• quality of life
• general health

In addition to being highly disabling, anxiety disorders are also associated with an increased risk for mortality from cardiac events. Thus co-morbid anxiety in bipolar illness deserves a special mention. In this chapter, we mainly focus on the clinical and pathophysiological implications of BD associated with co-morbid anxiety disorders and some medical conditions such as diabetes and cardiovascular disorders.

Co-morbidity

Co-morbidity is the co-occurrence of several disease states with different aetiologies. Bipolar disorder occurs so frequently with other Axis I and II psychiatric disorders and certain general medical conditions that clinicians should consider bipolarity as a marker for co-morbidity (see Table 12.2). Anxiety and substance-use disorders are a frequent co-morbidity in patients

Table 12.2 Co-morbid conditions

• BD is associated with
Anxiety:
• generalised anxiety disorder (GAD)
• phobias
• panic disorder (PD)
• obsessive–compulsive disorder (OCD)
• post-traumatic stress disorder (PTSD)
Medical:
• diabetes
• hypertension
• stroke
• coronary artery disease (CAD)
• metabolic syndrome (MBS)

with BD. However, other disorders such as conduct disorders, eating disorders and personality disorders also occur in some patients (McIntyre *et al.*, 2006). There are several medical conditions that are associated with bipolar illness, which include important conditions such as diabetes, cardiovascular disease (CVD) and metabolic syndrome (Krishnan, 2005; McIntyre *et al.*, 2005).

Co-morbid anxiety disorders

Anxiety has been described in association with bipolar states since the time of Kraepelin (Geraud, 1997; Akiskal, 2007). In a recent article, McIntyre and co-workers reviewed the occurrence of anxiety in BD in great detail and suggested that anxiety disorders may be the most common co-morbid conditions (McIntyre *et al.*, 2006). Here, one has to differentiate between the symptoms of anxiety associated with BD versus anxiety as a disorder/syndrome. However, it is also important to note that the aetiology of all these disorders may not be completely disease-specific, as there is substantial co-morbidity among anxiety disorders as well. The effects of stress can include both a sense of helplessness associated with depression and various psychological and somatic symptoms of anxiety. These symptoms of anxiety are seen among different anxiety disorders, and this issue makes it even more difficult to dissect out the effects of anxiety symptoms associated with BD versus anxiety disorders. It is well known that major depressive disorder (MDD) is frequently associated with anxiety, which leads to frequent recurrences and treatment resistance in some cases. Anxiety co-morbidity also increases the illness severity in BD, including suicidality and a protracted course of the illness (Kessler *et al.*, 1997).

Readers are referred to an excellent review by McIntyre and co-workers (McIntyre *et al.*, 2006) as to the prevalence of anxiety disorder in general and the subtypes such as GAD, different types of phobia, PD and post-traumatic stress disorder (PTSD). These authors summarised the results of all the articles published in English in a MEDLINE search between 1966 and 2005. Based on this review, these authors concluded that anxiety disorders are very commonly associated with BD and that anxiety hastens the onset of BD. These authors report that anxiety appears to be a core dimension of BD in many patients. Further, they suggest that the association leads to poor insight, poor outcome and protracted course of BD. They also highlight the fact that anxiety disorder co-morbidity is associated with other disorders, such as substance abuse, which further complicates the management of BD.

Studies after 2005

In a recent longitudinal study of the course of major psychiatric disorders from illness onset, Salvatore and co-workers reported on the decade-long McLean–Harvard First Episode Project & International Consortium for Bipolar Disorder Research (Salvatore *et al.*, 2007). This study systematically followed up large numbers of patients with DSM-IV-TR (APA, 2000) bipolar or psychotic disorders from first hospitalisation (Salvatore *et al.*, 2007). One major finding among patients with BD type I was the presence of very high rates of suicidal behaviour and accidents occurring early, and substance-use co-morbidity associated with anxiety. They suggest that the long-term prognosis of BD type I is much less favourable than had been believed previously.

In another important investigation of co-morbid OCD, Zutshi and co-workers reported on clinical characteristics of bipolar OCD (Zutshi *et al.*, 2007). These authors compared the clinical characteristics of bipolar and non-bipolar OCD. Bipolar OCD individuals had an episodic course of OCD, high family loading for mood disorder and co-morbidity with depression, social phobia and GAD. They had less severe OCD and had a somewhat different symptom profile compared with that of non-bipolar OCD. The OCD pre-dated BD in 54% of the bipolar OCD individuals. In the remaining individuals, it started during the course of BD. Most bipolar OCD individuals reported worsening of OCD in depression and improvement in manic/hypomanic episodes. These authors suggest that the OCD in those with a primary diagnosis of BD is perhaps pathophysiologically related to BD rather than to OCD. This is strongly supported by the episodic course of OCD, high familial loading for mood disorders and worsening of OCD in depression with improvement in hypomanic/manic phases. Thus, future studies need to explore the OCD–bipolar co-morbidity in both OCD and bipolar samples. This may be an important factor in the treatment of OCD associated with BD. Other reports also suggest that patients with BD have multiple anxiety disorders (Levander *et al.*, 2007), which may worsen the prognosis even further (Lee & Dunner, 2007; Tamam, 2007). These findings further support those from the review of McIntyre and co-workers (McIntyre *et al.*, 2006).

Goldstein and Levitt have reported on youth-onset anxiety disorders. They are associated with the increased prevalence of subsequent BD type I among adults to identify risk factors for BD in this population (Goldstein & Levitt, 2007). This study used the National Epidemiologic Survey on Alcohol and Related Conditions to identify respondents with social phobia, PD or GAD onset in youth (<19 years) that was not preceded by a major depressive, manic or mixed episode. The sample comprised 1571 participants (572 males and 999 females). The prevalence of BD among participants with, versus without, these youth-onset anxiety disorders was examined along with other co-morbid variables, including conduct disorder, youth-onset substance-use disorders and family history of depression and/or alcoholism. Analyses were performed separately for males and females. This study showed that the prevalence of BD was significantly greater among adults with, versus without, primary youth-onset anxiety disorders for both males and females. Youth-onset anxiety disorders remained significantly associated with BD even after controlling for major depression. This finding remained significant for each of the specific anxiety disorders studied. Among males with youth-onset primary anxiety disorders, conduct disorder and family history of depression were associated with significantly increased risk for BD. Among females, conduct disorder and family history of alcoholism were associated with significantly increased risk for BD. The conclusion was that prevalence of BD was elevated among subjects with youth-onset primary anxiety disorders, particularly if co-morbid conduct disorder was present. This study emphasises the fact that the co-morbidity is one of a spectrum of related disorders rather than a single entity.

Another study examined anxiety co-morbidity in treatment-resistant bipolar patients (Lee *et al.*, 2007). These authors found that a majority had co-morbid PD, PTSD or OCD. They compared bipolar patients with and without a history of co-morbid anxiety disorders regarding several clinical variables, such as mean age, percentage of women, mean age of onset, history of suicide attempts, history of rapid cycling, history of substance abuse, family history and mood rating scales. Bipolar patients with co-morbid anxiety disorders were more significantly ill than bipolar patients without co-morbid anxiety disorders. For instance, patients with an anxiety disorder were more likely to have an earlier age of onset of illness, have higher (worse) ratings on the Hamilton anxiety rating scale, the 17-item Hamilton depression rating scale, the Montgomery–Asberg depression rating scale and the Beck depression inventory, and were more impaired on the global assessment of functioning scale. Co-morbid anxiety disorders were also associated with a more frequent history of substance abuse and higher ratings of suicidal ideation. Anxiety disorders negatively affected the prognosis of BDs. Levander and co-workers also examined the prevalence rate of anxiety co-morbidity in bipolar individuals with and without alcohol-use disorders (Levander *et al.*, 2007). Bipolar men and women who entered the Stanley Foundation Bipolar Network (SFBN) were divided into those participants meeting current or lifetime criteria for an alcohol-use disorder versus those participants who did not. Lifetime rates of co-morbid anxiety disorder were evaluated between groups. In this study, 46.5% met the criteria for an anxiety disorder. Panic disorder and OCD were the most common anxiety disorders in the alcohol-use and non-alcohol-use groups. Obsessive–compulsive disorder and specific phobias were significantly less prevalent in BD type I patients with alcohol use compared to those without. Bipolar women with alcohol use had a significantly higher rate of PTSD than those without. The higher rate of PTSD and lower rate of OCD in bipolar women with alcohol co-morbidity need to be studied further. All these studies suggest that the clinician should be very vigilant for co-morbid anxiety when treating BD, as the treatment options may differ. However, BD may also be an associated condition in different anxiety disorders.

A special reference to children and adolescents

Masi and co-workers published a series of reports examining the co-morbidity of anxiety associated with the bipolar spectrum (Masi *et al.*, 2001, 2004, 2007a, 2007b). Co-morbid anxiety is a frequent marker of early-onset BD. A substantial number of children diagnosed with BD have a co-morbid PD, and some of these patients may initially present with separation anxiety disorder. The presentation with PD appears to be more frequent in females. Sometimes other anxiety disorders such as OCD and social phobia are associated with BD. Masi and co-workers suggest that differential risk for PD among bipolar patients is indicative of heterogeneous subtypes of BD. Savino and co-workers have shown that anxious-bipolar co-morbidity occurs in patients with cyclothymic and hyperthermic temperaments (Savino *et al.*, 1993). These studies appear to have genetic implications for anxious-bipolar co-morbidity. The findings from the Systematic Treatment Enhancement Program for Bipolar Disorder (STEP-BD) study suggest that patients who experience an early onset of BD before the age of 13 years are more likely to have a lifetime diagnosis of PD, and this incidence appears to decrease with a later onset of BD (Perlis *et al.*, 2004). These findings emphasise the importance of screening younger BD patients for the concomitant symptoms of panic-anxiety. However, one should note that BD in children is preceded by several other conditions, such as attention deficit hyperactivity disorder (ADHD) and substance-abuse disorders (Rende *et al.*, 2007). The presentation is one of depression in the majority of cases.

Co-morbid medical conditions

There is an increased mortality among patients with BD due to suicide as well as several co-morbid medical conditions. Diabetes, metabolic syndrome (MBS) and other CVDs are common among patients with BD, and co-morbid anxiety probably carries an even greater risk for serious cardiovascular events.

Diabetes

Bipolar disorder is associated with obesity and unhealthy dietary habits, and thus the association of this condition with type 2 diabetes is not surprising. Some investigations suggest that patients who suffer from BD with co-morbid diabetes mellitus have a more severe course and outcome, lower quality of life and higher prevalence of other medical illnesses. The prevalence of diabetes mellitus in bipolar illness may be three times greater than in the general population. Some of the factors include abnormal glucose metabolism and other symptoms associated with MBS. All patients with BD should be screened for the development or co-occurrence of diabetes mellitus.

Metabolic syndrome

There are several definitions of MBS, and as new information becomes available, the definition of this syndrome may keep on changing. The metabolic syndrome is a cluster of abnormalities related to metabolism, including abdominal obesity, glucose intolerance, hypertension and dyslipidaemia. It is also associated with an increased risk of cardiovascular events. Since the initial description of this syndrome, there has been a general agreement that MBS is a major public health challenge worldwide. There are several definitions of this syndrome, which include those proposed by the World Health Organization, the European Group for the Study of Insulin Resistance, the National Cholesterol Education Program Adult Treatment Panel III, the American College of Endocrinology and American Association of Clinical Endocrinologists and the latest International Diabetes Federation definition, which includes ethnic-specific waist circumference cut-off points. In a recent study, the prevalence of this syndrome in patients with BD was 58% higher than that reported for the general Spanish population (Garcia-Portilla *et al.*, 2008). Although the prevalence rates may change in different populations, all the studies stress the importance of identifying these metabolic abnormalities in patients with BD so that serious vascular events may be prevented. Clinicians should be aware of this issue and appropriately monitor patients with BD for metabolic syndrome as part of the standard of care for these patients.

Cardiovascular disease and sudden death

Patients with BD are at a higher risk for CVD as a result of several conditions, including metabolic syndrome, hypertension and diabetes. There is also a strong link between anxiety and fatal CVD, particularly sudden death. Some studies link anxiety to cardiovascular and cerebrovascular disease. Though some of these studies included retrospective data, recent prospective studies have provided strong evidence for a relationship between anxiety and CVD (Kawachi *et al.*, 1994a, 1994b; Kubzansky *et al.*, 1998).

Pathophysiological significance

Imaging studies

Bipolar disorder

Abnormalities in brain activation using functional magnetic resonance imaging (fMRI) during cognitive and emotional tasks have been described in patients with BD (Yurgelun-Todd & Ross, 2006). These abnormalities include the frontal, subcortical and limbic regions. Several investigations suggest that mood state may be differentiated by lateralisation of brain activation in fronto-limbic regions. The interpretation of fMRI studies is limited by the choice of regions of interest, medication effects, co-morbidity and the task performance. It appears that there is a complex alteration in regions important for neural networks underlying cognition and emotional processing. However, new techniques and data are needed to resolve some of the above issues. Structural brain changes are apparent in some MRI studies of patients with mood disorders, but results are inconsistent. Recent studies suggest that small hippocampal volumes are apparent in patients with recurrent MDD, but not generally reported early in the course of adult-onset depression. Small hippocampal volumes are infrequently reported in BD. Changes in amygdala volumes are inconsistent in patients with MDD or BD. An extensive review of preclinical literature suggests that various psychotropic medications may have neurotrophic and neuroprotective effects, which makes it difficult to interpret some of these studies. In addition, patients' age, sex, age at onset of the illness, and treatment with different medications may affect the detection of regional brain volume changes in patients (Campbell & MacQueen, 2006).

Anxiety

Most animal studies of the effects of stress on the brain have been used as a model for anxiety. Brain areas involved in stress response, including prefrontal cortex, hippocampus and amygdala, appear to play a role in the symptoms of anxiety. Although some anxiety disorders may fit in with animal models of stress, both conceptually and in terms of imaging findings, it is too simplistic to assume that conditions such as PTSD, PD and OCD all have the same abnormalities in brain structure and function (Bremner, 2004).

Some of these abnormalities in different brain regions overlap between depressive and anxiety disorders, and it may be difficult to delineate how exactly co-morbid anxiety might change these abnormalities in brain function in BD.

Hypothalamic–pituitary–adrenal axis

Hypothalamic–pituitary–adrenal (HPA) axis function is usually measured by the responses to the combined dexamethasone/corticotrophin-releasing hormone (dex/CRH) test, the dexamethasone suppression test (DST) and basal cortisol levels. The HPA axis has been reported to be abnormal in BD. The dex/CRH test is abnormal in both remitted and non-remitted patients with BD. This measure of HPA axis dysfunction may be a potential trait marker in BD and thus possibly indicative of the pathophysiological process in this illness (Watson *et al.*, 2004).

Bipolar disorder and PD are episodic disorders of affect regulation, and the common pathophysiological mechanism is likely to involve deficits in amygdala-mediated, plasticity-dependent emotional conditioning, as suggested by some of the imaging studies. Bipolar and PDs have both been associated with abnormality in the amygdala and related structure. Abnormalities have been described with regard to serotonin, norepinephrine, brain-derived neurotrophic factor (BDNF) and corticotrophin-releasing factor (CRF) (MacKinnon & Zamoiski, 2006). Thus there may be some common pathophysiological mechanism related to bipolar and anxiety disorders.

Cardiovascular pathophysiology

Abnormalities in cardiovascular and respiratory function are inter-related and are associated with symptoms of anxiety in patients (see Table 12.3). As described above, patients with anxiety as well as BD are at a higher risk for significant cardiovascular events. Here, we discuss some of the factors that may be responsible.

Heart rate variability

We have previously summarised the findings linking decreased heart rate variability to anxiety and anxiety disorder (Yeragani, 2004). Our own investigations have shown decreased heart rate variability in patients with PD (Balon *et al.*, 1990, 1993; Yeragani *et al.*, 1998, 2000a; Rao & Yeragani, 2001). This is important due to the fact that decreased heart rate variability is associated with decreased vagal function, which results in increased cardiac mortality (Malliani *et al.*, 1991; Bigger *et al.*, 1992; Gottsater *et al.*, 2006). Some studies also suggest decreased cardiac vagal function in patients with BD (Cohen *et al.*, 2003). Several recent studies have shown that increased beat-to-beat QT interval variability is associated with increased cardiac mortality due to serious ventricular arrhythmias (Berger *et al.*, 1997; Atiga *et al.*, 1998; Berger, 2003). Our studies have shown an increase in QT interval variability in patients with anxiety as well as depression (Yeragani *et al.*, 2000b, 2003). In another study, we have shown that depression is associated with increased QT interval variability (Carney *et al.*, 2003). Thus, there appears to be a common pathophysiological mechanism in both these conditions.

Table 12.3 Abnormal cardiovascular and respiratory function

- Abnormalities
  - Decreased heart rate variability
  - Increased QT interval variability
  - Increased blood pressure variability
  - Respiratory irregularity
  - Atherosclerosis

Blood pressure variability

Increased blood pressure variability is another non-invasive indicator of increased mortality due to vascular disease and end organ damage (Mancia *et al.*, 1997). We have shown that patients with PD have increased blood pressure variability (Yeragani *et al.*, 2004a, 2004b). We are not aware of such studies in patients with BD. Increased pulse wave velocity is associated with atherosclerosis. We have recently shown that there is increased pulse wave velocity in patients with anxiety (Yeragani *et al.*, 2006). Patients with BD are at a higher risk of developing hypertension (Newcomer, 2006). Hence, the effects of co-morbid anxiety should be carefully evaluated in these patients.

Respiratory irregularity

There are several studies suggesting respiratory irregularity in patients with anxiety using linear and non-linear techniques (Yeragani *et al.*, 2002). We have also shown that this abnormality disappears after treatment with paroxetine, which is a serotonin reuptake inhibitor (Yeragani *et al.*, 2004a, 2004b).

Treatment

Pharmacotherapy

Singh and Zarate (2006) have reviewed pharmacological treatment of co-morbidity in BD and concluded that there is only empirical evidence for the efficacy, and there are no controlled studies in this area of co-morbid anxiety in BD. Usually, the co-morbid symptoms respond to mood-stabiliser therapy, but there are always exceptions. Carbamazepine and lithium may not be effective in treating anxiety disorders, whereas lamotragine appears to be effective in treating symptoms of PTSD. Sodium valproate may prove more effective in mixed mania and rapid cycling, which are often seen with early-onset BD (Steele & Fisman, 1997). Sodium valproate may also help with symptoms of anxiety and agitation. Gabapentin can be effective in co-morbid social phobia and also in patients with prominent symptoms of anxiety (Yasmin *et al.*, 2001). Atypical antipsychotics such as risperidone may be effective in patients with co-morbid OCD and also PTSD. Quetiapine and olanzapine may be effective in OCD and also in the management of treatment-resistant depression (Kennedy & Lam, 2003). Antidepressants in general are effective in treating anxiety disorders, but the side effects need to be carefully considered.

Side effects of pharmacotherapy

All psychotropic agents have notable cardiac side effects. Other adverse effects from mood stabilisers include weight gain, dizziness and sedation. Often, patients with BD also receive typical and atypical antipsychotics to control manic symptoms. Antipsychotics can induce weight gain and, more importantly, can cause cardiac side effects including prolongation of the QT interval. This can lead to serious ventricular arrhythmias and sudden death. A substantial number of patients also need an antidepressant in addition to the mood stabilisers and/or antipsychotic medication for their anxiety/depression. Most of the antidepressant medications can also result in weight gain and may increase serum lipid levels (Yeragani *et al.*, 1989, 1990; Chokka *et al.*, 2006). Tricyclic antidepressants in particular can cause serious cardiac side effects, including heart block. Thus, patients receiving the above combinations are prone to develop type 2 diabetes and MBS. Another important side effect from antidepressants is the emergent manic symptoms, or even the induction of rapid cycling, which further complicates treatment.

As described above, patients with bipolar illness are already more prone to develop diabetes, CVD and MBS. Hence, pharmacological treatment of these patients has to be carefully evaluated to avoid serious cardiovascular problems.

Psychosocial therapies

Pharmacological interventions alone do not provide sufficient benefit for some individuals with BD. Studies have shown that psychoeducation and cognitive-behavioural psychotherapy, along with other psychosocial approaches, are very useful in the long-term management of patients with BD. These techniques are especially useful to prevent relapse in BD patients.

Summary

Bipolar disorder is associated with several psychiatric, as well as medical, co-morbid conditions, which have a considerable negative impact on the long-term prognosis of this disorder. Quite a few patients with BD have multiple co-morbid anxiety disorders as well as substance-abuse disorder. As described in this article, the economic burden due to the co-morbid conditions is substantial, and early recognition and treatment are of paramount importance. Some of the underlying pathophysiological factors include a dysfunction of the HPA axis and cardiovascular autonomic dysfunction. All these factors may contribute to increased cardiovascular mortality. Some of the atypical antipsychotics appear to be effective in the treatment of BD and also co-morbid anxiety. Treating anxiety with the least cardiotoxic drugs is important in addition to careful monitoring for the development of MBS. Medical co-morbid conditions such as diabetes, CVD and MBS should receive prompt medical attention while treating patients with BD. Finally, psychosocial approaches are just as important, to foster treatment compliance and to improve lifestyle of these patients.

References

Akiskal, H.S. (2007). The emergence of the bipolar spectrum: validation along clinical-epidemiologic and familial-genetic lines. *Psychopharmacology Bulletin*, **40**, 99–115.

APA (2000). *The Diagnostic and Statistical Manual of Mental Disorders*, 4th edition. Arlington, VA: American Psychiatric Association. www.psychiatryonline.com/resourceTOC.aspx?resourceID=1 [accessed 23 December 2009].

Atiga, W.L., Calkins, H., Lawrence, J.H., Tomaselli, G.F., Smith, J.M., Berger, R.D. (1998). Beat-to-beat repolarization lability identifies patients at risk for sudden cardiac death. *Journal of Cardiovascular Electrophysiology*, **9**, 899–908.

Balon, R., Pohl, R., Yeragani, V.K., Berchou, R., Gershon, S. (1990). Monosodium glutamate and tranylcypromine administration in healthy subjects. *Journal of Clinical Psychiatry*, **51**, 303–6.

Balon, R., Yeragani, V.K., Pohl, R., Ramesh, C. (1993). Sexual dysfunction during antidepressant treatment. *Journal of Clinical Psychiatry*, **54**, 209–12.

Berger, R.D. (2003). QT variability. *Journal of Electrocardiology*, **36** (Suppl.), 83–7.

Berger, R.D., Kasper, E.K., Baughman, K.L., Marban, E., Calkins, H., Tomaselli, G.F. (1997). Beat-to-beat QT interval variability: novel evidence for repolarization lability in ischemic and nonischemic dilated cardiomyopathy. *Circulation*, **96**, 1557–65.

Bigger Jr, J.T., Fleiss, J.L., Steinman, R.C., Rolnitzky, L.M., Kleiger, R.E., Rottman, J.N. (1992). Frequency domain measures of heart period variability and mortality after myocardial infarction. *Circulation*, **85**, 164–71.

Bremner, J.D. (2004). Brain imaging in anxiety disorders. *Expert Review of Neurotherapeutics*, **4**, 275–84.

Campbell, S. & MacQueen, G. (2006). An update on regional brain volume differences associated with mood disorders. *Current Opinion in Psychiatry*, **19**, 25–33.

Carney, R.M., Freedland, K.E., Stein, P.K., *et al.* (2003). Effects of depression on QT interval variability after myocardial infarction. *Psychosomatic Medicine*, **65**, 177–80.

Chokka, P., Tancer, M., Yeragani, V.K. (2006). Metabolic syndrome: relevance to antidepressant treatment. *Journal of Psychiatry and Neuroscience*, **31**, 414.

Cohen, H., Kaplan, Z., Kotler, M., Mittelman, I., Osher, Y., Bersudsky, Y. (2003). Impaired heart rate variability in euthymic bipolar patients. *Bipolar Disorders*, **5**, 138–43.

Garcia-Portilla, M.P., Saiz, P.A., Benabarre, A., *et al.* (2008). The prevalence of metabolic syndrome in patients with bipolar disorder. *Journal of Affective Disorders*, **106**, 197–201.

Geraud, M. (1997). [Emil Kraepelin and bipolar disorder: invention or over-extension?]. *L'Encephale*, **23**, 12–19.

Goldstein, B.I. & Levitt, A.J. (2007). Prevalence and correlates of bipolar I disorder among adults with primary youth-onset anxiety disorders. *Journal of Affective Disorders*, **103**, 187–95.

Gottsater, A., Ahlgren, A.R., Taimour, S., Sundkvist, G. (2006). Decreased heart rate variability may predict the progression of carotid atherosclerosis in type 2 diabetes. *Clinical Autonomic Research*, **16**, 228–34.

Kauer-Sant'Anna, M., Frey, B.N., Andreazza, A.C., *et al.* (2007). Anxiety comorbidity and quality of life in bipolar disorder patients. *Canadian Journal of Psychiatry*, **52**, 175–81.

Kawachi, I., Colditz, G.A., Ascherio, A., *et al.* (1994a). Prospective study of phobic anxiety and risk of coronary heart disease in men. *Circulation*, **89**, 1992–7.

Kawachi, I., Sparrow, D., Vokonas, P.S., Weiss, S.T. (1994b). Symptoms of anxiety and risk of coronary heart disease. The Normative Aging Study. *Circulation*, **90**, 2225–9.

Kennedy, S.H. & Lam, R.W. (2003). Enhancing outcomes in the management of treatment resistant depression: a focus on atypical antipsychotics. *Bipolar Disorders*, **5** *(Suppl.* 2), 36–47.

Kessler, R.C., Rubinow, D.R., Holmes, C., Abelson, J.M., Zhao, S. (1997). The epidemiology of DSM-III-R bipolar I disorder in a general population survey. *Psychological Medicine*, **27**, 1079–89.

Krishnan, K.R. (2005). Psychiatric and medical comorbidities of bipolar disorder. *Psychosomatic Medicine*, **67**, 1–8.

Kubzansky, L.D., Kawachi, I., Weiss, S.T., Sparrow, D. (1998). Anxiety and coronary heart disease: a synthesis of epidemiological, psychological, and experimental evidence. *Annals of Behavioral Medicine*, **20**, 47–58.

Lee, J.H. & Dunner, D.L. (2007). The effect of anxiety disorder comorbidity on treatment resistant bipolar disorders. *Depression and Anxiety*, **25**, 91–7.

Levander, E., Frye, M.A., McElroy, S., *et al.* (2007). Alcoholism and anxiety in bipolar illness: differential lifetime anxiety comorbidity in bipolar I women with and without alcoholism. *Journal of Affective Disorders*, **101**, 211–17.

MacKinnon, D.F. & Zamoiski, R. (2006). Panic comorbidity with bipolar disorder: what is the manic-panic connection? *Bipolar Disorders*, **8**, 648–64.

Malliani, A., Pagani, M., Lombardi, F., Cerutti, S. (1991). Cardiovascular neural regulation explored in the frequency domain. *Circulation*, **84**, 482–92.

Mancia, G., Di Rienzo, M., Parati, G., Grassi, G. (1997). Sympathetic activity, blood pressure variability and end organ damage in hypertension. *Journal of Human Hypertension*, **11** (Suppl. 1), S3–S8.

Masi, G., Perugi, G., Millepiedi, S., *et al.* (2007a). Bipolar co-morbidity in pediatric obsessive-compulsive disorder: clinical and treatment implications. *Journal of Child and Adolescent Psychopharmacology*, **17**, 475–86.

Masi, G., Perugi, G., Millepiedi, S., *et al.* (2007b). Clinical and research implications of panic-bipolar comorbidity in children and adolescents. *Psychiatry Research*, **153**, 47–54.

Masi, G., Perugi, G., Toni, C., *et al.* (2004). Obsessive-compulsive bipolar comorbidity: focus on children and adolescents. *Journal of Affective Disorders*, **78**, 175–83.

Masi, G., Toni, C., Perugi, G., Mucci, M., Millepiedi, S., Akiskal, H.S. (2001). Anxiety disorders in children and adolescents with bipolar disorder: a neglected comorbidity. *Canadian Journal of Psychiatry*, **46**, 797–802.

McIntyre, R.S., Konarski, J.Z., Misener, V.L., Kennedy, S.H. (2005). Bipolar disorder and diabetes mellitus: epidemiology, etiology, and treatment implications. *Annals of Clinical Psychiatry*, **17**, 83–93.

McIntyre, R.S., Soczynska, J.K., Bottas, A., Bordbar, K., Konarski, J.Z., Kennedy, S.H. (2006). Anxiety disorders and bipolar disorder: a review. *Bipolar Disorders*, **8**, 665–76.

Merikangas, K.R., Akiskal, H.S., Angst, J., *et al.* (2007). Lifetime and 12-month prevalence of bipolar spectrum disorder in the National Comorbidity Survey replication. *Archives of General Psychiatry*, **64**, 543–52.

Newcomer, J.W. (2006). Medical risk in patients with bipolar disorder and schizophrenia. *Journal of Clinical Psychiatry*, **67** (Suppl. 9), 25–30.

Perlis, R.H., Miyahara, S., Marangell, L.B., *et al.* (2004). Long-term implications of early onset in bipolar disorder: data from the first 1000 participants in the systematic treatment enhancement program for bipolar disorder (STEP-BD). *Biological Psychiatry*, **55**, 875–81.

Rao, R.K. & Yeragani, V.K. (2001). Decreased chaos and increased nonlinearity of heart rate time series in patients with panic disorder. *Autonomic Neuroscience*, **88**, 99–108.

Rende, R., Birmaher, B., Axelson, D., *et al.* (2007). Childhood-onset bipolar disorder: evidence for increased familial loading of psychiatric illness. *Journal of the American Academy of Child and Adolescent Psychiatry*, **46**, 197–204.

Salvatore, P., Tohen, M., Khalsa, H.M., Baethge, C., Tondo, L., Baldessarini, R.J. (2007). Longitudinal research on bipolar disorders. *Epidemiologiae Psichiatria Sociale*, **16**, 109–17.

Savino, M., Perugi, G., Simonini, E., Soriani, A., Cassano, G.B., Akiskal, H.S. (1993). Affective comorbidity in panic disorder: is there a bipolar connection? *Journal of Affective Disorders*, **28**, 155–63.

Simon, N.M., Otto, M.W., Wisniewski, S.R., *et al.* (2004). Anxiety disorder comorbidity in bipolar disorder patients: data from the first 500 participants in the Systematic Treatment Enhancement Program for Bipolar Disorder (STEP-BD). *American Journal of Psychiatry*, **161**, 2222–9.

Singh, J.B. & Zarate Jr, C.A. (2006). Pharmacological treatment of psychiatric comorbidity in bipolar disorder: a review of controlled trials. *Bipolar Disorders*, **8**, 696–709.

Steele, M. & Fisman, S. (1997). Bipolar disorder in children and adolescents: current challenges. *Canadian Journal of Psychiatry*, **42**, 632–6.

Tamam, L. (2007). [Comorbid anxiety disorders in bipolar disorder patients: a review]. *Turk Psikiyatri Dergisi*, **18**, 59–71.

Watson, S., Gallagher, P., Ritchie, J.C., Ferrier, I.N., Young, A.H. (2004). Hypothalamic-pituitary-adrenal axis function in patients with bipolar disorder. *British Journal of Psychiatry*, **184**, 496–502.

Yasmin, S., Carpenter, L.L., Leon, Z., Siniscalchi, J.M., Price, L.H. (2001). Adjunctive gabapentin in treatment-resistant depression: a retrospective chart review. *Journal of Affective Disorders*, **63**, 243–7.

Yeragani, V.K. (2004). New linear and nonlinear techniques to study cardiovascular autonomic function in anxiety. In: Velotis, C.M. (Ed.). *Progress in Anxiety Disorder Research*, 1st edition. New York: Nova Science, pp. 57–88.

Yeragani, V.K., Balon, R., Pohl, R., Berchou, R., Merlos, B., Weinberg, P. (1990). Lactate induced panic and beta-2 adrenergic activation. *Pharmacopsychiatry*, **23**, 198.

Yeragani, V.K., Balon, R., Pohl, R., Ramesh, C. (1989). Imipramine treatment and increased serum cholesterol levels. *Canadian Journal of Psychiatry*, **34**, 845.

Yeragani, V.K., Mallavarapu, M., Radhakrishna, R.K., Tancer, M., Uhde, T. (2004a). Linear and nonlinear measures of blood pressure variability: increased chaos of blood pressure time series in patients with panic disorder. *Depression and Anxiety*, **19**, 85–95.

Yeragani, V.K., Rao, R., Tancer, M., Uhde, T. (2004b). Paroxetine decreases respiratory irregularity of linear and nonlinear measures of respiration in patients with panic disorder. A preliminary report. *Neuropsychobiology*, **49**, 53–7.

Yeragani, V.K., Nadella, R., Hinze, B., Yeragani, S., Jampala, V.C. (2000a). Nonlinear measures of heart period variability: decreased measures of symbolic dynamics in patients with panic disorder. *Depression and Anxiety*, **12**, 67–77.

Yeragani, V.K., Pohl, R., Jampala, V.C., Balon, R., Ramesh, C., Srinivasan, K. (2000b). Increased QT variability in patients with panic disorder and depression. *Psychiatry Research*, **93**, 225–35.

Yeragani, V.K., Radhakrishna, R.K., Tancer, M., Uhde, T. (2002). Nonlinear measures of respiration: respiratory irregularity and increased chaos of respiration in patients with panic disorder. *Neuropsychobiology*, **46**, 111–20.

Yeragani, V.K., Sobolewski, E., Igel, G., *et al.* (1998). Decreased heart-period variability in patients with panic disorder: a study of Holter ECG records. *Psychiatry Research*, **78**, 89–99.

Yeragani, V.K., Tancer, M., Seema, K.P., Josyula, K., Desai, N. (2006). Increased pulse-wave velocity in patients with anxiety: implications for autonomic dysfunction. *Journal of Psychosomatic Research*, **61**, 25–31.

Yeragani, V.K., Tancer, M., Uhde, T. (2003). Heart rate and QT interval variability: abnormal alpha-2 adrenergic function in patients with panic disorder. *Psychiatry Research*, **121**, 185–96.

Yurgelun-Todd D.A. & Ross A.J. (2006). Functional magnetic resonance imaging studies in bipolar disorder. *CNS Spectrums*, **11**, 287–97.

Zutshi, A., Kamath, P., Reddy, Y.C. (2007). Bipolar and nonbipolar obsessive–compulsive disorder: a clinical exploration. *Comprehensive Psychiatry*, **48**, 245–51.

Chapter 13

Bipolar disorder co-morbid with addictions

Jose M. Goikolea and Eduard Vieta

Introduction

Bipolar disorder (BD) is commonly associated with different kinds of co-morbidities. Addictions, mainly substance abuse or dependence, are one of the most common.

Substance-use disorder (SUD) is defined by DSM-IV-TR (APA, 2000) as a maladaptive pattern of a substance use in a recurrent way and with significant adverse consequences associated with the use of these substances. Co-morbidity with SUD is complex, as each disorder has an impact on the outcome of the other, and multiple co-morbidities are frequent.

In clinical practice, when BD complicates with SUD, the outcome is seriously compromised in all senses: clinical, functional and therapeutic. These patients require intensive treatment, ideally in the same setting, where the whole patient can be treated by the same therapeutic team.

Pharmacologically, it is difficult to make sound recommendations given the scarcity of randomised, controlled trials. To date, only valproate and lithium have one positive trial of these characteristics. More studies are needed to assess the efficacy and safety of current drugs. Even in the absence of these data, current practice should be improved as SUDs are pharmacologically undertreated and benzodiazepines are overprescribed in this population.

Specific psychotherapies that integrate psychological treatment of both disorders have shown positive results and are likely to be the most efficient treatment for co-morbidity. Despite the clinical severity of these patients, a subgroup in which SUD precedes onset of BD shows a milder form of illness with a better prognosis.

Epidemiology

Most epidemiological studies have confirmed the strong association between BD and SUD.

Table 13.1 Lifetime prevalence rates of co-morbid SUD in the Epidemiologic Catchment Area (ECA) and National Comorbidity Survey Replication (NCS-R) studies

	BD type I (%)	BD type II (%)	Subthreshold BP
ECA			
Any SUD	60.7	48.1	–
Alcohol dependence	31.5	20.8	–
Drug dependence	27.6	11.7	–
Alcohol-use disorders	46.2		–
Other drug-use disorders	40.7		–
NCS-R			
Alcohol abuse	56.3	36.0	33.2
Alcohol dependence	38.0	19.0	18.9
Drug abuse	48.3	23.7	22.9
Drug dependence	30.4	8.7	9.5
Any substance	60.3	40.4	35.5

The Epidemiologic Catchment Area (ECA) study found a 56.1% lifetime prevalence for any SUD in the whole bipolar sample, 60.7% for BD type I and 48.1% for BD type II. Other prevalence rates found in the study are shown in Table 13.1. These co-morbidity rates were the highest among any Axis I disorders (Regier *et al.*, 1990). As part of the ECA study, Helzer and Pryzbeck (1998) showed that mania and alcohol-use disorders are far more likely to occur together (odds ratio, 6.2) than would be expected by chance. Among all other diagnoses, only antisocial personality disorder is more likely than mania to be related to alcohol-use disorders (Helzer & Pryzbeck, 1998).

According to the National Comorbidity Survey (NCS), the risk of alcohol dependence among bipolar

Practical Management of Bipolar Disorder, eds. Allan H. Young, I. Nicol Ferrier and Erin E. Michalak. Published by Cambridge University Press.

subjects was tenfold higher compared with the general population. The risk for psychoactive substance dependence was eightfold higher for bipolar patients (Kessler *et al.*, 1997).

At the 2001–2002 National Epidemiologic Survey on Alcohol and Related Conditions (*n*=43 093), lifetime prevalence of DSM-IV-TR bipolar I disorder (BD type I) was as high as 3.3%. The 12-month prevalence was 23.6% for alcohol-use disorder and 12.9% for other SUDs. Lifetime prevalence was 58 and 37.5%, respectively. In this study BD type I was found to be highly and significantly related to substance use, as well as to anxiety, and personality disorders (Grant *et al.*, 2005).

At the National Comorbidity Survey Replication (NCS-R) study, not only BD types I and II, but also subthreshold BD, prevalence and their co-morbidities were assessed. Co-morbidity with SUDs was substantially higher in BD type I than in type II or subthreshold BD. Lifetime prevalences are shown in Table 13.1 (Merikangas *et al.*, 2007).

In clinical samples, the prevalence of co-morbid addictions varies greatly depending on different factors, such as the country or the type of setting (primary, tertiary, etc.) (see Table 13.2). The Systematic Treatment Enhancement Program for Bipolar Disorder (STEP-BD) found in a sample of 1000 US patients a lifetime SUD prevalence of 48%, with 8% meeting criteria for a current alcohol-use disorder, and 5% meeting criteria for a current non-alcohol substance-use diagnosis (Simon *et al.*, 2004). It should be emphasised that although the lifetime prevalence of SUD in BD is high, SUDs wax and wane in the course of BD and many patients with a lifetime history of SUD recover or have periods of recovery from their SUD, current SUD prevalence rates being much lower. For example, in another sample of 288 outpatients with BD, the lifetime history of SUD was 42%, whereas only 4% of individuals met the criteria for a current SUD (McElroy *et al.*, 2001). Similarly, Winokur found that although 37% of outpatients with BD had current alcohol dependence at study entry, only 5% had current alcohol dependence at 5-year follow-up (Winokur *et al.*, 1995).

The Zurich Cohort Study in Switzerland followed up 591 individuals, and found that those having manic symptoms – that is, BD type I – were at significantly greater risk for the later onset of alcohol abuse/dependence, cannabis use and abuse/dependence, and benzodiazepine use and abuse/dependence. Bipolar disorder type II predicted only alcohol abuse/dependence and benzodiazepine use and abuse/dependence (Merikangas *et al.*, 2008).

Table 13.2 Rates of SUD co-morbidity in clinical samples of bipolar patients

Study	Rate (%)
Brady, 1991 (inpatients)	30, alcohol or drug abuse
Cassidy, 2001 (manic inpatients)	48.5, alcohol abuse 43.9, drug abuse
Keller, 1986	5, alcohol abuse – manic 8, alcohol abuse – depressive 13, mixed or rapid cycling
McElroy, 2001	42, lifetime SUD 4, current SUD
STEP-BD (Simon, 2004)	48, lifetime SUD 8, current alcohol-use disorder 5, current other substance-use disorder
Strakowski, 1998 (psychotic mania)	45, alcohol or drug abuse
Tondo, 1999 (major affective disorder)	32, SUD 33, SUD in BD type I 28, SUD in BD type II
Vieta, 2000, 2001	19.3, SUD in BD type I 14.7, alcohol-use disorder in BD type I 4.65, other substance-use disorder in BD type I 25, SUD in BD type II 20, alcohol-use disorder in BD type II
Winokur, 1995	37, current alcohol dependence 5, current alcohol dependence at 5 years

If bipolar patients are prone to co-morbidity with SUDs, the opposite is also true: that is, among patients with SUDs, prevalence of BD is higher than in the general population, although co-morbidity rates do not reach such high levels. Cocaine abusers seem to be more likely to also suffer BD. Table 13.3 shows some co-morbidity data in clinical samples of patients suffering SUDs. However, these data should be taken cautiously, as some studies assess only mania history (BD type I), whereas others evaluate bipolar spectrum disorders.

Male sex, young age and low educational level have been identified as sociodemographic risk factors for substance abuse in patients with BD (Sonne & Brady, 1999; Tondo *et al.*, 1999). Alcohol-use disorders are more prevalent in bipolar men than in women, as in the general population. A study conducted with 267 outpatients enrolled in the Stanley Foundation Bipolar Network found that 49% of men compared with 29% of women with BD met the criteria for lifetime alcoholism. However, the relative risk of suffering alcoholism was greater for women with BD (odds ratio, 7.35) than

Table 13.3 Rates of bipolar co-morbidity in clinical samples of patients diagnosed with SUDs

Study (bipolar subtype assessed)	% of patients
Gawin & Kleber, 1986 (bipolar spectrum)	Cocaine abusers: 20%
Hesselbrock, 1985 (mania)	Hospitalised alcoholics: 2%
Mirin, 1999 (bipolar disorder or cyclothimia)	Stimulant abusers: 17.5% Hypnotic abusers: 6.8% Opiate abusers: 5.4%
Nunes, 1989 (bipolar spectrum)	Cocaine abusers: 30%
Ross, 1988 (lifetime mania)	Alcohol and drug dependents: 1.9%
Rousanville, 1991 (bipolar spectrum)	Cocaine abusers: 31%
Weiss, 1986 (bipolar spectrum)	Cocaine abusers: 23%

for men with BD (odds ratio, 2.77) when compared with the general population (Frye *et al.*, 2003).

It is also clear that suffering SUD is associated with a higher risk of suffering a second or third co-morbidity, both in terms of another SUD or other kinds of co-morbidity, such as personality disorders or anxiety disorders (Merikangas *et al.*, 2007). A recent Finnish study with 90 BD type I and 101 BD type II participants, found that Axis II co-morbidity, and specifically cluster B personality disorder co-morbidity, significantly increased the likelihood of additional co-morbidity including SUDs (Mantere *et al.*, 2006).

In addition, the second co-morbidity is usually associated with a worse outcome than the first one. As an example, in a secondary analysis of a randomised, placebo-controlled trial that assessed valproate in 56 patients with BD type I and current active alcohol dependence, 48% of the patients also reported co-morbid marijuana abuse. These patients were younger, had fewer years of education and had a higher number of additional psychiatric co-morbidities. Their alcohol and other drug use pattern was more severe and they were more likely to have an additional cocaine-abuse disorder. In this trial, the placebo-treated marijuana abuse group had the worst alcohol-use outcome (Salloum *et al.*, 2005b).

The Stanley Foundation Bipolar Network compared bipolar patients with and without alcohol-use disorders regarding co-morbidity with anxiety disorders. Bipolar women with alcohol-use disorders had a significantly higher rate of post-traumatic stress disorder (PTSD), whereas co-morbid obsessive–compulsive disorder was less prevalent (Levander *et al.*, 2007).

Some data suggest that specific drugs may be more associated with specific co-morbid disorders. In a retrospective assessment, generalised anxiety disorder was more prevalent in bipolar patients with alcohol dependence, whereas higher rates of PTSD and antisocial personality disorder were found in bipolar patients with cocaine dependence (Mitchell *et al.*, 2007).

Pathogenesis

Different hypotheses have been suggested to explain the high rates of co-morbidity between BD and SUDs. Most likely, they all explain to a certain degree the co-morbidity phenomenon.

1. Consumption of psychoactive substances might trigger mania or depression in a genetically vulnerable subject. This can be a case of a secondary mania or depression, but sometimes the substance can trigger a full affective disorder.
2. Symptoms of hypomania or mania include excessive involvement in pleasurable activities that have a high potential for painful consequences. Substance misuse is a good example.
3. Genetic diathesis: some studies have found vulnerability genes for BD overlapping with vulnerability genes for SUD.
4. Anxiety: from both the genetic and clinical points of view, anxiety may be a relevant mediator of the link between BD and SUD. Anxiety disorders are highly co-morbid with BD, and patients with anxiety are more prone to SUD. Indeed, some substances have clear anxiolytic effects.
5. Common pathophysiology. Both disorders share some pathophysiological mechanisms. Disturbances in neurotransmitter systems, especially dopaminergic pathways, or adaptations in post-receptor signalling pathways, including regulation of neural gene expression, could be involved in the aetiology of both disorders. For example, chronic cocaine administration in animals has been considered a pertinent model for refractory mood disorder. Acute low-dose cocaine administration in animals produces an important model for the euphoric and psychomotor components of hypomania, and higher chronic doses may model the dysphoric and psychotic components of mania. Also, the phenomenon of kindling may present an interesting model to explain this co-morbidity. Post (2007) has suggested that increased rates of recurrence of BD

in patients with co-morbid alcohol abuse may be due to a kindling-like effect produced by alcohol.

6. Social diathesis: psychosocial impairment due to BD, unemployment and other adverse situations might lead to marginality that increases the risk of substance misuse.
7. Self-medication: this hypothesis postulates that BD patients would abuse substances as a way to relieve symptoms of their primary disease (i.e. stimulants to reverse the symptoms of depression, etc.) or side effects of the pharmacological treatment (i.e. nicotine reduces blood levels of several drugs and, therefore, reduces dose-dependent side-effects). However, this possible 'self-medication' effect is controversial and has not been clearly supported by empirical research. Although it may be real in some cases, it is also true that bipolar patients are more prone to use cocaine or stimulants in the manic phase rather than in the depressive one.
8. Personality features. High impulsivity and high scores in novelty-seeking (Cloninger temperament and character inventory (TCI)) are usually found in bipolar patients. These features will increase the risk of substance use.

In summary, it remains unclear whether one of the above hypotheses, a combination of these or a yet-unidentified mechanism is responsible for the strong association between BD and SUDs that is seen in clinical and community populations (Tohen *et al.*, 1998).

Clinical picture and course

Effect of co-morbid substance-use disorders on outcome in bipolar disorder

Most studies agree that co-morbidity with SUD is associated with a range of complications that make BD more severe and difficult to treat, and involves a worsening of the outcome.

Patients with BD who experience substance abuse are more likely to suffer mixed episodes than patients with BD only (Himmelhoch *et al.*, 1976; Keller *et al.*, 1986; Goldberg *et al.*, 1999a). Recovery from mixed episodes is slower, but even when controlling for this variable, substance abuse itself involves slower recovery from affective episodes (Strakowski *et al.*, 1998) as well as a lower rate of remission during hospitalisation (Goldberg *et al.*, 1999a). These patients are more likely to require hospitalisations throughout the course of the illness (Brady & Sonne, 1995; Feinman & Dunner, 1996; Cassidy *et al.*, 2001). A substantial increase in the affective morbidity, in terms of more mood episodes, more rapid cycling and more days ill, has also been described (Keller *et al.*, 1986; Feinman & Dunner, 1996; Strakowski *et al.*, 1998; Strakowski *et al.*, 2007). Goldberg and Whiteside found that a history of substance abuse and/or dependence increased the risk for antidepressant-induced mania, with an odds ratio of 6.99 (Goldberg & Whiteside, 2002).

Co-morbidity with substance abuse is one of the main variables associated with a higher severity of manic episodes (Nolen *et al.*, 2004), and in acutely manic patients those with current alcohol-use disorder show an increased number of manic symptoms, impulsivity and high-risk behaviour, including violence (Salloum *et al.*, 2002). In fact, studies support the clinical impression of increased aggression and more criminal actions in bipolar patients with co-morbid SUD. In a study conducted in Los Angeles County (USA), 66 arrested people were identified as diagnosed with BD type I. Most of them were manic (74.2%) and psychotic (59%) at the time of their arrest. And as many as 50 of these patients (75.8%) had a co-morbid substance abuse (Quanbeck *et al.*, 2004).

Different studies have linked suicidality with co-morbidity with SUD in BD (Feinman & Dunner, 1996; Goldberg *et al.*, 1999b; Tondo *et al.*, 1999). This was equally true for both BD type I and II patients (Vieta *et al.*, 2000; Vieta *et al.*, 2001).

As expected from all these data, co-morbid patients show poorer global functioning and worse occupational status (Feinman & Dunner, 1996; Strakowski *et al.*, 2000). One of the variables that might act as a mediator for a worse outcome is poor compliance, which has repeatedly been shown to be frequent in these patients (Strakowski *et al.*, 1998; Colom *et al.*, 2000; González-Pinto *et al.*, 2006).

Even nicotine dependence, very prevalent in many psychiatric disorders and specifically in BD, has been correlated with several negative outcomes. In an assessment of 399 outpatients, having ever smoked was associated with earlier age at onset of first depressive or manic episode, lower global assessment of functioning scores, higher clinical global impressions–bipolar disorder scale scores, lifetime history of a suicide attempt and lifetime co-morbid disorders: anxiety disorders, alcohol abuse and dependence and substance abuse and dependence (Ostacher *et al.*, 2006).

Obviously, all these characteristics make patients with BD and SUD extremely difficult to manage.

Regarding the influence of different substances, a study has shown a correlation between the duration of alcohol abuse during follow-up and the time patients experienced depression, whereas the duration of cannabis abuse correlated with the duration of mania (Strakowski *et al.*, 2000). Cannabis use may be well correlated to an increase in manic, mixed and psychotic morbidity (Strakowski *et al.*, 2007; Pacchiarotti *et al.*, 2009).

Effect of bipolar disorder on substance-use disorder

Although not a general rule, bipolar patients often increase substance use in mania, rather than in depression. When manic, some patients increase their drinking, although most of them (approximately three-quarters) do not change their alcohol consumption, and only rarely do they decrease their drinking. By contrast, during depression they are as likely to decrease as to increase their drinking (approximately 15% each), although, again, most do not change their alcohol use (Strakowski *et al.*, 2000).

Bipolar first or substance-use disorder first: different subtypes

In some patients, onset of bipolar illness precedes substance abuse. In others, substance abuse precedes BD. *Primary* and *secondary* have been used to specify which disorder started first. But these terms are no longer used, as they may invoke a causality relationship. Causality requires a temporal sequence, but the complex relationship between these two disorders cannot be explained in such a simple way, as if the first disorder was the cause of the second. In any case, the fact is that different studies agree that patients whose BD starts after SUD is already present ('SUD first') are a distinct subgroup.

Winokur studied bipolar patients with co-morbid alcoholism, and compared the two groups with the different onset patterns. There were no significant differences in family history between the groups. Patients with 'bipolar first' showed a more severe illness, with more affective episodes after the initial recovery, shorter median time to first affective relapse, and lower likelihood of recovery during follow-up. This led the authors to suggest that patients with 'alcoholism first' suffer a less severe subtype of illness that may have needed the added insult of alcoholism to make it manifest (Winokur *et al.*, 1995).

More recent data support Winokur's suggestion. Strakowski *et al.* found that bipolar patients who experienced alcohol abuse before onset of BD ('alcohol first') showed a less severe form of affective illness: patients in this group were more likely to recover and recover more quickly than patients in which onset of BD preceded alcohol abuse ('bipolar first') or even than bipolar patients without alcohol-use disorder. Affective symptomatic recurrence curves were similar among groups, but the 'bipolar first' group spent more time with affective episodes and symptoms of an alcohol-use disorder during follow-up than the 'alcohol first' group (Strakowski *et al.*, 2005). The same group conducted a very similar study, this time in bipolar patients with co-morbid cannabis-use disorder after the first hospitalisation for mania, and got similar results. The group of 'cannabis first' had better recovery than the 'bipolar first' group, or even than the 'no cannabis use' group, although when adjusted for potential mediator variables these results did not persist. The 'cannabis first' group had an older mean age at hospitalisation for first manic episode (23 years), followed by the 'no cannabis' group (18 years) and by the 'first bipolar' group (16 years). In addition, patients with bipolar first had more mixed and manic episodes than the other groups, and more severe dependence than the 'cannabis first' group. However, the effects of the sequence of onsets of bipolar and cannabis-use disorders were less pronounced than those observed in co-occurring alcohol and BDs (Strakowski *et al.*, 2007).

The Vieta group has found similar results in their sample with a retrospective methodology. Patients with previous onset of the SUD ('SUBP') were less likely to have a depressive onset (compared with 'NSUBP', which meant no co-morbidity or co-morbidity starting after the onset of bipolar illness), although both groups were similar for lifetime polarity of mood episodes. The SUBP group had a lower number of lifetime depressive and/or hypomanic episodes, despite a worse compliance. There were no significant differences regarding number of manic episodes. Despite this fact, the SUBP group showed a higher lifetime prevalence of psychotic symptoms. As this was the only outcome that was worse in this group, the authors suggested it was a direct effect of the substance abuse (Pacchiarotti *et al.*, 2009).

In conclusion, these studies tell us that co-morbidity with alcohol or cannabis worsens the outcome of 'bipolar first' when compared with 'no co-morbid' patients, but the 'substance first' group seems to be a different subtype with lower vulnerability to the illness, which makes a better outcome possible.

Treatment

Pharmacological

Clinical trials on the efficacy and safety of drugs for dual bipolar patients are very scarce. Different reasons can explain this fact. First, design difficulties are obvious, as two different kinds of outcome need to be assessed: that is, both affective and substance-use outcomes. In addition, complexity of clinical presentation of bipolar patients confers many different situations to be assessed: patients in different types of episode (manic, hypomanic, mixed, depressed or euthymic), using different drugs – frequently more than one – and each drug modifying the psychopathological status of the patient, and with frequently more co-morbidities, makes the picture too complex to be captured in a trial design. Second, if these patients are difficult to engage in regular follow-up, it becomes much more difficult for them to consent to participation in clinical trials. Finally, pharmaceutical companies do not start these trials, as little benefit is perceived compared to the necessary costs, as indications are not regarded in the field.

Some data show that psychopharmacological treatment of these patients should be clearly improved. For example, data from the STEP-BD have shown that even in specialised settings for bipolar patients, the use of drugs more specific for the SUD, such as disulfiram, methadone, naltrexone and buprenorphine, is extremely low, with only 0.4% (4/955) receiving at least one of these four drugs (Simon *et al.*, 2004). Besides, benzodiazepines are probably overprescribed in this population. Benzodiazepine use among New Hampshire Medicaid beneficiaries with psychiatric disorders was assessed, to find that 75% of bipolar patients with co-morbid SUD (vs 58% of bipolar patients without co-morbid SUD) were prescribed benzodiazepines during a 5-year period. Although a local study, these data are likely to reflect prescription practice in most Western countries, which is contrary to clinical guidelines, due to the higher risk of abuse in this population (Clark *et al.*, 2004).

Valproate is one of the few drugs that have been shown in a randomised, double-blind design to improve the outcome of substance abuse, specifically alcohol use, in bipolar patients. In this trial, 59 bipolar patients with co-morbid alcohol dependence, treated with lithium and experiencing a concurrent acute mood episode (manic, mixed or depressed), were randomised to adjunctive valproate or placebo for 24 weeks. The group treated with adjunctive valproate showed a reduction in heavy alcohol drinking and in the number of drinks consumed when drinking heavily, as well as an increase in the number of days to relapse (Figure 13.1). Higher valproate serum concentration significantly correlated with improved alcohol-use outcomes. Manic and depressive symptoms improved equally in both groups. There were no serious drug-related adverse events, and the level of gamma-glutamyl transpeptidase was significantly higher in the placebo group (Salloum *et al.*, 2005a).

Clinical features of bipolar patients with co-morbid SUD, such as more mixed episodes or more rapid cycling, as well as some retrospective data (Goldberg *et al.*, 1999a) have suggested that these patients may respond better to anticonvulsants rather than to lithium, but this hypothesis has not yet been proven. In fact, lithium has shown positive results in double-blind randomised conditions, although in a short-term study limited to adolescents. Lithium was superior to placebo, both in terms of affective symptoms and substance use in 25 bipolar adolescents with secondary substance-dependency disorders (Geller *et al.*, 1998). On the other hand, two small open trials with lithium in cocaine abusers with bipolar spectrum disorders showed contradictory results. A first trial with nine patients suggested efficacy of lithium in this subgroup of patients (Gawin & Kleber, 1984). However, a second trial with ten cocaine abusers showed no benefit of this drug (Nunes *et al.*, 1990).

Recently, a randomised, double-blind, placebo-controlled add-on trial of quetiapine in bipolar patients with co-morbid alcohol-use disorders, found this drug to be effective for depressive symptoms but not in reduction of alcohol use (Brown *et al.*, 2008), despite previous positive results in open trials (Brown *et al.*, 2002).

In fact, several drugs have also shown positive results, both in terms of affective outcomes and substance-use outcome, in open trials. But, as no randomisation nor control group was present in these studies, these results should be taken cautiously, and the drugs need further assessment in this dual population. The case of quetiapine is a good example. Some of these drugs, such as lamotrigine (Brown *et al.*, 2003, 2006b) or aripiprazole (Brown *et al.*, 2005), are efficacious in

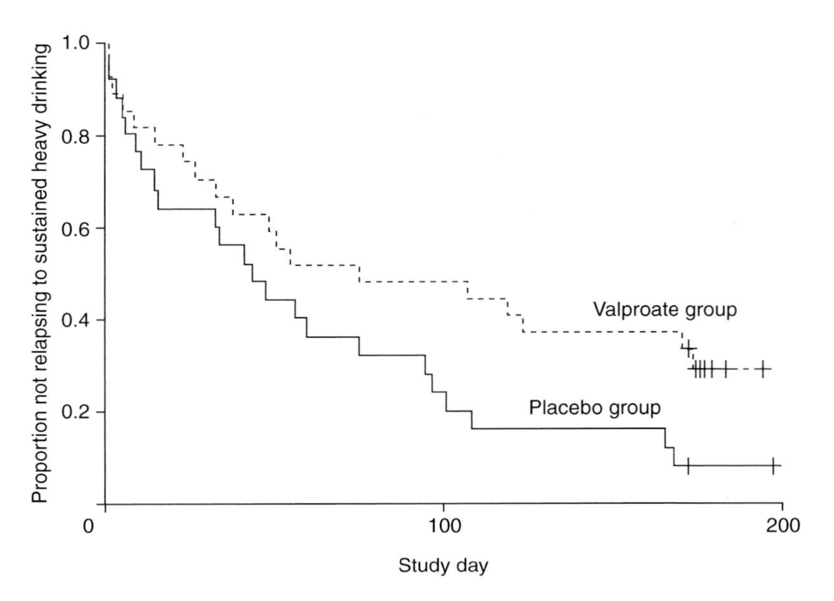

Figure 13.1 Kaplan–Meier survival curve for time to relapse to sustained heavy drinking by treatment group.

BD. Naltrexone has also been shown to improve affective symptoms and alcohol use in alcohol-dependent bipolar patients in a similar open design (Brown *et al.*, 2006a).

A different approach has been used with citicoline, a drug that modulates phospholipid metabolism and neurotransmitter levels and appears to improve cognition in some central nervous system disorders. In a 12-week, randomised, placebo-controlled, parallel-group, add-on trial in 44 bipolar outpatients with a history of cocaine dependence, citicoline was associated with improvement in some aspects of declarative memory and in cocaine use (6.41 times lower odds of testing positive for cocaine at exit), without worsening of affective symptoms (Brown *et al.*, 2007).

Finally, given the scarcity of methodologically sound data, it seems wise to use drugs that have proved to be efficacious in substance abuse populations and, at least, do not worsen the outcome of BD. Topiramate or acamprosate for alcohol dependence are two examples.

Psychotherapy

Dual patients require treatment of both BD and SUD. Group psychoeducation has proved to be effective in relapse prevention of BD, and treatment of SUD has traditionally used psychotherapy – especially group therapy. On this ground, Roger Weiss and his team in Harvard have specifically designed an *integrated group therapy*. A model that consists of 20 weekly hour-long meetings has been tested. It employs a cognitive behavioural relapse prevention model that integrates treatment by focusing on similarities between recovery and relapse processes in BD and SUD. A foundation of integrated group therapy is that the same types of thoughts and behaviours that facilitate recovery from one disorder will enhance the likelihood of recovery from the other disorder. Conversely, similar types of thoughts and behaviours can impede recovery from both disorders. Most sessions deal with topics relevant to both disorders. The interaction of the two disorders is emphasised, for example, by focusing on the adverse impact of substance abuse on the course of BD. This therapy has been compared in a randomised controlled design with an active treatment, such as group drug counselling, which is one of the most common treatments these patients can receive in a substance-abuse treatment programme and is primarily focused on substance use. Sixty-two patients were included: 80.6% diagnosed with BD type I, 16.1% with BD type II and 3.2% with BD not otherwise specified. Overall, substance use decreased during treatment. Integrated group therapy patients had fewer days of substance use, and reported approximately half as many days of substance use during both treatment and follow-up. During follow-up, substance use remained lower in the integrated group therapy group. When drug and alcohol use were separated, decreased alcohol use, not drug use, accounted for most of the differences in substance use between the groups. This result may have reflected a greater severity of alcohol problems and thus more room for improvement and perhaps greater motivation among patients to address their drinking (Weiss *et al.*, 2007).

Global considerations: towards an integrated treatment

Unfortunately, the dual bipolar patient suffers the consequences of the division between mental health and substance-use treatment services in most countries. Treating both conditions by one global team will undoubtedly ensure the most efficient and convenient treatment for these difficult-to-treat patients. In the future, specialised units involving experts on both conditions providing integrated care for the most-difficult-to-treat patients should coexist with community settings in which the less severe patients and those achieving remission could be followed up. With the progress of pharmacotherapy and the development of specific psychosocial interventions, with strong educational and preventative orientation towards both mood episode recurrences and addictions, we can expect better outcomes for patients with co-morbid BD and substance abuse and perhaps a reduction in the incidence of these two highly prevalent conditions.

References

APA (2000). *The Diagnostic and Statistical Manual of Mental Disorders*, 4th edition. American Psychiatric Association, Arlington, VA, USA. www.psychiatryonline.com/resourceTOC.aspx?resourceID=1 [accessed 23 December 2009].

Brady, K., Casto, S., Lydiard, R.B., Malcolm, R., Arana, G. (1991). Substance abuse in an inpatient psychiatric sample. *American Journal of Drug and Alcohol Abuse*, **17**, 389–97.

Brady, K.T. & Sonne, S.C. (1995). The relationship between substance abuse and bipolar disorder. *Journal of Clinical Psychiatry*, **56**, 19–24.

Brown, E.S., Beard, L., Dobbs, L., Rush, A.J. (2006a). Naltrexone in patients with bipolar disorder and alcohol dependence. *Depression and Anxiety*, **23**, 492–5.

Brown, E.S., Garza, M., Carmody, T.J. (2008). A randomized, double-blind, placebo-controlled add-on trial of quetiapine in outpatients with bipolar disorder and alcohol use disorders. *Journal of Clinical Psychiatry*, **69**, 701–5.

Brown, E.S., Gorman, A.R., Hynan, L.S. (2007). A randomized, placebo-controlled trial of citicoline add-on therapy in outpatients with bipolar disorder and cocaine dependence. *Journal of Clinical Psychopharmacology*, **27**, 498–502.

Brown, E.S., Jeffress, J., Liggin, J.D., Garza, M., Beard, L. (2005). Switching outpatients with bipolar or schizoaffective disorders and substance abuse from their current antipsychotic to aripiprazole. *Journal of Clinical Psychiatry*, **66**, 756–60.

Brown, E.S., Nejtek, V.A., Perantie, D.C., Bobadilla L. (2002). Quetiapine in bipolar disorder and cocaine dependence. *Bipolar Disorders*, **4**, 406–11.

Brown, E.S., Nejtek, V.A., Perantie, D.C., Orsulak, P.J., Bobadilla, L. (2003). Lamotrigine in patients with bipolar disorder and cocaine dependence. *Journal of Clinical Psychiatry*, **64**, 197–201.

Brown, E.S., Perantie, D.C., Dhanani, N., Beard, L., Orsulak, P., Rush, A.J. (2006b). Lamotrigine for bipolar disorder and comorbid cocaine dependence: a replication and extension study. *Journal of Affective Disorders*, **93**, 219–22.

Cassidy, F., Ahearn, E.P., Carroll, B.J. (2001). Substance abuse in bipolar disorder. *Bipolar Disorders*, **3**, 181–8.

Clark, R.E., Xie, H., Brunette, M.F. (2004). Benzodiazepine prescription practices and substance abuse in persons with severe mental illness. *Journal of Clinical Psychiatry*, **65**, 151–5.

Colom, F., Vieta, E., Martínez-Arán, A., Reinares, M., Benabarre, A., Gastó, C. (2000). Clinical factors associated with treatment noncompliance in euthymic bipolar patients. *Journal of Clinical Psychiatry*, **61**, 549–55.

Feinman, J.A. & Dunner, D.L. (1996). The effect of alcohol and substance abuse on the course of bipolar affective disorder. *Journal of Affective Disorders*, **37**, 43–9.

Frye, M.A., Altshuler, L.L., McElroy, S.L., *et al.* (2003). Gender differences in prevalence, risk, and clinical correlates of alcoholism comorbidity in bipolar disorder. *American Journal of Psychiatry*, **160**, 883–9.

Gawin, F.H. & Kleber, H.D. (1984). Cocaine abuse treatment: open pilot trial with desipramine and lithium carbonate. *Archives of General Psychiatry*, **41**, 903–9.

Gawin, F.H. & Kleber, H.D. (1986). Abstinence symptomatology and psychiatric diagnosis in cocaine abusers. Clinical observations. *Archives of General Psychiatry*, **43**, 107–13.

Geller, B., Cooper, T.B., Sun, K., *et al.* (1998). Double-blind and placebo-controlled study of lithium for adolescent bipolar disorders with secondary substance dependency. *Journal of the American Academy of Child and Adolescent Psychiatry*, **37**, 171–8.

Goldberg, J.F., Garno, J.L., Leon, A.C., Kocsis, J.H., Portera, L. (1999a). A history of substance abuse complicates remission from acute mania in bipolar disorder. *Journal of Clinical Psychiatry*, **60**, 733–40.

Goldberg, J.F., Garno, J.L., Portera, L., Leon, A.C., Kocsis, J.H., Whiteside, J.E. (1999b). Correlates of suicidal ideation in dysphoric mania. *Journal of Affective Disorders*, **56**, 75–81.

Goldberg, J.F. & Whiteside, J.E. (2002). The association between substance abuse and antidepressant-induced mania in bipolar disorder: a preliminary study. *Journal of Clinical Psychiatry*, **63**, 791–5.

González-Pinto, A., Mosquera, F., Alonso, M., *et al.* (2006). Suicidal risk in bipolar I disorder patients and adherence to long-term lithium treatment. *Bipolar Disorders*, **8**, 618–24.

Grant, B.F., Stinson, F.S., Hasin, D.S., *et al.* (2005). Prevalence, correlates, and comorbidity of bipolar I disorder and axis I and II disorders: results from the National Epidemiologic Survey on Alcohol and Related Conditions. *Journal of Clinical Psychiatry*, **66**, 1205–15.

Helzer, J.E. & Pryzbeck, T.R. (1998). The co-occurrence of alcoholism with other psychiatric disorders in the general population and its impact on treatment. *Journal of Studies on Alcohol*, **49**, 219–24.

Hesselbrock, M.N., Meyer, R.E., Keener, J.J. (1985). Psychopathology in hospitalized alcoholics. *Archives of General Psychiatry*, **42**, 1050–5.

Himmelhoch, J.M., Mulla, D., Neil, J.F., Detre, T.P., Kupfer, D.J. (1976). Incidence and significance of mixed affective states in a bipolar population. *Archives of General Psychiatry*, **33**, 1062–6.

Keller, M.B., Lavori, P.W., Coryell, W., *et al.* (1986). Differential outcome of pure manic, mixed/cycling, and pure depressive episodes in patients with bipolar illness. *JAMA*, **255**, 3138–42.

Kessler, R.C., Crum, R.M., Warner, L.A., Nelson, C.B., Schulenberg, J., Anthony, J.C. (1997). Lifetime co-occurrence of DSM-III-R alcohol abuse and dependence with other psychiatric disorders in the National Comorbidity Survey. *Archives of General Psychiatry*, **54**, 313–21.

Levander, E., Frye, M.A., McElroy, S.L., *et al.* (2007). Alcoholism and anxiety in bipolar illness: differential lifetime anxiety comorbidity in bipolar I women with and without alcoholism. *Journal of Affective Disorders*, **101**, 211–17.

Mantere, O., Melartin, T.K., Suominen, K., Rytsälä, H.J., Valtonen, H.M., Arvilommi, P. (2006). Differences in Axis I and Axis II comorbidity between bipolar I and II disorders and major depressive disorder. *Journal of Clinical Psychiatry*, **67**, 584–93.

McElroy, S.L., Altshuler, L.L., Suppes, T., *et al.* (2001). Axis I psychiatric comorbidity and its relationship to historical variables in 288 patients with bipolar disorder. *American Journal of Psychiatry*, **158**, 420–6.

Merikangas, K.R., Akiskal, H.S., Angst, J., *et al.* (2007). Lifetime and 12-month prevalence of bipolar spectrum disorder in the National Comorbidity Survey replication. *Archives of General Psychiatry*, **64**, 543–52.

Merikangas, K.R., Herrell, R., Swendsen, J. , Rösler, W., Ajdacic-Gross, V., Angst, J. (2008). Specificity of bipolar spectrum conditions in the comorbidity of mood and substance use disorders: results from the Zurich cohort study. *Archives of General Psychiatry*, **65**, 47–52.

Mirin, S.M., Weiss, R.D., Griffin, M.L., Michael, J.L. (1991). Psychopathology in drug abusers and their families. *Comprehensive Psychiatry*, **32**, 36–51.

Mitchell, J.D., Brown, E.S., Rush, A.J. (2007). Comorbid disorders in patients with bipolar disorder and concomitant substance dependence. *Journal of Affective Disorders*, **102**, 281–7.

Nolen, W.A., Luckenbaugh, D.A., Altshuler, L.L., *et al.* (2004). Correlates of 1-year prospective outcome in bipolar disorder: results from the Stanley Foundation Bipolar Network. *American Journal of Psychiatry*, **161**, 1447–54.

Nunes, E.V., McGrath, P.J., Wagner, S., Quitkin, F.M. (1990). Lithium treatment for cocaine abusers with bipolar spectrum disorders. *TheAmerican Journal of Psychiatry*, **147**, 655–7.

Nunes, E.V., Quitkin, F.M., Klein, D.F. (1989). Psychiatric diagnosis in cocaine abuse. Psychiatry Research, **28**, 105–14.

Ostacher, M.J., Nierenberg, A.A., Perlis, R.H., *et al.* (2006). The relationship between smoking and suicidal behavior, comorbidity, and course of illness in bipolar disorder. *Journal of Clinical Psychiatry*, **67**, 1907–11.

Pacchiarotti, I., Marzo, S.D., Colom, F., Sánchez-Moreno, J., Vieta, E. (2009). Bipolar disorder preceded by substance abuse: a different phenotype with not so poor outcome? *World Journal of Biological Psychiatry*, **10**, 209–16.

Post, R.M. (2007). Kindling and sensitization as models for affective episode recurrence, cyclicity, and tolerance phenomena. *Neuroscience and Biobehavioral Reviews*, **31**, 858–73.

Quanbeck, C.D., Stone, D.C., Scott, C.L., McDermott, B.E., Altshuler, L.L., Frye, M.A. (2004). Clinical and legal correlates of inmates with bipolar disorder at time of criminal arrest. *Journal of Clinical Psychiatry*, **65**, 198–203.

Regier, D.A., Farmer, M.E., Rae, D.S., *et al.* (1990). Comorbidity of mental disorders with alcohol and other drug abuse. Results from the Epidemiologic Catchment Area (ECA) Study. *JAMA*, **264**, 2511–18.

Ross, H.E., Glaser, F.B., Germanson, T. (1988). The prevalence of psychiatric disorders in patients with alcohol and other drug problems. *Archives of General Psychiatry*, **45**, 1023–31.

Rousanville, B.J., Antón, S.F., Carroll, K., Budde, D., Prusoff, B.A., Gawin, F. (1991). Psychiatric diagnoses of treatment-seeking cocaine abusers. *Archives of General Psychiatry*, **48**, 43–51.

Salloum, I.M., Cornelius, J.R., Daley, D.C., Kirisci, L. (2005a). Efficacy of valproate maintenance in patients with bipolar disorder and alcoholism: a double-blind placebo-controlled study. *Archives of General Psychiatry*, **62**, 37–45.

Salloum, I.M., Cornelius, J.R., Douaihyu, A., Kirisci, L., Daley, D.C., Kelly, T.M. (2005b). Patient characteristics and treatment implications of marijuana abuse among bipolar alcoholics: results from a double blind, placebo-controlled study. *Addictive Behaviors*, **30**, 1702–8.

Salloum, I.M., Cornelius, J.R., Mezzich, J.E., Kirisci, L. (2002). Impact of concurrent alcohol misuse on symptom presentation of acute mania at initial evaluation. *Bipolar Disorders*, **4**, 418–21.

Simon, N.M., Otto, M.W., Weiss, R.D., *et al.* (2004). Pharmacotherapy for bipolar disorder and comorbid conditions: baseline data from STEP-BD. *Journal of Clinical Psychopharmacology*, **24**, 512–20.

Sonne, S.C. & Brady, K.T. (1999). Substance abuse and bipolar comorbidity. *Psychiatric Clinics of North America*, **22**, 609–27.

Strakowski, S.M., DelBello, M.P., Fleck, D.E., Arndt, S. (2000). The impact of substance abuse on the course of bipolar disorder. *Biological Psychiatry*, **48**, 477–85.

Strakowski, S.M., DelBello, M.P., Fleck, D.E., *et al.* (2005). Effects of co-occurring alcohol abuse on the course of bipolar disorder following a first hospitalization for mania. *Archives of General Psychiatry*, **62**, 851–8.

Strakowski, S.M., DelBello, M.P., Fleck, D.E., *et al.* (2007). Effects of co-occurring cannabis use disorders on the course of bipolar disorder after a first hospitalization for mania. *Archives of General Psychiatry*, **64**, 57–64.

Strakowski, S.M., Sax, K.W., McElroy, S.L., Kec Jr, P.E., Hawkins, J.M., West, S.A. (1998). Course of psychiatric and substance abuse syndromes co-occurring with bipolar disorder after a first psychiatric hospitalisation. *Journal of Clinical Psychiatry*, **59**, 465–71.

Tohen, M., Greenfield, S.F., Weiss, R.D., Zarate, C.A., Vagge, L.M. (1998). The effect of comorbid substance use disorders on the course of bipolar disorder: a review. *Harvard Review of Psychiatry*, **6**, 133–41.

Tondo, L., Baldessarini, R.J., Hennen, J., *et al.* (1999). Suicide attempts in major affective disorder patients with comorbid substance use disorders. *Journal of Clinical Psychiatry*, **60**, 77–84.

Vieta, E., Colom, F., Corbella, B., *et al.* (2001). Clinical correlates of psychiatric comorbidity in bipolar I patients. *Bipolar Disorders*, **3**, 253–8.

Vieta, E., Colom, F., Martínez-Arán, A., Benabarre, A., Reinares, M., Gastó, C. (2000). Bipolar II disorder and comorbidity. *Comprehensive Psychiatry*, **41**, 339–43.

Weiss, R.D., Griffin, M.L., Kolodziej, M.E., *et al.* (2007). A randomized trial of integrated group therapy versus group drug counselling for patients with bipolar disorder and substance dependence. *American Journal of Psychiatry*, **164**, 100–7.

Weiss, R.D., Mirin, S.M., Michael, J.L., Sollogub, A.C. (1986). Psychopathology in chronic cocaine abusers. *American Journal of Drug and Alcohol Abuse*, **12**, 17–29.

Winokur, G., Coryell, W., Akiskal, H.S., *et al.* (1995). Alcoholism in manic-depressive (bipolar) illness: Familial illness, course of illness, and the primary-secondary distinction. *American Journal of Psychiatry*, **152**, 365–72.

Chapter 14

Practical management of cyclothymia

Giulio Perugi and Dina Popovic

The relationship between severe melancholic and manic states and attenuated or subclinical forms of mood disorders has been recognised since antiquity. The use of the term 'cyclothymia' referred to a mood disorder is due to Ewald Hecker, 1877 (Koukopoulos, 2003), a pupil of Ludwig Kahlbaum. His accurate clinical descriptions and his profound knowledge of psychopathology were the forerunners of modern descriptions of cyclothymia and bipolar disorder (BD) type II.

Cyclothymia officially became a part of contemporary nosography with Kraepelin's definition of manic-depressive illness (what we call bipolar disorder nowadays). In addition to traditional forms of mania and melancholy, Kraepelin included within manic-depressive illness some attenuated depressive conditions alternating with episodes of manic excitement of lower intensity (hypomania). He also described long-lasting depressive, manic (hyperthymic), cyclothymic and irritable temperamental traits (which he referred to as 'basic states').

Thanks to the contribution of Hagop S. Akiskal, the diagnosis of 'cyclothymic disorder' was included in DSM-III-TR (APA, 1980) in the chapter 'Mood disorders', and subsequently ICD-10 (WHO, 2007) has followed this trend. Cyclothymia received a large empirical validation as a bipolar spectrum disorder, and for this reason it remained classified in DSM-IV-TR (APA, 2000) as an Axis I mood disorder, alongside bipolar I disorder, bipolar II disorder and 'not otherwise specified'. Specific symptoms were not provided, except for reduced intensity of mood swings and protracted duration (more than 2 years).

A source of confusion comes from the fact that some of the core characteristics of cyclothymia, such as affective instability, mood reactivity and extreme emotionality, are reported in DSM-IV-TR among the criteria of dramatic cluster of personality disorders. Many of the criteria proposed by the manual for hystrionic and borderline personality disorders seem to describe some aspects of cyclothymia from a different perspective. The tendency to include many of the characteristics of cyclothymia in personality disorders limits the probability of understanding the existing relationship between stable temperamental dysregulation and major mood episodes in bipolar patients.

Akiskal (Akiskal, 1981) developed the temperamental perspective of cyclothymia, incorporating the Kraepelinian concept of 'basic states' as constitutional expression of manic-depressive illness. The criteria for cyclothymic temperament proposed by Akiskal *et al.* (1998) reflect the classic descriptions and require the presence, throughout much of the patient's life and starting from childhood/adolescence, of sudden mood swings and three out of five opposed conditions of each of the following two sets. The first set includes: (1) hypersomnia vs decreased need for sleep; (2) introverted self-absorption vs uninhibited people-seeking; (3) taciturnity vs talkativeness; (4) unexplained tearfulness vs buoyant jocularity; (5) psychomotor inertia vs restless pursuit of activities. The second set includes: (1) lethargy and somatic discomfort vs eutonia; (2) dulling of senses vs keen perceptions; (3) slow-witted vs sharpened thinking; (4) shaky self-esteem alternating between low self-confidence and overconfidence; (5) pessimistic brooding vs optimism and carefree attitudes. The bipolar nature of cyclothymic temperament is supported by a series of studies that have highlighted the strong propensity of these subjects to switch towards hypomania and/or mania when treated with antidepressants and to have a positive family history of BD (Angst & Marneros 2001; Koukopoulos *et al.*, 2003).

In both DSM-IV-TR and ICD-10, most of these 'attenuated' forms have not been recognised, and BD is identified, restrictively, with descriptions of classic manic-depressive illness, or with forms in which the depression is associated with 'hypomanic' episodes. Hypomania is described as a period of mood elevation of lower intensity when compared with mania,

Practical Management of Bipolar Disorder, eds. Allan H. Young, I. Nicol Ferrier and Erin E. Michalak. Published by Cambridge University Press.

lasting for at least 4 days. This classification is the result of Leonhard, Angst and Winokur's unipolar/bipolar distinction. Introduced in the mid twentieth century, it has favoured a clear separation between depressive and bipolar disorders. In recent decades, this distinction has shown important limitations. In particular, Akiskal helped to spread the idea of the existence of a continuum between bipolar and unipolar forms (Akiskal & Pinto, 1999).

Recently, the concept of 'soft' bipolar spectrum has been further enriched by epidemiological data reported by Angst & Marneros (2001), which showed a high prevalence (around 5% of the general population) of attenuated and short-lasting hypomanic episodes, almost always associated with depressive fluctuations. Based on these observations, the 4-day threshold proposed by DSM-IV-TR for the definition of hypomanic episode has been criticised. Independently from the definition, attenuated bipolar forms in general, and cyclothymia in particular, seem to represent the most common phenotype of mood disorders. Despite its epidemiological relevance, cyclothymia remains understudied from clinical and therapeutic points of view, and most of the research on BD is focused on BD type I.

Epidemiological and clinical aspects

Among mood disorders, cyclothymia is the one that has received less attention in epidemiological studies, and only recently prevalence rates in the general population have become available. This is surprising, given the frequency with which the disorder is found in clinical practice. Recent studies conducted in Switzerland by Angst reported lifetime prevalence rates ranging between 5 and 8% for brief episodes of hypomania associated with short-lasting depression, with a preponderance of about 2:1 among women (Angst & Marneros 2001).

Studies of clinical populations have shown how depression that occurs on a cyclothymic background is the most common manifestation of BD, found in around 50% of depressed patients seen in psychiatric outpatient settings (Hantouche *et al.*, 1998). This figure was also confirmed in general practice (Manning *et al.*, 1997), where it is assumed that observed cases are less severe.

The clinical presentation of cyclothymia is particularly rich in psychopathological manifestations, which are mainly expressed as behavioural disorders and impaired interpersonal relationships. In this sense, the DSM-IV-TR description, essentially based on the presence of mood symptoms, can be misleading.

In cyclothymia, the intensity of mood swings is generally limited, although in some cases major affective episodes of both polarities may appear. Cyclothymic individuals have continuous and irregular 'highs' and 'lows' of mood for extended periods of time; mood switches are often abrupt, while interposed periods of relative mood stability are infrequent. The unpredictability of mood swings is a major cause of distress, as it weakens self-esteem and produces a considerable instability in terms of vocation, behaviour and relations.

The hypomanic episodes are not easily identifiable, in particular if considering prolonged duration, as stipulated by DSM-IV-TR diagnostic criteria. In many cases, elated phases last for hours and only less frequently for more than 1 or 2 days The mood swings have a circadian component with biphasic characteristics, such as lethargy alternated with euphoria, reduced verbal productivity alternated with excessive loquacity, low self-esteem alternated with excessive trust in one's abilities. Major difficulties involve interpersonal relationships; relatives and friends point out how the individual often appears hostile towards people around them. In many cases they have explosions of rage following minor disputes, which have the effect of triggering 'avalanche' reactions with destructive consequences on their interpersonal life.

Mood reactivity is a specific characteristic of cyclothymia, and it influences the overall clinical presentation. Patients often present, as a stable trait since adolescence, a particular kind of increased sensitivity to environmental stimuli. They react to positive events by becoming quickly joyful, enthusiastic and dynamic and taking initiative (sometimes with excessive euphoria and impulsiveness); then in the case of 'negative' events (real or experienced as such) they become distressed, while experiencing feelings of deep prostration, sadness and extreme fatigue. Even minor disappointments can precipitate distress, at times complicated by crushing reactions, with the implementation of self-harming gestures.

Exaggerated positive and negative emotional reactions can be triggered by any sort of external stimuli, either psychological (for example falling in love vs sentimental disappointments), environmental (for example meteorological changes or changes of time zone), physical (for example immobility vs hyperactivity) or chemical (for example medications and alcohol or drugs).

Table 14.1 Psychological aspects of cyclothymia

Interpersonal sensitivity (judgement, criticism, rejection)
Separation anxiety
Affective dependency
Pathological jealousy
Obsessive need to please
Compulsive need for compliments and emotional rewards
Novelty-seeking mixed with harm avoidance
Hyper-control
Compulsive behaviours
Shaky self-esteem, low self-confidence to overconfidence
Lack of future projection

Psychological dysfunctions related to cyclothymia are summarised in Table 14.1. Interpersonal sensitivity is strongly related to mood reactivity (Perugi *et al.*, 2003). This feature, partially related to weak self-esteem, determines high sensitivity to judgement, criticism and rejection by others. In cyclothymic subjects, susceptibility to rejection and disapproval by others prevail. They are promptly offended and sensitive to the possibility of being wounded, with feelings of hostility and anger towards those who evoke these reactions and who are considered responsible for their suffering. When emotional reactions are very intense, sensitivity may favour the onset of a more-or-less transient tendency to interpret and overvalue ideas.

Interpersonal sensitivity and mood reactivity seem to be two strictly related characteristics, representing two different aspects, cognitive and affective, of the same psycho(patho)logic dimension (Perugi *et al.*, 2003). It is difficult to determine which of these two elements is primarily altered and in which way they affect each other. However, mood instability of the cyclothymic type also seems strongly related to interpersonal sensitivity and mood reactivity, suggesting the existence of a common background.

The fear of being rejected, turned away or disapproved of sometimes determines the tendency to please others with submissive behaviour and excessive dedication, which may result in conducts of 'pathological altruism'. The oscillation between complacency and excessive feelings of anger/hostility has a negative impact on sentimental relationships, family or employment, which become difficult and unstable.

When individuals with such mood reactivity switch towards exhilaration, they often seek sentimental and interpersonal relationships; by contrast, when dysphoric they tend to isolate themselves from others. Indeed, the youth of many of these patients can be a continuous succession of tempestuous short and intense sentimental relationships with partners who are often unsuitable or unlikely. What appears to afflict these patients most is their periodic swinging between behavioural inhibition and activation, which prompts them towards interpersonal relations, a situation to which they respond afterwards with an avalanche of hardly manageable emotional reactions – creating a path full of existential dramas and tragedies.

The great reactivity of cyclothymics favours impulsive, sensation-seeking and self-stimulating behaviour, which amplifies during elated phases (Perugi *et al.*, 2002). In many cases we assist the emergence of true impulse-control disorders, such as pathological gambling and compulsive sexuality in men, and compulsive buying and binge-eating in women.

Cyclothymia also represents a fertile ground for drug abuse (Maremmani *et al.*, 2006). On one hand sensation-seeking behaviour; on the other a high sensibility to substances encourages the use of alcohol, stimulants and cocaine, but also hypnotics and sedatives. In some subjects, for environmental reasons, mood instability and impulsivity combined with substance abuse can favour the emergence of antisocial conduct, with accompanying legal problems.

Owing to this great variety of pathological behaviours and to the coexistence of contradictory psychopathologic elements such as anxiety and impulsiveness, cyclothymic subjects can also meet the DSM-IV-TR criteria for cluster B 'dramatic' or 'emotional' personality disorders. Histrionic and borderline disorders have broad symptomatic areas in common with cyclothymia, such as 'emotional instability' and excessive 'mood reactivity'. In many cases, the distinction between cyclothymia and histrionic or borderline personality disorders mainly depends on the perspective and moment of observation rather than on true clinical differences (Levitt *et al.*, 1990).

Psychiatric co-morbidity

A growing number of observations show that mood instability, typical of cyclothymia, is the common factor underlying a wide range of co-morbidity and complications associated with bipolarity, including anxiety

(Perugi *et al.*, 1999), impulsivity (McElroy *et al.*, 1996), suicidality (Rihmer and Pestality, 1999) and drug abuse (Maremmani *et al.*, 2006). The co-morbid disorders often represent the reason for which these subjects require psychiatric intervention. They are often referred to psychiatrists for anxiety, binge-eating, substance abuse or other behavioural problems rather than for the mood instability, which is frequently egosyntonic and considered part of the 'normal' character of the individual (Perugi *et al.*, 1999, 2002).

Patients with cyclothymia frequently report panic attacks, anxiety and varying degrees of phobic avoidance, up to extended agoraphobia. Panic attacks can be triggered by separation events or the influence of certain substances such as stimulants or cannabis. In some cases panic attacks start on the acme of periods of hyperactivity or excitement and sometimes mark the switch from an elated to a depressive phase (Perugi *et al.*, 2001).

Recently MacKinnon *et al.* (MacKinnon *et al.*, 2003; MacKinnon & Pies, 2006) carried out a series of clinical and family studies on bipolar patients with rapid mood switches, which are similar in many ways to those affected by cyclothymia. They found that the presence of rapid mood fluctuations is associated with a high familial load for mood and anxiety disorders, early onset, marked suicidal risk and co-morbidity with panic disorder.

These findings are consistent with the results of the studies on the characteristics of BD in children and adolescents, in whom high familial loading, co-morbidity with multiple anxiety disorders and rapid circadian switches have been reported (Masi *et al.*, 2007). Co-morbidity with panic disorder and rapid switches seems to define a particular familial subtype of BD, characterised by early onset and cyclothymic instability (MacKinnon *et al.*, 2002; Masi *et al.*, 2007).

Some recent observations indicate how separation anxiety can often be associated with cyclothymic type mood instability (Pini *et al.*, 2005). Intense depressive reactions to real or imagined loss or separation seem to be present since childhood and, in many cases, remain in adulthood.

Some cyclothymic individuals present social anxiety. When social anxiety is associated with cyclothymia it confers particularly favourable grounds for alcohol or substance misuse. Hypomanic switches triggered by treatment with antidepressants are extremely frequent (Himmelhoch *et al.*, 1998).

Obsessive–compulsive symptoms can be associated with cyclothymia, as well as impulse-control disorders of various kinds. McElroy *et al.* (1996) highlighted the correlation between bipolar spectrum disorders and some impulsive behaviours such as those related to control of aggression and sexual instincts, paraphilias and pathological gambling. Impulse-control disorders have many affinities with BD, in terms of symptoms, co-morbidity and response to mood stabilisers. Both are characterised by egosyntonic, harmful or dangerous but rewarding, behaviour, impulsiveness, poor insight and emotional instability. High rates of BD were found in patients with impulse-control disorders, and vice versa.

Impulse-control disorders and cyclothymia present a large overlap on the co-morbidity spectrum with other mental disorders, including anxiety disorders, alcohol and substances abuse and eating disorders. In cyclothymic individuals, mood instability and impulsiveness are interrelated and characterise particular periods of behavioural disinhibition, poor insight and marked instability between tension, dysphoria and satisfaction.

Eating disorders, especially those that include impulsive conducts towards food, such as bulimia, bingeing-purging anorexia and binge-eating disorder, can be considered a particular subtype of impulse-control disorder. The frequent association between eating and mood disorders is well documented, especially with unipolar depression, whereas the literature on co-morbidity with BD is less extensive (Godart *et al.*, 2005; Blinder *et al*, 2006; Perugi *et al.*, 2006). However, several familial studies have suggested a correlation with BD type II and cyclothymic forms (McElroy *et al.*, 2005). The association seems to be more common in bulimic patients who present serious and chronic forms.

In a more speculative way, it is possible to hypothesise that, in cyclothymia, sensation-seeking and self-stimulating behaviour might involve, more or less consciously, any type of potentially addictive substance or activity, such as food, alcohol, drugs, physical exercise, work, travelling, the Internet, sex. In some cases persistent addictive behaviour can be the major source of distress, overshadowing the underlying mood instability.

To date, the relationship between borderline personality disorder (BPD) and bipolar spectrum disorders is questioned. Some of the diagnostic criteria for BPDs have, in fact, a strong emotional connotation: unjustified rage, emotional instability, suicidal tendencies, unstable relationships. In various surveys on borderline patients, a high prevalence of cyclothymia and/

or attenuated bipolar spectrum disorders has been reported and vice versa, in cyclothymic patients the prevalence of borderline personality traits is very high (Levitt *et al.*, 1990; Stone, 1990).

In a German study in which 'sub-affective personality disorders' were rigorously assessed, patients with BPD and those with cyclothymic-irritable temperament presented considerable overlap of clinical presentations (Sab *et al.*, 1993; Stone, 2006). In addition, borderline personality was a predictive factor for antidepressant-induced hypomania (Levy *et al.*, 1998; Stone, 2006). Also follow-up studies have suggested a close correlation of BPD with mood disorders, considering the number of young borderline patients that developed BD type I, or, more frequently, BD type II, over the years (Gunderson *et al.*, 2006).

Essential elements of affective deregulation of these patients are extreme reactivity and mood lability, which together with interpersonal sensitivity may be the substrate for some of the common elements between cyclothymia, atypical depression and BPD. Unstable work adjustment, sexual promiscuity, substance abuse, impulsiveness and strong self-harming tendencies could be interpreted as the result of long-lasting affective instability and excessive mood reactivity. The suicide attempts and other self-harming acts might reflect desperation and feelings of hopelessness; unstable relations might result from low self-esteem and excessive impulsivity. However, many authors with experience in this research area minimise the mood component in BPD and prefer to consider these patients' extreme emotional and behavioural lack of control as a result of physical and psychological abuse (Gunderson *et al.*, 2006; Stone, 2006). Several hereditary, biological and environmental factors influence this process in various ways. On one hand some give priority to constitutional factors, and on the other those who attribute the characteriologic disorder to development-related events.

Cyclothymic patients are considered more syntonic and outgoing, able to pursue socially acceptable objectives. Clinical experience suggests, however, that these two poles, rather than appertaining to two distinct categories, depend on the severity of impulsivity and mood symptoms and the presence or absence of an adequate 'goodness of fit'. This concept refers to the possibility that some environmental factors, as well as the expectations and environmental requests, can form an individual's personality and temperamental characteristics as well as his/her lifestyle. In this perspective it may be easier to understand how cyclothymia and its variants represent, on one hand, the background for antisocial and psychopathic behaviour and, on the other hand, those extraordinary qualities such as creativity and aptitude for leadership, that characterise some cyclothymic individuals. The latter is determined by variables independent from mood disorder, such as skill, talent and intelligence, but also opportunity, luck and external influences.

Course

Regarding the course of BD, similarly to manic and depressive symptomatological manifestations, mood swings seem to be distributed according to a spectrum ranging from highly unstable forms, with ultra-rapid cyclycity and circadian instability, to forms with low frequency of episodes. In cyclothymic individuals, short-lasting hypomanic and depressive phases alternate from adolescence, in most cases in a highly unstable way.

The depressive phases dominate the clinical presentation, with the interposition of periods of relative stability, irritability or occasional hypomania, but most frequently cyclothymics seek the aid of physicians for depressive symptoms. Depression frequently shows atypical features such as mood reactivity, hypersomnia, hyperphagia and marked fatigue (leaden paralysis), responsible for significant functional impairment (Davidson *et al.*, 1982; Perugi *et al.*, 1998; Benazzi, 1999; Perugi *et al.*, 2003). Severe manifestations such as psychotic symptoms and serious psychomotor disorders are generally rare.

Although cyclothymia can be observed in some patients with full-blown manic-depressive disorder (BD type I), more commonly it is associated with the BD type II pattern. In a French study on major depression, 88% of subjects with cyclothymic characteristics belonged to the BD II subtype (Hantouche *et al.*, 1998). Akiskal and Pinto have defined major depression in a cyclothymic background as 'Bipolar II-1/2 disorder' (Akiskal & Pinto, 1999), in order to distinguish it from bipolar II disorder, characterised by major depressive episodes alternating with protracted hypomania and free intervals (DSM-IV-TR bipolar II disorder; APA, 2000).

The NIMH study (Akiskal *et al.*, 1995) on originally unipolar patients who switched to BD type II during a long-term prospective follow-up has provided some important information. The variables that characterised at entry the patients with subsequent hypomanic switches were early age at onset, recurrent depression, high rates of divorce or separation, high rates of scholastic and/or job maladjustment, isolated 'antisocial acts' and drug abuse. In addition, the index depressive episode was characterised by such features as phobic anxiety, interpersonal sensitivity, separation anxiety, obsessive–compulsive symptoms, somatisation (subpanic symptoms), worsening in the evening, self pity, exacting, subjective or overt anger, jealousy, suspiciousness and ideas of reference. This pattern points to a broad array of cyclothymic and 'atypical' depressive symptoms with co-morbid anxious and impulsive features. Finally, some temperamental attributes of 'mood lability', 'energy activity' and 'daydreaming', already described by Kretschmer for 'cycloid temperament', have been proven specific to identify unipolar depressives who switched to hypomania (Kretschmer, 1936).

Cyclothymia seems to be the basic state of many BD type II depressions, in which the mood instability is supported by an intrinsic temperamental deregulation. Unfortunately, the major diagnostic systems (ICD-10 (WHO, 2007) and DSM-IV-TR (APA, 2000)) are oriented primarily on symptomatology, not recognising the role of traits and temperamental dispositions.

Some information on the long-term course of cyclothymia can be derived from a series of follow-up studies on patients with BPD treated as inpatients, published at the end of the 1980s by Michael Stone (for a review see Stone, 1990). Many of these patients, most of whom were in their twenties at the time of first observation, showed clinical presentations compatible with the diagnosis of BD type II or cyclothymia, as admitted by Stone himself (Stone, 2006). Approximately two-thirds of these patients presented, after 10 or 25 years, a 'mild symptomatology with good overall performance, and few significant interpersonal relationships'. Considering the other end of the clinical spectrum, from 3 to 9% of the same cohort had committed suicide. The age at initial assessment was lower, and the risk of suicide was greater. Favourable prognostic factors were represented by high intelligence, artistic talent and, in the case of patients who had co-morbid alcohol abuse, ability to follow rehabilitative treatments. By contrast, negative prognostic factors, with high risk of suicide, were represented by physical or sexual abuse by family members, the combination of antisocial traits and marked impulsiveness.

In several studies, the suicidal rates of cyclothymic patients are analogous to those of patients with BD or schizophrenia (Rihmer *et al.*, 1999), which is indicative of the disorder severity. However, most cases presented a better long-term prognosis than that of major psychosis. In fact, many participants begin to improve past the threshold of 40 years. This observation is another element in common with data on BPD.

Treatment strategies and practical management

Cyclothymia continues to be underestimated and is frequently misdiagnosed and inappropriately treated. Reasons for difficulties in treating cyclothymia are summarised in Table 14.2. Lack of consensus on the definition of bipolar spectrum and difficulty in the diagnosis of hypomania should be considered the major obstacles to the correct diagnosis. A complex clinical picture, lack of clear-cut episodes, extremely rich co-morbidity, early onset and overlap with personality disorders can also be responsible for the diagnostic delay. Complicated patient–physician relationships and weak response to conventional approaches produce further difficulties in the recognition of cyclothymia. Even when properly identified, there is no consensus on the treatment.

Treatment of cyclothymic and BD type II depression is surprisingly understudied, especially in comparison with mania and unipolar depression. The lack of adequate research in this area is even more astonishing, considering that depressive episodes and symptoms prevail in most bipolar patients seen in clinical practice (Judd & Akiskal, 2003) and are associated with a significant risk of suicide and high mortality from different causes.

Agents such as lithium, valproate and carbamazepine have been studied much more thoroughly in BD type I. Similarly, only a few controlled studies have focused on the efficacy of antidepressants in monotherapy. The relative risk of (hypo)manic switches or of rapid cycle induction further complicates the treatment of cyclothimic depression.

The rates of (hypo)manic switches among bipolar depressives treated with antidepressants reported in the literature range from 10 to 70% (Möller & Grunze, 2000; Thase & Sachs, 2000). The extreme variability of these estimates is due to the differences in sampling procedures, treatment setting and symptomatologic evaluation. Some data derive from randomised controlled trials, others from naturalistic studies; in some research, patients are treated with antidepressants in monotherapy, in others, these drugs are combined with mood stabilisers.

Although it would seem reasonable to limit the assessment of the risk of antidepressant-induced

Table 14.2 Reasons for difficulties in treating cyclothymia

Long delay for correct diagnosis
Complex clinical picture
Lack of clear-cut episodes
Rich co-morbidity
Young age of onset
Overlap with personality disorders
Influence on development and building self-esteem
Multitude of psychological dysfunctions
Complicated patient–physician relationships
Frequent exposure to antidepressants

mania to randomised, placebo-controlled trials, one limitation of this approach is that such trials are not primarily designed to assess this issue (they are mainly designed to assess efficacy). Furthermore, side effects are frequently underestimated in clinical trials because of numerous exclusion criteria that screen out those with risk factors for serious side effects. In the case of antidepressant-induced mania, some potential risk factors, such as substance abuse (Goldberg & Whiteside, 2002), are common exclusion criteria, which could lead to a serious underestimation.

New-generation antidepressants such as selective serotonin reuptake inhibitors (SSRIs), serotonin-norepinephrine reuptake inhibitors (SNRIs) and bupropion may result in a lower number of (hypo)manic 'switches' than do tricyclic antidepressants (TCAs); however, this conclusion mostly derives from pooled analysis (Peet, 1994), wherein a rate of 4% with SSRIs, versus 12% with TCAs and 4% with placebo, is reported. The data included in the pooled analysis were based on selected randomised trials of unipolar patients and, therefore, are limited to the few BD type II patients included in these samples. Moreover, the information on (hypo) manic switches was not explored with specific instruments, but was culled from an evaluation of the adverse events. We know from experience with SSRI-related sexual dysfunction that this approach can seriously underestimate side effects because of measurement bias. It is likely that egosyntonic hypomania, very frequent with SSRIs and SNRIs in clinical practice, is recorded as full remission instead of adverse reaction.

In contrast to these findings, the contemporary literature is extremely rich in case reports and series reporting (hypo)manic switches with SSRIs, not only in patients with major depression, but also in patients treated for dysthymia, anxiety and eating and impulse-control disorders (for a review, see Perugi & Akiskal, 2002). Two well-conducted observational studies (Bottlender *et al.*, 1998; Henry *et al.*, 2001) suggest that antidepressant-induced (hypo)mania is only partially prevented by concomitant treatments with mood stabilisers, and furthermore that switch rates with SSRIs appear to be notably higher in real-world patients (about 20%) than in randomised studies.

Recently, Post *et al.* (2001, 2003) reported data on hypomanic switches in patients with bipolar depression included in a double-blind, randomised trial comparing adjunctive bupropion, sertraline and venlafaxine with mood stabilisers. Of the patients included in the acute phase of the treatment (10 weeks), 14% presented manic (6%) or hypomanic (8%) switches. Among the patients who entered in the long-term maintenance phase (1 year), the rate of switching increased to 33% (20% hypomanic, 13% manic). No statistical difference was reported among the three antidepressants utilised; however, because of the absence of placebo control, it was very difficult to reach a firm conclusion on the differences between the rate of spontaneous switches and those induced by antidepressant treatments.

Antidepressants can worsen bipolar illness in two ways. Acutely, they can cause hypomanic or manic switch, while in the long-term they can cause rapid cycling or cycle acceleration, the latter with a serious worsening of the course of the illness, often with longer and more frequent depressive episodes. Because of its long-term complex nature, it is difficult to assess definitively cycle acceleration in randomised settings. In fact, to our knowledge there is only one study that has attempted to do so (Wehr *et al.*, 1988), and it supports an association between antidepressant use and cycle acceleration. Observational evidence can be cited both for and against this association, as is the case with observational evidence for many topics.

With this background, some conclusions seem reasonable. Antidepressants may be acutely effective for the current major depressive episodes in cyclothymics, but long-term continuation is ineffective at best and harmful at worst. If one accepts these conclusions, then use of antidepressants would be short-term, with discontinuation after recovery from the acute major depressive episode (perhaps within 2–6 months). This indeed was the suggestion of the most recent revision of the American Psychiatric Association treatment guidelines for bipolar disorder (APA, 2002), severely criticised by some clinicians in Europe (Möller & Grunze, 2001).

However, it could potentially be possible that using antidepressants could lead to cycle acceleration, whereas stopping them could lead to depressive relapses (Altshuler *et al.*, 2001). The two possibilities are not contradictory. If depressive relapse *frequently* followed antidepressant discontinuation, however, clinicians would be in a major clinical dilemma: how to identify patients who, with long-term use of antidepressants, develop an increasing frequency of episodes and rapid cycling and those who, when antidepressant medications are discontinued, tend to produce depressive recurrences. Those patients with a history of past switching on antidepressants are likely to do it again, so it is a good idea to avoid antidepressants in such patients.

In practice it is generally best to utilise mood stabilisers as the preferred choice of treatment and reserve antidepressants for non-responders. In particular, lithium and lamotrigine appear to have antidepressant properties. Other mood stabilisers, such as valproate, may also be acutely effective in some bipolar patients. A possible option is represented by quetiapine, which has been reported to be effective in the treatment of acute BD types I and II depression in several randomised, controlled trials on large clinical populations (Hirschfeld *et al.*, 2006; Thase *et al.*, 2006). Long-term efficacy and side effects such as sedation and weight gain might limit its utility in the treatment of cyclothymia. Antidepressants should be added only in the most severe or resistant forms, always in combination with mood stabilisers; high dosages of antidepressants and their presumably synergistic combination should be avoided. Their long-term use, even with mood stabilisers, is justified in rare cases only. Even in those studies frequently cited to support long-term antidepressant use (Altshuler *et al.*, 2003), only about 15% of patients with bipolar depression appear to benefit from long-term antidepressant use. In many patients it is advisable to combine two mood stabilisers (lithium plus valproate, lithium plus olanzapine, lamotrigine plus quetiapine) before adding an antidepressant.

With this approach, the acute depressive symptoms of most patients can be treated and mood-stabilising combinations continued in the long term. When patients are not acutely suicidal, such approaches emphasising mood stabilisers can be taken. Use of antidepressants may be necessary in very severely suicidal patients, or hospitalised patients, in which more rapid recovery is desired; care should be taken to discontinue antidepressants in most patients after the resolution of acute major depressive episodes. Clinical experience suggests that some patients may need long-term antidepressant treatment, but that group is a minority (15–20% of the bipolar population) (Ghaemi & Goodwin, 2001), and even then, in our experience, many of those patients develop an oscillating course of mood swings that may be mild but is generally not consistently euthymic.

The treatment of cyclothymic patients is often complicated by the presence of co-morbidity. In many cases, the concomitant anxiety, impulse-control or eating disorders represent the major complaints and require specific treatment. Most of the controlled trials on BD exclude patients with co-morbid drug abuse, anxiety and impulse-control disorders, and vice versa; as a consequence, the empirical basis for treating patients with complex co-morbidity is almost exclusively derived from open clinical experience. This is a deplorable situation, because the most common patients treated in everyday clinical practice are cyclothymic–BD type II with complex co-morbidity. 'Pure' BD is an abstract concept that is never encountered in the real world.

Bipolar patients with high anxiety ratings are less likely to respond to lithium (Young *et al.*, 1993). Controlled data suggest valproate may be more effective than lithium in mania associated with depressive features, even when the depressive features are mild (Swann *et al.*, 1997). As anxiety symptoms are often seen in mixed states and may even be related to depression in mania, future investigations should evaluate anxiety features as possible predictors of response to valproate (and other antimanic agents) in mania. Actually, valproate has been used successfully in the treatment of panic disorder (Lum *et al.*, 1991). In addition, in an open-label study, Calabrese and Delucchi (1990) reported that rapid-cycling BD patients with co-morbid panic attacks experienced reduction in their panic symptoms with valproate treatment. Benzodiazepines – such as clonazepam – are relatively safe and well tolerated when used in combination with mood stabilisers. However, long-term benzodiazepine use may be problematic because of the development of tolerance, physical dependence and withdrawal phenomena. Gabapentin – which has been shown to be effective in panic disorder (Pande *et al.*, 1999) and social phobia – seems to be helpful when anxiety disorders or alcohol abuse are co-morbid (Perugi *et al.*, 2002).

Antidepressant-induced (hypo)manic symptoms have been reported to occur in the course of the treatment of virtually all anxiety disorders, including obsessive–compulsive, panic disorder/agoraphobia and social phobia (Sholomskas, 1990; Himmelhoch *et al.*, 1998; Steiner, 1991). When treating co-morbid bipolar and anxiety disorders, it is imperative to begin treatment with a mood stabiliser *first.*

In the treatment of co-morbid cyclothymia and panic disorder/agoraphobia, it appears reasonable to utilise as first choice mood stabilisers that have been shown to possess some antipanic efficacy, such as valproate. In patients with persistent and disabling anxiety, combination with small dosages of SSRI or TCA (e.g. paroxetine or trimipramine) can be considered. Paroxetine may be considered an optimal choice, because it is effective in all anxiety disorders as well as major depression, and in addition has proven efficacy in BD type II depression in combination with mood stabilisers (Young *et al.*, 2000).

Less information is available for co-morbid cyclothymia and social phobia. A small positive study (Pande *et al.*, 1999) on gabapentin in social phobia suggests the possible efficacy of this drug for both mood and anxiety disorders. The combination of mood stabilisers – such as lithium, valproate and carbamazepine with SSRIs, reversible inbibitors of monoamine oxidase (type A) (RIMAs) or monoamine oxidase inhibitors (MAOIs) – might reduce the number of switches in these patients. However, there is virtually no information on long-term outcome of patients treated with drug combinations for co-morbid BD and social anxiety.

Patients with cyclothymia and co-morbid obsessive–compulsive disorder are among the most difficult to treat. Although no mood stabiliser has been shown to exert any anti-obsessive–compulsive activity, highly effective anti-obsessive pharmacological treatments (e.g. high doses of clomipramine or SSRIs) are likely to trigger (hypo)manic switches and to increase mixed states in bipolar spectrum patients (Akiskal *et al.*, 2003; Perugi *et al.*, 2003). A combination of different mood stabilisers (e.g. lithium plus anti-epileptics) is often necessary. However, many of these patients present residual obsessive–compulsive symptomatology and very severe manic or mixed episodes (aggressive, hostile mood) that may require hospitalisation. In some cases, SSRI augmentation with low-dose atypical antipsychotics (e.g. risperidone, olanzapine, aripiprazole) can be considered on an empirical basis. Indeed, even if in some cases, atypical antipsychotics have been reported to exacerbate obsessions and compulsions (Baker *et al.*, 1992), in others, these drugs display a mood-stabilising, anti-aggressive activity, which permits an efficacious approach to obsessive–compulsive symptoms (McDougle *et al.*, 2000).

As for non-pharmachological intervention, cyclothymia should be considered a distinct form of bipolarity, which requires a specific approach. Most cyclothymic patients do not match the model of BD type I in psychoeducation groups. By contrast, they want to share their own experiences with other patients and need an appropriate 'format' of pychoeducation, based on an adapted 'model'. The classical description of BD characterised by manic and depressive episodes followed by periods of remission, with different algorithms for the treatment of different episodes, does not apply to cyclothymia, where depression and excitement are strongly related and inter-episodic mood instability is the rule. In these patients, free intervals and long-lasting remissions are very rare, and for this reason the psychoeducational intervention should start as soon as possible. It should be focused on the patient becoming an 'expert' on cyclothymia, getting familiar with its psychological dysfunction and behavioural problems and developing self-skills to reduce them and the impact of the illness. Currently the correct evaluation of the role and the effectiveness of this type of non-pharmachological intervention requires better-designed prospective observations.

General principles in the practical management of patients with cyclothymia are summarised in Table 14.3. In defining outcome measures, a 'primacy' of hypomania should be established, avoiding the frequent error of considering everything that is not typical (hypo)mania as depression. The therapeutic intervention should focus not only on major affective episodes but also on basic mood dysregulation, which underlies most psychological dysfunction and behavioural problems in these patients. Both the pharmacological and psychoeducational interventions should target specific goals such as excitement, depression, co-morbidity, impulsivity, hostility, mood-reactivity, interpersonal sensitivity and risk-taking behaviour. Moreover, in order to increase treatment adherence, psychoeducation should be started from the beginning. Finally, regarding the pharmacological approach, mood stabilisers or antimanic drugs should be used before antidepressants and, when the latter are utilised, particular attention should be devoted to the possibility of hypomanic switches or cycle acceleration.

Table 14.3 Practical management of cyclothymia

Refine diagnosis
Primacy of hypomania
Systematic error: 'everything that is not typical (hypo)mania is depression'
Excitement, irritability, inner agitation, and impulsivity are fundamental dimensions of soft bipolarity
Conception of the basic illness
Episodes are probably the 'wings' emerging from temperaments that could be the 'roots'
Mood episodes, psychological dysfunctions and behavioural problems result from a 'clash' between basic temperament and environment (importance of 'patient's life systems')
Focus treatment on specific clinical targets
Depression
Co-morbidity
Specific dimensiion: impulsivity, hostility, hyperreactivity, interpersonal sensitivity, risk-taking behaviour, excitement, inner tension
Start psychoeducation from the beginning
Use mood stabilisers or antimanic drugs before antidepressants
Be vigilant when using antidepressants
Hypomanic switches
Cycle acceleration
Prolonged excitement (protracted mixed states)

Conclusion

In clinical settings, 30 to 50% of all individuals who seek help for depression are deemed to be affected by BD type II after careful screening, and many of these are affected by cyclothymia. These data emerge both from academic centres specialising in mood disorders and from public and private outpatient facilities. The proportion of depressed patients who can be classified as cyclothymic grows significantly if the 4-day threshold for the hypomanic episode proposed by the DSM-IV-TR is reconsidered. The average length of a hypomanic episode in the general population seems to be 2 days; in many cyclothymics, elated episodes are shorter than 1 day and are often associated with environmental stimuli or substance misuse.

Many cyclothymic patients are diagnosed as affected by personality disorders, especially those with frequent relapses, severe impulsivity and extreme mood instability. The presence of 'borderline' traits in cyclothymic patients seems to derive from an important mood deregulation, where the interpersonal sensitivity, motivational and emotional instability have an important effect on patients' personal history since childhood.

Finally, cyclothymia as a clinical syndrome with early onset and protracted course represents the common denominator of a complex co-morbidity with anxiety and impulse-control disorders, which these patients manifest from the beginning of their adult life. Alcohol and substance misuse can be interpreted as related to self-stimulation and sensation-seeking, although it may further worsen impulsiveness and mood instability.

The treatment and clinical management of cyclothymia present the greatest challenge today. Antidepressant use may be problematic for a large number of patients suffering from cyclothymic depression. Mood stabilisers are the first choice, with the addition of antidepressants only in the most resistant cases. Atypical antipsychotics, in particular olanzapine or quetiapine, should be considered in non-responders or when psychotic or mixed features are present. Unfortunately, the use of classical mood stabilisers, such as lithium, carbamazepine and valproate, is sometimes limited by side effects. The new anticonvulsants with reduced side effects might represent promising therapeutic tools. In particular, lamotrigine seems to be efficacious in the prevention of rapid cycling and depressive recurrences.

The frequent coexistence of cyclothymia with anxiety, impulse-control, eating and substance-use disorders is a substantial therapeutic challenge pertaining to a large number of patients. The use of psychotropic combinations is rendered necessary because of the syndromic complexity and the contrasting effects of pharmacological treatments. The identification of differential patterns of co-morbidity may provide important information for distinguishing more homogeneous clinical subtypes of affective disorders from the therapeutic point of view. The pattern of complex relationships among these disorders requires better-designed prospective observations. This is also true for specific psychological approaches and psychoeducation focused not only on prevention of the mood episode but also on the complex co-morbidity and the basic temperamental dysregulation.

References

Akiskal, H.S. (1981). Subaffective disorders: dysthymic, cyclothymic and bipolar II disorders in the 'borderline' realm. *The Psychiatric Clinics of North America*, **4**, 25–46.

Akiskal, H.S. & Pinto, O. (1999). The evolving bipolar spectrum. Prototypes I, II, III, and IV. *Psychiatric Clinics of North America*, **22**, 517–34.

Akiskal, H.S., Hantouche, E.G., Allilaire, J.F., *et al.* (2003). Validating antidepressant-associated hypomania (bipolar III): a systematic comparison with spontaneous hypomania (bipolar II). *Journal of Affective Disorders*, **73**, 65–74.

Akiskal, H.S., Maser, J.D., Zeller, P.J., *et al.* (1995). Switching from 'unipolar' to bipolar II. An 11-year prospective study of clinical and temperamental predictors in 559 patients. *Archives of General Psychiatry*, **52**, 114–23.

Akiskal, H.S., Placidi, G.F., Maremmani, I., *et al.* (1998). TEMPS-I: delineating the most discriminant traits of the cyclothymic, depressive, hyperthymic and irritable temperaments in a nonpatient population. *Journal of Affective Disorders*, **51**, 7–19.

Altshuler, L., Kiriakos, L., Calcagno, J., *et al.* (2001). The impact of antidepressant discontinuation versus antidepressant continuation on 1-year risk for relapse of bipolar depression: a retrospective chart review. *Journal of Clinical Psychiatry*, **62**, 612–16.

Altshuler, L., Suppes, T., Black, D., *et al.* (2003). Impact of antidepressant discontinuation after acute bipolar depression remission on rates of depressive relapse at 1-year follow-up. *American Journal of Psychiatry*, **160**, 1252–62.

American Psychiatric Association. (2002). Practice guideline for the treatment of patients with bipolar disorder (revision). *American Journal of Psychiatry*, **159**, 1–50.

Angst, J., Arlington, VA, & Marneros, A. (2001). Bipolarity from ancient to modernities, birth and rebirth. *Journal of Affective Disorders*, **67**, 3–19.

APA (1980). *The Diagnostic and Statistical Manual of Mental Disorders*, 3rd edition. Arlington, VA: American Psychiatric Association.

APA (2000). *The Diagnostic and Statistical Manual of Mental Disorders*, 4th edition. American Psychiatric Association, Arlington, VA: www.psychiatrylonline.com/resourceTOC.aspx?resourceID=1 [accessed 23 December 2009].

Baker, R.W, Chengappa, K.N., Baird, J.W., Steingard, S., Christ, M.A., Schooler, N.R. (1992). Emergence of obsessive compulsive symptoms during treatment with clozapine. *Journal of Clinical Psychiatry*, **53**, 439–42.

Benazzi, F. (1999). Prevalence of bipolar II disorder in atypical depression. *European Archives of Psychiatry and Neurological Sciences*, **249**, 62–5.

Blinder, B.J., Cumella, E.J., Sanathara, V.A. (2006). Psychiatric comorbidities of female inpatients with eating disorders. *Psychosomatic Medicine*, **68**, 454–62.

Bottlender, R., Rudolf, D., Strauss, A., Möller, H.J. (1998). Antidepressant-associated maniform states in acute treatment of patients with bipolar I depression. *European Archives of Psychiatry and Clinical Neuroscience*, **248**, 296–300.

Calabrese, J.R. & Delucchi, G.A. (1990). Spectrum of efficacy of valproate in 55 patients with rapid-cycling bipolar disorder. *American Journal of Psychiatry*, **147**, 431–4.

Davidson, J.R., Miller, R.D., Turnbull, C.D., Sullivan, J.L. (1982). Atypical depression. *Archives of General Psychiatry*, **39**, 527–34.

Ghaemi, S.N. & Goodwin, F.K. (2001). Long-term naturalistic treatment of depressive symptoms in bipolar illness with divalproex vs. lithium in the setting of minimal antidepressant use. *Journal of Affective Disorders*, **65**, 281–7.

Godart, N.T., Flament, M.F., Perdereau, F., Jeammet, P. (2005). Comorbidity between eating disorders and mood disorders: review. *L'Encéphale*, **31**, 575–87.

Goldberg, J.F . & Whiteside, J.E. (2002). The association between substance abuse and antidepressant-induced mania in bipolar disorder: a preliminary study. *Journal of Clinical Psychiatry*, **63**, 791–5.

Gunderson, J.G., Weinberg, I., Daversa, M.T., *et al.* (2006). Descriptive and longitudinal observations on the relationship of borderline personality disorder and bipolar disorder. *American Journal of Psychiatry*, **163**, 1173–8.

Hantouche, E.G., Akiskal, H.S., Lancrenon, S., *et al.* (1998). Systematic clinical methodology for validating bipolar-II disorder: data in mid-stream from a French national multi-site study (EPIDEP). *Journal of Affective Disorders*, **50**, 163–73.

Henry, C., Sorbara, F., Lacotte, J. (2001). Antidepressant-induced mania in bipolar patients: identification of risk factors. *Journal of Clinical Psychiatry*, **62**, 249–55.

Himmelhoch, J.M. (1998). Social anxiety, hypomania and the bipolar spectrum: data, theory and clinical issues. *Journal of Affective Disorders*, **50**, 203–13.

Hirschfeld, R.M., Weisler, R.H., Raines, S.R., Macfadden, W. (2006). Quetiapine in the treatment of anxiety in patients with bipolar I or II depression: a secondary analysis from a randomized, double-blind, placebo-controlled study. *Journal of Clinical Psychiatry*, **67**, 355–62.

Judd, L.L. & Akiskal, H.S. (2003). Depressive episodes and symptoms dominate the longitudinal course of bipolar disorder. *Current Psychiatry Reports*, **5**, 417–18.

Koukopoulos, A. (2003). Ewald Hecker's description of cyclothymia as a clinical mood disorder: its relevance to the modern concept. *Journal of Affective Disorders*, **73**, 199–205.

Kretschmer, E. (1936). *Physique and Character*. London: Kegan Paul.

Levitt, A.J., Joffe, R.T., Ennis, J., MacDonald, C., Kutcher, S.P. (1990). The prevalence of cyclothymia in borderline personality disorder. *Journal of Clinical Psychiatry*, **51**, 335–9.

Levy, D., Kimhi, R., Barak, Y., Aviv, A., Elizur, A. (1998). Antidepressant-associated mania: a study of anxiety disorders patients. *Psychopharmacology (Berl)*, **136**, 243–6.

Lum, M., Fontaine, R., Elie, R., Ontiveros, A., (1991). Probable interaction of sodium divalproex with benzodiazepines. *Progress in Neuro-psychopharmacology & Biological Psychiatry*, **15**, 269–73.

MacKinnon, D.F. & Pies, R. (2006). Affective instability as rapid cycling: theoretical and clinical implications for borderline personality and bipolar spectrum disorders. *Bipolar Disorders*, **8**, 1–14. Review.

MacKinnon, D.F., Zandi, P.P., Cooper, J., *et al.* (2002). Comorbid bipolar disorder and panic disorder in families with a high prevalence of bipolar disorder. *American Journal of Psychiatry*, **159**, 30–5.

MacKinnon, D.F., Zandi, P.P., Gershon, E., Nurnberger Jr, J.I., Reich, T., DePaulo J.R. (2003). Rapid switching of mood in families with multiple cases of bipolar disorder. *Archives of General Psychiatry*, **60**, 921–8.

Manning, J.S., Haykal, R.F., Connor, P.D., Akiskal, H.S. (1997). On the nature of depressive and anxious states in a family practice setting: the high prevalence of bipolar II and related disorders in a cohort followed longitudinally. *Comprehensive Psychiatry*, **38**, 102–8.

Maremmani, I., Perugi, G., Pacini, M., Akiskal, H.S. (2006). Toward a unitary perspective on the bipolar spectrum and substance abuse: opiate addiction as a paradigm. *Journal of Affective Disorders*, **3**, 1–12.

Masi, G., Perugi, G., Millepiedi, S., *et al.* (2007). Clinical and research implications of panic-bipolar comorbidity in children and adolescents. *Psychiatric Research Reports*, **153**, 47–54.

McDougle, C.J., Epperson, C.N., Pelton, G.H., Wasylink, S., Price, L.H. (2000). A double-blind, placebo-controlled study of risperidone addition in serotonin reuptake inhibitor-refractory obsessive-compulsive disorder. *Archives of General Psychiatry*, **57**, 794–801.

McElroy, S.L., Kotwal, R., Keck Jr, P.E., Aksikal, H.S. (2005). Comorbidity of bipolar and eating disorders: distinct or related disorders with shared dysregulations? *Journal of Affective Disorders*, **86**, 107–27. Review.

McElroy, S.L. Pope, H.G. Jr., Keck, P.E. Jr., Hudson, J.I., Phillips, K.A., Strakowski, S.M. (1996). Are impulse-control disorders related to bipolar disorder? *Comprehensive Psychiatry*, **37**, 229–40.

Möller, H.J. & Grunze, H. (2001). Are antidepressants less effective in the acute treatment of bipolar I compared to unipolar depression? *Journal of Affective Disorders*, **67**, 141–6.

Möller, H.J. & Grunze, H. (2000). Have some guidelines for the treatment of acute bipolar depression gone too far in the restriction of antidepressants? *European Archives of Psychiatry and Clinical Neuroscience*, **250**, 57–68.

Pande, A.C., Davidson, J.R., Jefferson, J.W., *et al.* (1999). Treatment of social phobia with gabapentin: a placebo controlled study. *Journal of Clinical Psychpharmacology*, **19**, 341–8.

Peet, M. (1994). Induction of mania with selective serotonin reuptake inhibitors and tricyclic antidepressants. *British Journal of Psychiatry*, **164**, 549–50.

Perugi, G. & Akiskal, H.S. (2002). The soft bipolar spectrum redefined: focus on the cyclothymic, anxious-sensitive, impulse-dyscontrol, and binge-eating connection in bipolar II and related conditions. *Psychiatric Clinics of North America*, **25**, 713–37.

Perugi, G., Akiskal, H.S., Lattanzi, L., *et al.* (1998). The high prevalence of 'soft' bipolar (II) features in atypical depression. *Comprehensive Psychiatry*, **39**, 63–71.

Perugi, G., Akiskal, H.S., Toni, C., Simonini, E., Gemignani, A. (2001). The temporal relationship between anxiety disorders and (hypo)mania: a retrospective examination of 63 panic, social phobic and obsessive-compulsive patients with comorbid bipolar disorder. *Journal of Affective Disorders*, **67**, 199–206.

Perugi, G., Toni, C., Akiskal, H.S. (1999). Anxious-bipolar comorbidity. Diagnostic and treatment challenges. *Psychiatric Clinics of North America*, **22**, 565–83.

Perugi, G., Toni, C., Frare, F, Travierso, M.C., Hantouche, E., Akiskal, H.S. (2002). Obsessive-compulsive-bipolar comorbidity: a systematic exploration of clinical features and treatment outcome. *Journal of Clinical Psychiatry*, **63**, 1129–34.

Perugi, G., Toni, C., Passino, M.C., Akiskal, K.K., Kaprinis, S., Akiskal, H.S. (2006). Bulimia nervosa in atypical depression: the mediating role of cyclothymic temperament. *Journal of Affective Disorders*, **92**, 91–7.

Perugi, G., Toni, C., Travierso, M.C., Akiskal, H.S. (2003). The role of cyclothymia in atypical depression: toward a data-based reconceptualization of the borderline-bipolar II connection. *Journal of Affective Disorders*, **73**, 87–98.

Pini, S., Abelli, M., Mauri, M., *et al.* (2005). Clinical correlates and significance of separation anxiety in patients with bipolar disorder. *Bipolar Disorders*, **7**, 370–6.

Post, R.M., Altshuler, L.L., Frye, M.A., *et al.* (2001). Rate of switch in bipolar patients prospectively treated with second-generation antidepressants as augmentation to mood stabilizers. *Bipolar Disorders*, **3**, 259–65.

Post, R.M., Leverich, G.S., Nolen, W.A., *et al.* (2003). A re-evaluation of the role of antidepressants in the treatment of bipolar depression: data from the Stanley Foundation Bipolar Network. *Bipolar Disorders*, **5**, 396–406.

Rihmer, Z. & Pestality, P. (1999). Bipolar II disorder and suicidal behavior. *Psychiatric Clinics of North America*, **22**, 667–73.

Sab, H., Herpertz, S., Steinmeyer, E.M. (1993). Subaffective personality disorders. *International Clinical Psychopharmacology*, **1** (Suppl. 1), 39–46.

Sholomskas, A.J. (1990). Mania in a panic disorder patient treated with fluoxetine. *American Journal of Psychiatry*, **147**, 1090–1.

Steiner, W. (1991). Fluoxetine-induced mania in a patient with obsessive-compulsive disorder. *American Journal of Psychiatry*, **148**, 1403–4.

Stone, M.H. (1990). The fate of borderline patients: successful outcome and psychiatric practice. In: *Diagnosis & Treatment of Mental Disorders*. New York: Guilford Publications.

Stone, M.H. (2006). Relationship of borderline personality disorder and bipolar disorder. *American Journal of Psychiatry*, **163**, 1126–8.

Swann, A.C., Bowden, C.L., Morris, D., *et al.* (1997). Depression during mania: treatment response to lithium or divalproex. *Archives of General Psychiatry*, **54**, 37–42.

Thase, M.E. & Sachs, G.S. (2000). Bipolar depression: pharmacotherapy and related therapeutic strategies. *Biological Psychiatry*, **48**, 558–72.

Thase, M.E., Macfadden, W., Weisler, R.H., *et al.*; BOLDER II Study Group (2006). Efficacy of quetiapine monotherapy in bipolar I and II depression: a double-blind, placebo-controlled study (the BOLDER II study). *Journal of Clinical Psychopharmacology*, **26**, 600–9.

Wehr, T.A., Sack, D.A., Rosenthal, N.E., Cowdry, R.W. (1988). Rapid cycling affective disorder: contributing factors and treatment responses in 51 patients. *American Journal of Psychiatry*, **145**, 179–84.

WHO (2007). *International Classification of Diseases*, 10th edition. Geneva: World Health Organization. http://apps.who.int/classifications/apps/icd/icd10online/ [accessed 14 December 2009].

Young, L.T., Cooke, R.G., Robb, J.C., Levitt, A.J., Joffe, R.T. (1993). Anxious and non-anxious bipolar disorder. *Journal of Affective Disorders*, **29**, 49–52.

Young, L.T., Joffe, R.T., Robb, J.C., MacQueen, M., Marriott, M., Patelis-Siotis, I. (2000). Double-blind comparison of addition of a second mood stabilizer versus an antidepressant to an initial mood stabilizer for treatment of patients with bipolar depression. *American Journal of Psychiatry*, **157**, 124–6.

Chapter 15

Circadian and sleep/wake considerations in the practical management of bipolar disorder

Greg Murray

Introduction

The neurobiology of bipolar disorder (BD) is not well understood, but there is consensus that instability of circadian rhythms is a significant pathway in the development and course of the disorder. Importantly, this pathway is modifiable by non-pharmacological means, and so management of circadian rhythms and sleep has become a core element of psychosocial interventions for BD. The aim of this chapter is to provide a clinician-friendly overview of strategies for addressing the circadian and sleep/wake vulnerabilities in BD. The interested reader can find more detail in comprehensive clinician guides (see Lam *et al.*, 1999; Newman *et al.*, 2002; Morin & Espie, 2003; Frank, 2005).

Background: circadian and sleep/wake processes in bipolar disorder

The endogenous circadian time-keeping system is adapted to optimise engagement with the Earth's cyclic environment by predicting critical events, particularly sunrise and sunset (Moore-Ede, 1986). In humans and other mammals, the pacemaker responsible for the generation of circadian rhythms is located in the hypothalamic suprachiasmatic nucleus (SCN) (Reppert & Weaver, 2002). The endogenous period generated in the SCN is close to, but generally not equal to, 24 h. The process by which the pacemaker is set to a 24-h period and kept in appropriate phase with seasonally shifting daylength is called 'entrainment'.

Entrainment occurs via zeitgebers (environmental events that affect pacemaker amplitude, phase and period), the most significant of which is the daily light/dark cycle caused by the planet's rotation. The SCN is also responsive to non-photic cues such as arousal, locomotor activity, social cues, feeding, sleep deprivation and temperature (Gooley & Saper, 2005). Of course in free-living humans, exposure to the primary zeitgeber (light) is also mediated by activity. As discussed below, the fact that circadian entrainment is therefore modifiable by volitional behaviour (directly through behaviours that act as zeitgebers, and indirectly through behaviours that mediate light exposure) is central to the management of circadian vulnerability in BD.

The daily cycle of sleep and wakefulness is the most prominent behavioural rhythm in humans and is partly under the control of the SCN. Under the accepted 'two-process' model, sleep propensity is a function of a circadian oscillatory process, and a homeostatic process that increases with wakefulness and decreases during sleep (Czeisler *et al.*, 2005). The sleep/wake cycle has been called the master circadian rhythm, because it modulates many behavioural and physiological rhythms that are timed by the circadian pacemaker (Turek *et al.*, 2005). Although strongly integrated with circadian function, the sleep/wake rhythm and its disorders warrant separate attention as causal factors in BD and its outcomes (Riemann *et al.*, 2002).

Circadian instability in the aetiology of bipolar disorder

There is broad agreement that circadian instability is implicated in the aetiology and course of BD (see, e.g., Ehlers *et al.*, 1988; Jones, 2001; Wirz-Justice *et al.*, 2005; Goodwin & Jamison, 2007; McClung, 2007). Although the mechanisms of the relationship are unclear, the general conclusion is supported by a wide range of clinical and basic research (see Box 15.1). Evidence suggests not only that circadian and sleep function co-vary with symptoms of BD, but also that manipulation of circadian and sleep parameters affects symptoms, and that these parameters may constitute trait-like (or endophenotypic) (Hasler *et al.*, 2006) features of BD vulnerability.

A final demonstration of the role of circadian function in BD is presented in the circadian case study below (Figure 15.1). The data in the figure are presented as the smoothed sequence of 103 d of standardised circadian

Practical Management of Bipolar Disorder, eds. Allan H. Young, I. Nicol Ferrier and Erin E. Michalak. Published by Cambridge University Press.

Box 15.1: Bipolar disorder, circadian function and sleep

- A number of circadian parameters (including body temperature, cortisol and thyrotropin) are altered in episodes of BD (see Jones, 2001).
- Effective treatments for BD (lithium, antidepressants, anxiolytics, electroconvulsive therapy) affect circadian function (Padiath *et al.*, 2004; Yin *et al.*, 2006).
- Episodes of BD can be precipitated by seasonal change, time zone travel and other zeitgeber challenges (see Murray, 2006).
- Bright light can induce symptoms of mania, and 'dark therapy' has been successful in stabilising patients with rapid-cycling BD (Benedetti *et al.*, 2007).
- Polymorphisms in circadian genes have been associated with symptoms of BD in pre-clinical and human studies (Mitterauer, 2000; Roybal *et al.*, 2007; Benedetti *et al.*, 2008).
- Brain regions and neurotransmitters putatively involved in mood disorder also influence circadian and sleep function (Nestler *et al.*, 2002).
- There is a circadian rhythm in normal mood state (Murray *et al.*, 2002a), alterations of which may be causally involved in mood episodes (Boivin *et al.*, 1997; Murray, 2007).
- The primary temperamental predisposition to BD, neuroticism, may be associated with circadian instability (Murray *et al.*, 2002b).
- Instability in sleep and activity rhythms continues outside episodes (Harvey *et al.*, 2005; Jones *et al.*, 2005), and variation in circadian activity rhythm predicts subclinical variation in mood (Murray *et al.*, 2007).
- Diagnostic criteria for major depressive, manic, hypomanic and mixed episodes include sleep changes (APA, 2000).
- Compared with individuals without sleep disturbance, individuals with sleep disturbance are >10 times more likely to experience mood disorder (Breslau *et al.*, 1996).
- Altered sleep often precedes deterioration in clinical state (Colombo *et al.*, 1999; Jackson *et al.*, 2003; Bauer *et al.*, 2006), and decreased need for sleep has been called the final common pathway to mania.
- Episodes of BD are associated with sleep polysomnographic changes. Sleep architecture changes also persist outside acute episodes and may constitute a trait marker for mood disorder (Benca, 2005).
- Deliberate sleep deprivation is a same-day powerful treatment for bipolar (and unipolar) depression (McClung, 2007). Maintenance of the therapeutic effect beyond the next sleep phase is a target of current research (Benedetti *et al.*, 2007).

activity amplitude (measured by wrist-worn actigraph) for a single patient with BD. The patient was participating in a long-term prospective investigation of actigraphy as a predictor of mood change (Murray *et al.*, 2007). At the end of the sequence in Figure 15.1, the participant experienced a manic relapse. In the days and weeks preceding relapse, a pattern of decreased circadian amplitude and erratic regulation is clearly visible.

Therapeutic process

At one level, the strategies described below are simply commonsense behavioural changes. But human behaviour is notoriously difficult to change (how many of us get the exercise we require?), so therapeutic process becomes a critical issue. The strategies can be applied as one-size-fits-all modules, and presented in group or even self-help formats (e.g. Basco, 2006). However, where resources allow, they will ideally be integrated into an individualised case formulation and treatment plan. A comprehensive case formulation approach is particularly critical in BD, given the high rates of co-morbidity and the distinct strengths that are commonly seen in this population (Zaretsky *et al.*, 2007; Murray, 2010).

Like all evidence-based psychosocial interventions for BD, practical management of circadian rhythms and sleep calls for a strongly collaborative therapeutic relationship (Berk *et al.*, 2004; Corsini & Wedding, 2005). Key elements are: (1) the clinician is warm, directive and concerned, a problem-solver and a coping model; (2) the patient is active in determining the specific targets of therapy, while the clinician is expert in proposing pathways to achieve these goals; (3) the patient and clinician therefore work as a collaborative team; (4) therapeutic emphasis is on measurable changes outside the therapy room; (5) therapeutic goals are supported by learning principles (e.g. change as incremental).

In practice, a collaborative therapeutic atmosphere can be generated by asking patients to give their opinions about treatment, and what they think might be effective on the basis of past experience. A collaborative treatment approach also means that sometimes the patient's, rather than the clinician's, treatment choice is followed. As noted by Basco and Rush, the potential for short-term difficulties this generates is likely to be outweighed by the long-term impacts on the working relationship (Basco & Rush, 1996).

Collaboration requires that the patient be acculturated to the model that the clinician is bringing to their case. The circadian system is well suited to a stress–vulnerability (or chronic-disease management)

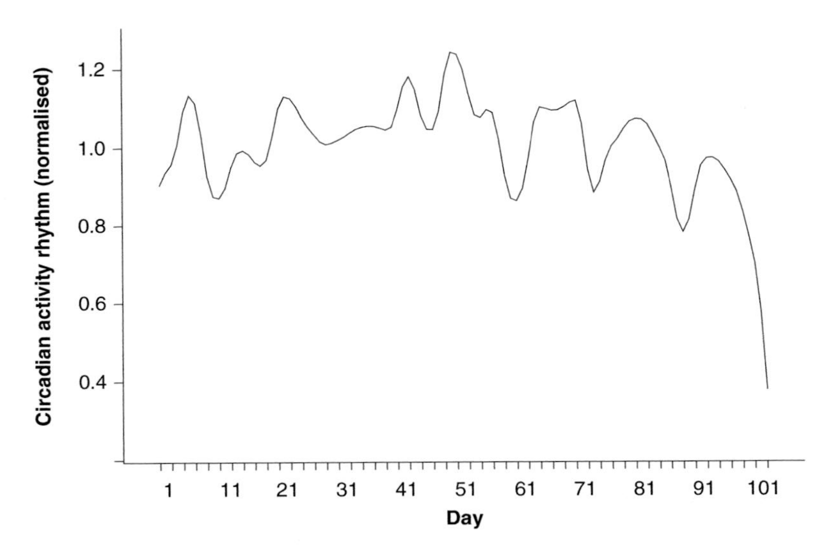

Figure 15.1 Case study of progressive circadian dysregulation in a manic prodrome.

framework for BD – the disorder is conceptualised not as an endogenous biological illness, nor as a response to overwhelming exogenous challenge, but as a dynamic interaction between the person's circadian function (vulnerability) and their life experiences (stress). The information in Box 15.1 can be used as part of psychoeducation about this aspect of the disorder.

A framework for integrating circadian rhythm management into other aspects of psychosocial treatment is the stress–vulnerability model shown in Figure 15.2, which is adapted from the models of Lam, Jones, Frank, Healy and others (Healy & Waterhouse, 1995; Lam *et al.*, 1999; Jones, 2001; Frank, 2005). The model in Figure 15.2 identifies points at which the core biological vulnerability of BD might be mitigated; the tools for achieving this are described next.

Management of circadian and sleep/wake rhythms in bipolar disorder

Circadian rhythm and sleep processes are fundamental to human biology, and alterations in these processes are strongly implicated in the development and course of BD. Not surprisingly, rhythm stabilisation is a widely recognised therapeutic goal (e.g. Miklowitz *et al.*, 2008) and a component of consensus treatment guidelines (e.g. Yatham *et al.*, 2006). The remainder of this chapter is devoted to describing the interventions that target this aspect of BD vulnerability. Techniques aimed at regularising social rhythms are covered first, followed by strategies targeting sleep quality. Both sets of techniques work by developing the patient's self-management skills, and therefore can be expected to have prophylactic as well as acute therapeutic effects.

Pharmacotherapy remains the first-line treatment for BD, and the psychosocial strategies described below must be seen as adjuncts to mood-stabilising medication. The evidence base for these adjunctive interventions is incomplete, but supporting data include research implicating circadian rhythms in the aetiology of BD (above), qualified support for social rhythm disruption in episode onset and the general effectiveness of adjunctive psychosocial interventions for BD, in which rhythm stabilisation plays a central role (Zaretsky *et al.*, 2007). The interventions described below are relevant across the subtypes of BD, but may be most effective when introduced earlier in the course of the disorder and at a time when functioning is not seriously impaired (Scott *et al.*, 2007).

The social zeitgeber hypothesis and social rhythm management

As noted above, the circadian system is primarily entrained by light, but is also sensitive to activity and social cues. In the classic BD handbook, Goodwin and Jamison observed that vulnerability to circadian dysregulation is a defining feature of the disorder, and that disruptions in social rhythms may be a key stressor for this diathesis (Goodwin & Jamison, 1990). Likewise, Ehlers, Frank and others (Ehlers *et al.*, 1988; Healy & Waterhouse, 1995; Frank, 2007) have argued that rhythmic features of the social environment (such as the timing of sleep, eating and exercise) are significant

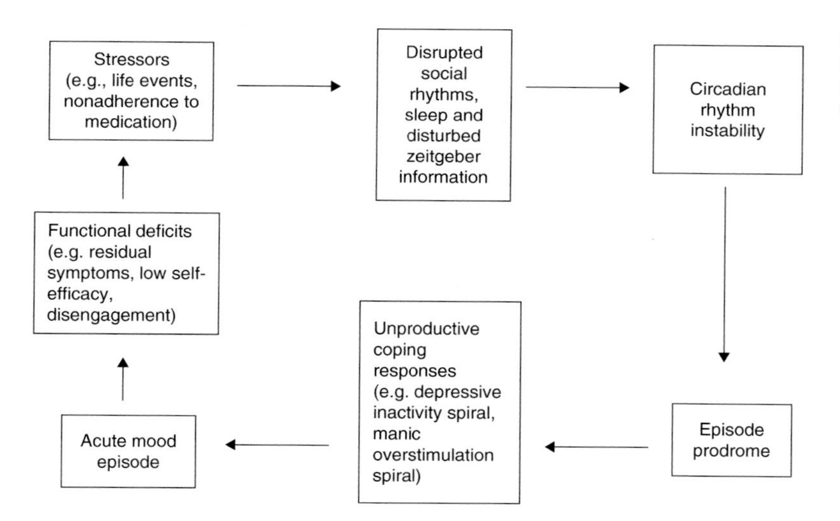

Figure 15.2 A stress–vulnerability model of circadian instability in bipolar disorder (after Lam, Jones, Frank, Healy and others).

components of human circadian entrainment, and disruption of these rituals may challenge the circadian clock.

The social zeitgeber hypothesis of depression proposes that major life events (such as loss of a significant relationship) not only have psychological meaning but also weaken zeitgeber information to the SCN through destabilisation of daily activities and light exposure. In biologically vulnerable individuals, this may trigger dysregulation of circadian rhythms and precipitate an episode of mood disorder (see Figure 15.2). Ehlers and colleagues further highlight the importance of social rhythm disruption in the term 'zeitstorer', a feature of the social environment that actively disturbs timing signals (Ehlers *et al.*, 1993). Shift work, travel across time zones, changes in family structure and more subtle events (such as couple conflict or promotion at work) can act as zeitstorers by disrupting links between the internal clock and chronobiologically significant time cues.

The social zeitgeber hypothesis has received some empirical support. In two studies, onset of manic episodes has been found to be associated with life events that involve disruption of social rhythms (Malkoff-Schwartz *et al.*, 1998, 2000). Interestingly, there is no strong evidence that social rhythm disruption precedes depressive symptoms in BD, and it has been suggested that trait-like circadian vulnerability may play a greater role in this more insidious phase of the disorder (Grandin *et al.*, 2006).

The clinical application of the social zeitgeber hypothesis is social rhythm therapy (Frank *et al.*, 1994), a largely behavioural psychotherapy aimed at helping patients maintain stability in their social rhythms and so reduce the risk of relapse. In the treatment of BD, social rhythm therapy is typically integrated with principles from interpersonal psychotherapy in a treatment known as interpersonal and social rhythm therapy (IPSRT) (Frank *et al.*, 1994), which has proven effective in two large studies. In the first (Frank *et al.*, 2005), IPSRT was compared with an intensive clinical management (ICM) condition. Patients entered the study while in acute manic, mixed or depressive episodes, and were allocated to either IPSRT or ICM as adjunctive treatment for both acute and maintenance phases. Time to stabilisation of acute episode did not differ between the groups, but irrespective of which maintenance condition patients were allocated to, those randomised to IPSRT in the acute phase had longer survival times to an affective episode. Significantly, at the end of the acute phase, social rhythmicity was higher in the IPSRT group, and the ability to increase regularity of routines during the acute phase was inversely proportional to the likelihood of relapse during the maintenance phase. Interpersonal and social rhythm therapy also had an effect on residual symptoms, and patients receiving IPSRT spent more time euthymic and less time depressed. The authors conclude that the findings support the idea of the prophylactic effect of IPSRT. In a second major investigation, IPSRT as an adjunct to mood stabilisers was shown to improve symptomatic and functional outcomes in bipolar depression relative to a brief psychoeducational intervention (Miklowitz *et al.*, 2007). Significantly, this study found no difference between IPSRT and two other psychosocial treatments (cognitive behavioural therapy and family-focused therapy). This absence of differential efficacy might be explained by the fact that improving sleep/wake regulation is a goal common to all psychosocial interventions (Frank, 2007).

Monitoring and assessing social rhythmicity

A key element of social rhythm management is measurement of social rhythm stability using the social rhythm metric (SRM). The latest version of this self-report diary instrument, the SRM-II-5 (Frank, 2005), assesses the regularity of the time at which the individual (1) gets out of bed, (2) has first contact with another person, (3) starts work, school or housework, (4) has dinner and (5) goes to bed. The social rhythm score is the proportion of events per week that fall within a target window of the ideal clock time of their occurrence and provides a quantitative point of reference for improving social rhythmicity. Beyond the social rhythm score, the clinician and patient should explore the weekly SRM reports: What is the relationship between social rhythmicity and daily mood? What role do others play in the patient's social rhythmicity? What barriers to regularising social rhythms can be identified?

Social rhythm stability can be affected by the patient's cognitions as well as behaviours. For example, reward sensitivity, autonomy and a hyperpositive sense of self may promote overstimulating behaviours, whereas low self-esteem may trigger and maintain depressive inactivity and sleep disturbance (Johnson & Tran, 2007). It is therefore useful to introduce the notion of cognitive mediation of affect as part of monitoring behaviours and mood on the SRM (see, e.g., Newman *et al.*, 2002). Cognitively mediated amplification of altered motivation and energy levels may be particularly significant in prodromes (see Figure 15.2), and hence a target for relapse prevention work.

Regularising social rhythms and titrating stimulation

Once the patient's social rhythm patterns, zeitgebers and zeitstorers are understood prospectively (via SRM reports) and retrospectively (using an illness timeline or life-charting method), the aim of social rhythm therapy is to strengthen basal social rhythmicity and pre-emptively address disruptions to social rhythms.

The importance of stability is well accepted by people diagnosed with BD, as is the bond between sleep quality and wellbeing (Michalak *et al.*, 2007). In this context, simple problem-solving (see, e.g., Falloon, 2000) is often effective in tackling current barriers and future threats to social rhythmicity. For example, the circadian challenge of overseas travel can be minimised by early attention to flight times, sleep strategies and likely triggers for over-arousal or anxiety.

Chronic instability in social rhythms due to erratic work habits may lead to a review of the patient's lifestyle choices. The archetypal BD temperament encourages artistic pursuits (Goodwin & Jamison, 2007), but this occupational niche is particularly challenging to social rhythms (lack of a working day and working week, project-based activities leading to extreme variations in stimulation, etc.). The clinician can provide a 'clock's eye view' of these activities and encourage the patient to consider ways to proactively regulate stimulation levels (e.g. scheduling activities that provide modest daily stimulation irrespective of performance or project demands).

The SRM not only measures the *timing* of activities but also provides a record of the *quantity* of activity, social interactions and other stimulating events. Patients showing hypomanic symptoms should be encouraged to decrease social, cognitive and biological (particularly light) stimulation, or to schedule stimulating activities earlier in the day. Conversely, the behavioural activation techniques effective in unipolar depression (e.g., activity scheduling) (Cuijpers *et al.*, 2007) can be used to encourage additional stimulating and engaging activities during depressive phases. The complexity in BD is the potential for ascent into mania and the consequent need to carefully monitor and titrate stimulating activity.

Managing sleep in bipolar disorder

Poor sleep quality is a major barrier to social rhythm stability and a primary target of IPSRT (Frank, 2005). Sleep disturbances are also a significant detriment to quality of life in BD and a major clinical focus in their own right (Michalak *et al.*, 2007).

Common sleep disturbances

Insomnia (typically middle and late insomnia) is an almost universal concomitant of the depressive phase of BD. Given that depression is the most common clinical state in BD (Judd *et al.*, 2003), insomnia treatment is therefore commonly required. For many patients, bipolar depression can also be associated with hypersomnia, difficulty awakening and excessive daytime sleepiness. In manic and mixed episodes, by contrast, sleep-onset insomnia is frequent and often associated with highly distressing over-arousal and subjective inability to relax. For some patients, of course, hypomanic and manic mood states generate not insomnia but an elated, decreased need for sleep – when manic, patients may require only 3–4 h sleep to feel refreshed.

In one of the few quantitative investigations of sleep in euthymic BD (Harvey *et al.*, 2005), 70% of patients were found to exhibit clinically significant sleep disturbance, and 55% met diagnostic criteria for insomnia (except the overarching psychiatric disorder exclusion, Criterion D). Compared with good sleepers, BD patients exhibited decreased sleep efficiency (time asleep as a proportion of time in bed), lower daytime activity levels, misperceptions about sleep and more dysfunctional attitudes about sleep. Indeed, the quality and extent of sleep disturbance in BD patients were similar to those of a comparison group presenting for treatment of primary insomnia. Cognitive arousal at bedtime was a significant subjective concern in both BD and insomnia groups. Bipolar patients also reported concerns about achieving a stable sleeping routine. Many BD patients are familiar with the role of sleep in their episode prodromes (Lam & Wong, 2005), so it is not surprising that low self-efficacy around sleep is an ongoing source of worry in this population.

Brief assessment of sleep disturbances

Given the elevated prevalence of sleep disturbance in mood disorder populations, sleep should be routinely assessed in patients diagnosed with BD. This assessment can be simplified to a two-step process. First, the clinician can assess for the most common presenting symptoms of sleep disorders (insomnia and excessive daytime sleepiness). Second, additional symptoms such as snoring, observed apnoea and abnormal sleep behaviours should be queried. It is worth noting that obesity, a common medical co-morbidity in BD, increases the likelihood of breathing-related sleep disturbance.

Overall sleep quality can be usefully measured on the valid and reliable Pittsburgh sleep quality index (PSQI) (Buysse *et al.*, 1989). This 19-item questionnaire is retrospective over the previous month, and comprises seven component scores: (1) subjective sleep quality; (2) sleep onset latency; (3) sleep duration; (4) habitual sleep efficiency; (5) sleep disturbance; (6) use of sleep medications; and (7) daytime functioning. These component scores are summed to provide a global score that ranges from 0 to 21, with higher scores indicating worse sleep. A global score of >5 suggests clinically significant sleep disturbance on the PSQI, and as noted earlier, the majority of patients with BD can be expected to meet this criterion.

Improving sleep quality

Insomnia in patients with serious mental illness is often addressed with medication, and many patients' timely use of small doses of benzodiazepines or sedating antipsychotics is an effective part of manic relapse prevention (Russell, 2005). However, there is good evidence that the psychosocial techniques effective for sleep disturbance in non-psychiatric populations are also effective in patients with BD, and these should be offered as a first-line treatment (Smith *et al.*, 2005; Biancosino *et al.*, 2006; Smith & Perlis, 2006). These interventions are of five types (see Table 15.1) but overlap and are often generically termed cognitive behavioural therapy (CBT) for insomnia (see Morin & Espie, 2003).

As summarised in Table 15.1, many cognitive behavioural strategies aim to develop good sleep habits, which both facilitate normal sleep and minimise insomnia and hypersomnia. When applied to BD, the ultimate goal of these techniques is to regularise social rhythms, decrease sleep-driven relapse and increase self-efficacy around sleep. Moderate-to-large effect sizes have been demonstrated for the cognitive-behavioural treatment of chronic insomnia in sleep-disorder populations, and similar magnitudes of outcome can be expected in BD (Smith & Perlis, 2006).

There are a number of principles that the clinician should consider in introducing these strategies, as listed below.

1. Given the relationship between sleep and relapse, the onset of either insomnia or hypersomnia should cue the clinician to increase the frequency of appointments.
2. The choice of intervention should be based on the factors maintaining the sleep problem. If sleep disruption is due to unproductive cognitions about sleep, a cognitive strategy is indicated; if sleep disruption is being maintained by an association between bed and wakefulness, a sleep hygiene intervention is the first step.
3. The phase of the disorder impacts on sleep targets. As mood becomes elevated, patients may underestimate the need for sleep in comparison to task completion, or they may use exhaustion or alcohol in lieu of an appropriate bedtime routine. By contrast, attention to the daily wake phase (using the behavioural activation techniques

Table 15.1 A summary of cognitive–behavioural strategies for improving sleep quality and treating common sleep disturbances

Approach	Strategies	Rationale	Potential issues and qualifications
Stimulus control	Five principles of stimulus control: • go to bed only when sleepy • get out of bed when unable to sleep (perform a non-arousing distracting activity and return to bed when sleepy; repeat if necessary) • use the bed for sleep only (or sex or brief reading) • arise around the same time every morning • no napping during the day	Re-associates the bed with sleep (particularly if insomnia has become a chronic problem); establishes a consistent sleep/wake schedule	Daytime sleepiness in the depressive phase may be difficult to overcome, and so napping may need to be moderated rather than proscribed
Sleep restriction	Limit time in bed to the amount of time asleep (e.g. if person reports 6 h sleep for 8 h in bed, initial prescribed sleep time is 6 h, progressively increased until optimal duration achieved)	The inclination for poor sleepers to increase the amount of time spent in bed is unproductive; mild sleep deprivation increases sleep drive; reduces sleep anticipatory anxiety	Even mild sleep deprivation may be a risk for hypo/manic swings; careful monitoring is required; *this intervention may be unsuitable for patients with a history of rapid manic switches*
Relaxation training	• progressive muscle relaxation • abdominal breathing • meditation/guided imagery • exogenous relaxation (e.g. hot bath, chamomile tea or hot milk half an hour before bed) • direct patient to one of the useful self-help guides for anxiety management (e.g. Davis *et al.*, 2000)	Decreases physical arousal; controls negative thoughts that maintain arousal	Relaxation is an aversive state for some people with chronic anxiety; it is necessary to trial relaxation strategies in-session to gauge response; distraction (get up and perform a basic task, clean up the dishes, do a crossword, knitting) is an alternative cognitive technique for these patients
Cognitive approaches	In therapy, open a discussion around beliefs about sleep and insomnia. People suffering insomnia typically exaggerate both the degree of their sleep disturbance and the seriousness of the consequences for next-day functioning Teach strategies for coping with negative thoughts at bedtime (e.g. writing down concerns to be addressed in the morning). Remember: rumination at night is often masked as problem-solving	Changes unproductive beliefs about sleep and consequences for wellbeing; decreases monitoring of daytime sleepiness and negative thoughts around bed; decreases arousal at night	Important not to 'disagree' or argue with patient's beliefs and attitudes; more useful to explore the consequences of holding unproductive beliefs ('It's a disaster if I don't sleep') vs more productive beliefs ('It would be good to have an uninterrupted night's sleep, but I can sleep poorly without serious consequences')

Sleep hygiene education	Psychoeducation about the role of lifestyle regularity, exercise, diet, substances and light exposure in sleep. Provide information about normative need for sleep and individual differences in preference and need. Four components of sleep hygiene: **1. Develop a wake-up routine** Wake-up time around 07:00–07.30 h is a goal for many people; get up as soon as you wake up, even if you feel that you haven't had enough sleep; when you get up, get outside into the sunlight and do some physical activity (the body clock is particularly sensitive to light and exercise) **2. During the day** Napping should be avoided; physical and social activity in moderation **3. Develop a bedtime routine** Go to bed around 22:30–23:00 h; avoid alcohol around bedtime; arousing stimuli (smoking, drugs, TV, the Internet, movies or emotional discussions with family members) can make getting to sleep more difficult; don't take sleeping pills for more than a week at a time; add in a wind-down time (30 min or so) between any work and bed **4. In bed** Keep the sleeping environment cool, dark and quiet and remove stimulating cues (such as computer and artistic materials)	Informs patient about the nature of sleep and its connection to daytime activities: wake and sleep are opposite sides of the same coin	If atypical symptoms predominate in the depressive phase, waking at 07.30 h is an over-ambitious goal; wake time can be successively shifted earlier across 2 or 3 weeks. The bedtime routine may be a particular focus during hypomanic periods, or for patients whose clinical picture includes prominent ascents into mania. The wake-up routine is particularly relevant for residual depressive symptoms

mentioned above) is key for sleep problems in bipolar depression.

4. Acknowledge that behaviour change is difficult, incremental and requires reinforcement. New behaviours should be seen as experiments; 'failures' are data to be fed back into the formulation.
5. The clinician must address the information-processing deficits commonly associated with BD. Patients require a clear, engaging rationale for each strategy, and should be provided with written instructions and monitoring sheets to support behavioural change.
6. Recognise individual differences: reading in bed and evening exercise can be too stimulating for many patients, but may form part of a trusted bedtime ritual for others.
7. Consider the social context of the sleep problem. For example, the night-time behaviour of others in the family may be a chronic challenge to sleep and social rhythm stability. Similarly, progression of BD across the lifespan is often associated with poorer living conditions, and physical safety is a prerequisite for restful sleep.
8. Interventions should be integrated with the patient's existing wellbeing strategies. A patient is unlikely to adhere to scheduled bedtime if it curtails their sole pleasure of late-night entertainment on TV.

Summary

The present chapter can be summarised in three take-home points. First, circadian rhythms and their moderation are crucial in the development, course and treatment of BD. Second, the principles of social rhythm management and sleep quality improvement provide evidence-based direction for the patient and clinician to address this core vulnerability. In a significant proportion of patients, these interventions can be expected to decrease symptom burden, enhance quality of life and improve prognosis. Finally, by focusing on the ubiquitous and adaptive 24-h cycle of engagement/disengagement, we underscore what is shared between people who do, and people who do not, face the challenge of BD. It is thus a normalising and empowering framework from which to work.

References

APA (2000). *The Diagnostic and Statistical Manual of Mental Disorders,* 4th edition. American Psychiatric Association, Arlington, VA, USA. www.psychiatryonline.com/resourceTOC.aspx?resourceID=1 [accessed 23 December 2009].

Basco, M.R. (2006). *The Bipolar Workbook: Tools for Controlling Your Mood Swings.* New York: Guilford Press.

Basco, M.R. & Rush, J.A. (1996). *Cognitive-Behavioral Therapy for Bipolar Disorder.* New York, NY: Guilford Press.

Bauer, M., Grof, P., Rasgon, N., Bschor, T., Glenn, T., Whybrow, P.C. (2006). Temporal relation between sleep and mood in patients with bipolar disorder. *Bipolar Disorders,* **8**, 160–7.

Benca, R.M. (2005). Mood disorders. In: Kryger, M.H., Roth, T., Dement, W.C. (Eds). *Principles and Practice of Sleep Medicine,* 4th edition. New York: Saunders.

Benedetti, F., Barbini, B., Colombo, C., Smeraldi, E. (2007). Chronotherapeutics in a psychiatric ward. *Sleep Medicine Reviews,* **11**, 509–22.

Benedetti, F., Radaelli, D., Bernasconi, A., *et al.* (2008). Clock genes beyond the clock: CLOCK genotype biases neural correlates of moral valence decision in depressed patients. *Genes, Brain, and Behavior,* **7**, 20–5.

Berk, M., Berk, L., Castle, D. (2004). A collaborative approach to the treatment alliance in bipolar disorder. *Bipolar Disorders,* **6**, 504–18.

Biancosino, B., Rocchi, D., Dona, S., Kotrotsiou, V., Marmai, L., Grassi, L. (2006). Efficacy of a short-term psychoeducational intervention for persistent non-organic insomnia in severely mentally ill patients. A pilot study. *European Psychiatry,* **21**, 460–2.

Boivin, D.B., Czeisler, C.A., Dijk, D.J., *et al.* (1997). Complex interaction of the sleep-wake cycle and circadian phase modulates mood in healthy subjects. *Archives of General Psychiatry,* **54**, 145–52.

Breslau, N., Roth, T., Rosenthal, L., Andreski, P. (1996). Sleep disturbance and psychiatric disorders: a longitudinal epidemiological study of young adults. *Biological Psychiatry,* **39**, 411–18.

Buysse, D.J., Reynolds III, C.F, Monk, T.H., Berman, S.R., Kupfer, D.J. (1989). The Pittsburgh Sleep Quality Index: a new instrument for psychiatric practice and research. *Psychiatry Research,* **28**, 193–213.

Colombo, C., Benedetti, F., Barbini, B., Campori, E., Smeraldi, E. (1999). Rate of switch from depression into mania after therapeutic sleep deprivation in bipolar depression. *Psychiatry Research,* **86**, 267–70.

Corsini, R.J. & Wedding, D. (eds) (2005). *Current Psychotherapies.* Itasca, Ill: F.E. Peacock.

Cuijpers, P., Van Straten, A., Warmerdam, L. (2007). Behavioral activation treatments of depression: a meta-analysis. *Clinical Psychology Review,* **27**, 318–26.

Czeisler, C.A., Buxton, O.M., Khalsa, S.B. **S.** (2005). The human circadian timing system and sleep-wake regulation. In: Kryger, M.H., Roth, T., Dement, W.C.

(Eds). *Principles and Practice of Sleep Medicine*, 4th edition. Philadelphia, PA: Elsevier.

Davis, M., Eshelman, E.R., McKay, M. (2000). *The Relaxation and Stress Reduction Workbook*. Oakland, CA: New Harbinger.

Ehlers, C.L., Frank, E., Kupfer, D.J. (1988). Social zeitgebers and biological rhythms. *Archives of General Psychiatry*, **45**, 948–52.

Ehlers, C.L., Kupfer, D.J., Frank, E., Monk, T.H. (1993). Biological rhythms and depression: the role of zeitgebers and zeitstorers. *Depression*, **1**, 285–93.

Falloon, I.R. (2000). Problem solving as a core strategy in the prevention of schizophrenia and other mental disorders. *Australian and New Zealand Journal of Psychiatry*, **34** (Suppl.), S185–90.

Frank, E. (2005). *Treating Bipolar Disorder: A Clinician's Guide to Interpersonal and Social Rhythm Therapy*. New York: Guilford Press.

Frank, E. (2007). Interpersonal and social rhythm therapy: a means of improving depression and preventing relapse in bipolar disorder. *Journal of Clinical Psychology*, **63**, 463–73.

Frank, E., Kupfer, D.J., Ehlers, C.L., *et al.* (1994). Interpersonal and social rhythm therapy for bipolar disorder: Integrating interpersonal and behavioural approaches. *Behavior Therapy*, **17**, 143–9.

Frank, E., Kupfer, D J., Thase, M.E., *et al.* (2005). Two-year outcomes for interpersonal and social rhythm therapy in individuals with bipolar I disorder. *Archives of General Psychiatry*, **62**, 996–1004.

Goodwin, F. & Jamison, K. (1990). *Manic-depressive Illness*. New York: Oxford University Press.

Goodwin, F.K. & Jamison, K.R. (2007). *Manic-depressive Illness: Bipolar Disorders and Recurrent Depression*. New York: Oxford University Press.

Gooley, J.J. & Saper, C.B. (2005). Anatomy of the mammalian circadian system. In: Kryger, M.H., Roth, T., Dement, W.C. (Eds). *Principles and Practice of Sleep Medicine*, 4th edition. Philadelphia, PA: Elsevier.

Grandin, L.D., Alloy, L.B., Abramson, L.Y. (2006). The social zeitgeber theory, circadian rhythms, and mood disorders: review and evaluation. *Clinical Psychology Review*, **26**, 679–94.

Harvey, A.G., Schmidt, D.A., Scarna, A., Semler, C.N., Goodwin, G.M. (2005). Sleep-related functioning in euthymic patients with bipolar disorder, patients with insomnia, and subjects without sleep problems. *The American Journal of Psychiatry*, **162**, 50–7.

Hasler, G., Drevets, W.C., Gould, T.D., Gottesman, I.I., Manji, H.K. (2006). Toward constructing an endophenotype strategy for bipolar disorders. *Biological Psychiatry*, **60**, 93–105.

Healy, D. & Waterhouse, J.M. (1995). The circadian system and the therapeutics of the affective disorders. *Pharmacology and Therapeutics*, **65**, 241–63.

Jackson, A., Cavanagh, J., Scott, J. (2003). A systematic review of manic and depressive prodromes. *Journal of Affective Disorders*, **74**, 209–17.

Johnson, S. & Tran, T. (2007). Bipolar disorder: what can psychotherapists learn from the cognitive research? *Journal of Clinical Psychology*, **63**, 425–32.

Jones, S.H. (2001). Circadian rhythms, multilevel models of emotion and bipolar disorder: an initial step towards integration? *Clinical Psychology Review*, **21**, 1193–209.

Jones, S.H., Hare, D.J., Evershed, K. (2005). Actigraphic assessment of circadian activity and sleep patterns in bipolar disorder. *Bipolar Disorders*, **7**, 176–86.

Judd, L.L., Akiskal, H.S., Schettler, P.J., *et al.* (2003). A prospective investigation of the natural history of the long-term weekly symptomatic status of bipolar II disorder. *Archives of General Psychiatry*, **60**, 261–9.

Lam, D. & Wong, G. (2005). Prodromes, coping strategies and psychological interventions in bipolar disorders. *Clinical Psychology Review*, **25**, 1028–42.

Lam, D.H., Jones, S.H., Hayward, P., Bright, J.A. (1999). *Cognitive Therapy for bipolar Disorder: a Therapist's Guide to Concepts, Methods and Practice*. New York: Wiley.

Malkoff-Schwartz, S., Frank, E., Anderson, B., *et al.* (1998). Stressful life events and social rhythm disruption in the onset of manic and depressive bipolar episodes. *Archives of General Psychiatry*, **55**, 702–7.

Malkoff-Schwartz, S., Frank, E., Anderson, B.P., *et al.* (2000). Social rhythm disruption and stressful life events in the onset of bipolar and unipolar episodes. *Psychological Medicine*, **30**, 1005–16.

McClung, C.A. (2007). Circadian genes, rhythms and the biology of mood disorders. *Pharmacology & Therapeutics*, **114**, 222–32.

Michalak, E.E., Murray, G., Young, A.H., Lam, R.W. (2007). Quality of life impairment in bipolar disorder. In: Ritsner, M. (Ed.). *Quality of Life Impairment in Schizophrenia, Mood and Anxiety Disorders: From Brain Functions to Clinical Practice*. New York: Springer.

Miklowitz, D.J., Goodwin, G.M., Bauer, M.S., Geddes, J.R. (2008). Common and specific elements of psychosocial treatments for bipolar disorder: a survey of clinicians participating in randomized trials. *Journal of Psychiatric Practice*, **14**, 77–85.

Miklowitz, D.J., Otto, M.W., Frank, E., *et al.* (2007). Psychosocial treatments for bipolar depression: a 1-year randomized trial from the Systematic Treatment Enhancement Program. *Archives of General Psychiatry*, **64**, 419–26.

Mitterauer, B. (2000). Clock genes, feedback loops and their possible role in the etiology of bipolar disorders: an integrative model. *Medical Hypotheses*, **55**, 155–9.

Moore-Ede, C.M. (1986). Physiology of the circadian timing system: predictive versus reactive homeostasis. *American Journal of Physiology*, **250**, R737–52.

Morin, C.M. & Espie, C.A. (2003). *Insomnia: a Clinical Guide to Assessment and Treatment*. New York: Kluwer Academic/Plenum.

Murray, G. (2006). *Seasonality, Personality and the Circadian Regulation of Mood*. New York: Nova Science.

Murray, G. (2007). Diurnal mood variation in depression: a signal of disturbed circadian function? *Journal of Affective Disorders*, **102**, 47–53.

Murray, G. (2010). Individual cognitive behavioural therapy for bipolar disorder. *E-Journal of Applied Psychology*, in press.

Murray, G., Allen, N.B., Trinder, J. (2002a). Mood and the circadian system: investigation of a circadian component in positive affect. *Chronobiology International*, **19**, 1151–69.

Murray, G., Allen, N.B., Trinder, J., Burgess, H. (2002b). Is weakened circadian rhythmicity a characteristic of neuroticism? *Journal of Affective Disorders*, **72**, 281–9.

Murray, G., Judd, F., Bullock, B. (2007). Circadian and sleep/wake variables as predictors of relapse in bipolar disorder. *Bipolar Disorders*, **9** (Suppl. 2), 11.

Nestler, E.J., Barrot, M., Dileone, R.J., Eisch, A.J., Gold, S.J., Monteggia, L.M. (2002). Neurobiology of depression. *Neuron*, **34**, 13–25.

Newman, C.F., Leahy, R.L., Beck, A.T., Reilly-Harrington, N.A., Gyulai, L. (2002). *Bipolar Disorder: a Cognitive Therapy Approach*. London: American Psychological Association.

Padiath, Q.S., Paranjpe, D., Jain, S., Sharma, V.K. (2004). Glycogen synthase kinase 3B as a likely target for the action of lithium on circadian clocks. *Chronobiology International*, **21**, 43–55.

Reppert, S.M. & Weaver, D.R. (2002). Coordination of circadian timing in mammals. *Nature*, **418**, 935–41.

Riemann, D., Voderholer, U., Berger, M. (2002). Sleep and sleep-wake manipulations in bipolar depression. *Neuropsychobiology*, **45** (Suppl. 1), 7–12.

Roybal, K., Theobold, D., Graham, A., *et al.* (2007). Mania-like behavior induced by disruption of CLOCK. *Proceedings of the National Academy of Sciences of the United States of America*, **104**, 6406–11.

Russell, S. (2005). *A Lifelong Journey: Staying Well With Manic Depression/Bipolar Disorder*. Melbourne: Michelle Anderson Publishing.

Scott, J., Colom, F., Vieta, E. (2007). A meta-analysis of relapse rates with adjunctive psychological therapies compared to usual psychiatric treatment for bipolar disorders. *The International Journal of Neuropsychopharmacology*, **10**, 123–9.

Smith, M.T. & Perlis, M.L. (2006). Who is a candidate for cognitive-behavioral therapy for insomnia? *Health Psychology*, **25**, 15–19.

Smith, M.T., Huang, M I., Manber, R. (2005). Cognitive behavior therapy for chronic insomnia occurring within the context of medical and psychiatric disorders. *Clinical Psychology Review*, **25**, 559–92.

Turek, F.W., Dugovic, C., Laposky, A.D. (2005). Master circadian clock, master circadian rhythm. In: Kryger, M.H., Roth, T., Dement, W.C. (Eds). *Principles and Practice of Sleep Medicine*, 4th edition. Philadelphia, PA: Elsevier.

Wirz-Justice, A., Benedetti, F., Berger, M. A., *et al.* (2005). Chronotherapeutics (light and wake therapy) in affective disorders. *Psychological Medicine*, **35**, 939–44.

Yatham, L.N., Kennedy, S.H., O'Donovan, C., *et al.* (2006). Canadian Network for Mood and Anxiety Treatments (CANMAT) guidelines for the management of patients with bipolar disorder: update 2007. *Bipolar Disorders*, **8**, 721–39.

Yin, L., Wang, J., Klein, P.S., Lazar, M.A. (2006). Nuclear receptor Rev-erbalpha is a critical lithium-sensitive component of the circadian clock. *Science*, **311**, 1002–5.

Zaretsky, A.E., Rizvi, S., Parikh, S.V. (2007). How well do psychosocial interventions work in bipolar disorder? *Canadian Journal of Psychiatry*, **52**, 14–21.

Chapter 16

A clinician's guide to psychosocial functioning and quality of life in bipolar disorder

Erin E. Michalak and Greg Murray

Introduction

Bipolar disorder (BD) is a complex, chronic condition that can be characterised by a wide variety of symptoms and marked variability in outcome. An individual with BD can experience episodes of depression, hypomania or mania and, indeed, can experience a mixture of emotional states or cycle rapidly between them. Variability also occurs in terms of the frequency and length of episodes experienced by individuals with BD over their lifetime, the severity and type of symptoms encountered and the degree of inter-episode recovery achieved. The presence of subsyndromal features between episodes, however, is the rule rather than the exception (e.g. Post *et al.*, 2003a). Also, it is now appreciated that depression is the predominant mood symptom in BD. For example, Judd and colleagues' seminal prospective studies (Judd *et al.*, 2002, 2003) of the natural course of BD (for average time frames of over a decade) showed that individuals with BD type I experienced symptoms of depression for approximately 31% of weeks, compared with 10% of weeks for hypo/manic symptoms (Judd *et al.*, 2003). Individuals with BD type II experienced depression for a staggering 52% of weeks, in comparison with just over 1% of weeks for hypo/manic symptoms. Similar findings have been reported for the Stanley Foundation Bipolar Network (SFBN) cohort (for further details regarding SFBN methodology, see Leverich *et al.*, 2001; Post *et al.*, 2001), where individuals with BD reported approximately threefold more depressive symptoms compared with hypomanic or manic over a 12-month period (Post *et al.*, 2003b; Altshuler *et al.*, 2006).

Assessing outcome in bipolar disorder: a shifting paradigm

Outcome in BD has traditionally been determined by the assessment of externally assessed clinical information, such as rates of relapse, the number of times a person is hospitalised or degree of symptom reduction on clinician-rated assessment scales. More recently, however, there has been discussion about the need for additional forms of assessment to measure response to treatment in this population. For example, Keck Jr (2004) has suggested that 'functional outcomes are more meaningful measures of response to treatment for bipolar disorder than are scores on various psychiatric rating scales' (Keck Jr, 2004, p. 25).

Psychosocial functioning describes a person's ability to perform the tasks of daily life and to engage in mutual relationships with other people in ways that are gratifying to the individual and others, and that meet the needs of the community in which the person lives. A complementary measurement paradigm that is attracting increasing interest in the BD field concerns the assessment of quality of life (QoL). The World Health Organization has described QoL as 'individuals' perception of their position in life in the context of the culture and value systems in which they live and in relation to their goals, expectations, standards and concerns' (WHOQOL Group, 1995). This broad conceptualisation of QoL has been distinguished by some from the concept of 'health-related quality of life' (HRQOL), which is said to refer specifically to those aspects of an individual's life that are impacted on by their health or ill health. From our point of view, it is difficult to imagine life domains that are unaffected by an individual's state of physical or emotional wellbeing, so we choose to use the term 'QoL' over 'HRQOL'. Furthermore, although the terms 'functioning' and 'QoL' are sometimes used interchangeably in the literature, many researchers agree that functional status and QoL are not identical constructs (Coons *et al.*, 2000). Psychosocial functioning tends to refer to the assessment of a small number of behavioural domains, and functioning is generally considered to be a contributing factor to QoL. Further, the majority of psychosocial functioning scales are objective, rather than subjective

Practical Management of Bipolar Disorder, eds. Allan H. Young, I. Nicol Ferrier and Erin E. Michalak. Published by Cambridge University Press.

measures. In comparison, the assessment of QoL is usually (but not always) conducted subjectively, and QoL scales tend to assess wellbeing in a wider variety of domains than do functioning scales.

Having provided brief definitions of psychosocial functioning and QoL (as we construe the constructs), we will now turn to providing a rationale for incorporating such scales into clinical practice.

A rationale for assessing psychosocial functioning and quality of life in clinical practice

A strong argument can be made for the importance of including functioning and/or QoL measures when assessing outcome in BD clinically. Our rationale is based on several key observations.

1. Individuals with BD may give higher priority to improvements in functioning or QoL than improvements in symptoms.
2. Symptoms and functioning are not as closely related as we might assume.
3. There is evidence that deficits in functioning/QoL can *precede* episode onset and *predict* subsequent episodes.
4. Treatment interventions with equivalent efficacy in symptom reduction can have differential effects in improving functioning/QoL.

With respect to point 1, it seems (from our clinical experience at least) that peoples' goals for treatment tend to be orientated towards markers of functional recovery, or global life quality. For example, when we initiate a course of cognitive behavioural therapy (CBT) with an individual with BD, we will typically begin with a session examining the person's individual goals for the course of treatment. No client to date has responded to this goal-setting exercise by stating that they would like to see their mean depression score drop by 50%! Instead, people tend to say things like: 'I need to stay in work; otherwise this bout of depression will succeed in consuming me'. Or, 'I need to keep my marriage together'. Or, 'I want to be able to have fun and enjoy my life like I used to'. These are the types of goal by which people with mood disorders judge the success of our pharmacological and psychosocial treatment interventions, and we need to routinely incorporate the assessment of such outcomes into clinical practice, in addition to continuing to assess symptomatic outcomes. It is reasonable to expect that treatment alliance will also benefit when clinicians share this more holistic viewpoint.

In our as-yet-unpublished 'Self-management strategies in bipolar disorder' study, we have been using qualitative methods to investigate definitions of wellness in 'high-functioning' individuals with BD (level of functioning is determined on the basis of a comprehensive psychosocial functioning assessment scale, the multidimensional scale for independent functioning (MSIF) (Jaeger *et al.*, 2003)). When asked to define what 'functioning well' means to them, participants in the study have tended to provide complex descriptions of good functioning that go beyond regulation of symptoms. For example, they describe their need to feel self-confident, maintain their ability to have fun, have healthy social relationships, enjoy life, meet their goals and maintain their creativity.

With respect to point 2, it is important to be aware that the relationship between symptoms and functioning/QoL is not always clear-cut; some individuals with BD appear to function well despite relatively severe symptoms, whereas others who could be deemed dysfunctional present with few symptoms. A relatively large body of evidence suggests that marked functional deficits often remain after symptomatic recovery in BD. For example, one study of first-episode patients (Tohen *et al.*, 2000) reported that 98% of the sample achieved syndromal recovery within 2 years, compared with only 38% achieving functional recovery (defined as the proportion of patients who regained occupational and living situations equivalent to those they held before their episode), suggesting that the speed with which a person restores their functioning after an episode is influenced by more than disease state alone. Further, inter-episode functioning also frequently remains compromised in the condition (MacQueen *et al.*, 2001). Available research indicates that 25–35% of individuals with BD continue to experience partial impairment in their work and social functioning; a similar proportion will exhibit extreme functional problems (Dion *et al.*, 1988; Harrow *et al.*, 1990; Tohen *et al.*, 1990; Goldberg *et al.*, 1995).

With respect to point 3, there is now growing evidence that deficits in functioning/QoL can occur before the onset of a full episode. Research in the schizophrenia field, for example, has shown that individuals who are considered to be at risk for a first episode of psychosis (the sample were deemed to be in a 'putative early prodromal state') exhibited poorer QoL than healthy controls, and lower affective QoL than the first-episode patient comparison sample, suggesting that subjective QoL can be compromised before the onset of first positive schizophrenia symptoms (Bechdolf *et al.*, 2005).

From a clinical standpoint, then, there is some early evidence that diminishing functioning or QoL could be a harbinger of hard times to come in some individuals who are at risk for developing a mental illness. A recent study in the BD literature has indicated that impaired psychosocial functioning can predict subsequent depressive, but not manic, symptoms in individuals with BD (Weinstock & Miller, 2008).

With respect to point 4, there is now evidence that pharmacological treatments with equivalent efficacy in relation to symptom relief can have very different impacts on psychosocial functioning or QoL. For example, Shi and colleagues randomised 453 patients with acute mania to olanzapine or haloperidol (Shi *et al.*, 2002). Although remission rates in the two intervention groups were similar at 6 and 12 weeks, olanzapine was associated with greater improvement in health functioning at both time points and greater impact on work functioning at 12 weeks. Omitting the measurement of functioning or QoL may, therefore, result in the omission of important clinical data.

Having provided a rationale for incorporating the assessment of functioning and/or QoL into clinical practice, we will now move on to review some of the literature on psychosocial functioning in BD, focusing upon one area that may be particularly amenable to clinical intervention: the domain of work. If the reader is interested in exploring the scientific literature pertaining to other domains of functioning (for example, social or marital life), a review chapter can be found in Johnson and Leahy's book, *Psychological Treatment of Bipolar Disorder* (Hammen & Cohen, 2004).

Work and bipolar disorder

Although the symptoms of BD have been recognised for centuries, attention has only fairly recently been directed towards improving our understanding of the impact of the condition upon psychosocial functioning. One important but often overlooked (or poorly captured) area of functioning is an individual's ability to work. At a basic level, this can be interpreted simply as a person's ability to obtain and maintain paid employment. At a more complex level, it can refer to an individual's ability to engage in work, paid or unpaid, that they perceive to be meaningful, personally satisfying and in keeping with their educational achievements, expectations, skills or vocational aspirations. Although there is a far larger body of research addressing the relationship between employment and schizophrenia, several quantitative studies have now been conducted in bipolar populations, and have generally indicated that BD can have a profoundly negative impact upon occupational functioning.

In a review published in 2004, Dean and colleagues identified 14 quantitative studies that had assessed work impairment in patients with BD (Dean *et al.*, 2004). The studies were heterogeneous, alternatively assessing degree of work impairment by rates of long-term unemployment, occupational functioning, absenteeism due to emotional or physical problems and reduced work performance. For example, several studies indicated that rates of employment are low in people with BD in comparison to those observed in the general population, and those observed in patients with other affective disorders. In one study that prospectively followed individuals (n=67) for 6 months following hospitalisation for an episode of mania, 43% were employed during this time, with only 21% working at their expected level of employment, although 80% of the sample were considered to be symptom-free or mildly symptomatic (Dion *et al.*, 1988). Another prospective study reported that only 42% of a sample of 73 patients with BD were in continuous employment over a 1.7-year observation period (Harrow *et al.*, 1990). When examining degree of work impairment by type of BD, research has indicated that BD type II is associated with similar levels of impairment to those of BD type I (Ruggero *et al.*, 2007). The same study reported that one-third of people with BD type II had missed over a year of work in the last 5 years due to the illness (although there was more absenteeism from work in individuals with BD type I, this difference appeared to be accounted for by more frequent episodes of hospitalisation).

Several variables have been shown to be predictive of poor work functioning in patients with BD, including demographic or clinical characteristics and lack of social support. For example, Dickerson and colleagues examined variables associated with employment status in individuals (n=117) diagnosed with BD types I and II, including demographic variables, cognitive functioning, symptom severity and course of illness (Dickerson *et al.*, 2004). Multivariate analysis indicated that current employment status was significantly associated with level of cognitive functioning, severity of symptoms, history of psychiatric hospitalisation and level of maternal education. In other research, substance misuse and personality trait scores have been shown to be associated with degree of impairment in work functioning (Loftus & Jaeger, 2006), although not at a multivariate level. Another interesting study of individuals (n=52) with

BD type I found that the presence of a strong, supportive relationship was actually a stronger predictor of work functioning than clinical status (recent or current symptoms or number of previous hospitalisations) (Hammen *et al.*, 2000). Similar results have been reported in other studies; level of current psychiatric symptoms does not reliably predict occupational functioning (Dion *et al.*, 1988; Kusznir *et al.*, 1996). Instead, for some people, there appears to be a marked time lag between recovery from a mood episode and return to the workforce, if indeed that return occurs at all.

In order to better understand the effects of BD upon employment, it is important to take into consideration the ways in which these effects are experienced, and a small number of qualitative studies have now addressed this issue. In a notable study, Tse and Yeats assessed the factors related to successful employment in a relatively large sample (*n*=67) of people with BD in New Zealand (Tse & Yeats, 2002). Two main factors were found to determine readiness to re/join the workforce: (1) recovery from their acute episode of BD; and (2) goodness of fit between the individual, their job, the support available to them and wider contextual components. Specifically, having a sense of determination, good professional qualifications, a good work record, faith in God and good illness management skills were important individual factors in determining whether vocational integration was successful or not. Work factors related to the meaning and value the person derived from their occupation, and the nature and structure of the job (for example, whether there was a balance between routine and flexibility). Perceived support consisted of either support within the workplace (from managers, colleagues or in terms of entitlement to leave) or wider social support (friends and family, and professional or community support groups). Finally, wider contextual components included factors such as lack of perceived stigma towards people with psychiatric illness and appropriate governmental policies and legislation. Importantly, Tse and Yeats note that 'being employed should not be viewed as the end of the rehabilitation process in itself. Achieving an employment status can potentially act as a catalyst to prompt the person concerned to further advance his/her career pursuits and recovery from BD' (Tse & Yeats, 2002).

In our own more recent qualitative study, we reported on the relationship between BD and work in participants in a variety of employment situations, ranging between those with no employment history through to those in highly skilled professional positions (Michalak *et al.*, 2007b). Five main themes emerged from the data: lack of continuity in work history, loss, illness management strategies in the workplace, stigma and disclosure in the workplace, and interpersonal problems at work. Potential barriers for people with BD in the workplace and possible solutions for clinicians are provided in Table 16.1.

Having provided an overview of some of the research that has examined psychosocial functioning in BD, and having taken an in-depth look at one aspect of functioning (the work domain), we will now turn to examine research into QoL in BD.

QoL in bipolar disorder: an overview

The following section is intended to provide a succinct overview of some of the existing research into QoL in BD. For more a more theoretical analysis of the body of literature around QoL in BD, see (Michalak *et al.*, 2007a). Issues pertaining specifically to the relationship between bipolar depression and QoL are reviewed comprehensively in Michalak *et al.* (2008).

Five systematic reviews of research into QoL in BD have been published to date – three aimed to be comprehensive (Namjoshi & Buesching, 2001; Dean *et al.*, 2004; Michalak *et al.*, 2005a;) and two focused specifically upon clinical trial data (Michalak & Murray, 2008; Revicki *et al.*, 2005). In the first of the reviews, Namjoshi and colleagues identified ten studies assessing QoL published before 1999 (Namjoshi & Buesching, 2001). Building on this work, Dean and colleagues (Dean *et al.*, 2004) described studies published before November 2002 that had assessed QoL, work impairment or health-care costs and utilisation in patients with BD. This review applied a broad definition of QoL, including studies utilising scales that assessed single domains of functioning, such as the global assessment of functioning (GAF) scale (Endicott *et al.*, 1976). Using these criteria, the authors identified 65 studies, allowing them to conclude that: (1) the QoL of patients with BD is similar to that of patients with unipolar depression and equal to or lower than that observed in patients with chronic medical conditions; and (2) treatment interventions for BD have been shown to have a beneficial impact on life quality. In the most recent of the (comprehensive) reviews, we identified 28 studies when using relatively tight

Table 16.1 Improving functioning in the workplace: tips for the clinician

Issue	Possible role for clinician
Individual believes s/he has little perceived control over moods in the workplace	Assist the individual in developing a strategy for recognising and acting upon prodromes in the workplace
The individual is ambivalent about/nonadherant with medications for BD due to a sense of loss for periods of increased productivity at work during highs	Conduct cost–benefit analysis with individual; for example, encourage them to consider: (1) Is the increased productivity worthwhile if the episode escalates into a full mania? (2) Are there potential physiological/psychosocial consequences of experiencing repeated mood episodes? (3) Is the increased productivity always associated with good quality, appropriate work? (4) Is the individual's creativity/productivity definitely hampered by mood-stabilising medications?
BD may impact upon individual's ability to engage in what they view as meaningful work	Work with individual to take a creative approach towards identifying rewarding and meaningful forms of employment (for example, volunteer work, self-employment, competitive employment, sheltered work, transitional work or casual work (Tryssenaar, 1998))
Individual feels hopeless about their career future	Encourage client to remember their long-term work goals; although initial work placements may appear demeaning or unrewarding, these may be impermanent phases that will be replaced by more satisfying and positive employment opportunities as their stamina, strengths and capabilities improve over time
Individual is overwhelmed by the negative impacts of their BD upon their employment	Ask individual to consider whether there are any positive impacts of BD upon their work life; for example, Has having BD opened up new avenues of employment? Does having the condition give them greater empathy for colleagues struggling with mental health problems? Do they consider themselves more resilient than they were before their diagnosis?
Individual is struggling with symptoms of BD in the workplace	Explore possible management strategies to impede cascade into increasing severity of episode; for example, Is it likely to be beneficial if the individual reduces their workload, changes their work activities, enlists the help of trusted co-workers or seeks further support from their health-care/social support system?
Individual is concerned about stigma in the workplace	Explore potential coping strategies; for example, waiting for a period of time before disclosing diagnosis in a new job, garnering support from a trusted co-worker

inclusion criteria (Michalak *et al.* 2005a). The studies were heterogeneous; several undertook to assess QoL during different phases of the disorder, for example, cross-sectional research that compared perceived QoL in euthymic, manic or depressed patients with BD. Other studies compared QoL in BD samples and other populations (both psychiatric and medical groups). Finally, we identified a small number of studies that had used a QoL instrument to assess outcome in trials of treatment inventions for the condition. At the time of the review, the studies were also of variable scientific quality. Methodological shortcomings included small sample sizes, cross-sectional designs, unusual diagnostic methods or poorly differentiated diagnostic groups, use of poorly validated QoL instruments and lack of control for clinical variables.

Two reviews have assessed data from clinical trials that have incorporated QoL assessment scales. In the first, Revicki and colleagues focused on pre-2003 trials, finding just three studies that had included QoL outcome assessments (Revicki *et al.*, 2005). More recently, we conducted a review of studies published up to June 2006, identifying ten clinical trials, eight studies of pharmacological treatment interventions and two of psychosocial interventions (Michalak & Murray, 2008). On the basis of this review, we concluded that:

1. QoL outcome measures appear to provide additional important information over that provided by symptomatic measures
2. there is some evidence that QoL in BD improves relatively slowly after treatment
3. emerging research into adjunctive psychosocial treatments suggests that even relatively brief interventions may have effects on QoL
4. existing literature on QoL in BD is relatively immature. In particular, there is no current consensus about which QoL instruments are most appropriate for use as outcome measures (at the time of the review there was no disorder-specific measure to assess QoL in BD).

The following section will briefly review the literature on QoL in BD that may be of particular interest to the clinician, addressing the following four specific questions.

1. How impaired is QoL in BD?
2. Does QoL vary across mood state in BD?
3. Can QoL be impacted by treatment interventions for BD?
4. What do qualitative data tell us about QoL in BD?

How impaired is QoL in bipolar disorder?

Several studies have attempted to ascertain the degree of impairment in QoL experienced by individuals with BD.

Unsurprisingly, QoL in bipolar populations appears to fall far below that observed in general population samples, at least in the realms of emotional or psychosocial wellbeing. For example, one study using the Medical Outcomes Study SF-36 (Ware *et al.*, 1993), the most widely used QoL measure in this population to date, compared scores between individuals with BD (n=44) with previously reported norms for a general population sample (Arnold *et al.*, 2000). The SF-36 contains eight subscales that assess physical functioning, social functioning, role limitations (physical), role limitations (emotional), pain, mental health, general health and vitality. The results of the study indicated that QoL was significantly compromised in people with BD in all SF-36 domains except physical functioning.

In other research, Yatham and colleagues compared QoL in individuals with BD type I (n=920) who were either currently depressed, or had experienced a recent episode of depression with general population norms (Yatham *et al.*, 2004). Scores were significantly lower across all scale domains in the bipolar group compared with norms reported for the general US population, with markedly lower scores in the mental health, vitality, social functioning and role emotional domains. Yatham and colleagues went on to compare these scores with those derived from seven large studies of QoL in patients with unipolar depression (Yatham *et al.*, 2004). Scores on four domains (general health, social functioning, role-physical and role-emotional) were lower than those observed in unipolar depression. By contrast, the unipolar samples tended to exhibit higher scores in the bodily pain domain.

Several studies have compared QoL in BD with QoL in other psychiatric populations. For example, the NEMESIS study compared SF-36 scores in 136 adults with BD with that observed in a variety of other psychiatric disorders (ten Have *et al.*, 2002). Participants with BD showed significantly more impairment in most SF-36 domains compared with other participants. For example, in the domain of mental health, participants with BD type I experienced significantly lower scores than people with other mood, anxiety or substance-use disorders. Other research (Atkinson *et al.*, 1997; Chand *et al.*, 2004; Goldberg & Harrow, 2005) has compared

QoL in patients with schizophrenia with that observed in patients with BD, but has to date generated mixed results (see Michalak *et al.*, 2007a for a detailed review of this body of research).

Clinical take home message

Studies using a range of measures have generally confirmed that QoL is, in a range of domains and to a marked extent, lower among patients with BD than in the normal population. Beyond this commonsense finding, there is some evidence that QoL is poorer in BD than in other mood and anxiety disorders.

Does QoL vary across mood state in BD?

The course of BD can include a range of abnormal mood states, including the co-occurrence of extremes of mania and depression. It is reasonable to expect that QoL would be negatively affected by the depressive episodes of BD, and as noted above, the relatively large study of Yatham and colleagues did find remarkably low QoL among individuals with BD who were either depressed or had experienced a recent episode of depression (Yatham *et al.*, 2004). Other recent research has indicated that QoL may be more impaired in individuals with BD type II than in those with BD type I, further underscoring the deleterious effect of depression in BD upon life quality (Maina *et al.*, 2007).

The complex issue of QoL across clinical states in BD has recently been addressed more comprehensively in a cross-sectional study of the first 2000 participants enrolled in the Systematic Treatment Enhancement Program for Bipolar Disorder (STEP-BD) (Zhang *et al.*, 2006). This is a large multicentre prospective, naturalistic study that features several embedded randomised-controlled trials measuring QoL via the quality of life enjoyment and satisfaction questionnaire (Q-LES-Q) (Endicott *et al.*, 1993) and the SF-36 (Ware *et al.*, 1993). In addition, the study assessed a comprehensive battery of possible confounders, including demographic and socioeconomic factors, family history, psychiatric comorbidity, clinical characteristics, personality, social support, negative life events and attributional style. The primary finding from the analysis was that depressive symptoms were strongly associated with poorer emotional QoL, even after relevant confounding variables were controlled for. Conversely, apparent 'supranormal' QoL reported in patients with hypomania or mania (compared with that of euthymic patients) disappeared after statistical control. Related to this are results from an earlier study that found negative QoL impacts of 'elevated' mood states. Vojta and colleagues administered a short form of the SF-36 to individuals with BD who were in a manic/hypomanic episode (n=16), depressive episode (N=26), mixed episode (N=14) or who were euthymic (n=30), finding that individuals with mania/hypomania showed significantly lower QoL scores than those who were euthymic, and depressed or mixed episodes were associated with significantly poorer QoL again (Vojta *et al.*, 2001).

Clinical take home message

Although manic or hypomanic symptoms distinguish BD from unipolar depression in the current classification, much of the morbidity and mortality in BD appears to be a consequence of the depressive phase of the disorder. Quality of life in individuals with BD is most impaired during episodes of depression, but also appears to be lowered in manic, and possibly even hypomanic, states.

Can quality of life be impacted by treatment interventions for bipolar disorder?

In a review of the literature (Michalak *et al.*, 2005a), we identified ten treatment outcome studies that had incorporated a QoL outcome measure: eight clinical trials that examined pharmacological interventions for BD; and two studies that assessed non-pharmacological interventions. The following section will review key findings from these studies.

Namjoshi and colleagues have conducted a series of studies examining the efficacy of olanzapine as a treatment for BD. In the first of these, they evaluated the impact of acute (3-week) treatment with olanzapine or placebo and long-term (49-week open-label) treatment of BD type I (manic/mixed) (Namjoshi *et al.*, 2002). During the acute-phase treatment period, treatment-related improvements in QoL were apparent only for the physical functioning domain of the SF-36. Improvement in other aspects of QoL (specifically, pain, vitality, general health and social functioning) occurred during the open-label treatment period, indicating that olanzapine may have a relatively rapid effect in terms of improving physical functioning in patients with acute mania, but other QoL domains may be slower to respond to treatment. Shi and colleagues compared the treatment effects of olanzapine and haloperidol in individuals with acute mania (n=453) (Shi *et al.*, 2002; Tohen *et al.*, 2003). During acute treatment, significantly greater improvement in five of the SF-36 domains was apparent in the olanzapine group. Superiority of olanzapine over haloperidol persisted over the study's 6-week continuation phase, during which time improvements in work and household functioning also became apparent. Other research has examined the effects of adding olanzapine to lithium or valproate in patients with BD (*n*=224) (Namjoshi *et al.*, 2004). Combination therapy was associated with better outcome in several QoL domains compared with lithium or valproate monotherapy. The SF-36 and a broader QoL scale were used in a study comparing the benefits of olanzapine alone versus an olanzapine–fluoxetine combination or placebo (Shi *et al.*, 2004). Compared with placebo, patients who received olanzapine showed greater improvement at 8 weeks in SF-36 mental health summary scores, and three domain scores. The combination group fared significantly better in terms of QoL improvement than did the olanzapine-alone group.

The Q-LES-Q has been administered at baseline (hospital discharge) and at 6 and 12 weeks in a comparison of divalproex sodium and olanzapine in the treatment of acute mania (Revicki *et al.*, 2003). No significant treatment effects were detected in Q-LES-Q scores in the study, although only 52 (43%) of the 120 patients randomised to either divalproex or olanzapine completed the QoL instrument. The authors reported an association between weight gain and poorer change scores in the physical, leisure and general activities domains of the Q-LES-Q at 6 weeks (but not at 12 weeks). Negative correlations were reported between increased weight (at 6 weeks) and overall life satisfaction, physical health, mood, general activities and satisfaction with medication on the Q-LES-Q. More recent research has also reported an association between increased weight and lower QoL (as measured by the SF-36 and a weight-related QoL scale) in patients with BD (Kolotkin *et al.*, 2008).

Finally, QoL data have now been published from the BOLDER study (Calabrese *et al.*, 2005; Endicott *et al.*, 2007). This was a large, 8-week, multicentre, double-blind, randomised, fixed-dose, placebo-controlled monotherapy study of quetiapine (600 or 300 mg/d) versus placebo in outpatients with DSM-IV-TR bipolar I or bipolar II disorder, with or without rapid cycling, in a major depressive episode. The study administered the 16-item short form of the Q-LES-Q (Q-LES-Q SF) at baseline and weeks 4 and 8 to assess QoL, finding 12- and 11-point increases in Q-LES-Q SF scores at last assessment in the high- and low-dose groups, respectively, compared with a 7-point change in the placebo group (i.e. significantly greater improvement in QoL in both groups after 8 weeks of treatment compared with placebo). The 12-point change in Q-LES-Q SF score observed in the BOLDER study is in keeping with outcome data from other depression studies (Kocsis *et al.*, 1997; Miller *et al.*, 1998), indicating that QoL in patients with BD is amenable to relatively rapid change, even when study inclusion criteria are broadened to include greater diagnostic heterogeneity, such as rapid cycling and BD type II patients. In a more recent publication (Endicott *et al.*, 2007), the minimal clinically important difference (MCID) for the Q-LES-Q SF was calculated; for a moderate effect size (0.5), a minimum score of 7 points on the Q-LES-Q would be required. For a large effect size (0.8), a score of 10 would be necessary. Both quetiapine groups showed effect sizes in the moderate range (0.4–0.5), indicating that the intervention had a clinically meaningful impact upon QoL, as well as a statistically significant impact. The MCID, determined by a 1-point decrease on the clinical global impression–severity scale (CGI-S) (Guy, 1976), was 15 points on the Q-LES-Q, or a decrease of 12 points to be assessed as 'minimally improved' on the same scale.

Although pharmacology forms the bedrock of treatment for BD, there is a clear need for other treatment modalities that augment the effects of medication in this complex condition. Over the last two decades, we have seen an upsurge of interest in examining the role of psychological interventions as an adjunct to the pharmacological treatment of BD, with most of this research examining the efficacy of psychotherapy as a treatment intervention. Several multi-modal psychotherapeutic interventions for BD have been developed, such as family-focused therapy (FFT) (Miklowitz *et al.*, 1988; Miklowitz & Goldstein, 1997), interpersonal and social rhythm therapy (Frank *et al.*, 2000) and CBT (Otto *et al.*, 2003). Surprisingly few (Patelis-Siotis *et al.*, 2001; Dogan & Sabanciogullari, 2003; Michalak *et al.*, 2005b) studies of psychosocial treatment interventions for BD, however, have used QoL measures to assess outcome. This might be an important oversight, because it could be in functional/QoL outcomes that psychosocial interventions make their strongest contribution.

Clinical take home message

Research using QoL as an outcome measure has indicated that QoL measures provide additional important information over that provided by symptom measures. Although based on a small number of studies, there is some evidence that QoL improves relatively slowly after treatment (perhaps paralleling functional recovery).

Table 16.2 Provisional recommendations for monitoring psychosocial functioning/QoL in BD in clinical practice

- Use assessment scales that have demonstrated validity in this population.
- In the absence of a disorder-specific scale for QoL in BD, it is appropriate to use QoL scales developed for individuals with unipolar depression or other psychiatric conditions. The Q-LES-Q is the most extensively used scale of this type; it can be recommended for use in patients with BD; it is available for clinical use in long and short forms (see Chapter X for reprint of the short form) and shows sound psychometric properties, including responsiveness.
- When assessing functioning/QoL, also assess symptoms of depression and hypomania or mania using valid, clinically appropriate assessment scales.
- While you might expect an individual's symptoms of depression or hypomania to respond to treatment over a relatively short period of time, changes in some functioning/QoL domains (e.g. social or occupational wellbeing) take longer to occur. Therefore assess functioning/QoL over longer periods of time than you would traditionally assess symptomatic outcome.
- Best practice suggests annual monitoring of functioning/QoL in individuals whose condition is stable to observe the interaction between developmental stages and the illness itself.

What do qualitative data tell us about QoL in BD?

As part of development of a disorder-specific scale to assess QoL in BD, we conducted a series of qualitative interviews to identify the ways in which BD impacts upon QoL (Michalak *et al.*, 2006). We sought the views of a representative sample of people diagnosed with both BD type I and BD type II. Clinical characteristics of the sample ranged widely, from individuals who had been clinically stable for several years through to inpatients who were recovering from a severe episode of depression or mania. The results of the study provided some interesting initial data concerning the impact of BD upon QoL. The majority of participants described how the condition had had a profoundly negative effect upon their life quality, often having serious and enduring effects on their ability to find meaningful work, become educated, maintain their independence, have healthy social and intimate relationships and have a coherent sense of self. Having said this, we also interviewed a number of people who were functioning exceptionally well despite their diagnosis; a minority of people even espoused the view that their condition had opened up new doors of opportunity for them – for example, in terms of positively changing their career paths or social networks. On the whole, however, even these individuals described having undergone several years of hardship and adjustment before getting 'back on track'.

Respondents described a wide variety of factors that influenced their QoL, including, but not limited to, side effects of medications, occupation, education, physical functioning, environment, health-care factors, leisure activities, routine and sexuality. Some of the factors raised (for example, independence, identity, stigma and disclosure, and spirituality) are not frequently examined in relation to QoL, yet they appear to have a significant impact upon peoples' ability to live their lives to the full in the context of BD. We are continuing to develop the QoL.BD in close consultation with individuals with BD in the hope of maximising the validity of the resulting scale.

Clinical take home message

Qualitative research has indicated that QoL in BD is impacted by a wide range of influences over and above the degree of symptoms experienced. It is important to invite patients to consider these factors when developing treatment plans or assessing the impact of treatment interventions.

Incorporating psychosocial functioning and QoL scales into clinical practice

Chapter X of this book makes specific recommendations for psychosocial functioning and QoL scales that are appropriate for use in routine clinical practice – in particular, recommending the use of the Q-LES-Q SF (note that the short form of the Q-LES-Q has the same content as the General Activities section of the regular Q-LES-Q) to assess QoL and the MIRECC version of the GAF (Niv *et al.*, 2007) or the functioning assessment short test (Rosa *et al.*, 2007) to assess functioning in individuals with BD. There are currently no consensus guidelines for the clinical measurement of outcomes for psychosocial functioning or QoL in BD. Drawing on our own clinical experience, research and related literature, however, we are able to make provisional recommendations for monitoring psychosocial functioning or QoL in clinical practice (Table 16.2).

Concluding remarks

There has been a recent upsurge of interest in defining and measuring psychosocial functioning and QoL in BD. Although research is at an early stage, there is no doubt that symptom measures alone constitute a limited assessment of BD outcomes, and more valid understandings (scientifically and clinically) are achieved with the addition of QoL measures. Existing research has revealed, for example, the marked negative impact of BD on QoL, a disjunction between symptom level and functional outcome in BD, and the apparent primacy of depressive symptoms over hypo/mania in BD QoL outcomes. The emerging body of research clearly suggests that it is both feasible and important to assess functioning and QoL in patients with this complex psychiatric condition. For the practising clinician, routinely adding a QoL measure to outcome monitoring will enrich understanding of patient progress, with consequent benefits for tailoring treatment regimens and for the therapeutic alliance.

References

Altshuler, L.L., Post, R.M., Black, D.O., *et al.* (2006). Subsyndromal depressive symptoms are associated with functional impairment in patients with bipolar disorder: results of a large, multisite study. *Journal of Clinical Psychiatry*, **67**, 1551–60.

Arnold, L.M., Witzeman, K.A., Swank, M.L., McElroy, S.L., Keck Jr, P.E. (2000). Health-related quality of life using the SF-36 in patients with bipolar disorder compared with patients with chronic back pain and the general population. *Journal of Affective Disorders*, **57**, 235–9.

Atkinson, M., Zibin, S., Chuang, H. (1997). Characterizing quality of life among patients with chronic mental illness: a critical examination of the self-report methodology. *American Journal of Psychiatry*, **154**, 99–105.

Bechdolf, A., Pukrop, R., Kohn, D., *et al.* (2005). Subjective quality of life in subjects at risk for a first episode of psychosis: a comparison with first episode schizophrenia patients and healthy controls. *Schizophrenia Research*, **79**, 137–43.

Calabrese, J.R., Keck Jr, P.E., Macfadden, W., *et al.* (2005). A randomized, double-blind, placebo-controlled trial of quetiapine in the treatment of bipolar I or II depression. *American Journal of Psychiatry*, **162**, 1351–60.

Chand, P.K., Mattoo, S.K., Sharan, P. (2004). Quality of life and its correlates in patients with bipolar disorder stabilized on lithium prophylaxis. *Psychiatry and Clinical Neurosciences*, **58**, 311–18.

Coons, S.J., Rao, S., Keininger, D.L., Hays, R.D. (2000). A comparative review of generic quality-of-life instruments. *Pharmacoeconomics*, **17**, 13–35.

Dean, B.B., Gerner, D., Gerner, R.H. (2004). A systematic review evaluating health-related quality of life, work impairment, and healthcare costs and utilization in bipolar disorder. *Current Medical Research and Opinion*, **20**, 139–54.

Dickerson, F.B., Boronow, J.J., Stallings, C.R., Origoni, A.E., Cole, S., Yolken, R.H. (2004). Association between cognitive functioning and employment status of persons with bipolar disorder. *Psychiatric Services (Washington, D.C.)*, **55**, 54–8.

Dion, G.L., Tohen, M., Anthony, W.A., Waternaux, C.S. (1988). Symptoms and functioning of patients with bipolar disorder six months after hospitalization. *Hospital & Community Psychiatry*, **39**, 652–7.

Dogan, S. & Sabanciogullari, S. (2003). The effects of patient education in lithium therapy on quality of life and compliance. *Archives of Psychiatric Nursing*, **17**, 270–5.

Endicott, J., Nee, J., Harrison, W., Blumenthal, R. (1993). Quality of Life Enjoyment and Satisfaction Questionnaire: a new measure. *Psychopharmacology Bulletin*, **29**, 321–6.

Endicott, J., Rajagopalan, K., Minkwitz, M., Macfadden, W. (2007). A randomized, double-blind, placebo-controlled study of quetiapine in the treatment of bipolar I and II depression: improvements in quality of life. *International Clinical Psychopharmacology*, **22**, 29–37.

Endicott, J., Spitzer, R.L., Fleiss, J.L., Cohen, J. (1976). The global assessment scale. A procedure for measuring overall severity of psychiatric disturbance. *Archives of General Psychiatry*, **33**, 766–71.

Frank, E., Swartz, H.A., Kupfer, D.J. (2000). Interpersonal and social rhythm therapy: managing the chaos of bipolar disorder [In Process Citation]. *Biological Psychiatry*, **48**, 593–604.

Goldberg, J.F. & Harrow, M. (2005). Subjective life satisfaction and objective functional outcome in bipolar and unipolar mood disorders: a longitudinal analysis. *Journal of Affective Disorders*, **89**, 79–89.

Goldberg, J.F., Harrow, M., Grossman, L.S. (1995). Course and outcome in bipolar affective disorder: a longitudinal follow-up study. *American Journal of Psychiatry*, **152**, 379–84.

Guy, W. (1976). *ECDEU Assessment Manual for Psychopharmacology: Publication ADM 76-338.* Rockville, MD: US Department of Health, Education and Welfare.

Hammen, C. & Cohen, A.N. (2004). Psychosocial functioning. In: Johnson, S.L. & Leahy, R.L. (Eds).

Psychological Treatment of Bipolar Disorder. New York: The Guilford Press, pp. 17–34.

Hammen, C., Gitlin, M., Altshuler, L. (2000). Predictors of work adjustment in bipolar I patients: a naturalistic longitudinal follow-up. *Journal of Consulting and Clinical Psychology*, **68**, 220–5.

Harrow, M., Goldberg, J.F., Grossman, L.S., Meltzer, H.Y. (1990). Outcome in manic disorders. A naturalistic follow-up study. *Archives of General Psychiatry*, **47**, 665–71.

Jaeger, J., Berns, S.M., Czobor, P. (2003). The multidimensional scale of independent functioning: a new instrument for measuring functional disability in psychiatric populations. *Schizophrenia Bulletin*, **29**, 153–68.

Judd, L.L., Akiskal, H.S., Schettler, P.J., *et al.* (2002). The long-term natural history of the weekly symptomatic status of bipolar I disorder. *Archives of General Psychiatry*, **59**, 530–7.

Judd, L.L., Akiskal, H.S., Schettler, P.J., *et al.* (2003). A prospective investigation of the natural history of the long-term weekly symptomatic status of bipolar II disorder. *Archives of General Psychiatry*, **60**, 261–9.

Keck Jr, P.E. (2004). Defining and improving response to treatment in patients with bipolar disorder. *Journal of Clinical Psychiatry*, **65** (Suppl. 15), 25–9.

Kocsis, J.H., Zisook, S., Davidson, J., *et al.* (1997). Double-blind comparison of sertraline, imipramine, and placebo in the treatment of dysthymia: psychosocial outcomes. *American Journal of Psychiatry*, **154**, 390–5.

Kolotkin, R.L., Corey-Lisle, P.K., Crosby, R.D., *et al.* (2008). Impact of obesity on health-related quality of life in schizophrenia and bipolar disorder. *Obesity* (Silver Spring, Md.), **16**, 749–54.

Kusznir, A., Scott, E., Cooke, R.G., Young, L.T. (1996). Functional consequences of bipolar affective disorder: an occupational therapy perspective. *Canadian Journal of Occupational Therapy*, **63**, 313–22.

Leverich, G.S., Nolen, W.A., Rush, A.J., *et al.* (2001). The Stanley Foundation Bipolar Treatment Outcome Network. I. Longitudinal methodology. *Journal of Affective Disorders*, **67**, 33–44.

Loftus, S.T. & Jaeger, J. (2006). Psychosocial outcome in bipolar I patients with a personality disorder. *Journal of Nervous and Mental Disease*, **194**, 967–70.

MacQueen, G.M., Young, L.T., Joffe, R.T. (2001). A review of psychosocial outcome in patients with bipolar disorder. *Acta Psychiatrica Scandinavica*, **103**, 163–70.

Maina, G., Albert, U., Bellodi, L., *et al.* (2007). Health-related quality of life in euthymic bipolar disorder patients: differences between bipolar I and II subtypes. *Journal of Clinical Psychiatry*, **68**, 207–12.

Michalak, E.E. & Murray, G.W. (2008). Using quality of life as an outcome measure in patients with bipolar disorder. *Aspects of Affect*, 2.

Michalak, E.E., Murray, G., Young, A., Lam, R.W. (2007a). Quality of life impairment in bipolar disorder. In: Ritsner M. & Awad A.G. (Eds). *Quality of Life Impairment in Schizophrenia, Mood and Anxiety Disorders. New Perspectives on Research and Treatment*. Dordrechht, The Netherlands: Springer, pp. 253–74.

Michalak, E.E., Murray, G., Young, A.H., Lam, R.W. (2008). Burden of bipolar depression: impact of disorder and medications on quality of life. *CNS Drugs*, **22**, 389–406.

Michalak, E.E., Yatham, L.N., Kolesar, S., Lam, R.W. (2006). Bipolar disorder and quality of life: a patient-centered perspective. *Quality of Life Research*, **15**, 25–37.

Michalak, E.E., Yatham, L.N., Lam, R.W. (2005a). Quality of life in bipolar disorder: a review of the literature. *Health and Quality of Life Outcomes*, **3**, 72.

Michalak, E.E., Yatham, L.N., Maxwell, V., Hale, S., Lam, R.W. (2007b). The impact of bipolar disorder upon work functioning: a qualitative analysis. *BipolarDisorders*, **9**, 126–43.

Michalak, E.E., Yatham, L.N., Wan, D.D., Lam, R.W. (2005b). Perceived quality of life in patients with bipolar disorder. Does group psychoeducation have an impact? *Canadian Journal of Psychiatry*, **50**, 95–100.

Miklowitz, D.J. & Goldstein, M.J. (1997). *Bipolar Disorder: A Family-focused Treatment Approach*, New York: Guilford Press.

Miklowitz, D.J., Goldstein, M.J., Nuechterlein, K.H., Snyder, K.S., Mintz, J. (1988). Family factors and the course of bipolar affective disorder. *Archives of General Psychiatry*, **45**, 225–31.

Miller, I.W., Keitner, G.I., Schatzberg, A.F., *et al.* (1998). The treatment of chronic depression, part 3: psychosocial functioning before and after treatment with sertraline or imipramine. *Journal of Clinical Psychiatry*, **59**, 608–19.

Namjoshi, M.A. & Buesching, D.P. (2001). A review of the health-related quality of life literature in bipolar disorder. *Quality of Life Research*, **10**, 105–15.

Namjoshi, M.A., Rajamannar, G., Jacobs, T., *et al.* (2002). Economic, clinical, and quality-of-life outcomes associated with olanzapine treatment in mania. Results from a randomized controlled. *Journal of Affective Disorders*, **69**, 109–18.

Namjoshi, M.A., Risser, R., Shi, L., Tohen, M., Breier, A. (2004). Quality of life assessment in patients with bipolar disorder treated with olanzapine added to lithium or valproic acid. *Journal of Affective Disorders*, **81**, 223–9.

Niv, N., Cohen, A.N., Sullivan, G., Young, A.S. (2007). The MIRECC version of the Global Assessment of Functioning scale: reliability and validity. *Psychiatric services (Washington, D.C.)*, **58**, 529–35.

Otto, M.W., Reilly-Harrington, N., Sachs, G.S. (2003). Psychoeducational and cognitive-behavioral strategies in the management of bipolar disorder. *Journal of Affective Disorders*, **73**, 171–81.

Patelis-Siotis, I., Young, L.T., Robb, J.C., *et al.* (2001). Group cognitive behavioral therapy for bipolar disorder: a feasibility and effectiveness study. *Journal of Affective Disorders*, **65**, 145–53.

Post, R.M., Denicoff, K.D., Leverich, G.S., *et al.* (2003a). Morbidity in 258 bipolar outpatients followed for 1 year with daily prospective ratings on the NIMH life chart method. *Journal of Clinical Psychiatry*, **64**, 680–90.

Post, R.M., Leverich, G.S., Altshuler, L.L., *et al.* (2003b). An overview of recent findings of the Stanley Foundation Bipolar Network (Part I). *Bipolar Disorders*, **5**, 310–19.

Post, R.M., Nolen, W.A., Kupka, R.W., *et al.* (2001). The Stanley Foundation Bipolar Network. I. Rationale and methods. *British Journal of Psychiatry, Supplement*, **41**, S169–76.

Revicki, D.A., Matza, L.S., Flood, E., Lloyd, A. (2005). Bipolar disorder and health-related quality of life: review of burden of disease and clinical trials. *Pharmacoeconomics*, **23**, 583–94.

Revicki, D.A., Paramore, L.C., Sommerville, K.W., Swann, A.C., Zajecka, J.M. (2003). Divalproex sodium versus olanzapine in the treatment of acute mania in bipolar disorder: health-related quality of life and medical cost outcomes. *Journal of Clinical Psychiatry*, **64**, 288–94.

Rosa, A.R., Sanchez-Moreno, J., Martinez-Aran, A., *et al.* (2007). Validity and reliability of the Functioning Assessment Short Test (FAST) in bipolar disorder. *Clinical Practice and Epidemiology in Mental Health*, **3**, 5.

Ruggero, C.J., Chelminski, I., Young, D., Zimmerman, M. (2007). Psychosocial impairment associated with bipolar II disorder. *Journal of Affective Disorders*, **104**, 53–60.

Shi, L., Namjoshi, M.A., Swindle, R., *et al.* (2004). Effects of olanzapine alone and olanzapine/fluoxetine combination on health-related quality of life in patients with bipolar depression: secondary analyses of a double-blind, placebo-controlled, randomized clinical trial. *Clinical Therapeutics*, **26**, 125–34.

Shi, L., Namjoshi, M.A., Zhang, F., *et al.* (2002). Olanzapine versus haloperidol in the treatment of acute mania: clinical outcomes, health-related quality of life and work status. *International Clinical Psychopharmacology*, **17**, 227–37.

ten Have, M., Vollebergh, W., Bijl, R., Nolen, W.A. (2002). Bipolar disorder in the general population in The Netherlands (prevalence, consequences and care utilisation): results from The Netherlands Mental Health Survey and Incidence Study (NEMESIS). *Journal of Affective Disorders*, **68**, 203–13.

The WHOQOL Group (1995). The World Health Organization Quality of Life assessment (WHOQOL): position paper from the World Health Organization. *Social Science & Medicine*, **41**, 1403–9.

Tohen, M., Goldberg, J.F., Gonzalez-Pinto Arrillaga, A.M., *et al.* (2003). A 12-week, double-blind comparison of olanzapine vs haloperidol in the treatment of acute mania. *Archives of General Psychiatry*, **60**, 1218–26.

Tohen, M., Hennen, J., Zarate Jr, C.M., *et al.* (2000). Two-year syndromal and functional recovery in 219 cases of first-episode major affective disorder with psychotic features. *American Journal of Psychiatry*, **157**, 220–8.

Tohen, M., Waternaux, C.M., Tsuang, M.T. (1990). Outcome in mania. A 4-year prospective follow-up of 75 patients utilizing survival analysis. *Archives of General Psychiatry*, **47**, 1106–11.

Tryssenaar, J. (1998). Vocational exploration and employment and psychosocial disabilities. In: *Psychosocial Occupational Therapy: A Holistic Approach*. San Diego, CA: Singular Publishing Group, pp. 351–73.

Tse, S. & Yeats, M. (2002). What helps people with bipolar affective disorder succeed in employment: a grounded theory approach. *Work*, **19**, 47–62.

Vojta, C., Kinosian, B., Glick, H., Altshuler, L., Bauer, M.S. (2001). Self-reported quality of life across mood states in bipolar disorder. *Comprehensive Psychiatry*, **42**, 190–5.

Ware, J. E., Snow, K.K., Kosinski, M., Gandek, B. (1993). *SF-36 Health Survey: Manual and Interpretation Guide*. Boston, MA: The Health Institute.

Weinstock, L.M. & Miller, I.W. (2008). Functional impairment as a predictor of short-term symptom course in bipolar I disorder. *Bipolar Disorders*, **10**, 437–42.

Yatham, L.N., Lecrubier, Y., Fieve, R.R., Davis, K.H., Harris, S.D., Krishnan, A.A. (2004). Quality of life in patients with bipolar I depression: data from 920 patients. *Bipolar Disorders*, **6**, 379–85.

Zhang, H., Wisniewski, S.R., Bauer, M.S., Sachs, G.S., Thase, M.E. (2006). Comparisons of perceived quality of life across clinical states in bipolar disorder: data from the first 2000 Systematic Treatment Enhancement Program for Bipolar Disorder (STEP-BD) participants. *Comprehensive Psychiatry*, **47**, 161–8.

Chapter 17

Service delivery of integrated care for bipolar disorder

Richard Morriss

Service delivery for bipolar disorder (BD) provides the platform for the evidence-based assessment and management of people with BD. It is important to consider the needs of the atypical patient as well as the typical one, and services must be able to manage patients who pose the highest risk of adverse outcomes. Services also have to consider issues such as access to care, acute treatment versus continuing care, rehabilitation back to normal function, including work, and education of patients and families. There are issues of boundaries within the health service between primary care and secondary mental health care, secondary general mental health care and mood or bipolar specialist services and specialist teams that control resources of use to some but not all BD patients, e.g. assertive outreach teams. There are issues of boundaries related to the transition from child to working-age adult services, and working-age to older adult services. There are issues concerning care provided by publicly funded versus privately funded services, and between self-help, voluntary services and professionally delivered services.

There are issues that influence service delivery that are likely to vary considerably from one country to another: among these are the boundary between health and social care, the services provided to the patient as a right, the issue of the patient's personal responsibilities for managing themselves and seeking care, the protection of the patient and others, the legal and professional framework, personal choice (of the patient and more contentiously the health professional), political influence, the existing service configuration, and resources (financial, personnel) available. Rarely do commissioners and service providers have the luxury of completely designing services from scratch; usually they are utilising and reshaping existing resources, sometimes with a diminishing, rather than an increasing, budget. It should also be borne in mind that the various movers and shakers that decide upon service delivery perceive evidence of effectiveness and uncertainty differently: researchers look for statistical evidence of effectiveness and access in typical populations; commissioners examine evidence from cost-effectiveness and population perspectives; managers look at organisational and resource needs; clinicians rely on case series experience and consider services to be effective when they meet the needs of atypical cases rather than focus on typical cases; patients appraise evidence from personal experience; and politicians utilise evidence from ideological, economic and pressure group and media perspectives. Implementation in a service is more complex than running a randomised controlled trial (RCT) of a single intervention because, unlike an RCT, patients are rarely excluded, treatment is multimodal, outcomes are not always clear-cut and the service has multiple, sometimes barely compatible, aims, such as containment of risk and cost, staff welfare and compliance with legal and professional standards, as well as patient outcome.

There is a need academically to consider the delivery of services for BD patients for a number of reasons.

1. There is a 'chasm' between the services that are delivered and the care that it is possible to provide, even within existing resources. As well as not implementing evidence-based interventions, costly, ineffective and harmful interventions are sometimes implemented. The reasons for this are worth examining.
2. There are issues to do with access to services for people with BD. Even with extensive publicity, psychoeducation for BD was taken up by only half the patients with BD who might have benefited from it in the Texas Algorithm Project (Toprac *et al.*, 2006). There is evidence that people with BD have trouble accessing care for physical health problems. For instance, people with BD have a higher standardised mortality ratio from cardiovascular disease than the general population but are less likely to access specialist cardiovascular services.

Practical Management of Bipolar Disorder, eds. Allan H. Young, I. Nicol Ferrier and Erin E. Michalak. Published by Cambridge University Press.

3. There may be systematic failings in service delivery to people with BD or subgroups of patients with BD. For instance, medicine has tended to specialise more in terms of service delivery, but at least 50% of patients with BD have another co-morbid medical, psychiatric or substance-use disorder in the course of their BD. A specialist service for BD may neglect the co-morbid condition, whereas a service for the co-morbid disorder may neglect BD. Research suggests that in some co-morbidities such as substance-use disorders, the management of BD and substance-use disorders is intimately connected, such that the management of one condition directly affects the outcome of the other. In service-delivery terms, management of such co-morbidity requires joint assessment and management of both conditions, rather than separate treatment of each.
4. It is essential to understand the ingredients of a treatment that cannot be varied if it is going to be effective and which elements can be varied. Mental health services have to make new interventions fit into an existing configuration of services, so inevitably there are some differences between the intervention carried out in an RCT and the intervention delivered in a service. Table 17.1 shows the features of an intervention that need to be considered before it is applied in a service. Services have to adapt to deliver fresh treatment approaches and often have to use the same resources that were previously available in terms of personnel, finance, facilities and organisational structures. The requirements for change in these resources and additional requirements for training, supervision and new technology need to be understood. Such an approach improves commissioning and reduces variation in service delivery of treatment.

This chapter cannot address all these issues, because influences on service delivery vary so much nationally and within each locality in each country. However, we will first review the service needs that are universal, evidence of effectiveness using service-delivery models that have been evaluated, and the theory and evidence base for implementing service change.

Universal needs of patients with bipolar disorder from services

Bipolar disorder usually has an onset in childhood or young adulthood. Typically patients with BD will experience clusters of episodes of mania and depression interspersed with quiescent periods characterised by no symptoms or symptoms of mild depression. However, bipolar episodes will occur through adulthood into later life. There is a high risk of suicide and self-harm associated with the mood episodes and an increased mortality from cardiovascular disease and stroke. Function is often impaired even when remission from a manic or depressive episode has been achieved. There are a number of effective drug and psychological treatments, but both types of treatment are effective in particular phases of the illness and need careful monitoring. There are frequently substance use and other psychiatric and physical co-morbidities. According to the World Health Organization and World Bank (Murray & Lopez, 1996), opportunity costs for BD are high because patients who look after themselves well and receive a high standard of care from health services have a much more benign illness and contribute to society; by contrast, those that do not look after themselves well and do not receive a high standard of care may need looking after by society. Even from this brief description of the condition, there are several implications for care delivered by health services: (1) continuity of care is going to be required throughout a patient's life; (2) however, when the illness is in a quiescent period, primary care rather than mental health services could provide this continuity; (3) as patients with BD become older they will interact with different groups of professionals serving different age groups ('vertical care'); (4) the service has to be a platform for the delivery of a range of drug and psychological treatments, but the service may also need to provide crisis or inpatient care, rehabilitation back into work, education and social roles, physical care monitoring, and treatments for substance use and other co-morbid mental health problems.

Providing all of these functions would be a challenge for any specialist BD service and may be prohibitively expensive. Therefore even in specialist centres for BD, core mental health professionals with expertise and knowledge in BD are integrated with groups of health professionals who can fulfil specific but occasional needs of BD patients, such as the provision of inpatient care and rehabilitation services. More rural services that cover a large geographical area are likely to be much less specialised and may have no core mental health professionals with specialist expertise in BD. Nevertheless, the principle of providing continuity of care and a diverse range of care options will apply in

Table 17.1 Assessment of the key ingredients of service delivery of a treatment

Ingredient	Question	Source of evidence
Population	Which patients is the intervention suitable for?	Survey of characteristics of patients in service compared to research demonstrating effectiveness
Setting	Does the intervention require a specialist setting, e.g. inpatient care?	Assessment of risks and benefits of giving in specialist versus non-specialist setting
Access	What proportion of the patients who would benefit are likely to access?	Audit
Key tasks	What tasks must be delivered and when?	Functional analysis of the intervention
Competencies of therapists	Who has the competency to deliver tasks and are they supported to do this? Do they need additional training and supervision?	Functional analysis of tasks
Attitudes and culture	What are the attitudes of staff, patients and carers to this intervention?	Focus group, survey or interview with key stakeholders
Resources	What resources are available and how are they most effectively used?	Identification of costs, savings and budget to implement and sustain service delivery
Key outcomes	What outcomes are expected and will be audited against?	Specific, measurable, attainable, relevant and timely
Risks	What can go wrong and what are the consequences?	Functional risk assessment, audit and review of adverse incidents or complaints
Choice	How might offering this intervention promote or decrease patient or health professional choice?	Interview with key stakeholders

both settings. A further issue of growing importance in BD is the increasing demands of patients for self-management, the exercise of patient choice in treatment and the growing appreciation that lifestyle issues play an important part in service delivery. Table 17.2 shows the care that patients utilised to stay well with BD for at least two years (Russell & Brown, 2005). Ideally services will promote these opportunities for self-management.

Specific service-delivery models

Collaborative care model

The collaborative care model was originally devised to promote the management of chronic medical illness and has been effective in improving outcome in primary care for patients with unipolar depression. The five elements of care that are characteristic of the collaborative care model are: (1) the use of evidence-based planned care; (2) the definition of provider roles within an overall monitored system of health care; (3) patient and carer self-management supported and enhanced by health providers; (4) clinical information systematically obtained on the patient; (5) decision support provided by more specialist care providers to less specialist carer providers, including if necessary assessment and management by more specialist care providers for a period of time.

This model has been applied to the management of BD in secondary care mental health services. A nurse acts as the care manager coordinating care. Group psychoeducation is delivered to promote patient self-management. Medication is delivered by psychiatrists by means of a treatment algorithm. Additional treatment is obtained for the patient as required by the nurse, who can discuss the case with, or get the patient seen by, a psychiatrist. Two RCTs using the collaborative care model treatment in BD have been carried out. They both show reductions in the amount of time patients spent in mania and improvements in function compared with treatment as usual, but there were no improvements in depression outcomes (Bauer *et al.*, 2006a; Simon *et al.*, 2006). Although, the treatment effects were not large, they were sustained over periods of 2–3 years (Table 17.3).

In the Bauer *et al.* RCT (Bauer *et al.*, 2006b), a psychiatrist and mental health nurse coordinator (0.5 and 1.0 full-time equivalent for 90–100 BD patients) were employed in an outpatient setting. There were no out-of-office-hours services or community outreach

Table 17.2 Coping strategies of BD patients who stayed well for two years

1. Acceptance of diagnosis
2. Mindfulness
3. Education
4. Identify trigger factors
5. Recognise warning signs
6. Manage sleep & stress
7. Making lifestyle changes
8. Medication
9. Access support
10. Stay well plans

Taken from Russell & Brown, 2005.

Table 17.3 Summary of results of collaborative care RCTs for BD

Bauer *et al.* (2006a)	Over three years: - 6.2 week reduction in time in episode - 4.5 week reduction in mania - no reduction in depression - modest effect (0.3)[a] on overall function, including effects on work, parental and extended family roles - effects sustained over three years - modestly reduced costs in intervention group because of reduced psychiatric and general medical and surgical inpatient costs
Simon *et al.* (2006)	Modest improvements in mania symptoms over every quarter for 2 years Effect confined to patients symptomatic at baseline No improvement in depression Higher costs of intervention service and antipsychotic use not offset by reduced inpatient care costs

[a] Below NICE arbitrary threshold for clinically significant effectiveness (0.5).

services provided. All treatment of medical and psychiatric co-morbidity, primary care health problems and psychotherapy were provided by usual providers and were continued if started already. The Bipolar Disorders Program itself consisted of group psychoeducation under the direction of the nurse care coordinator at the beginning of the intervention. The group psychoeducation stressed the identification of personal symptom profiles, early warning symptoms and triggers for illness episodes. Each patient was encouraged to develop a personal cost–benefit analysis, and group feedback from other patients helped to improve coping responses and develop collaborative action plans with the service providers. A practice guideline was developed for the psychiatrist and nurse care coordinator stressing identification and early, full-dose treatment of bipolar episodes, the recognition of early warning signs and subsyndromal symptoms. The medication algorithm specified classes of medications to use to allow for patient-centered collaborative decision-making based on efficacy and side effects. The organisation of care itself through the nurse care coordinator recognised three types of contacts: (1) 'backbone scheduled care' – regularly scheduled appointments for monitoring regardless of clinical status; (2) 'demand-responsiveness services' are requested by patients for issues that cannot wait for the next scheduled appointment and are dealt by the nurse on the same day by telephone or the next day face to face; (3) 'outreach and inreach contacts' include prompt follow-up for missed appointments and liaison with other care providers, such as emergency room contacts, inpatient services or other specialties.

The nurse care coordinator also facilitated information flow to the psychiatrist by providing patient assessments, implementing reminders for guideline-based monitoring and tracking laboratory results.

The intervention in the Simon *et al.* RCT (Simon *et al.*, 2006) was provided by nurse care managers under weekly supervision from a psychiatrist, who worked with the patient's existing psychiatrists and psychotherapists. The caseload was similar to the Bauer *et al.* intervention (1.0 full-time equivalent for 95 BD patients). The intervention itself had five elements: (1) the nurse care manager reviewed each patient's treatment history and developed a collaborative treatment plan, including current medications, expected frequency of follow-up visits, early warning signs of mood episodes, coping strategies for responding to warning signs and identification of a care partner (family member or significant other); (2) the nurse care manager telephones each patient monthly to complete structured clinician ratings of current symptoms, current medication use and medication adverse effects, and patients also self-monitored their mood; (3) the structured information collected by nurse care managers was fed back to treating psychiatrists and psychotherapists, who decided upon further management; (4) a similar group psychoeducation intervention to the Bauer intervention (the Life Goals Program) was delivered weekly for five weeks and then twice per month for up to 48 sessions; (5) as required, nurse care managers made additional telephone contacts to provide general support, encourage group participation and facilitate in-person follow-up care.

Table 17.4 'Normalisation process model' applied to the implementation of service delivery of integrated care for BD, from May 2006

1. Interactional workability
Shared expectations by professionals between each other and the patient about the nature of the work, including the time taken and its goals
2. Relational integration
Credibility of knowledge and practice and the level of expertise that is required
3. Skill-set workability
Allocation of work within the health service
4. Contextual integration
Allocation of resources, infrastructure and control, and integration into existing patterns of activity

The telephone monitoring and care management were supported by an Internet-based computer application integrating contact tracking, structured assessment and standardised feedback reports to providers, including graphic displays of mood symptoms and medication use, and algorithm-based recommendations for medication adjustment covering around 7000 scenarios.

Although both interventions are extensive, it is claimed that the Bauer *et al.* intervention (Bauer *et al.*, 2006b) re-uses the time of a nurse and a psychiatrist who were already available and does not require extensive training or Internet-based resources. The Simon *et al.* intervention (Simon *et al.*, 2006) requires an extensive psychoeducation programme and Internet-based resource, although nurse resources are comparable to the Bauer *et al.* (2006b) intervention. Neither intervention had any impact on depressive symptoms, possibly indicating the need to incorporate cognitive behaviour, social rhythm or family interventions, which all may have some impact on depressive symptoms. Although research evidence in BD exists for a nurse-led collaborative care model, there are still relatively few routine clinical services that are run this way. The interventions require a considerable change from usual practice in terms of both organisation of care and the method of practice, constituting barriers to implementation (May, 2006) (Table 17.4). Possibly the clinical benefits on outcome are not sufficiently great for services to make such substantial changes to practice.

Extended role of the lithium clinic

A traditional form of service delivery, especially in Europe, is the lithium clinic, an outpatient clinic or sometimes day-hospital service, usually provided by psychiatrists and coordinated by a senior nurse. Many of these started when most patients with BD were treated with lithium in the 1960s and 1970s. They often provide psychoeducation about BD and specifically about lithium through leaflets or sometimes more intensively in group psychoeducation sessions that are open to patients. Sometimes carers are invited to these sessions or are offered separate sessions. The clinic coordinates the investigations required for the safe management of lithium and keeps a database of patients, ensuring that regular reviews and investigations are carried out in accordance with guidelines for lithium monitoring. There is usually good liaison with specialist medical services required for the management of medical complications such as renal physicians and endocrinologists for thyroid problems, and chemical pathology laboratories. Research has shown that patients who attend such clinics and adhere to lithium treatment (Ahrens *et al.*, 1995; Kallner *et al.*, 2000) have better mental and physical health outcomes, including reduced mortality rates, than those who do not. However, such results may be explained by selection, as patients who look after themselves better are more likely to attend outpatient clinics regularly and adhere to lithium treatment. Now that alternatives to lithium are readily available, most lithium clinics are evolving into specialist affective services or specialist BD services. For many years lithium clinics have not just looked after patients with BD, they have also looked after some patients with unipolar depression who are taking lithium augmentation of antidepressants, a practice more commonly found in Europe than in some other parts of the world such as North America.

There are a number of disadvantages to lithium clinics. First, they are focused on a specific form of management rather than a condition. There is a danger that the service may give less emphasis to the management of patients who can no longer take lithium. Second, such clinics may not manage non-affective psychiatric co-morbidities such as substance-use disorders well, although they often look after core mood disorder psychopathology and physical health better than many generic health services. Third, lithium clinics rarely have the capacity for outreach work for those patients who do not attend regularly and do not adhere well to medication. In some services, there are co-working arrangements between lithium clinics and community outreach teams who can visit patients at home and other specialist psychiatric services such as substance-use disorder services. With the development of joint working protocols and an attitude of reciprocity between the services, it is possible to overcome the disadvantages of lithium clinics.

Specialist affective disorder services

Specialist affective disorder or BD services are a common form of service delivery. The service is usually led by a senior psychiatrist, but it is often multidisciplinary. Such services may comprise an outpatient clinic or both outpatient and inpatient services, sometimes with out-of-hours provision and some capacity for community outreach. Many services provide some degree of psychoeducation for patients and/or carers, although the form this takes is highly variable. In many privately run services and in Germany, courses of psychoeducation start during inpatient care and will continue for the outpatient, providing the patient can access the sessions. Such service provision arises because health purchasers buy packages of care including psychoeducation, although there is evidence that psychoeducation is ineffective if it is started during an acute episode (Morriss *et al.*, 2007).

Sometimes specialist affective disorder services are assessment and advice services, but usually they provide a one-stop service for integrated care, including some access to psychological treatment. The provision of psychological treatment except psychoeducation is rarely provided universally and is usually reserved for those patients who are thought to particularly benefit from it. Sometimes medication is delivered using treatment algorithm or protocol approaches; in other services, the opposite view is taken, particularly if the service is seen as providing a service for patients who have failed to benefit from routine general psychiatric care. Thus, these services share many of the features of lithium clinics, except they are focused on conditions rather than a particular form of management. They often adopt some of the excellent procedural practices of lithium clinics, such as the systematic organisation of routine follow-up consultations and investigations, and close working relationships with medical specialists and laboratories. They are staffed by experts in mood disorders who are up to date in continuing professional education, and there is a tradition of consulting on patients who have failed to make progress with routine treatment from general adult psychiatrists.

The main disadvantages of specialist services are: (1) in routine practice, many patients do not receive a diagnosis of BD when they should because all the expertise in mood disorders is concentrated in the specialist affective disorders service; (2) specialist mood disorder services may not offer much choice in service-delivery terms, such as the option for home treatment or day-hospital care as an alternative to inpatient admission for acute mood episodes; (3) if they provide inpatient services and are staffed by many well-paid experts, they are expensive to run unless the specialist mood disorder services are targeted at patients who most require specialist expertise. The willingness of the specialist affective disorders services to train, provide advice and supervise generic specialist mental health and primary care staff in the assessment of patients who may have BD will overcome lack of expertise in diagnosis and the management of mild cases with BD. Negotiation of protocols and a willingness for reciprocity between services can tackle other disadvantages. For instance, a crisis resolution and home treatment team (CRHTT) might be persuaded to offer alternative care at home for patients with mania under the specialist affective disorders service. Protocols are devised to help the CRHTT decide when it is safe and appropriate to manage a patient with mania at home, what medication it should offer, and when the specialist affective disorders service might take over care and advise if complications arise.

Supporting generic community mental health teams

In most services in most countries, specialist affective disorders services have not been created, although there may be individual practitioners who have developed specialist expertise in BD. Unfortunately, data the author has personally collected in the UK shows that non-medical staff in generic community mental health teams do not draw a distinction in either assessment or management between patients with BD and any other form of serious mental illness. The model for most practice appeared to be schizophrenia, and the assumption was that unless patients were persistently symptomatic and functionally impaired then they did not require follow-up. The only specific diagnostic and management input came from a psychiatrist whose interest was restricted to the assessment of current mood and medication for the current mood state. Access to psychological treatment was very restricted, and psychoeducation was not available except on the Internet through an independent patient-led charity. Therefore, not only were the patient's longitudinal needs not fully assessed and met, but also patients were likely to be discharged even though there was a high likelihood of a recurrence of an acute mood episode in the next 12 months. Research carried out in other countries suggests that in generic mental health services, the service provided to patients with BD is much the same or even worse. Perhaps in part to fill the void, patients with BD have formed a network of patient-led organisations that not only provide information but provide services themselves and lobby for better care, e.g. MDF in the UK, and Bright Blue in Australia.

In many countries there have been attempts to improve the standard of care through the development of national treatment guidelines, e.g. NICE in the UK (National Collaborating Centre for Mental Health, 2006), CANMAT (Yatham *et al.*, 2006) in Canada, and the American Psychiatric Association (APA, 2002) in the USA. However, there is a well-established body of work that shows that even government- or service-provider-supported national treatment guidelines are likely to change practice only involving small increments of knowledge, such as drug treatment (Morriss, 2008). Service delivery requiring acquisition of new skills, a change in attitude and organisational change must have local champions and a range of additional inputs, such as training, support, investment and managerial and patient support. A further problem with paper guidelines is that in a fast-moving research field such as BD, recommended practice may require constant revision.

Nevertheless, generic mental health services in many countries are now sophisticated and potentially could provide an excellent service for patients with BD if they could work in partnership with clinicians and patients with specialist expertise in BD. The expert clinicians and expert patients would become official advisors to the commissioners, managers and staff of generic services on the organisation and delivery of care. They would be involved in training, advising and supervising other staff. Such a model has been piloted in North West England, where a senior mental health nurse with experience in both the education of staff and managing patients with BD was trained by a psychiatrist and clinical psychologist with expertise in BD, with further advice from an expert patient in terms of self-management. The nurse trained all members of community mental health teams and supervised at least one case in a programme for BD, known as Enhanced Relapse Prevention (ERP) (Lobban *et al.*, 2007). Without such expertise, the chasm between the service provided to the majority of patients with BD and what is possible to achieve is likely to grow wider over time.

Working on the interface with primary care

In the Netherlands, a country with universal health care and comprehensive geographically organised primary care and mental health services, approximately 50% of patients with BD type I or II disorder had no recent contact with specialist mental health services and 25% never had contact with specialist mental health services (ten Have *et al.*, 2002). These figures illustrate that an integrated service care model for a patient with BD has to include consideration of the role of primary care services as well as specialist mental health services. One way of doing this is by establishing a joint care protocol between primary care and secondary care for BD, a procedure that is becoming common among managed health-care services across the world. Where such formal agreements are not made or encouraged, informal agreement on such protocols would improve equity of access to care and overall standards of care from the patient's perspective.

Clearly, there is a need to educate and support primary care services in relation to diagnosing BD but there may be a role for primary care in terms of managing milder cases of BD who have infrequent episodes accompanied by little functional impairment and low risk of adverse events such as suicide. Primary care may also manage patients with more serious BD who have remained symptomatically and functionally well for some time (at least two years, in my view). Unfortunately, primary care is often required to provide the main source of care of patients with more serious BD who have recently been acutely ill or have continuing functional impairment. In cases that are under the management of specialist mental health services, primary care can still play a part in integrated service delivery, e.g. in supporting carers, addressing other important health needs such as smoking cessation, or where protocols have been drawn up by local health providers in relation to physical health and medication monitoring.

There are collaborative care model services established for the assessment and management of unipolar depression across the primary care and secondary care divide. One way of supporting primary care might be to adapt protocols used for such collaborative care so that questions on the presence or absence of mania and hypomania are included in primary care assessment. If it is assumed that all suspected cases of BD are assessed by a secondary care mental health specialist, then lines of referral to such clinicians and shared management procedures will then need to be drawn.

In developing countries, support workers with a short basic training in the assessment and management of mental health problems often provide initial assessments of patients with mood disorders in primary care. It is important that such support workers are given enough training and support to detect depressive, manic, hypomanic and mixed episodes, and that protocols are in place for the support worker to refer to specialists who are able to assess and manage patients with BD.

Assertive outreach teams

One form of service delivery that can usefully complement specialist mood disorder or more generic mental health services providing most secondary health care for BD is assertive outreach (AO). It should be used for those patients that cause the greatest concern because the risks associated with their mental health are great yet their adherence to any form of treatment or follow-up is low. Typically such multidisciplinary services aim to stabilise patients within a period of around 12 months before returning them to more routine secondary care at the end of this period. Given the technical expertise required now for the optimal management of difficult-to-treat BD, it would particularly make sense for there to be consultation by experts in the assessment and management of BD with the staff of AO teams.

Early intervention teams

Increasingly, first- or second-onset episodes of severe mood disturbance and psychotic symptoms in adolescents and young adults are assessed and managed through early intervention teams. They promote engagement of younger people by meeting the patient in more informal and, from the young person's perspective, more acceptable settings. The aim is to initiate treatment early so that the illness trajectory of the patient is set on a more benign course given that patients with early onset BD are more likely to have substance-use disorder and other psychiatric co-morbidity, make more suicide attempts and have a worse functional outcome than patients with later-onset BD. Such services usually have a strong emphasis on psychoeducation and are multidisciplinary. In clinical practice, the distinction between the first episode of bipolar affective disorder and the first episode of schizophrenia is not easy to make but may require very different treatment approaches. Early intervention teams are primarily focused on early-onset schizophrenia, so again input from specialists in the assessment and management of BD, as well as specific psychoeducation for BD, would be helpful.

Determining whether service change will be implemented

Interventions that are clinically cost-effective such as the collaborative care models for BD are frequently not implemented, even when there is research evidence to support them. Why? Interventions are less likely to be implemented if they require changes to usual practice that go beyond merely applying new knowledge and require changes in attitude, skills and organisational change. The ease with which a service is able to make organisational changes to practice can be predicted from the model shown in Table 17.4 (May, 2006). Using this model, the Bauer collaborative care model in the US Veterans Administration health-care system might be implemented, but there may still be barriers to its generalised adoption, as shown below.

1. *Interactional workability* – some health professionals such as psychiatrists and some patients used to more traditional care would not expect such a direct role of the nurse in planning care.
2. *Relational integration* – the intervention did not produce improvements in depression and relatively modest improvements in time spent in mania.
3. *Skill set workability* – requires a different allocation of work between the nurse care coordinator and psychiatrist.
4. *Contextual integration* – resources and infrastructure are similar.

In the UK publicly funded mental health service in which the author works, nurses work in mobile community teams and spend little time in outpatient settings, where only the psychiatrist works. Psychoeducation is rare. The barriers to implementing a Bauer-type intervention would be considerable, because the expectations of patients and professionals are for nurses to work in the community. An outpatient-based service with no community outreach would be seen as ignoring the needs of some of the most vulnerable patients, who cannot easily, or will not, go to the outpatient clinic. In the UK, the ERP form of service delivery by nurses working in the community from community mental health team bases (Lobban *et al.*, 2007) may be a more acceptable intervention should there be evidence of its effectiveness. However, in US services such as the Veterans Administration, there would have to be compelling evidence of the clinical and cost-effectiveness of ERP before clinic-based nurses would start to practice in the community.

Conclusion

Service delivery for BD needs to provide continuity of care through the lifespan as well as providing acute care and rehabilitation. Primary and secondary care mental health services have a role to play. The needs of people with BD are considerable and vary quite widely from person to person, and also from country to country. Care needs to be integrated, and the needs of people with BD are unlikely to be met entirely, even by a dedicated specialist service for BD. A number of service models have been considered, but only collaborative care models have been fully evaluated. The results of these randomised controlled trials may not be sufficient to stimulate the organisational changes that would be required in existing services.

References

Ahrens, B., Grof, P., Moller, H.J., Muller-Oerlinghausen, B., Wolf, T. (1995). Extended survival of patients on long-term lithium treatment. *Canadian Journal of Psychiatry*, **40**, 241–6.

APA. (2002). Practice guideline for the treatment of patients with bipolar disorder (revision). *American Journal of Psychiatry*, **159** (Suppl. 4), 1–50.

Bauer, M.S., McBride, L., Williford, W.O., *et al.*: Coauthors for the Cooperative Studies Program 430 Study Team (2006a). Collaborative care for bipolar disorder: Part II. Impact on clinical outcome, function, and costs. *Psychiatric Services*, **57**, 937–45.

Bauer, M.S., McBride, L., Williford, W.O., *et al.*: Coauthors for the Cooperative Studies Program 430 Study Team (2006b). Collaborative care for bipolar disorder: Part II. Impact on clinical outcome, function, and costs. *Psychiatric Services*, **57**, 927–36.

Kallner, G., Lindelius, R., Petterson, U., Stockman, O., Tham, A. (2000). Mortality in 497 patients with affective disorders attending a lithium clinic or after having left it. *Pharmacopsychiatry*, **33**, 8–13.

Lobban, F., Gamble, C., Kinderman, P., *et al.* (2007). Enhanced relapse prevention for bipolar disorder – ERP trial. A cluster randomised controlled trial to assess the feasibility of training care coordinators to offer enhanced relapse prevention for bipolar disorder. *BMC Psychiatry*, **7**, 6.

May, C. (2006). A rational model for assessing and evaluating complex interventions in health care. *BMC Health Services Research*, **6**, 86.

Morriss, R.K. (2008). Implementing clinical guidelines for bipolar disorder. *Psychology and Psychotherapy*. **81**, 437–58.

Morriss, R.K., Faizal, M.A., Jones, A.P., Williamson, P.R., Bolton, C., McCarthy, J.P. (2007). Interventions for helping people recognise early signs of recurrence in bipolar disorder. *Cochrane Database of Systematic Reviews*, **1**, CD004854.

Murray, C.J.L., Lopez, A.D. (Eds) (1996). *The Global Burden of Disease: a Comprehensive Assessment of Mortality and Disability from Diseases, Injuries, and Risk Factors in 1990 and Projected to 2020.* Cambridge, MA: Harvard School of Public Health on behalf of the World Health Organisation and the World Bank.

National Collaborating Centre for Mental Health. (2006). *Bipolar Disorder: the Management of Bipolar Disorder in Adults, Children and Adolescents, in Primary and Secondary Care.* London: NICE. www.guidance.nice.org.uk/CG38 [accessed 3 January 2010].

Russell, S.J. & Browne, J.L. (2005). Staying well with bipolar disorder. *Australian and New Zealand Journal of Psychiatry*, **39**, 187–93.

Simon, G.E., Ludman, E.J., Bauer, M. S., Unutzer, J., Operskalski, B. (2006). Long-term effectiveness and cost of a systematic care program for bipolar disorder. *Archives of General Psychiatry*, **63**, 500–8.

ten Have, M., Vollebergh, W., Bijl, R., Nolen, W.A. (2002). Bipolar disorder in the general population in The Netherlands (prevalence, consequences and care utilisation): results from The Netherlands Mental Health Survey and Incidence Study (NEMESIS). *Journal of Affective Disorders*, **68**, 203–13.

Toprac, M.G., Dennehy, E.B., Carmody, T.J., *et al.* (2006). Implementation of the Texas Medication Algorithm Project patient and family education program. *Journal of Clinical Psychiatry*, **67**, 1362–72.

Yatham, L.N., Kennedy, S.H., O ' Donovan, C., *et al.*: Guidelines Group, CANMAT (2006). Canadian Network for Mood and Anxiety Treatments (CANMAT) guidelines for the management of patients with bipolar disorder: update 2007. *Bipolar Disorders*, **8**, 721–39.

Chapter 18

Training and assessment issues in bipolar disorder: a clinical perspective

Sagar V. Parikh and Erin E. Michalak

Common training/skill problems in bipolar disorder encountered by clinicians

Making a diagnosis of bipolar disorder

Bipolar disorder (BD) is diagnosed by the occurrence of a single manic episode (for BD type I) or by the occurrence of a hypomanic episode and at least one full major depressive episode (for BD type II) (APA, 2000). Although the clinical presentation of acute mania is often dramatic and unmistakable, the disorder is often not recognised, even by experienced clinicians (Perlis, 2005; Parikh, 2007; Kamat, 2008). Errors in diagnosis flow from many causes, with perhaps two reasons explaining much of the diagnostic confusion: (1) patients in acute mania often have progressed in symptom expression beyond classical descriptions of mania into an agitated or psychotic state that is initially impossible to clarify, particularly if it is compounded by acute substance use and non-cooperativeness with the treatment team; and (2) patients experiencing hypomania or mild–moderate expressions of mania rarely present for treatment, coming in only in the context of an acute depression, when memory is poor of a possible previous hypomania (Meyer & Meyer, 2009). The fact that the disorder has so many presentations – including being mistaken for 'just' substance abuse when that co-morbidity is prominent, or for personality disorder when there are frequent mild episodes, or for schizophrenia when psychotic symptoms are very prominent – invites clinicians to make a diagnosis at one point in time that is plausible, but in fact incorrect. Two studies that highlight the difficulty in delay include the fact that correct diagnosis often takes 10 years of contact with the health-care system (Lish *et al.*, 1994; Hircshfeld *et al.*, 2003), and in research surveys using well-regarded interview instruments such as the composite international diagnostic interview, BD is often the disorder with the poorest inter-rater reliability (Kessler *et al.*, 1997).

Success in diagnosis may be optimised by using the Robins and Guze (1970) approach to diagnosis, which is further delineated by Guze (1992) and most recently supported by the report from the International Society of Bipolar Disorders (Ghaemi *et al.*, 2008). In this approach, diagnosis is made along five dimensions: (1) acute clinical presentation, (2) laboratory tests, (3) course/outcome, (4) response to treatment and (5) family history. As the acute clinical presentation is not usually conclusive, and as there are no reliable laboratory tests for the diagnosis, the remaining factors merit consideration. For instance, when a 30-year-old person presents for evaluation for the first time, if the history indicates that there was a severe depression at age 15, some uncharacterically wild behaviour for a few weeks at 19, and a couple of depressions since, with good functioning between episodes even without treatment, that course alone suggests a recurrent mood disorder such as BD. Similarly, if a person with a previously diagnosed fairly prominent borderline personality disorder undergoes a 6-month trial of a mood stabiliser with a dramatic improvement in mood and functioning, again the possibility of BD looms large. Finally, a mood disorder in a first-degree relative also elevates the chance that the 30-year-old being assessed now also has a mood disorder. The absence of these suggestive features does *not* rule out BD, but such features do provide soft support for the diagnosis.

A key approach in validating the diagnosis involves obtaining collateral history, with the permission of the patient, from family members, particularly spouses. Although such information may be biased, the confirmation of repeated or prolonged instances of increased energy with decreased sleep, for instance, may be a key confirmatory piece of information when BD is already suspected.

Practical Management of Bipolar Disorder, eds. Allan H. Young, I. Nicol Ferrier and Erin E. Michalak. Published by Cambridge University Press.

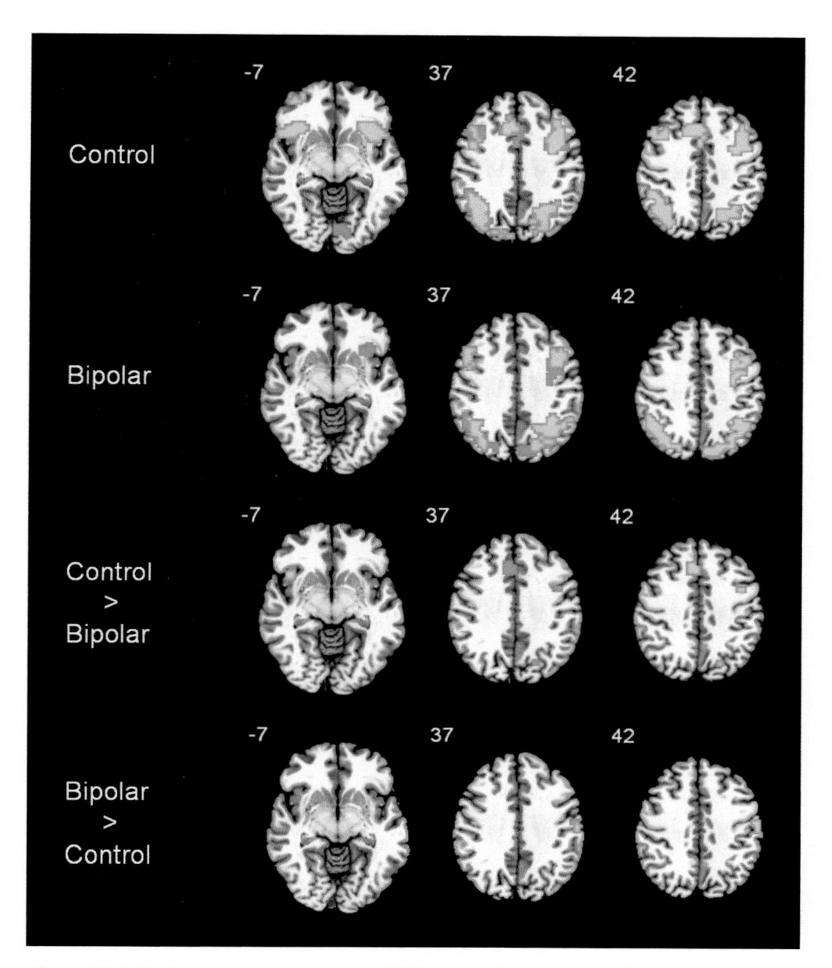

Figure 20.1 Task-dependent responses in fMRI: two-back task (adapted from Monks *et al.*, 2004). These brain maps represent regions of BOLD response during performance of a two-back task by euthymic bipolar (*n*=12) and control (*n*=12) participants. Yellow clusters = regions with significant activation during two-back condition.

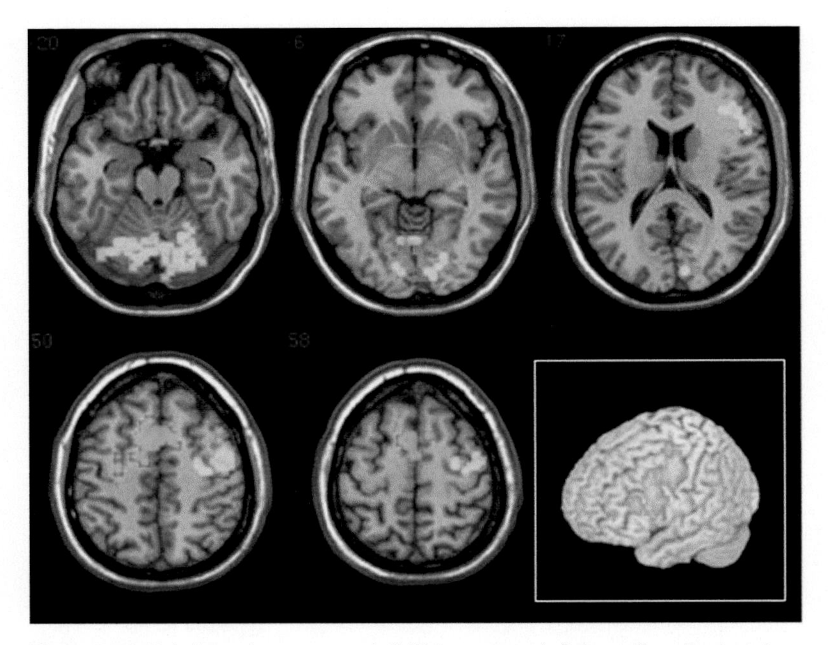

Figure 20.2 Task-dependent responses in fMRI: language tasks (adapted from Curtis *et al.*, 2007). During the same scanning session as Monks *et al.* (2004), participants also performed tasks of language production. In addition to baseline tasks, participants undertook four active tasks designed to investigate components of language. Yellow voxels represent regions of relatively greater response in the bipolar subjects and blue voxels represent regions of relatively greater response in the control participants.

Finally, diagnosis may be aided by a willingness to review and change diagnosis over time. A thorough initial assessment may suggest unipolar depression, but specific experience with that patient over time may yield observations of hypomania that were previously seen by the patient and family as simply good cheer. Often, a young individual hospitalised for a first psychotic episode and discharged with a probable diagnosis of schizophrenia may show no cognitive decline, good interpersonal functioning and perhaps an occasional depression over the next few years – this more closely fits the course of BD rather than schizophrenia. Such new information adds to and revises the diagnostic perspective. It is important for the clinician to be open-minded about changing the diagnosis, rather than feel beholden to the original diagnosis simply because it was made by thorough initial assessment or by distinguished colleagues.

Identifying substance-abuse issues

Most individuals with BD experience a time when substance misuse becomes a serious problem (Brown, 2005). Alcohol abuse, particularly in women, is especially common in BD. Drugs that are stimulating, such as cocaine and amphetamines, are also commonly abused, and in fact confound the diagnosis of BD. Given the ubiquity of substance abuse, the clinician should be thorough in initially asking about substances, and persistent in evaluating for substances in the context of acute episodes. Particularly in acute mania, it is helpful to conduct urine drug screens for substances and examine blood indices such as liver function tests, especially for the enzyme GGT as a marker of alcohol abuse.

There is no commonly accepted, widely validated tool for identifying substance abuse; simply asking in two different ways about substance misuse often is sufficient. Examples of good screening questions include: (1) Has there ever been a time in your life when you were using alcohol, street drugs or prescription drugs to excess?; or (2) Have you ever had any medical problems (liver problems, bleeding, etc.) or legal problems (public drunkenness, driving while intoxicated) related to drugs or alcohol? Such questions should always be accompanied by a straightforward summary of current consumption of alcohol, marijuana or other items. Many individuals consume marijuana in modest quantities on a daily basis, and so will not report this as a problem with 'excess substance use'.

Identifying co-morbid anxiety disorders

According to many studies, close to 90% of individuals with BD have a significant anxiety disorder at some point (Keller, 2006). Furthermore, the presence of such anxiety disorders – when not addressed – worsens the treatment outcome of BD (El-Mallakh & Hollifield, 2008). This issue is covered in detail in Chapter 12 of this book. Given the ubiquity of such anxiety disorders, simple screening questions for each of the major anxiety disorders is warranted. Specifically, consider asking the following questions: (1) Do you often feel tense or anxious?; and (2) Do you worry a lot about things? If either response is positive, continue with questions about panic disorder (sudden fear with many physical symptoms such as pounding heartbeat, breathing problems, sweating, etc.), social phobia (fear in social situations such as public speaking, eating in public, etc.) and obsessive–compulsive disorder, as these three disorders are most frequent. Further information about key assessment scales for anxiety disorders is available at the end of this chapter.

Clarifying 'personality' issues

Personality disorders are a controversial area for many reasons. The DSM-IV-TR characterises personality disorders as 'an enduring pattern of inner experience and behaviour that deviates markedly from the expectations of the individual's culture, in either cognition, affectivity, interpersonal functioning or impulse control' (APA, 2000). Many individuals with BD have such difficulties, but it is also known that mood episodes accentuate underlying personal styles to the extent that the individual experiences problems in functioning. Hence, proper diagnosis of a personality disorder should only be done after some months of stable mood, and should focus heavily on actual performance in interpersonal and vocational functioning and behaviour during such periods of stable mood (Fan & Hassell, 2008). A common question arises when borderline personality disorder and BD are being entertained as competing diagnostic possibilities. In this situation, careful adherence to the diagnostic principles cited above in the 'Making a diagnosis' section is necessary. Additionally, a key question should be: Are the antecedents to the personality disorder seen here? It is well recognised that the vast majority of individuals with borderline personality disorder have experienced some degree of severe emotional, physical or sexual abuse as children. Current or recent interpersonal behaviours may suggest borderline personality disorder, but such a diagnosis would be difficult to defend in the absence of a previous history of abuse, particularly if the individual also had long periods of good relationships and job stability.

Measuring side effects and adherence with treatment

Given the heavy reliance on pharmacotherapy in psychiatry, it is disappointing that no widely accepted scale exists to reliably enumerate and quantify side effects. In part, this stems from the reality that each patient is truly different, and brings his or her own history of awareness of bodily sensations, specific health issues and attitudes towards symptoms and treatment. Furthermore, it is impossible in a brief medical visit to systematically review all body systems. Therefore, prudence suggests the following approach: identify three to four common side effects from any specific medication, and definitely any side effect that is potentially severe (could result in hospitalisation or death), even if rare. Ask about the common side effects in subsequent visits, and add a general question about 'any other changes in physical functioning or symptoms'. Psychotherapy, too, has side effects, including feeling upset for hours or days after a difficult issue is explored – this should be mentioned. Just as with medication, 'withdrawal' symptoms are also experienced when psychotherapy is being discontinued. Patients should be informed at the start about how long the proposed psychotherapy might last (6 months, 16 sessions, etc.) and notified periodically as time passes (at the half-way point and in each of the last few sessions). Ideally, psychotherapy, like medications, should be tapered, with longer periods between the last few sessions.

Adherence to treatment is a key challenge in all medical disorders (Osterberg & Blaschke, 2005; Haynes *et al.*, 2008) and poignant in BD, as individuals often seek either to re-experience a hypomania or simply be free of the burden of chronic medication. Stating that medication adherence is often difficult in BD before asking about the person's specific use of medication often reduces any shame or reluctance in admitting missed medication. Although numerous scales exist, two simple manoeuvres often suffice: (1) asking how often someone missed a medication since the last visit, along a continuum of 'almost never, sometimes, often, almost always'; and (2) blood monitoring for certain medications. Alternatively, clinicians can encourage the use of 'daily mood charts' for BD, the majority of which include a question on medication adherence over the previous 24 h (see, for example, the 'personal calendar' available via the Depression and Bipolar Support Alliance website www.dbsalliance.org [accessed 3 January 2010]).

As everyone becomes non-adherent sometimes, it is helpful during clinical encounters to maintain a spirit of non-judgemental tolerance and pleasantly encourage proper use of treatment. Periodic revisiting of the basic elements of psychoeducation, such as quoting the rates of unmedicated relapse (perhaps 6% per month; thus over two-thirds of patients will have an episode within a year) (Suppes *et al.*, 1991, 2000) and clarifying the perceived benefits of *non*-adherence are useful. The focus on advantages of non-adherence allows the patient to more clearly articulate to themselves as well as the clinician what the goal of non-adherence is, and often provokes a more vigorous review of the pros and cons by the patient in private, without clinician hectoring.

Assessment scales for clinical practice

It has long been recognised in BD research that measuring symptom severity across time is helpful in evaluating the course of treatment for psychiatric conditions. For example, all published BD clinical trials involve measuring outcome by means of scales that focus on symptoms of interest. Systematically rating symptoms of BD is also an essential feature of psychological treatments such as cognitive–behavioural therapy, interpersonal and social rhythm therapy, and family therapy. The use of assessment scales, however, has not historically been a routine aspect of patient care in clinical practice. In part, this may be because many clinicians are not routinely trained in the use of assessment scales. Additionally, the nature of clinical practice with the pressure of high patient flow makes it difficult to incorporate yet more tasks into every clinical encounter.

Several recent developments, however, have emphasised that using assessment scales should be a priority for clinicians. First, evidence-based medicine has become the prevailing clinical framework for mental health. Evidence-based medicine promotes the use of evidence-based guidelines for clinical interventions and many of these guidelines offer treatment options based on scores from assessment scales. Second, there is increasing emphasis on patient self-education and self-management, which includes self-monitoring of symptoms. Third, there is increasing recognition of the importance of residual or subsyndromal symptoms as predictors of poor outcome in BD (Judd *et al.*, 2008). Residual symptoms may not be detected unless an assessment scale is used. Finally, a cornerstone of evidence-based medicine involves measuring the effectiveness of one's clinical practice. It is no longer sufficient to evaluate patient or practice outcomes by asking general questions about clinical status.

Of course, there are important caveats and questions to consider when using assessment scales to assess clinical outcome. What is the scale designed to measure? How effective is it at carrying out that task? What is the interval of assessment (today, the past week, the past month, etc.)? Is the scale aministered by a clinician or self-reported? Many scales require training for proper administration. Copyright issues dictate that some scales must be purchased for clinical use. Other scales are in the public domain and can be used freely. Users of self-rating scales must consider the unique characteristics of the individual with BD – whether they can read the language and understand the questions, whether there is there any cognitive impairment and whether their current mood state could result in them over- or under-endorsing symptoms, etc. Users of clinician-administered scales should consider issues such as inter-rater reliability and whether scoring conventions and rules are followed. Unstructured interviews are usually the least reliable among different raters, whereas structured or semi-structured interviews increase reliability by providing standardised questions for patients to answer. Explicit and clear anchor points for each item also improve reliability of assessment scales.

In summary, the therapeutic objective for the treatment of BD is full recovery, which includes the full remission of symptoms, a return to pre-morbid psychosocial functioning and acceptable quality of life (QoL). Assessment scales are useful to assess clinical symptoms, monitor response to treatment and return of functioning, promote self-management strategies, detect residual symptoms and ensure that side effects are not limiting treatment. Incorporating assessment scales into routine clinical practice means that treatment decisions can be made based on the best available information. For clinicians, the use of brief clinician-rated scales and/or self-rated scales can improve the quality and efficiency of clinical assessments. For patients with BD, systematically tracking outcomes can provide valuable feedback on the effect of clinical interventions as an important component of self-management programmes and evidence-based psychosocial treatments. In this way, assessment scales can serve to enhance the therapeutic alliance and to promote adherence to both psychological and pharmacological treatment interventions. The following section will describe some key assessment scales for BD that have high utility for clinical practice.

Diagnosing bipolar disorder

Tools to assist with diagnosing BD include the following.

Mood disorders questionnaire

Bipolar spectrum disorders, particularly BD type II, are under-diagnosed in primary care and psychiatric patient populations. The mood disorders questionnaire (MDQ) (Hirschfeld, 2000) is a brief 13-item self-report questionnaire designed to screen for bipolar spectrum disorders (BD types I and II, cyclothymia and BD 'not otherwise specified'). In a yes/no format, the scale screens for lifetime history of DSM-IV-TR mania/hypomania. The MDQ is an easy-to-administer screening tool with good psychometric properties and high clinical utility.

Scoring

The screen is considered positive when seven or more symptoms have occurred, several within the same time period, causing moderate to severe problems.

Assessing symptoms and severity of bipolar disorder

Recommended scales for assessing symptoms and severity of BD include the following.

Hamilton depression rating scale

The Hamilton depression rating scale (Hamilton, 1967) (HDRS; also known as the HAM-D) is the most widely used clinician-administered depression assessment scale. The original version contains 17 items (HDRS17) pertaining to symptoms of depression experienced over the past week. Although the scale was designed for completion after an unstructured clinical interview, there are now semi-structured interview guides available (Williams, 1988). The HDRS was originally developed for hospital inpatients – thus the emphasis on melancholic and physical symptoms of depression. A later 21-item version (HDRS21) included four items intended to subtype the depression but that are sometimes, incorrectly, used to rate severity. The scale takes approximately 20 min to administer and is in the public domain. There are numerous versions of varying lengths (e.g. the HDRS17, HDRS21, HDRS29, HDRS8, HDRS6, HDRS24 and HDRS7). The HAM-D7, also referred to as the Toronto HAM-D7, performs as well as the HDRS and the Montgomery–Asberg depression rating scale (MADRS) in tracking change over time.

Scoring

The method for scoring varies by version. For the HDRS17, a score of 0–7 is generally accepted to be within the normal range (or in clinical remission), whereas a score of 20 or higher (indicating at least moderate severity) is usually required for entry into a clinical trial.

Montgomery–Asberg depression rating scale

The MADRS (Montgomery, 1979) is a 10-item clinician-administered scale designed to assess depressive symptoms, particularly change following treatment with antidepressant medication. The scale is frequently used in pharmaceutical clinical trials to monitor change in response to treatment. The MADRS can be used 'for any time interval between ratings, be it weekly or otherwise, but this must be recorded'. The scale places greater emphasis upon psychological symptoms of depression (for example, sadness, tension, pessimistic thoughts and suicidal thoughts) than somatic in comparison with the HDRS. It takes approximately 10 min to administer, and a patient-rated version of the scale (the MADRS-S) has been developed.

Scoring

Items are rated on a 0–6 scale, yielding a total possible score of 60, where higher scores indicate greater depressive symptomatology. A score of ≤10 has been suggested as a remission criterion.

Young mania rating scale

The Young mania rating scale (YMRS) (Young, 1978) is an 11-item clinician-rated scale designed to assess severity of manic symptoms. Considered by some to be the gold standard of mania rating scales, the instrument is widely used in both clinical and research settings. Information for assigning scores is gained from the patient's subjective reported symptoms over the previous 48 h and from clinical observation during the interview. The scale is appropriate for assessing both baseline severity of manic symptoms and response to treatment in patients with BD types I and II. However, the YMRS does not assess concomitant depressive symptoms and should be administered in conjunction with a depression rating scale such as the HDRS or the MADRS in patients with symptoms of depression or those experiencing a mixed episode.

Scoring

Four of the YMRS items are rated on a 0–8 scale, with the remaining five items being rated on a 0–4 scale. Anchor-points are provided to help the clinician determine severity. A score of ≤12 indicates remission of symptoms.

Clinical global impression

Among the most widely used of brief assessment tools in psychiatry, the clinical global impression (CGI) (Guy, 1976) is a three-item clinician-rated scale that measures illness severity (CGI-S), global improvement or change (CGI-C) and therapeutic response. The illness severity and improvement sections of the instrument are used more frequently than the therapeutic response section, in both clinical and research settings. The CGI has proved to be a robust measure of efficacy in many clinical drug trials and can be quick and easy to administer, provided that the clinician knows the patient well. A CGI for BD (CGI-BD) is available.

Scoring

The CGI is rated on a 7-point scale, with the severity of illness scale using a range of responses from 1 (normal) through to 7 (among the most severely ill patients). The CGI-C scores range from 1 (very much improved) through to 7 (very much worse). Treatment response ratings should take account of both therapeutic efficacy and treatment-related adverse events and range from 0 (marked improvement and no side effects) to 4 (unchanged or worse, and side effects outweigh the therapeutic effects). Each component of the CGI is rated separately; the instrument does not yield a global score.

Recommended scales for assessing anxiety disorders include the following.

Beck anxiety inventory

The Beck anxiety inventory (BAI) (Beck, 1988) is a widely used 21-item self-report measure designed to assess severity of anxious symptoms over the past week. Each item describes a common symptom of anxiety, such as heart pounding or racing, inability to relax and dizziness. The scale was designed to discriminate depression from anxiety, and emphasises the more somatic, panic-type symptoms of anxiety, rather than symptoms of generalised anxiety, such as worry, sleep disturbance or poor concentration. The BAI is a reliable and widely used screen for somatic anxiety symptoms that is sensitive to treatment response, although it is not appropriate for the assessment of generalised anxiety disorder. The scale is not in the public domain (see www.pearsonassessments.com for copyright issues [accessed 13 January 2010]).

Scoring

Items are scored on a scale of 0 (not at all) to 3 (severely: I could barely stand it), with a score range of 0–63. Scores of 0–7 represent minimal anxiety, 8–15 mild anxiety, 16–25 moderate anxiety and 26–63 severe anxiety.

Hospital anxiety and depression scale

The hospital anxiety and depression scale (HADS) (Zigmond & Snaith, 1983) is a brief 14-item self-report instrument designed to screen for presence and severity of symptoms of depression and anxiety over the past week in medical patients. The instrument possesses a 7-item depression subscale (HADS-D) and a seven-item anxiety subscale (HADS-A), both of which omit somatic symptoms in an attempt to reduce the likelihood of false-positive diagnoses. The HADS represents a brief and useful screening tool for symptoms of depression and anxiety in patients with physical illness. It also appears to be sensitive to change. The scale is not in the public domain (see www.gl-assessment.co.uk/health_and_psychology/resources/hospital_anxiety_scale/hospital_anxiety_scale.asp for further details).

Scoring

Items are scored on a 0–3 scale: HADS-D and HADS-A subscale scores (range 0–21) are derived by summing the seven items on each scale (the scale developers warn against deriving a total score for the HADS). For both subscales, scores in the range of 0–7 are considered normal, 8–10 mild, 11–14 moderate and 15–21 severe.

Assessing psychosocial functioning and quality of life

Chapter 16 of this book draws attention to the importance of assessing psychosocial functioning and QoL in clinical settings. Psychosocial functioning refers to a person's ability to perform tasks of daily life and to engage in relationships with other people in ways that are gratifying to the individual and others, and that meet the needs of the community in which the person lives. Quality of life is usually considered a higher-order construct, and refers to the person's wellbeing across a wider variety of domains, such as emotional, social, occupational and physical life. Quality of life is also highly individual and subjective; what may be essential in determining one person's life quality may be unimportant to another. Factors such as these make QoL challenging to measure, but it nevertheless remains an important aspect of patient wellbeing to capture. When used clinically, QoL scales permit the patient to assess the impact of treatment interventions upon areas of their lives that may be of particular importance to them, such as their ability to engage in or enjoy their chosen leisure activities, or assess satisfaction with intimate or family relationships. Traditional symptomatic assessment scales may miss this valuable 'fine-grain' information, which can greatly enrich the clinical picture and help the health-care professional better assess the effects of treatment upon the patient's life as a whole. Several book chapters and review articles have now summarised existing research into psychosocial functioning and QoL in BD (Namjoshi & Buesching, 2001; Revicki *et al.*, 2005; Michalak *et al.*, 2005, 2008;). Here, we recommend key assessment scales for psychosocial functioning and QoL that are, in our opinion, particularly suitable for clinical use (selected scales are relatively succinct, have been validated in BD and are in the public domain).

Recommended scales for assessing psychosocial functioning include the following.

Global assessment of functioning

The global assessment of functioning (GAF) (Endicott, 1976), which constitutes Axis V of the DSM-IV-TR classification system (APA, 2000), assigns a numeric value on a 1–100 scale to psychosocial functioning. The clinician is required to assess the functioning of the patient disregarding impairment arising from physical or environmental limitations using information from any clinical source (e.g. clinical assessment, collateral, medical records). The GAF has been used extensively to assess baseline levels of psychosocial functioning and to predict and evaluate outcome in a wide variety of patient populations. The main strengths of the GAF are its brevity, ease of administration, high reliability and sensitivity to change. Limitations include the fact that it confounds symptoms and functioning; hence the development of the MIRECC version of the GAF (Niv *et al.*, 2007), which measures occupational functioning, social functioning and symptom severity on three subscales.

Scoring

The GAF is scored on a 1–100 scale, where 1 represents the hypothetically most impaired patient and 100 the hypothetically healthiest patient. The scale is divided into ten equal 10-point intervals (e.g. 1–10, 11–20) that have clear anchor points; the use of intermediate scores is encouraged. Specific scoring conventions are available for the MIRECC version of the GAF (Niv *et al.*, 2007).

Functioning assessment short test

The functioning assessment short test (FAST) scale (Rosa *et al.*, 2007) is a 24-item, clinician-administered instrument designed to assess several areas of functioning in individuals with psychiatric disorders, in particular BD. The scale assesses psychosocial functioning over the previous 2 weeks in six specific areas: autonomy, occupational functioning, cognitive functioning, financial issues, interpersonal relationships and leisure time. It showed good psychometric properties in an initial evaluation in patients with BD by its developers, and is brief, taking approximately 6 min to administer. The scale is in the public domain.

Scoring

All FAST items are scored on 4-point scale, where 0 = no difficulty, 1 = mild difficulty, 2 = moderate difficulty and 3 = severe difficulty. Item scores are summed to form an overall functioning score.

Recommended scales for assessing quality of life include the following.

Quality of life enjoyment and satisfaction questionnaire

The quality of life enjoyment and satisfaction questionnaire (Q-LES-Q) (Endicott, 1993) is a 93-item self-report measure of QoL that was developed in a population of outpatients with depression. The scale possesses eight subscales (physical health, work, school, household duties, subjective feelings, leisure activities, social relationships and general activities), although completion of the work, school and household duties sections is optional. The Q-LES-Q is one of the most widely used QoL scales in BD populations. The full version of the scale takes approximately15–20 min to complete – a short form (the Q-LES-Q SF), which corresponds to the General Activities section of the long form of the scale, is available and shows sound psychometric properties.

Scoring

Items are scored on a 1–5 scale; raw scores for the subscales and the total score are converted to percentages of maximum possible scores, where higher scores indicate better QoL.

Training in bipolar disorder diagnosis, treatment and rating scales

Several types of training are useful in developing mastery of BD.

Basic clinical exposure

Significant exposure to patients with BD over the course of a year provides a balanced exposure to different episodes, presentations and the pace of recovery. Inpatient settings allow for ample exposure to the severity of mania, to the perplexing face of a mixed episode and the intense agony of bipolar depression. However, most episodes are not severe enough to warrant hospitalisation. In fact, about 50% of the time, the individual is entirely well from a bipolar perspective, and perhaps another 30% of the time has subsyndromal symptoms (Judd *et al.*, 2002, 2003). Less than 20% of the time is spent in full episodes, and an even smaller percentage in hospitalisations, so the full picture of the challenges of the illness require outpatient exposure. It is suggested that three to five manic inpatients, three depressed bipolar inpatients and having outpatient contacts of at least 20 visits from various patients over the course of a year might begin to provide some true sense of the vagaries of the illness. Connection with a mood disorders clinic for one half day per week for 6 months, with participation in both new patient assessments and follow-ups, would provide a good clinical base for the typical health professional.

Key tools – rating scales, diagnostic instruments

Rating scales are often popularly seen as intimidating, possibly difficult to learn and potentially disruptive to the clinical interview. In reality, most clinicians who choose to use rating scales are able to harmoniously blend them into routine clinical practice. Furthermore, it has been our experience that patients often see the use of the rating scale as rendering the clinical visit more valid and scientific, the parallel to having a blood pressure measured at a family physician's office. It is helpful to limit such scales to those that are truly useful clinically, such as a single mania scale during a high, and a single depression instrument during an episode of depression. Examples were discussed earlier in this chapter.

Training in rating scales and diagnostic instruments is sometimes difficult to find but is often fairly straightforward. From our experience in teaching a variety of instruments, the average health professional can learn several scales in a one-day workshop. The design of the workshop, however, must reflect a combination of some didactic component, and an active 'doing' component. For example, in workshops on mania rating scales, we find teaching one mania scale with 30 min of didactic overview, followed by scoring of the rating scale by watching one video clip of a manic patient and one simulated interview of a patient provides solid background. After each rating scale exercise, the instructor discusses the score item by item with the class. Our experience suggests that using the rating scale on five occasions, with feedback and comparison, provides reasonable inter-rater reliability. We find it useful in research studies, to conduct a rating of a standardised video clip every 3–6 months, to ensure validity and correct anchoring. Individuals seeking training in rating scales are advised to look at the annual meetings of major associations of psychiatrists or psychologists, as well as contacting well-recognised centres for mood disorder treatment.

Continuing education

Issues of maintenance of competence are increasingly prominent in professional societies. Some organisations, such as the American Board of Psychiatry and Neurology, now require individuals to perform a number of tasks, including writing an exam, to maintain board certification, 10 years after initial certification. Given the speed with which treatment advances are being made in BD, and the efforts to expand the boundaries of the disorder, both in terms of criteria and in terms of special populations such as children, it is reasonable to expect clinicians to regularly receive new training in BD. The educational research literature highlights key features of effective continuing education: (1) interactive, rather than purely didactic education; (2) use of case-based scenarios; (3) teaching in which a skill is practised with some opportunity for feedback; (4) opportunities to discuss challenges or patient scenarios from one's own practice; (5) provision of tools to use in the practice, such as rating scales; and (6) education that incorporates larger elements of the treatment context – for instance interprofessional education, in which individuals understand and learn from people in other professions who also collaborate in treating BD (Davis *et al.*, 2008).

Keeping such parameters in mind, the clinician is therefore advised to do a reflective exercise on learning needs first. What are the principal problems for me – are they diagnostic, treatment, or other? What has worked effectively for me in terms of educational experiences in the past? Finally, a chart review is recommended, of the last five BD patients seen, to identify what problems arose – diagnosis, communication, treatment alliance, treatment refractoriness, lack of knowledge about specific treatments (medication, psychotherapy, community resources or other). Armed with the chart review and self-reflection, the clinician can make a more effective choice about suitable educational next steps. Most importantly, it should be stressed that the most common form of education – a simple lecture, whether in grand rounds or a major international conference – is considered the least likely to change practitioner behaviour. Attendance at such a lecture may be vital to identifying the need for further education, much like a newspaper headline alerts us to an interesting story, but the learning cannot end with the lecture. Active searching for interactive education, for preceptorships, for training via specialised centres, etc. may be done by directly emailing local mood disorder experts and by searches on the Internet, as well as contacting major organisations such as the International Society for Bipolar Disorders (www.isbd.org [accessed 3 January 2010]).

References

APA. (2000). *The Diagnostic and Statistical Manual of Mental Disorders, 4th edition.* Arlington, VA: American Psychiatric Association. www.psychiatryonline.com/resourceTOC.aspx?resourceID=1 [accessed 23 December 2009].

Beck, A.T., Epstein, N., Brown, G., Steer, R.A. (1988). An inventory for measuring clinical anxiety: psychometric properties. *Journal of Consulting and Clinical Psychology*, **56**, 893–7.

Brown, E.S. (2005). Bipolar disorder and substance abuse. *The Psychiatric Clinics of North America*, **28**, 415–25.

Davis, N., Davis, D., Bloch, R. (2008). Continuing medical education: AMEE Education Guide No 35. *Medical Teacher*, **30**, 652–66.

Endicott, J., Nee, J., Harrison, W., Blumenthal, R. (1993). Quality of Life Enjoyment and Satisfaction Questionnaire: a new measure. *Psychopharmacology Bulletin*, **29**, 321–6.

Endicott, J., Spitzer, R.L., Fleiss, J.L., Cohen, J. (1976). The global assessment scale: a procedure for measuring overall severity of psychiatric disturbance. *Archives of General Psychiatry*, **33**, 766–71.

El-Mallakh, R.S. & Hollifield, M. (2008). Comorbid anxiety in bipolar disorder alters treatment and prognosis. *Psychiatric Quarterly*, **79**, 139–50.

Fan, A.H. & Hassell, J. (2008). Bipolar disorder and comorbid personality psychopathology: a review of the literature. *Journal of Clinical Psychiatry*, **69**, 1794–803.

Ghaemi, S.N., Bauer, M., Cassidy, F., *et al.* (2008). Diagnostic guidelines for bipolar disorder: a summary of the International Society for Bipolar Disorders Diagnostic Guidelines Task Force Report. *Bipolar Disorders*, **10**, 117–28.

Guy, W. (1976). *ECDEU Assessment Manual for Psychopharmacology, revised edition*. Rockville, MD: US Department of Health, Education, and Welfare, Public Health Service, Alcohol, Drug Abuse and Mental Health Administration, NIMH Psychopharmacology Research Branch, Division of Extramural Research Programs, pp. 534–7.

Guze, S.B. (1992). *Why Psychiatry is a Branch of Medicine*. New York: Oxford University Press.

Guze, S. & Robins, E. (1970). Establishment of diagnostic validity in psychiatric illness: its application to schizophrenia. *American Journal of Psychiatry*, **126**, 983–7.

Hamilton, M. (1967). Development of a rating scale for primary depressive illness. *British Journal of Social and Clinical Psychology*, **6**, 278–96.

Haynes, R.B., Ackloo, E., Sahota, N., *et al.* (2008). Interventions for enhancing medication adherence. *Cochrane Database of Systematic Reviews*, **2**, CD000011.

Hirschfeld, R.M., Lewis, L., Vornik, L.A. (2003). Perceptions and impact of bipolar disorder: how far have we really come? Results of the national depressive and manic-depressive association 2000 survey of individuals with bipolar disorder. *Journal of Clinical Psychiatry*, **64**, 161–74.

Hirschfeld, R. M., Williams, J.B., Spitzer, R.L., *et al.* (2000). *The American Journal of Psychiatry*, **157**, 1873–5.

Judd, L.L., Akiskal, H.S., Schettler, P.J., *et al.* (2002). The long-term natural history of the weekly symptomatic status of bipolar I disorder. *Archives of General Psychiatry*, **59**, 530–7.

Judd, L.L., Akiskal, H.S., Schettler, P.J., *et al.* (2003). A prospective investigation of the natural history of the long-term weekly symptomatic status of bipolar II disorder. *Archives of General Psychiatry*, **60**, 261–9.

Judd, L. L., Schettler, P.J., Akiskal, H.S., *et al.* (2008). Residual symptom recovery from major affective episodes in bipolar disorders and rapid episode relapse/recurrence. *Archives of General Psychiatry*, **65**, 386–94.

Kamat, S.A., Rajagopalan, K., Pethick, N., *et al.* (2008). Prevalence and humanistic impact of potential misdiagnosis of bipolar disorder among patients with major depressive disorder in a commercially insured population. *Journal of Managed Care Pharmacy*, **14**, 631–42.

Keller, M.B. (2006). Prevalence and impact of comorbid anxiety and bipolar disorder. *Journal of Clinical Psychiatry*, **67** (Suppl. 1), 5–7.

Kessler, R.C., Rubinow, D.R., Holmes, C., *et al.* (1997). The epidemiology of DSM-III-R bipolar I disorder in a general population survey. *Psychological Medicine*, **27**, 1079–89.

Lish, J.D., Dime-Meenan, S., Whybrow, P.C., *et al.* (1994). The National Depressive and Manic-depressive Association (DMDA) survey of bipolar members. *Journal of Affective Disorders*, **31**, 281–94.

Meyer, F. & Meyer, T.D. (2009). The misdiagnosis of bipolar disorder as a psychotic disorder: some of its causes and their influence on therapy. *Journal of Affective Disorders*, **112**, 174–83.

Michalak, E. E., Murray, G., Young, A.H., Lam, R.W. (2008). Burden of bipolar depression: impact of disorder and medications on quality of life. *CNS Drugs*, **22**, 389–406.

Michalak, E. E., Yatham, L.N., Lam, R.W. (2005). Quality of life in bipolar disorder: a review of the literature. *Health and Quality of Life Outcomes*, **3**, 72.

Montgomery, S. A., Asberg, M. (1979). A new depression scale designed to be sensitive to change. *British Journal of Psychiatry*, **134**, 382–9.

Namjoshi, M. A. & Buesching, D.P. (2001). A review of the health-related quality of life literature in bipolar disorder. *Quality of Life Research*, **10**, 105–15.

Niv, N., Cohen, A.N., Sullivan, G., Young, A.S. (2007). The MIRECC version of the Global Assessment of Functioning scale: reliability and validity. *Psychiatric Services (Washington, D.C.)*, **58**, 529–35.

Osterberg, L. & Blaschke, T. (2005). Adherence to medication. *New England Journal of Medicine*, **353**, 487–97.

Parikh, S.V. (2007). *Bipolar disorder. Therapeutic Choices*, 5th edition. Ottawa: Canadian Pharmacists Association.

Perlis, R.H. (2005). Misdiagnosis of bipolar disorder. *American Journal of Managed Care*, **11** (Suppl. 9), S271–4.

Revicki, D. A., Matza, L.S., Flood, E., Lloyd, A. (2005). Bipolar disorder and health-related quality of life: review of burden of disease and clinical trials. *Pharmacoeconomics*, **23**, 583–94.

Rosa, A. R., Sanchez-Moreno, J., Martinez-Aran, *et al.* (2007). Validity and reliability of the Functioning Assessment Short Test (FAST) in bipolar disorder. *Clinical Practice and Epidemiology in Mental Health*, **3**, 5.

Suppes, T., Baldessarini, R.J., Faedda, G.L., *et al.* (1991). Risk of recurrence following discontinuation of lithium treatment in bipolar disorder. *Archives of General Psychiatry*, **48**, 1082–8.

Suppes, T., Dennehy, E.B., Gibbons, E.W. (2000). The longitudinal course of bipolar disorder. *Journal of Clinical Psychiatry*, **61** (Suppl. 9), 23–30.

Williams, J. B. (1988). A structured interview guide for the Hamilton Depression Rating Scale. *Archives of General Psychiatry*, **45**, 742–7.

Young, R. C., Biggs, J.T., Ziegler, V.E., Meyer, D.A. (1978). A rating scale for mania: reliability, validity and sensitivity. *British Journal of Psychiatry*, **133**, 429–35.

Zigmond, A. S. & Snaith, R.P. (1983). The Hospital Anxiety and Depression Scale. *Acta Psychiatrica Scandinavica*, **67**, 361–70.

Chapter 19

Practical guide to brain imaging in bipolar disorder

I. Nicol Ferrier and Adrian J. Lloyd

Introduction

The research literature on bipolar affective disorder abounds with both structural and functional imaging studies, but imaging is little used to guide the clinical management of this illness. Are there imaging techniques that should be more routinely applied to clinical practice or, if not, why is this the case when these methods are so widely employed as research tools? This chapter will briefly consider the imaging findings in the research literature to give context to the subsequent discussion of the applicability of imaging to routine clinical practice.

Main imaging techniques and findings in bipolar affective disorder

Structural imaging (computerised tomorgraphy [CT] and magnetic resonance imaging [MRI]) give information on changes in appearance and volume of brain structures, yielding data on generalised and localised atrophy or hypertrophy or altered tissue composition (for example, increased white matter image intensity on T2-weighted sequences that can correspond to altered water content, demyelination and/or be associated with vascular change). In addition, CT and MRI can be useful in identifying structural lesions, including vascular malformations, tumours and infarcts. The findings in structural brain imaging studies in bipolar disorder (BD) have been contradictory across many studies, probably owing to differences in populations and methodologies. The only regional volumetric change that stands up to meta-analysis is that of lateral ventricular enlargement (Soares & Mann, 1997a; Haldane & Frangou, 2004; McDonald *et al.*, 2004b). However, the range of volumetric changes that have been reported to be associated with BD with sufficient frequency to warrant further consideration include white matter volume reduction (Bruno *et al.*, 2004; Davis *et al.*, 2004; Haznedar *et al.*, 2005), reduced anterior cingulate gyrus volume (Haldane & Frangou, 2004; Sassi *et al.*, 2004; McDonald *et al.*, 2004a), possibly increased amygdala volume in adults (Altshuler *et al.*, 2000; Brambilla *et al.*, 2003; Frangou, 2005) but reduced volume in adolescents (Chen *et al.*, 2004; Blumberg *et al.*, 2005), increased striatal volumes (Strakowski *et al.*, 2002) and possibly cerebellar changes (Soares & Mann, 1997a; DelBello *et al.*, 1999). Changes in some of these areas, particularly those within fronto-subcortical circuits, are supported by data from other methodologies, such as detailed neuropsychological assessment (Bearden *et al.*, 2001; Thompson *et al.*, 2005). There is a strong evidence base for the association of MRI white matter hyperintensities with BD, possibly with the frontal lobes being most affected, and also limited evidence exists for increased hyperintensities in the subcortical grey nuclei, particularly the putamen (Dupont *et al.*, 1995; Altshuler *et al.*, 1995; Soares & Mann, 1997a; Bearden *et al.*, 2001; Moore *et al.*, 2001; Harrison, 2002; Kempton *et al.*, 2008; Lloyd *et al.*, 2009). Specific neuropathological data on the nature of white matter hyperintensities are lacking in bipolar subjects, and any conclusions with regard to potential cellular changes are by extrapolation from other conditions such as unipolar depression, where a relationship between hyperintensities and vascular pathology has been established in late-life depression (Thomas *et al.*, 2002a).

Magnetic resonance diffusion tensor imaging yields a dataset that allows the integrity of white matter to be considered by examining the way in which protons (essentially water) diffuse within the tissue and the degree to which this diffusion is constrained by surrounding cellular components (i.e. the diffusion is asymmetrical, or *anisotropic*). Early findings are indicative of disruption to the integrity of white matter tracts with increased diffusivity and reduced anisotropy in both the deep white matter and the corpus callosum (Adler *et al.*, 2004; Yurgelun-Todd *et al.*, 2007; Bruno *et al.*, 2008; Wang F *et al.*, 2008).

In functional imaging, glucose metabolism has been shown to be reduced in the prefrontal regions and

Practical Management of Bipolar Disorder, eds. Allan H. Young, I. Nicol Ferrier and Erin E. Michalak. Published by Cambridge University Press.

increased in the amygdala in depressed phases. Increases in activity in the anterior cingulate and head of caudate with positron emission tomography (PET) and single photon emission computed tomography imaging have been reported in BD. Brain activation with specific tasks and measured with functional MRI (fMRI) has shown both increased and decreased responses to word generation tasks in different specific areas of the prefrontal cortex and increased response to emotional tasks in the amygdala in bipolar patients (Soares & Mann, 1997b; Haldane & Frangou, 2004; Strakowski *et al.*, 2005).

Some of the changes summarised above suggest a possible developmental aspect to BD, e.g. amygdala volume that may be reduced in adolescents but increased in adults. Also anterior cingulate and white matter changes exist early in the course of illness and may predate its onset (Strakowski *et al.*, 2005), whereas other regions appear to degenerate with repeated affective episodes (e.g. alterations in lateral ventricle volumes and inferior prefrontal regions) and may therefore represent the effects of illness progression (Goodwin *et al.*, 2008). A 4-year follow-up study of patients and age-matched controls suggested a reduction in memory function and loss of grey matter volume in the medial temple cortex related to illness intensity (Moorhead *et al.*, 2007). This is evidence in favour of a correlation between illness course, brain change and cognition, but does not establish the direction of causality. In any event, the above changes, even those found early in the course of illness and in drug-naive patients, are not of sufficient reliability or robust enough to provide us with diagnostic or prognostic markers.

Given that the methodologies listed above are contributing so valuably to our understanding of the neurobiology of BD, it is perhaps surprising, as indicated in the introduction to this chapter, that they do not routinely form part of the investigative armamentarium in this illness. The key reason for this lies in the fact that the differences defined in controlled research studies show significant separation between sufficiently large samples of control and bipolar subjects, but with a large degree of overlap in the distributions of the two groups. The various measurements made lie on a continuum within each group, and thus the investigations are of little help diagnostically when used in individuals.

Where does this leave imaging in routine clinical management of BD – are there circumstances in which it should be employed?

There are two broad areas of practice where imaging may be of value:

1. where there are symptoms or signs suggestive of an underlying organic aetiology to the presentation that requires definition and that might also require specific treatment
2. where imaging data might help predict outcome and prognosis and be of value in directing treatment modalities or in giving valuable information to a patient or their carers about the nature of the illness.

Imaging for diagnostic purposes

As stated above, imaging is not of help in the differential diagnosis of BD from other 'functional' psychiatric illnesses. A useful question to ask is: Epidemiologically, what proportion of new presentations of BD are organic in nature? The concept of this relationship between physical conditions and elevated mood was introduced as 'secondary mania' by Krauthammer and Klerman (1978), and this has been supported by others since (Stone, 1989; Shulman, 1997).

There are numerous case reports of mania associated with neurological disorders of many types, including infections, malignancy and neuroendocrine problems, but perhaps the most commonly reported link is with cerebrovascular disease (e.g. Starkstein & Robinson, 1989). Pathology of the right cerebral hemisphere and also that affecting the basal ganglia, orbitofrontal regions and temporal lobe is reported as being associated with mania (Starkstein & Robinson, 1989; Shulman, 1997).

Evidence suggests that manic syndromes are associated more often with neurological conditions when they have their onset later in life. One study reported association with neurological disorders in up to 25% of individuals with onset of manic symptoms after age 50 (Snowdon, 1991). It has been reported that neurological disorders are associated more often with late-onset mania than they are with late-onset unipolar depression (Krishnan, 2002). However, there are difficulties with these data. Such studies generally consider manic symptoms rather than BD as defined by accepted classifications of psychiatric illnesses and are generally cross-sectional and retrospective (Krishnan, 2002). One study that examined subjects with clearly defined BD found that although the proportion of late-onset mania associated with organic conditions (2.8%) exceeded that of early-onset mania (1.2%), the absolute proportions were substantially smaller than those found when more broadly defined manic symptoms are considered (Almeida & Fenner, 2002). Mania can have a temporal relationship with physical problems in up to 12% of patients (Jacoby, 1997), but this does not necessarily mean that there is direct causality. One specific group of disorders that might seem worthy of consideration in terms of importance in psychiatry and with possible neuroimaging changes among older patients is the dementias. However, symptoms of the disinhibition that can be a feature of frontal lobe syndromes in dementia are much more common rather than the true manic syndrome. In reviewing this topic, Shulman did, however, note a case report of Lewy body dementia associated with a manic presentation (Shulman, 1997).

Therefore, it would appear that neuroimaging is only justifiable diagnostically if there is some pre-existing reason from the history or examination to suggest an underlying neurological problem that requires further investigation in its own right. It would not be appropriate currently to recommend routine neuroimaging to investigate BD diagnostically in other circumstances. This conclusion is in agreement with the UK NICE clinical guideline (CG) 38 for BD (National Collaborating Centre for Mental Health, 2006) and the technology appraisal (TA) 136 on structural neuroimaging in first-episode psychosis (including mania) (National Collaborating Centre for Mental Health, 2008). The CG suggests only undertaking imaging in BD when an underlying organic cause is suspected. The TA highlights the risks of routinely undertaking imaging, including: (1) problems for those with metallic implants, etc.; (2) anxiety of false positives owing to the relatively common incidental findings of no medical consequence; (3) the 5–10% rate of intolerance of scanning owing to claustrophobia with MRI; and (4) radiation exposure and reactions to contrast dyes with CT. The TA also gives a best estimate from the literature of the proportion of cases in which routine neuroimaging might have a direct effect on clinical management in first-episode psychosis – around 5% for MRI and around 0.5% for CT. In addition to the difficulties of adverse effects and false positives, the NICE TA concludes that routine neuroimaging in this group of patients is not cost-effective in terms of cost per quality-adjusted life year (QALY) unless it is assumed that detection of an abnormality would lead to effective treatment and the resolution of both the neurological condition and psychiatric symptoms – an outcome that is unlikely in many cases in clinical practice. Other well-recognised guidelines relating to BD from the American Psychiatric Association and the British Association for Psychopharmacology do not consider the appropriateness or otherwise of neuroimaging (APA, 2002; Goodwin, 2003).

Imaging for prognostic reasons

In contrast to the diagnostic assessment of acute presentations of BD, consideration may be given to imaging in circumstances where there is marked resistance to treatment or where there is a clinical history suggestive of neurological conditions that might affect outcome and warrant reassessment in view of the severity of psychiatric symptoms. In unipolar depression, a clear association has been defined between white matter MRI hyperintensities and vascular pathology and outcome (Thomas *et al.*, 2002a, 2002b), and although such a clear relationship has not been established in BD, there is evidence of an association between the illness outcome (particularly in illness of greater severity) and MRI white matter lesions (Moore *et al.*, 2001; Pompili 2007; Regenold 2008) and between white matter lesions and vascular disease (Fazekas *et al.*, 1993; Liao *et al.*, 1997; de Groot *et al.*, 2000; de Leeuw *et al.*, 2000). Findings of white matter lesions or vascular-type change on neuroimaging are not yet predictive of a need for particular treatment regimes, but may be of value in terms of information that can be discussed with sufferers and their carers that may aid the understanding of their illness and the reasons for relative treatment resistance. The neuroimaging findings could lead to the clinical decision to try additional therapies, but this is not, as yet, evidence based. It is likely that such circumstances would be uncommon and consideration of using imaging in this way correspondingly infrequent.

Conclusions

There would not currently appear to be a place for routine use of clinical neuroimaging in the assessment and treatment of patients presenting with symptoms suggesting BD. However, the important differential of an underlying organic cerebral disorder must be given consideration in the formulation of such presentations, and where history or physical examination suggest this aetiology, imaging can have an important role in investigating it further. Information from imaging investigations may be helpful in explaining the nature of illness to those who experience it but, as yet, is not able to guide specific management plans in the absence of defineable intracranial pathology or further research evidence.

References

Adler, C.M., Holland, S.K., Schmithorst, V., *et al.* (2004). Abnormal frontal white matter tracts in bipolar disorder: a diffusion tensor imaging study. *Bipolar Disorders*, **6**, 197–203.

Almeida, O.P.& Fenner, S. (2002). Bipolar disorder: similarities and differences between patients with illness onset before and after 65 years of age. *International Psychogeriatrics*, **14**, 311–22.

Altshuler, L.L., Bartzokis, G., Grieder, T., *et al.* (2000). An MRI study of temporal lobe structures in men with bipolar disorder or schizophrenia. *Biological Psychiatry*, **48**, 147–62.

Altshuler, L.L., Curran, J.G., Hauser, P., *et al.* (1995). T2 hyperintensities in bipolar disorder: magnetic resonance imaging comparison and literature meta-analysis. *American Journal of Psychiatry*, **152**, 1139–44.

APA (2002). Practice guideline for the treatment of patients with bipolar disorder (revision). *American Journal of Psychiatry*, **159**, 1–50.

Bearden, C.E., Hoffman, K.M., Cannon, T.D. (2001). The neuropsychology and neuroanatomy of bipolar affective disorder: a critical review. *Bipolar Disorders*, **3**, 106–50.

Blumberg, H.P., Fredericks, C., Wang, F., *et al.* (2005). Preliminary evidence for persistent abnormalities in amygdala volumes in adolescents and young adults with bipolar disorder. *Bipolar Disorders*, **7**, 570–6.

Brambilla, P., Harenski, K., Nicoletti, M., *et al.* (2003). MRI investigation of temporal lobe structures in bipolar patients. *Journal of Psychiatric Research*, **37**, 287–95.

Bruno, S.D., Barker, G.J., Cercignani, M., *et al.* (2004). A study of bipolar disorder using magnetization transfer imaging and voxel-based morphometry. *Brain*, **127**, 2433–40.

Bruno, S., Cercignani, M., Ron, M.A. (2008). White matter abnormalities in bipolar disorder: a voxel-based diffusion tensor imaging study. *Bipolar Disorders*, **10**, 460–8.

Chen, B.K., Sassi, R., Axelson, D., Hatch, J.P., *et al.* (2004). Cross-sectional study of abnormal amygdala development in adolescents and young adults with bipolar disorder. *Biological Psychiatry*, **56**, 399–405.

Davis, K.A., Kwon, A., Cardenas, V.A., Deicken, R.F. (2004). Decreased cortical gray and cerebral white matter in male patients with familial bipolar I disorder. *Journal of Affective Disorders*, **82**, 475–85.

de Groot, J.C., de Leeuw, F.E., Oudkerk, M., *et al.* (2000). Cerebral white matter lesions and cognitive function: the Rotterdam Scan Study. *Annals of Neurology*, **47**, 145–51.

DelBello, M.P., Strakowski, S.M., Zimmerman, M.E., *et al.* (1999). MRI analysis of the cerebellum in bipolar disorder: a pilot study. *Neuropsychopharmacology*, **21**, 63–8.

de Leeuw, F.E., de Groot, J.C., Oudkerk, M., *et al.* (2000). Aortic atherosclerosis at middle age predicts cerebral white matter lesions in the elderly. *Stroke*, **31**, 425–9.

Dupont, R.M., Jernigan, T.L., Heindel, W., *et al.* (1995). Magnetic resonance imaging and mood disorders. Localization of white matter and other subcortical abnormalities. *Archives of General Psychiatry*, **52**, 747–55.

Fazekas, F., Kleinert, R., Offenbacher, H., *et al.* (1993). Pathologic correlates of incidental MRI white matter signal hyperintensities. *Neurology*, **43**, 1683–9.

Frangou, S. (2005). The Maudsley Bipolar Disorder Project. *Epilepsia*, **46**, 19–25.

Goodwin, G.M.; Consensus Group of the British Association for Psychopharmacology (2003). Evidence-based guidelines for treating bipolar disorder: recommendations from the British Association Psychopharmacology. *Journal of Psychopharmacology*, **17**, 149–73.

Goodwin, G.M., Martinez-Aran, A., Glahn, D.C., Vieta, E. (2008). Cognitive impairment in bipolar disorder: neurodevelopment or neurodegeneration? An ECNP expert meeting report. *European Neuropsychopharmacology*, **18**, 787–93.

Haldane, M.& Frangou, S. (2004). New insights help define the pathophysiology of bipolar affective disorder: neuroimaging and neuropathology findings. *Progress in Neuro-psychopharmacology & Biological Psychiatry*, **28**, 943–60.

Harrison, P.J. (2002). The neuropathology of primary mood disorder. *Brain*, **125**, 1428–49.

Haznedar, M.M., Roversi, F., Pallanti, S., *et al.* (2005). Fronto-thalamo-striatal gray and white matter volumes and anisotropy of their connections in bipolar spectrum illnesses. *Biological Psychiatry*, **57**, 733–42.

Jacoby, R. (1997). Manic illness. In: Jacoby, R. & Oppenheimer, C. (Eds). *Psychiatry in the Elderly*. Oxford University Press.

Kempton M.J., Geddes J.R., Ettinger U., *et al.* (2008). Meta-analysis, database, and meta-regression of 98 structural imaging studies in bipolar disorder. *Archives of General Psychiatry*, **65**, 1017–32.

Krauthammer, C.& Klerman, G.L. (1978). Secondary mania: manic syndromes associated with antecedent physical illness or drugs. *Archives of General Psychiatry*, **35**, 1333–9.

Krishnan, K.R. (2002). Biological risk factors in late life depression. *Biological Psychiatry*, **52**, 185–92.

Liao, D., Cooper, L., Cai, J., *et al.* (1997). The prevalence and severity of white matter lesions, their relationship with age, ethnicity, gender, and cardiovascular disease risk factors: the ARIC Study. *Neuroepidemiology*, **16**, 149–62.

Lloyd, A.J., Moore, P. B., Cousins, D. A., *et al.* (2009). White matter lesions in euthymic patients with bipolar disorder. *Acta Psychiatrica Scandinvica*, **120**, 481–91.

Mcdonald, C., Bullmore, E.T., Sham, P.C., *et al.* (2004a). Association of genetic risks for schizophrenia and bipolar disorder with specific and generic brain structural endophenotypes. *Archives of General Psychiatry*, **61**, 974–84.

Mcdonald, C., Zanelli, J., Rabe-Hesketh, S., *et al.* (2004b). Meta-analysis of magnetic resonance imaging brain morphometry studies in bipolar disorder. *Biological Psychiatry*, **56**, 411–7.

Moore, P.B., Shepherd, D.J., Eccleston, D., *et al.* (2001). Cerebral white matter lesions in bipolar affective disorder: relationship to outcome. *British Journal of Psychiatry*, **178**, 172–6.

Moorhead, T. W., Mckirdy, J., Sussmann, J. E., *et al.* (2007). Progressive graymatter loss in patients with bipolar disorder. *Biological Psychiatry*, **62**, 894–900.

National Collaborating Centre for Mental Health (2006). *Bipolar Disorder: the Management of Bipolar Disorder in Adults, Children and Adolescents, in Primary and Secondary Care*. London: NICE. www.guidance.nice.org.uk/CG38 [accessed 3 January 2010].

National Collaborating Centre for Mental Health (2008). *Structural Neuroimaging in First-episode Psychosis*. London: NICE. www.guidance.nice.org.uk/TA136 [accessed 3 January 2010].

Pompili, M., Ehrlich, S., De Pisa, E., *et al.* (2007). White matter hyperintensities and their associations with suicidality in patients with major affective disorders. *European Archives of Psychiatry and Clinical Neuroscience*, **257**, 494–9.

Regenold, W.T., Hisley, K.C., Phatak, P., *et al.* (2008). Relationship of cerebrospinal fluid glucose metabolites to MRI deep white matter hyperintensities and treatment resistance in bipolar disorder patients. *Bipolar Disorders*, **10**, 753–64.

Sassi, R.B., Brambilla, P., Hatch, J.P., *et al.* (2004). Reduced left anterior cingulate volumes in untreated bipolar patients. *Biological Psychiatry*, **56**, 467–75.

Shulman, K.I. (1997). Disinhibition syndromes, secondary mania and bipolar disorder in old age. *Journal of Affective Disorders*, **46**, 175–82.

Snowdon, J. (1991). A retrospective case-note study of bipolar disorder in old age. *British Journal of Psychiatry*, **158**, 485–90.

Soares, J.C.& Mann, J.J. (1997a). The anatomy of mood disorders-review of structural neuroimaging studies. *Biological Psychiatry*, **41**, 86–106.

Soares, J.C.& Mann, J.J. (1997b). The functional neuroanatomy of mood disorders. *Journal of Psychiatric Research*, **31**, 393–432.

Starkstein, S.E.& Robinson, R.G. (1989). Affective disorders and cerebral vascular disease. *British Journal of Psychiatry*, **154**, 170–82.

Stone, K. (1989). Mania in the elderly. *British Journal of Psychiatry*, **155**, 220–4.

Strakowski, S.M., DelBello, M.P., Adler, C.M. (2005). The functional neuroanatomy of bipolar disorder: a review of neuroimaging findings. *Molecular Psychiatry*, **10**, 105–16.

Strakowski, S.M., DelBello, M.P., Zimmerman, M.E., *et al.* (2002). Ventricular and periventricular structural volumes in first- versus multiple-episode bipolar disorder. [comment]. *American Journal of Psychiatry*, **159**, 1841–7.

Thomas, A.J., O'brien, J.T., Davis, S., *et al.* (2002a). Ischemic basis for deep white matter hyperintensities in major depression: a neuropathological study. *Archives of General Psychiatry*, **59**, 785–92.

Thomas, A.J., Perry, R., Barber, R., *et al.* (2002b). Pathologies and pathological mechanisms for white matter hyperintensities in depression. *Annals of the New York Academy of Sciences*, **977**, 333–9.

Thompson, J.M., Gallagher, P., Hughes, J.H., *et al.* (2005). Neurocognitive impairment in euthymic patients with bipolar affective disorder. *British Journal of Psychiatry*, **186**, 32–40.

Wang, F., Kalmar, J.H., Edmiston, E., *et al.* (2008). Abnormal corpus callosum integrity in bipolar disorder: a diffusion tensor imaging study. *Biological Psychiatry*, **64**, 730–3.

Yurgelun-Todd, D.A., Silveri, M.M., Gruber, S.A., *et al.* (2007). White matter abnormalities observed in bipolar disorder: a diffusion tensor imaging study. *Bipolar Disorders*, 9, 504–12.

Chapter 20

From neuroscience to clinical practice

Annie J. Kuan, Vivienne A. Curtis and Allan H. Young

Introduction

I have seen you in various stages of undress
I have seen you through various states of madness
I have seen your refractions
And I did not recognize you
I have seen you through various states of madness
How high are your highest of heights
How low are your lows
I have seen you in the eyes
Of a thousand other stranger faces
I have seen you in unlikely and unfamiliar places
I have seen you be reckless in matters of love
I have seen by degrees the boiling point come and go
What lies at the end of <this> long and dark twisted road
How high are your highest of heights
How low are your lows
(Reprinted with permission from *Various Stages* by Tony Dekker [©Harbour Songs 2005] – from *Bodies and Minds* by Great Lake Swimmers.)

Bipolar disorder (BD) is a chronic illness characterised by recurrent episodes of depression and mania or hypomania, with a prevalence rate of 1–2% (Sajatovic, 2005). It is a highly disabling mood disorder interfering with cognition, behaviour, daily activities such as education and work, quality of life and interpersonal relationships (Zarate *et al.*, 2000). It also results in high mortality rates (Angst *et al.*, 2002). The common age of onset is between 17 and 21 years (Perlis *et al.*, 2004), with BD type I affecting men and women equally but BD type II appearing to be more common in women (Leibenluft, 1996). Bipolar individuals may experience depressive symptoms 40–50% of the time (Judd *et al.*, 2002), and have a mean recurrence rate of 50% at one year after an index manic episode and a relapse rate of 70–90% at 5 years (Bowden & Krishnan, 2004). Compared with a healthy population, those with BD have a significantly higher utilisation of health services and increased need for welfare or disability benefits, and higher rate of suicidal behaviour (Judd & Akiskal, 2003). It is estimated that a bipolar woman after age 25 without appropriate treatment may lose up to 14 years of productive functioning (e.g. employment, schooling, family activity), 12 years of normal health and possibly 9 years of life (US DHEW Medical Practice Project, 1979). In 2000, BD ranked in the top ten causes of non-fatal illnesses in the world, accounting for 2.5% of total years lost to disability (YLD) globally (Ayuso-Mateos, 2002).

One of the biggest challenges that academics and clinicians face is the translation of research evidence into clinical practice and significant benefits for the patients. Within the academic community there is now an emphasis on clinically relevant and meaningful research, but the impact of neuropsychological deficits on day-to-day functioning is rarely discussed. The media and the Internet are important conduits for the dissemination of research findings; thus patients and their caregivers are increasingly well informed. They will often question how the research that they may read about relates to them, specifically: the relationship between hormones and mood; cognitive functioning between episodes; the physical consequences of BD; illness progression; cognitive predictors of family risk; detection of emotions; the effects of medications on thinking; the relationship between cognition and psychosocial functioning; and psychotherapies and their effects on the brain.

In this chapter we will address these questions and, in doing so, attempt to put current neuroscientific knowledge into a clinical context.

How do hormones affect moods?

Both men and women can have alterations in mood as a consequence of hormonal dysregulation.

Whereas for some conditions, such as thyroid dysfunction, the mood change is secondary to end-organ dysfunction, there is also a large body of evidence to suggest that mood disorders are linked to dysregulation in the hypothalamic-pituitary-adrenal (HPA) axis. This disturbance can affect not only mood but also information processing and related memory loss (Belanoff

Practical Management of Bipolar Disorder, eds. Allan H. Young, I. Nicol Ferrier and Erin E. Michalak. Published by Cambridge University Press.

et al., 2001; Young, 2004). Although first reported in unipolar depression, abnormal HPA axis functioning has been demonstrated in BD (Daban *et al.*, 2005), and the extent of abnormality appears to be equal in remitted and symptomatic patients, suggesting that the HPA axis dysfunction may be an enduring marker in bipolar illness (Watson *et al.*, 2004).

Disturbance in HPA, particularly elevated cortisol levels, can cause a neurocognitive impairment (Belanoff *et al.*, 2001), and the close inter-relationship between hypercortisolaemia, residual mood symptoms and cognitive impairment is a source of confounding in many studies. By investigating the relationship between salivary cortisol and neuropsychological function in a matched euthymic bipolar and control groups, Thompson *et al.* (2005) found that impairments in verbal and executive function persist in the euthymic phase of bipolar illness. In these participants there were no differences in basal cortisol levels, but the bipolar group demonstrated impairments in attention, executive function, immediate memory and verbal and visuospatial declarative memory compared with the control group. This suggests that residual mood symptoms and, at least by this measure, cortisol output were not the source of the cognitive deficits.

As there is a suggestion that the HPA dysfunction may cause part of the cognitive impairments, Watson *et al.* (2006) examined the correlation between glucocorticoid receptor (GR) function and neuropsychological function in euthymic bipolar patients versus controls by performing the dexamethasone suppression test (DST) and measures of verbal declarative memory and working memory.

The results show that those with BD made significantly more erroneous responses relative to controls on the cognitive measures and that the working memory commission errors were positively correlated to post-dexamethasone cortisol levels in the bipolar patients. This relationship was not demonstrated in the controls, suggesting that the persistent impairment in executive performance seen in mood disorders is related to abnormal GR function. This implies that the GR is an important modulator of neurocognition and mood in bipolar illness (Young, 2004).

Hormonal state can also affect clinical presentations for women with BD. In addition to differences in clinical course between genders (women having a greater tendency for BD type II, depressive symptoms and a rapid cycling course), events unique to women (particularly hormonal variability across the menstrual cycle, contraception and pregnancy, lactation) can impact on the management of mood disorders.

The influence of the reproductive cycle on the course of BD is discussed in Chapter 10 of this book.

Will thinking be clear between episodes?

Bipolar disorder can affect all aspects of daily functioning, including cognitive skills – particularly attention, working memory, learning and executive function. Although there are demonstrated patterns of cognitive dysfunction in BD, there is no definitive cognitive profile that classifies bipolar patients owing to the range of psychosocial influences and demographic variables, selection criteria, cognitive overlap with other disorders, medications and illness severity and chronicity (Osuji & Cullum, 2005; Bearden *et al.*, 2001). Studies have attempted to address these problems at the source by using bipolar and control populations without co-morbid substance abuse, across the range of clinical presentations and early in their course of illness.

Rubinsztein *et al.* (2000) investigated neuropsychological function in a well-defined clinically remitted group of bipolar outpatients without a history of alcohol or drug dependence. The patients were between the ages of 18 and 60 and had to have been in remission for at least four months. Compared with healthy controls, these remitted patients with reported good social adaptation still displayed generalised deficits in memory and executive functioning. There was no difference between the remitted group and controls in impairment of accuracy on executive functioning tasks, but executive response latency was increased and visuospatial recognition memory was impaired.

Malhi *et al.* (2004) reviewed the literature comparing the neuropsychological profile of the three phases of BD (depression, mania or hypomania and euthymia) and concluded that the most common deficits were in executive functioning, attention and memory, which do not appear to be confined to any illness phase. In a study that controlled for estimated pre-morbid IQ, poorer performance in executive functioning and verbal memory was observed compared with controls across all bipolar phases (Martinez-Aran *et al.*, 2004a).

Cognitive impairment has also been reported in young euthymic patients with bipolar spectrum disorder in the areas of executive functioning and verbal memory compared with remitted young adults with recurrent major depression and matched controls

(Smith *et al.*, 2006). The remitted bipolar spectrum patients consistently showed poorer performance compared with the other groups, but the euthymic major depressed patients were in between the euthymic bipolar group and the controls in the cognitive dysfunction realm. Smith and colleagues suggest that these deficits in the bipolar spectrum group are an influence of bipolar predisposition more so than the severity of the unipolar illness (Smith *et al.*, 2006).

A recent meta-analysis reviewed the cognitive impairments in euthymic bipolar patients (Robinson *et al.*, 2006) and provided strong support for residual deficits in executive functioning and verbal learning in remitted bipolar illness. One key finding of this meta-analysis is that executive functions are not all impaired equally but rather more selectively. This review also concluded that these impairments appear to relate to illness or treatment factors and are cross-cultural, and not due to pre-morbid IQ differences.

Taken together, these findings suggest that cognitive impairment endures in remission and is independent of clinical state in BD. Neuroimaging provides us with a tool to investigate neuroanatomical correlates of these neuropsychological impairments, and may provide us with clues as to origin of the variability in results between study cohorts. A collaborative study between the Institute of Psychiatry in London, UK and the Department of Psychiatry at the Royal Victoria Infirmary in Newcastle, UK investigated this with a series of functional magnetic resonance imaging (fMRI) experiments in a cohort of euthymic BD type I patients and matched controls. Experiments were chosen that related to both neuropsychological and clinical observations. Thus, as disturbances in working memory are consistently reported in euthymic bipolar individuals and disorders of language production are clinically associated with relapse and can be indicative of a change in clinical state, participants undertook tasks of working memory and language production within the same scanning session. Monks *et al.* (2004) report the evidence for hypofrontality in the bipolar participants while performing tasks dependent on the central executive components of working memory (see Figure 20.1), whereas Curtis *et al.* (2007) report hyperfrontal responses during completion of language tasks (see Figure 20.2).

The region of prefrontal cortex identified within these studies is known to be under dopaminergic control, particularly sensitive to task demand, and activation varies with performance. These findings are in keeping with other studies of altered dopaminergic states and suggest that the altered dopaminergic function may be central to executive dysfunction in BD.

Are there any physical consequences of bipolar disorder?

Neurological soft signs including movement disorders have long been observed in schizophrenia and neurodegenerative disorders but have also been reported in bipolar patients in North America (Nashrallah *et al.*, 1983) as well as in India (Goswami *et al.*, 1998) and more recently in euthymic and symptomatic Ethiopian bipolar patients (Negash *et al.*, 2004). In one of the first published studies investigating neurological soft signs and neurocognitive functioning in remitted bipolar patients compared with healthy controls, Goswami *et al.* (2006) confirmed the presence of neurological soft signs in the euthymic bipolar group and significant impairment in executive functioning. Social disability was correlated with neurological soft sign severity whereas neurological soft signs were linked with executive function and verbal memory measures. Most interestingly, the neurocognitive impairments in Goswami's New Delhi euthymic bipolar sample are consistent with the results found by Thompson *et al.* (2005) in the UK, thus suggesting these cognitive deficits are independent of culture.

In a long-term prospective study of hospitalised affective disordered patients, Angst *et al.* (2002) found that inpatient men and women had increased mortality rates from suicide and circulatory diseases (cardiovascular for bipolar patients, cerebrovascular disorders for unipolar patients). The unipolar patients had significantly higher suicide rates than those with bipolar illness. Importantly, those in the unipolar and bipolar groups treated with long-term pharmacotherapy had a 2.5-fold reduction in suicide mortality compared with untreated patients. The difference in the survival analysis and standardised mortality ratios between treated and non-treated bipolar patients was higher compared with the treated versus non-treated unipolar patients. Increased mortality in cerebrovascular disorder was associated with neuroleptic treatment, whereas increased cardiovascular deaths were linked to a combination of antidepressants and neuroleptics. Regardless, in the studied inpatients, long-term medication therapy significantly decreased suicide rates, despite the fact that those treated were more severely ill. This emphasises the importance of long-term pharmacotherapy in mood disorders, as suicide prevention is one of the primary goals of treatment.

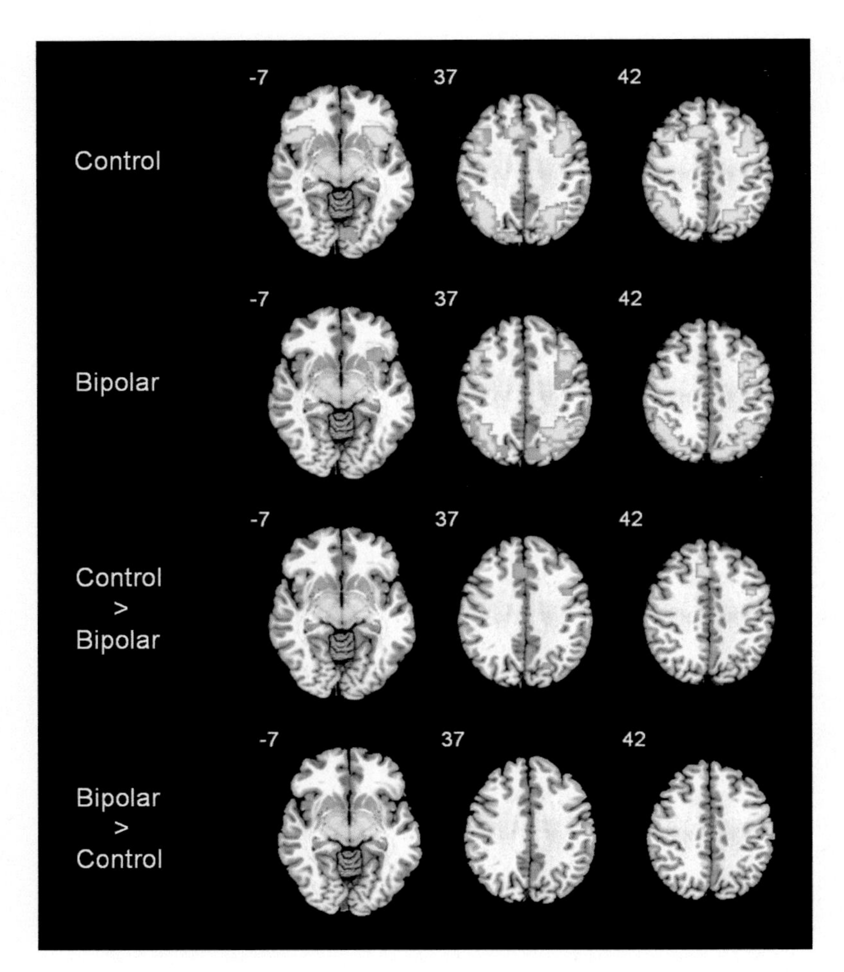

Figure 20.1 Task-dependent responses in fMRI: two-back task (adapted from Monks *et al.*, 2004). These brain maps represent regions of BOLD response during performance of a two-back task by euthymic bipolar (n=12) and control (n=12) participants. Yellow clusters = regions with significant activation during two-back condition. See colour plate section.

Will things get worse after each illness?

Cognitive deficits and bipolar illness course are associated and may progressively worsen with illness development, but these impairments may be present before illness onset, again suggesting neurocognitive impairment as a trait marker of bipolar illness.

Robinson and Ferrier (2006) undertook a systematic review of the relationship between neurocognitive performance and clinical outcome as well as the cognitive functioning in healthy first-degree relatives of bipolar probands. The reviewed evidence suggests that cognitive impairment in bipolar patients is related to illness variables (prior course of illness, number of manic episodes, number of hospitalisations and length of illness). Unaffected first-degree relatives had deficits in areas of executive function and verbal memory. For the patients, both depressive and manic episodes progressively impaired neurocognitive function, with mania most consistently associated with verbal declarative memory and some executive functions. The link between depressive episodes and a wider range of deficits was less consistently shown in this study, although other studies have suggested that cognitive functioning is related to the number of depressive episodes rather than to the number of manic episodes (Kessing, 1998), while others have reported that those with a longer and more severe duration of illness plus a higher number of affective episodes display more neuropsychological impairment (Denicoff *et al.*, 1999; Martinez-Aran *et al.*, 2004b). The causal relationship is unclear and has yet to be determined.

Tham *et al.* (1997) report that euthymic patients with recurring mood disorder had significantly more hospitalisations if they displayed cognitive deficits

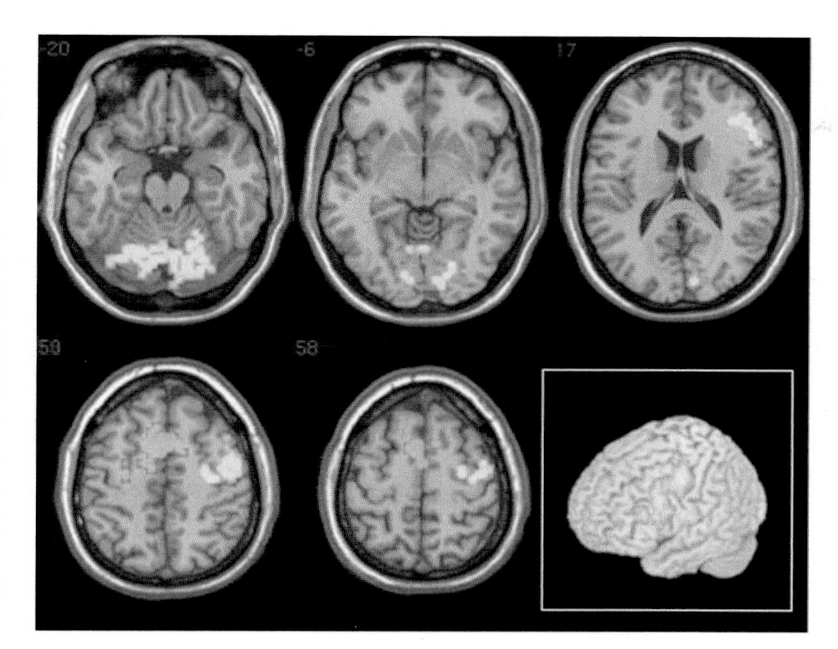

Figure 20.2 Task-dependent responses in fMRI: language tasks (adapted from Curtis *et al.*, 2007). During the same scanning session as Monks *et al.* (2004), participants also performed tasks of language production. In addition to baseline tasks, participants undertook four active tasks designed to investigate components of language. Yellow voxels represent regions of relatively greater response in the bipolar subjects and blue voxels represent regions of relatively greater response in the control participants. See colour plate section.

than those with non-impaired cognitive functioning, and Van Gorp *et al.* (1998) suggest a possible accumulative negative effect in the duration of bipolar illness on memory and executive function. Van Gorp and colleagues report that it is the collective duration of mania and depression experienced that has more effect on cognitive function than the number of episodes. However, this is in contrast to findings of Kerry *et al.* (1983), who found that cognitive functioning in those with more than 10 years' duration of bipolar illness was no different from that in those with less than 10 years' illness duration.

Further clarification regarding the relationship between prior course of illness and neuropsychological impairment in bipolar outpatients was examined by Denicoff *et al.* (1999). They found that cognitive impairment was associated with a longer duration and a more severe course of bipolar illness, as indicated by a higher number of prior episodes and hospitalisations. In particular, the longer the duration of illness, the poorer the cognitive performance on tests of attention and concentration. It is possible that the recurrent nature of this disorder may have an accumulating effect on the brain, which leads to greater cognitive dysfunction (Osuji & Cullum, 2005). Denicoff and colleagues have also suggested the possibility that those suffering form severe forms of BD have poorer cognitive aptitude at illness onset or that the illness episodes themselves hinder cognitive progression Denicoff *et al.* (1999).

Structural changes in the brains of bipolar patients have been reported (Bearden *et al.*, 2001) and some have found a correlation between cognitive deficits and cerebral structure. Altshuler (1993) suggests that the structural changes in the brain may be a result of having episodes of mania and depression, and that this recurrent destructive process impairs the efficiency of the central nervous system, in turn leading to cognitive deficits.

Can cognitive functioning tests tell if family members are at risk?

Ferrier *et al.* (2004) examined cognitive functioning in healthy first-degree relatives of those with BD compared with healthy matched controls. There was no difference in mood between the two groups and neurocognitive areas tested included psychomotor function, attention, executive function and declarative memory. The final analysis included only medication-free relatives without a current Axis I disorder (n=14). The healthy first-degree relatives showed significant impairments in visuospatial declarative memory and in visual and verbal span measures of working memory. As the deficits were not apparent in all areas of executive functioning, this suggests that the impairments are selective and not due to global cognitive decline. These results corroborate the literature regarding residual cognitive deficits in remitted BD and imply that this cognitive stamp may be a predictor of possible bipolar illness development.

In a similar study, Clark *et al.* (2005) compared the first-degree relative of BD type I sufferers with euthymic unipolar patients and healthy controls on a verbal learning test (California verbal learning test) and a visual discrimination task (intradimensional/extradimensional shift task). The BD type I relatives who had higher scores on the Hamilton depression rating scale had associated lower scores on the verbal learning test, suggesting a correlation between the two. The BD type I relatives and the euthymic unipolars had pronounced deficits on the shift task, but the total errors in each group did not correlate with mood scores. The significant impairment in both groups reflects a similar performance to that of remitted bipolar patients in a previous study (Clark *et al.*, 2002), suggesting impaired attentional shifting may be a putative endophenotype in mood disorder.

Will the ability to detect emotions remain intact?

Successful social behaviour entails the ability to differentiate various expressed facial emotions. It has been demonstrated that there is a generalised impairment in the perception of emotions in depressed patients (Persad & Polivy, 1993) and also in those with mania (Lembke & Ketter, 2002). As in the investigations of cognitive functioning in remitted bipolar patients, Harmer *et al.* (2002) assessed emotional processing in euthymic bipolar individuals. It would be expected that residual cognitive deficits would also impact emotional perception. They assessed the ability of remitted bipolar subjects to recognise the facial expressions of happiness, surprise, sadness, fear, anger and disgust, compared with healthy matched controls. The euthymic bipolar group demonstrated an enhanced recognition of facial disgust, but with no differences in the other expressions versus the controls, suggesting a selective process in the labelling of disgust in BD.

A more recent study examined whether differences in facial emotional recognition are dependent on mood state or whether this impairment is an underlying trait in BD (Venn *et al.*, 2004). Compared with healthy matched control individuals, the remitted bipolar group showed no difference in sensitivity to any specific emotion (happiness, surprise, sadness, fear, anger and disgust). However, a statistical trend was evident in the bipolar participants in the ability to recognise fear, but this was not significantly greater than that of other emotions. This study does not lend support for misinterpretation of facial emotions as an underlying trait in the development of bipolar illness. This proposes that errors in distinguishing facial expressions among depressed and manic patients may be mood-congruent influences. The discrepancy found in results between Venn's group and other investigators may be due to methodological differences, sample sizes, medication effects and collapsing of the bipolar subgroups.

Will the medications affect thinking?

Certain psychotropic medications are associated with negative neuropsychological performance in bipolar affective disorder, but it is difficult to tease out to what degree the cognitive impairments are a result of the illness and the result of the medication. This is also complicated by polypharmacy and duration of use, the presenting profile of bipolar illness, and the combinations of cognitive tests that measure different abilities.

Will problems with thinking affect psychosocial functioning?

It has been established that cognitive impairments can predict the level of community functioning in schizophrenia with a probability equal to or greater than psychotic symptoms (Kurtz *et al.*, 2005). Laes & Sponheim (2006) set out to examine whether this association is unique to schizophrenia only or if it is present in bipolar affective disorder as well. The association of community functioning with generalised cognitive performance, impairments in cognitive domains and symptomatology was investigated in schizophrenia, bipolar illness and healthy controls. Predictably, symptomatology was more strongly associated with degree of social functioning in the schizophrenia group than the bipolar group. They found that generalised cognitive functioning had little influence on community functioning in bipolar patients but had some bearing in the schizophrenia and control groups. However, poorer social functioning was associated with poorer planning and problem-solving in the bipolar patients, whereas difficulty in processing verbal-related material was linked to worse community functioning in the schizophrenia group.

Poor psychosocial functioning and course of bipolar illness are also linked with cognitive impairment (Zarate *et al.*, 2000; Martinez-Aran *et al.*, 2004b). Often, the individual's personal relationships and occupation are affected, as impairments in attention, working memory, learning and executive function may lead to poorer social adaptation. Martinez-Aran and colleagues (2004b) found that psychosocial functioning was associated more with cognitive measures than with clinical factors. Psychosocial functioning was worse when deficits were displayed in measures of working memory, verbal learning and memory. These memory difficulties may result in less compliance with treatment and in turn, lead to worsening of symptoms and poorer social behaviour.

Zarate *et al.* (2000) reviewed the literature on functional impairment and cognition in bipolar illness and found that the majority of long-term and short-term studies report a high percentage of impaired functional outcomes. It is estimated that 30–60% of bipolar patients do not regain complete functioning in the occupational or social realms (MacQueen *et al.*, 2001). The psychosocial impairments appear to persist for years, despite treatment and clinical remission, suggesting that symptomatic recovery occurs before functional recovery. Zarate and colleagues conclude that cognitive dysfunction may partially contribute to the low functional recovery rates in BD (Zarate *et al.*, 2000).

Dickerson *et al.* (2004) specifically studied variables associated with employment status among bipolar sufferers. These variables included cognitive functioning, symptom severity, demographic data and course of illness. They found that cognitive performance, particularly verbal memory abilities, was significantly associated with current employment status. Other significantly related factors with occupational outcome included total symptom severity, previous psychiatric hospitalisations and maternal educational level.

Martinez-Aran *et al.* (2004a) also found that poor neurological performance was associated with poor psychosocial functioning, duration of illness and the number of manic episodes, hospitalisations and suicide attempts. In another investigation, Martinez-Aran *et al.* (2004b) compared the neuropsychological test results of 40 euthymic bipolar patients with 30 healthy controls. After controlling for the effect of subclinical affective symptoms, age and pre-morbid IQ, the remitted bipolar group showed a worse performance than the controls in verbal memory, immediate recall, delayed recall and recognition. They again found that cognitive impairment was associated with poorer psychosocial functioning and worse clinical course, adding weight to the suggestion that the impairments may be linked to disease progression (Denicoff *et al.*, 1999).

The direction of causality between cognitive deficits and poor psychosocial outcome has yet to be fully established. Psychological treatments that emphasise cognitive rehabilitation or remediation may help decrease the influence of cognitive performance deficits in bipolar illness on overall daily functioning.

Will talking treatments help?

Pharmacotherapy is the primary treatment method for bipolar illness, but application of established psychological therapies as an adjunct provides optimal management for bipolar sufferers, particularly during depressive episodes. The four main psychological approaches are psychoeducation, cognitive behavioural therapy (CBT), interpersonal and social rhythm therapy (IPSRT) and family-focused therapy (FFT). Each of these therapies is highly structured by providing an overview of the patient's difficulties and promoting utilisation of learned skills to self-assist, and these therapies should be introduced early in the course of illness to stabilise daily functioning, improve medication compliance, identify relapse signs and minimise residual symptoms (Yatham *et al.*, 2009). Psychotherapeutic treatments alone without medication for acute mania are not generally useful, but they have some demonstrated efficacy for bipolar depression (Yatham *et al.*, 2009). A recent systematic review by Scott and Gutierrez (2004) concludes that bipolar sufferers receiving adjunctive psychological treatments had significantly fewer relapses compared with those treated with medication only or standard psychiatric treatment alone. The combination of psychotherapy and pharmacotherapy appears to contribute to better symptomatic and functional recovery. However, further research is needed as to which specific treatment is most effective during the different phases of the illness and for which particular patient profile. Psychosocial treatment is covered in detail in Chapter 6.

Can talking treatments change the way the brain works?

There are no known published systematic, prospective studies to date investigating the impact of psychosocial treatments on cognitive outcome in bipolar patients.

Oei and Sullivan (1999) reported on the recovery status of depression patients and changes in cognitive thoughts following a group CBT programme and found that mood and activity levels were consistently higher in the recovered group (Beck depression inventory [BDI] score of <10) than those in the non-recovered group (BDI score of ≥ 10 or more). The recovered group also had significantly lower scores regarding negative thoughts and negative expectations at completion of CBT than those in the non-recovered group, suggesting that negative cognition contributes to the recovery process in depression.

Cognitive remediation (CR) has been defined as a therapy that concentrates on improving cognitive functioning by using repeated practice of cognitive tasks or by applying training strategies for counteracting cognitive deficits (Pfammatter *et al.*, 2006). Glahn *et al.* (2006) report that even though the overall level of cognitive impairment in BD is less than that in schizophrenia, there is a partial overlapping of the memory systems involved. Glahn and colleagues suggest that clarifying the overlap and the intact functions may help in planning targeted CR programmes to improve patient outcomes (Glahn *et al.*, 2006). A meta-analysis on the efficacy of psychological therapy in schizophrenia conducted by Pfammatter *et al.* (2006) found that CR results in short-term, small-to-medium beneficial effects in cognitive functioning.

Wykes *et al.* (2002) examined whether or not concomitant brain activation changes occur following CR therapy in male patients with schizophrenia compared with a control therapy group and with a healthy control group. Those with schizophrenia were maintained on their antipsychotic medications throughout the study and either randomly assigned to the CR therapy or control therapy. Both therapies were 12 weeks long with 40 individual sessions consisting of occupational therapy activities (control therapy) or paper and pencil tasks involving practising information processing strategies in cognitive flexibility, working memory and planning (CR therapy). Functional magnetic resonance imaging was completed at baseline and post-therapy sessions along with tests of executive functioning. At completion of therapy, there were improvements in neuropsychological performance, but there was a definite advantage in those receiving CR in the memory domain. The fMRI data showed that both patient groups had increased brain activation, but this was particularly so in the CR group, especially in the fronto-cortical areas associated with working memory. There were no differences between the patient groups in symptoms or medications; thus it can be implied that the changes are associated with the psychological, rather than the pharmacological, treatment.

Thase *et al.* (1996) examined the relationship between HPA activity and response to CBT in unmedicated inpatients with major depression. Urinary free cortisol levels and DSTs were collected from patients, and each received individual CBT sessions over an average of 3 weeks (maximum 20 sessions). Response to the CBT treatment was evaluated relative to clinical severity of the depression and pre-therapy HPA parameters. The results showed a significant inverse association between pre-treatment urinary free cortisol excretion (increased) and response to inpatient CBT (poor), but CBT response was not strongly correlated with DST. The response to psychotherapy as it relates to HPA action was not explained by depression severity, as the pre- and post-treatment mood scores were not statistically significant. Thase and colleagues propose that because hypercortisolism is known to contribute to cognitive impairment in depression, the hypercortisolism may also impede one's ability to recall, incorporate and employ the cognitive strategies of psychotherapy (Thase *et al.*, 1996). Future therapeutic strategies should focus on treatment modalities that modulate HPA dysregulation and future research should investigate the effects in bipolar illness in relation to psychological interventions in order to tailor treatment options for individuals.

Conclusions

Bipolar illness is a chronic, recurrent lifelong mood disorder that affects behaviour, daily functioning and cognition. It is a devastating disorder that has been linked to increased mortality rates as a result of suicide and cardiovascular diseases. Until recently it was assumed that there were no neuropsychological deficits associated with euthymia in BD. However, it has now been demonstrated that residual cognitive impairments remain during bipolar remission, thus questioning the view that functional recovery is complete between illness episodes and suggesting that neurocognitive dysfunction may be a trait marker of BD. Neurological soft signs and HPA axis dysregulation have also been shown to contribute to cognitive deficits in the remitted phase of bipolar illness, implying that these may be early warning signs of illness development. The direction of causality between cognitive impairment and illness variables and between cognitive dysfunction and psychosocial outcome has yet to be fully established.

A combination of pharmacotherapy and adjunctive psychological therapy provides the best possible management for patients, especially during depressive episodes. Individual CBT has shown to be the most effective psychological intervention thus far for bipolar illness. However, the effect of psychological therapies on cognitive performance outcome is essentially unknown in BD.

Demographic and psychosocial factors, cognitive functioning, symptom severity and course of illness are some of the variables that have made it difficult to establish a best treatment strategy to achieve optimal long-term functioning and clinical outcome. The cause of cognitive impairment in bipolar illness needs to be elucidated, as well as the effects of existing pharmacological and psychological treatments on cognitive outcome. Treatments to improve neuropsychological functioning in the long term will help improve quality of life and clinical status of bipolar patients during and between episodes.

References

Altshuler, L.L. (1993). Bipolar disorder: are repeated episodes associated with neuroanatomic and cognitive changes? *Biological Psychiatry*, **33**, 563–5.

Angst, F., Stassen, H.H., Clayton, P.J., Angst, J. (2002). Mortality of patients with mood disorders: follow-up over 34–38 years. *Journal of Affective Disorders*, **68**, 167–81.

Ayuso-Mateos, J.L. (2002). *Global burden of bipolar disorder in the year 2000: GBD 2000 Working Paper*. Geneva: World Health Organization. www.who.int/healthinfo/global_burden_disease/en/ [accessed 11 January 2010].

Bearden, C.E., Hoffman, K.M., Cannon, T.D. (2001). The neuropsychology and neuroanatomy of bipolar affective disorder: a critical review. *Bipolar Disorders*, **3**, 106–50; Discussion 151–3.

Belanoff, J.K., Gross, K., Yager, A., Schatzberg, A.F. (2001). Corticosteroids and cognition. *Journal of Psychiatric Research*, **35**, 127–145.

Bowden, C.L. & Krishnan, A.A. (2004). Pharmacotherapy for bipolar depression: an economic assessment. *Expert Opinion on Pharmacotherapy*, **5**, 1101–7.

Clark, L., Iversen, S.D., Goodwin, G.M. (2002). Sustained attention deficit in bipolar disorder. *British Journal of Psychiatry*, **180**, 313–9.

Clark, L., Sarna, A., Goodwin, G.M. (2005). Impairment of executive function but not memory in first-degree relatives of patients with bipolar I disorder and in euthymic patients with unipolar depression. *American Journal of Psychiatry*, **162**, 1980–2.

Curtis, V.A., Thompson, J.M., Seal, M.L., *et al.* (2007). The nature of abnormal language processing in euthymic bipolar I disorder: evidence for a relationship between task demand and prefrontal function. *Bipolar Disorders*, **9**, 358–69.

Daban, C., Vieta, E., Mackin, P., Young, A.H. (2005). Hypothalamic-pituitary-adrenal axis and bipolar disorder. *The Psychiatric Clinics of North America*, **28**, 469–80.

Denicoff, K.D., Ali, S.O., Mirsky, A.F., *et al.* (1999). Relationship between prior course of illness and neuropsychological functioning in patients with bipolar disorder. *Journal of Affective Disorders*, **56**, 67–73.

Dickerson, F.B., Boronow, J.J., Stallings, C.R., Origoni, A.E., Cole, S., Yolken, R.H. (2004). Association between cognitive functioning and employment status of persons with bipolar disorder. *Psychiatric Services (Washington, D.C.)*, **55**, 54–8.

Ferrier, I.N., Chowdhury, R., Thompson, J.M., Watson, S., Young, A.H. (2004). Neurocognitive function in unaffected first-degree relatives of patients with bipolar disorder: a preliminary report. *Bipolar Disorders*, **6**, 319–22.

Glahn, D.C., Barrett, J., Bearden, C.E., *et al.* (2006). Dissociable mechanisms for memory impairment in bipolar disorder and schizophrenia. *Psychological Medicine*, **36**, 1085–95.

Goswami, U., Basu, S., Kahstgir, U., Chandrasekaran, R., Gangadhar, B. (1998). Neurobiological characterization of bipolar affective disorders: a focus on tardive dyskinesia and soft neurological signs in relation to serum dopamine beat hydroxylase activity. *Indian Journal of Psychiatry*, **40**, 201–11.

Goswami, U., Sharma, A., Khastigir, U., *et al.* (2006). Neuropsychological dysfunction, soft neurological signs and social disability in euthymic patients with bipolar disorder. *British Journal of Psychiatry*, **188**, 366–73.

Harmer, C.J., Grayson, L., Goodwin, G.M. (2002). Enhanced recognition of disgust in bipolar illness. *Biological Psychiatry*, **51**, 298–304.

Judd, L.L. & Akiskal, H.S. (2003). The prevalence and disability of bipolar spectrum disorders in the US population: re-analysis of the ECA database taking into account subthreshold cases. *Journal of Affective Disorders*, **73**, 123–31.

Judd, L.L., Akiskal, H.S., Schettler, P.J., *et al.* (2002). The long-term natural history of the weekly symptomatic status of bipolar I disorder. *Archives of General Psychiatry*, **59**, 530–7.

Kerry, R.J., McDermott, C.M., Orme, J.E. (1983). Affective disorders and cognitive performance: a clinical report. *Journal of Affective Disorders*, **5**, 349–52.

Kessing, L.V. (1998). Cognitive impairment in the euthymic phase of affective disorder. *Psychological Medicine*, **28**, 1027–38.

Kurtz, M.M., Moberg, P.J., Ragland, J.D., Gur, R.C., Gur, R.E. (2005). Symptoms versus neurocognitive test performance as predictors of psychosocial status in schizophrenia: a 1- and 4-year prospective study. *Schizophrenia Bulletin*, **31**, 167–74.

Laes, J.R. & Sponheim, S.R. (2006). Does cognition predict community function only in schizophrenia?: a study of schizophrenia patients, bipolar affective disorder patients, and community control subjects. *Schizophrenia Research*, **84**, 121–31.

Leibenluft, E. (1996). Women with bipolar illness: clinical and research issues. *American Journal of Psychiatry*, **153**, 163–73.

Lembke, A. & Ketter, T.A. (2002). Impaired recognition of facial emotion in mania. *American Journal of Psychiatry*, **159**, 302–4.

MacQueen, G.M., Young, L.T., Joffe, R.T. (2001). A review of psychosocial outcome in patients with bipolar disorder. *Acta Psychiatrica Scandinavica*, **103**, 163–70.

Malhi, G.S., Ivanovski, B., Szekeres, V., Olley, A. (2004). Bipolar disorder: it's all in your mind? The neuropsychological profile of a biological disorder. *Canadian Journal of Psychiatry*, vol. **49**, 813–9.

Martinez-Aran, A., Vieta, E., Colom, F., *et al.* (2004a). Cognitive impairment in euthymic bipolar patients: implications for clinical and functional outcome. *Bipolar Disorders*, **6**, 224–32.

Martinez-Aran, A., Vieta, E., Reinares, M., *et al.* 2004b. Cognitive function across manic or hypomanic, depressed, and euthymic states in bipolar disorder. *American Journal of Psychiatry*, **161**, 262–70.

Monks, P.J., Thompson, J.M., Bullmore, E.T., *et al.* (2004). A functional MRI study of working memory task in euthymic bipolar disorder: evidence for task-specific dysfunction. *Bipolar Disorders*, **6**, 550–64.

Nasrallah, H.A., Tippin, J., McCalley-Whitters, M. (1983). Neurological soft signs in manic patients: a comparison with schizophrenic and control groups. *Journal of Affective Disorders*, **5**, 45–50.

Negash, A., Kebede, D., Alem, A., *et al.* (2004). Neurological soft signs in bipolar I disorder patients. *Journal of Affective Disorders*, **80**, 221–30.

Oei, T.P. & Sullivan, L.M. (1999). Cognitive changes following recovery from depression in a group cognitive-behaviour therapy program. *Australian and New Zealand Journal of Psychiatry*, **33**, 407–15.

Osuji, I.J. & Cullum, C.M. (2005). Cognition in bipolar disorder. *The Psychiatric Clinics of North America*, **28**, 427–41.

Perlis, R.H., Miyahara, S., Marangell, L.B., *et al.*; STEP-BD Investigators (2004). Long-term implications of early onset in bipolar disorder: data from the first 1000 participants in the Systematic Treatment Enhancement Program for Bipolar Disorder (STEP-BD). *Biological Psychiatry*, **55**, 875–81.

Persad, S.M. & Polivy, J. (1993). Differences between depressed and nondepressed individuals in the recognition of and response to facial emotional cues. *Journal of Abnormal Psychology*, **102**, 358–68.

Pfammatter, M., Junghan, U.M., Brenner, H.D. (2006). Efficacy of psychological therapy in schizophrenia: conclusions from meta-analyses. *Schizophrenia Bulletin*, **32** (Suppl. 1), S64–80.

Robinson, L.J. & Ferrier, I.N. (2006). Evolution of cognitive impairment in bipolar disorder: a systematic review of cross-sectional evidence. *Bipolar Disorders*, **8**, 103–16.

Robinson, L.J., Thompson, J.M., Gallagher, P., *et al.* (2006). A meta-analysis of cognitive deficits in euthymic patients with bipolar disorder. *Journal of Affective Disorders*, **93**, 105–15.

Rubinsztein, J.S., Michael, A., Paykel, E.S., Sahakian, B.J. (2000). Cognitive impairment in remission in bipolar affective disorder. *Psychological Medicine*, **30**, 1025–36.

Sajatovic, M. (2005). Bipolar disorder: disease burden. *American Journal of Managed Care*, **11** (Suppl. 3), S80–4.

Scott, J. & Gutierrez, M.J. (2004). The current status of psychological treatments in bipolar disorders: a systematic review of relapse prevention. *Bipolar Disorders*, **6**, 498–503.

Smith, D.J., Muir, W.J., Blackwood, D.H. (2006). Neurocognitive impairment in euthymic young adults with bipolar spectrum disorder and recurrent major depressive disorder. *Bipolar Disorders*, **8**, 40–6.

Tham, A., Engelbrektson, K., Mathe, A.A., Johnson, L., Olsson, E., Aberg-Wistedt, A. (1997). Impaired neuropsychological performance in euthymic patients with recurring mood disorders. *Journal of Clinical Psychiatry*, **58**, 26–9.

Thase, M.E., Dube, S., Bowler, K., *et al.* (1996). Hypothalamic-pituitary-adrenocortical activity and response to cognitive behavior therapy in unmedicated, hospitalized depressed patients. *American Journal of Psychiatry*, **153**, 886–91.

Thompson, J.M., Gallagher, P., Hughes, J.H., *et al.* (2005) Neurocognitive impairment in euthymic patients with bipolar affective disorder. *British Journal of Psychiatry*, **186**, 32–40.

US DHEW Medical Practice Project (1979) *A State of the Service Report for the Office of the Assistant Secretary for the US Dept of Health, Education and Welfare*. Baltimore, MD: Policy Research.

van Gorp, W.G., Altshuler, L., Theberge, D.C., Wilkins, J., Dixon, W. (1998). Cognitive impairment in euthymic bipolar patients with and without prior alcohol dependence: a preliminary study. *Archives of General Psychiatry*, **55**, 41–6.

Venn, H.R., Gray, J.M., Montagne, B., *et al.* (2004). Perception of facial expressions of emotion in bipolar disorder. *Bipolar Disorders*, **6**, 286–93.

Watson, S., Gallagher, P., Ritchie, J.C., Ferrier, I.N., Young, A.H. (2004). Hypothalamic-pituitary-adrenal axis function in patients with bipolar disorder. *British Journal of Psychiatry*, **184**, 496–502.

Watson, S., Thompson, J.M., Ritchie, J.C., Nicol Ferrier, I., Young, A.H. (2006). Neuropsychological impairment in bipolar disorder: the relationship with glucocorticoid receptor function. *Bipolar Disorders*, **8**, 85–90.

Wykes, T., Brammer, M., Mellers, J., *et al.* (2002). Effects on the brain of a psychological treatment: cognitive remediation therapy: functional magnetic resonance imaging in schizophrenia. *British Journal of Psychiatry*, **181**, 144–52.

Yatham, L.N., Kennedy, S.H., Schaffer, A., *et al.* (2009). Canadian Network for Mood and Anxiety Treatments (CANMAT) and International Society for Bipolar Disorders (ISBD) collaborative update of CANMAT guidelines for the management of patients with bipolar disorder: update 2009. *Bipolar Disorders*, **11**, 225–55.

Young, A.H. (2004). Cortisol in mood disorders. *Stress*, **7**, 205–8.

Zarate Jr, C.A., Tohen, M., Land, M., Cavanagh, S. (2000). Functional impairment and cognition in bipolar disorder. *Psychiatric Quarterly*, **71**, 309–29.

Index

acamprosate 135
ACC *see* anterior cingulate cortex (ACC)
acute mania 15–20, 41, 44, 77, 80, 165, 169, 184 *see also* mania
  clinical subtypes of patients with 10
  diagnostic criteria for 11
  dosing and duration of treatment for 11–12
  emergency management of 35
  lithium, use of 88
  monotherapy for treatment of 36
  non-pharmacological treatment of 13
  pharmacological treatment of 13, 35–6, 75
    recommendations for 33, 36
  psychotherapeutic treatment of 205
  quality of life enjoyment and satisfaction questionnaire 170
  substance-abuse issues, impact of 185
  treatment algorithm for 14, 36, 37
ADHD *see* attention deficit hyperactivity disorder (ADHD)
adjunctive lamotrigine
  for BD type I depression treatment 25
  for second-line treatment of bipolar depression 39
adjunctive quetiapine 41
  for BD type I depression treatment 26
adolescent-onset bipolar disorder 73–4
adulthood, bipolar disorder in 73 *see also* late-life bipolar disorder
  care during transition to young adulthood 80
  clinical features 84–7
  depression 77
  management of 87–90
aetiopathogenesis, of bipolar disorder 9
AF *see* amniotic fluid (AF)
age of onset and clinical features, of late-life bipolar disorder 84–5
akathisia 18, 19, 110
alcohol-use disorders 122, 129, 130, 132
  alcohol abuse and 133
  bipolar patients with co-morbid 134
  bipolar women with 131
Alzheimer's disease 86
American Academy of Pediatrics 100, 116
American Association of Clinical Endocrinologists 114, 123
American Board of Psychiatry and Neurology 191
American Diabetes Association (ADA) 114
American Psychiatric Association (APA) 33, 113, 114, 115, 145, 181, 196
amisulpride 13, 37, 109, 115
  and antimanic treatment 19
amniotic fluid (AF) 99
anhedonia 66, 68
anterior cingulate cortex (ACC) 67
anticonvulsants 78, 113–14
  carbamazepine 16–17
  lamotrigine 109
  for late-life BD treatment 88–9
  other anticonvulsants with potential antimanic properties 17
  sodium valproate 109
  valproate 17
antidepressants 4, 36, 62, 142
  for BD type I depression treatment 26–7
  for BD type II depression treatment 28–9
  for late-life BD treatment 89–90
  for rapid cycling BD treatment 29
  tricyclic 99, 145
antidiabetic medications 107
antimanic
  agents 15, 20, 35, 36, 146
  medication 11, 19, 35
antipsychotics 4, 15, 35, 95, 108, 114–15
  and extrapyramidal side effects 110
  for late-life BD treatment 89
  therapy 113, 114
antisocial personality disorder 129, 131
anxiety disorders 1, 7, 9, 44, 50, 73, 74, 123, 124, 131, 142, 146, 147, 169
  common symptom of 120
  co-morbid 94, 120, 121, 122, 126, 131, 132, 185
  implications of BD associated with co-morbid 120
  personality disorders and 131
  pharmacotherapy for treatment of 125
  recommended scales for assessing 188
APA *see* American Psychiatric Association (APA)
aripiprazole 78, 79, 134
  for antimanic treatment 17–18
  for BD type I depression treatment 27–8, 41
asenapine 13
  for antimanic treatment 19, 35
assertive outreach (AO) teams 175, 182
atherosclerosis 125
attention deficit hyperactivity disorder (ADHD) 74, 101, 123
atypical antipsychotics 15, 26, 109–10, 125
  amisulpride 19
  aripiprazole 17–18
  asenapine 19
  olanzapine 18
  quetiapine 18
  risperidone 18–19
  ziprasidone 19
Axis I disorders 69, 94, 120, 129, 139, 203
Axis II disorders 48, 69, 120

BAI *see* Beck anxiety inventory (BAI)
Barcelona Bipolar Disorders Program 45, 59
Bauer collaborative care model 182
BDI *see* Beck depression inventory (BDI)
BDNF *see* brain-derived neurotrophic factor (BDNF)
Beck anxiety inventory (BAI) 188–9
Beck depression inventory (BDI) 68, 122, 205
Beck's cognitive model of mania 45

behavioural reward system (BRS) 68
behavioural rhythms 152
benzodiazepines 88
for antimanic treatment 20
for co-morbid SUD treatment 134
for improving sleep quality 157
as mood stabilisers 77
bimodal distribution of age of illness 84
binge-eating disorder 108, 142
binging-purging anorexia 142
bipolar depressive symptoms 38–40
bipolar disorder (BD) *see also individual entries*
care and stress upon family 80
clinical trials on efficacy and safety of drugs for 134
continuous circular 93
diagnostic issues in treatment of
adolescent-onset bipolar disorder 73–4
manic-like symptoms in pre-adolescent children 74
effect of co-morbid substance-use disorders on outcome in 132–3
episodes 176
imaging studies of 123–4
impact on psychosocial functioning and quality of life 9
interrelatedness of respect, rapport and recovery 5–6
maintenance treatment of
clinical management and practical considerations 41–2
prophylaxis 40
role of psychosocial treatments and 40–1
management of 1, 80–1
diagnosis 33–4
goals of treatment 34
treatment guidelines for 33
metabolic state in 106–8
non-adherence to medication, issue of 3
overview of 9
physical consequences of 201
with psychotic features 1
rates of recurrence of 131–2
recovery, personal medicine and treatment for 1–2
resource issues for 80
review of psychoeducation studies for 47–9
screening for 34
summary of comparative studies for 53–4
symptoms for 2
treatment for
pharmacological 134–5
psychotherapy 135–6
young people 80
types of *see* bipolar disorder (BD) type I; bipolar disorder (BD) type II
bipolar disorder (BD) type I 9, 15, 120, 130, 143, 165, 181, 204
illness in women 93
interpersonal and social rhythm therapy trials for 52
mood episode 51
pharmacotherapy
adjunctive lamotrigine 25
adjunctive quetiapine 26
antidepressants 26–7
aripiprazole 27–8
carbamazepine 25
ethyl-eicosapentaenoic acid (EPA) 27
lamotrigine 24–5
lithium 24
mifepristone 27
modafinil 27
olanzapine/fluoxetine combination (OFC) 26
pramipexole 27
quetiapine 25–6
valproate 25
QoL in 171
weight changes in patients 110
bipolar disorder (BD) type II 9, 120, 143, 166, 181
bipolar illness in women with 93
cyclothymia and 144
pharmacotherapy
antidepressants 28–9
pramipexole 28
quetiapine 28
QoL in 171
bipolar disorder patients
obesity and unhealthy dietary habits in 123
QoL of 171
response to electroconvulsive therapy 62–3
Bipolar Disorders Program 45, 59, 178
bipolar mood symptoms 51
bipolar spectrum disorders 130, 134, 139, 142, 187, 200
bipolar women
bipolar illness in
childhood abuse 94
co-morbid anxiety disorder and 94
course of 93–4
diagnostic delay 94
suicidality in 94
management of reproductive health issues in
breastfeeding 100
contraception 97–9
general and sexual health care 97
hyperprolactinaemia 95
hypothalamo-pituitary-gonadal axis dysfunction 94
polycystic ovary syndrome 95–7
pregnancy and puerperium 99–100
pre-pregnancy planning 99
prolactin-elevating medication 96
menstrual abnormalities in 95
mood symptoms in 93
panic disorder and social phobia 94
psychological and psychosocial issues for
development of self-identity in bipolar adolescent girl 101
mothers with bipolar disorder 101–2
self-identity and relationship issues in adults 101
rapid-cycling disorder in 94
blood–brain barrier 17, 19, 100
Body Mass Index (BMI) 42, 106, 107, 111, 114
borderline personality disorder (BPD) 68, 139, 141, 184, 185
co-morbid conditions for 121
correlation with mood disorders 143
impact of 120
relation with bipolar spectrum disorders 142
sub-affective personality disorders and 143
brain cells 64
brain imaging in bipolar disorder
imaging for
diagnostic purposes 195–6
prognostic reasons 196
imaging techniques and findings 194–5
brain-derived neurotrophic factor (BDNF) 124
breast cancer 6, 95
breastfeeding 99, 100, 117
bronchial carcinoma 95
BRS *see* Behavioural reward system (BRS)
bulimia 68, 96, 142
Bunney–Hamburg scale 25
bupropion 27, 29, 38, 40, 145

California verbal learning test 204
Canadian Network for Mood and Anxiety Treatments (CANMAT) 33, 34, 36–9, 181

recommendations for maintenance treatment of BD 41
cannabis 130, 133, 142
carbamazepine 19, 115
as anticonvulsants 109
for antimanic treatment 16–17
for BD type I and type II depression treatment 25
for late-life BD treatment 89
cardiac rhythm abnormalities 89
cardiovascular diseases (CVDs) 106, 111, 121
European guidelines for prevention of 112
relationship with bipolar disorder 107
and sudden death 123
cardiovascular pathophysiology 124
care and stress upon family, in early stages of illness 80
catatonia 12, 62
electroconvulsive therapy for 64
CBF *see* cerebral blood flow (CBF)
CBT *see* cognitive behavioural therapy (CBT)
central nervous system disorders 135
cerebellar atrophy 17
cerebral blood flow (CBF) 65, 67
CGI *see* clinical global impression (CGI)
CGI-S *see* clinical global impression–severity scale (CGI-S)
childhood abuse 94
children and adolescents 95, 122–3
characteristics of BD in 142
family-focused therapy for 51
management of physical health issues in 117
pharmacotherapy of acute mania in 75
prevalence of BD in 73
chlorpromazine 13, 16, 36, 63
for antimanic treatment 20
cingulate gyrus 66–7, 194
circadian and sleep/wake processes
circadian instability in aetiology of BD 152–3
management of circadian and sleep/wake rhythms 154
sleep management in bipolar disorder
assessment of sleep disturbances 157
common sleep disturbances 156–7
improvement of sleep quality 157–60
social zeitgeber hypothesis and social rhythm management 154–6
therapeutic process 153–4
circadian rhythms 45, 54, 152–5, 160
citicoline 135
clinical global impression as modified for bipolar illness (CGI-BP) depression 27
clinical global impression (CGI) 188
clinical global impression–severity scale (CGI-S) 26, 170
clinical practice, assessment scales for 186–7
clozapine 13, 14, 19, 36, 41, 100, 109, 110, 112, 114, 115, 117
for rapid cycling BD treatment 29
cocaine abuse 73, 130, 131, 134, 135, 141, 185
cognitive behavioural therapy (CBT) 16, 45–6, 53, 155–6, 164, 186, 205
Lam's programme for 46
review of trials 49–50
summary of trials 50–1
cognitive functioning tests 203–4
cognitive impairments, in euthymic bipolar patients 201
cognitive remediation (CR) 206
cognitive skills, affect of bipolar disorder on 200
collaborative care (CC) 53
model for management of BD 177–9
RCTs for bipolar disorder 178
co-morbid anxiety disorders 94, 120, 121, 126, 147, 185
in children and adolescents 122–3
co-morbid diabetes mellitus 123
co-morbid organic (neurological) disorders 16, 74
co-morbid SUD
benzodiazepine for treatment of 134
clinical features of bipolar patients with 134
lifetime prevalence rates of 129
rates in clinical samples of bipolar patients 130
co-morbidity
alcoholism 133
between BD and SUDs 131
bipolar disorder associated with 120–1
anxiety disorders 121
cardiovascular disease and sudden death 123
children and adolescents 122–3
diabetes 123
medical conditions 123
metabolic syndrome 123
studies after 2005 121–2
treatment 125
pathogenesis of 131–2
pathophysiological significance of
anxiety 124
bipolar disorder 123–4
blood pressure variability 125
cardiovascular pathophysiology 124
heart rate variability 124
hypothalamic-pituitary-adrenal (HPA) axis 124
respiratory irregularity 125
psychiatric 131
rates in clinical samples of patients diagnosed with SUDs 131
service-delivery terms in management of 176
with substance abuse 132–3
with SUDs 130, 132
computerised tomorgraphy (CT) 194
Confidential Enquiries into Maternal Deaths (CEMD) 94
congenital malformations (CMs) 99
contraception 97–9
coronary artery diseases 107, 121
corticotrophin-releasing factor (CRF) 124
CR *see* cognitive remediation (CR)
CRF *see* corticotrophin-releasing factor (CRF)
CRHTT *see* crisis resolution and home treatment team (CRHTT)
crisis management (CM) 51, 56
crisis resolution and home treatment team (CRHTT) 180
CT *see* computerised tomorgraphy (CT)
cyclothymia
BD type II depressions and 144
bipolar nature of 139
bipolar spectrum disorders and 142
characteristics of 139
and co-morbid obsessive–compulsive disorder 147
course of 143–4
cyclothymic-irritable temperament 143
drug abuse and 141
eating disorders and 142
epidemiological and clinical aspects 140–1
impulse-control disorders in 142
intensity of mood swings 140

obsessive–compulsive symptoms in 142
panic attacks 142
practical management of 148
psychiatric co-morbidity 141–3
psychological aspects of 141
reasons for difficulties in treating 145
social anxiety and 142
treatment strategies and practical management 144–7
cytochrome P450 enzyme 19

'daily mood charts' for BD 186
deep brain stimulation (DBS)
anatomical correlates of brain reward system as targets for 67–8
cingulate gyrus as target for 66–7
inferior thalamic peduncle 68–9
in psychiatric conditions 66
dementia 85, 86, 89, 90, 196
depression 9, 33
adjunctive therapy for 39
electroconvulsive therapy in treatment of 62–3
management of 37–8
clinical 39–40
pharmacological treatment 38–9
mortality rates 24
neurobiological models of 66
psychosocial impairment 24
symptoms 38
treatment algorithm for management of 39
treatment of 77
unipolar 184, 194
depressive psychosis 94
dexamethasone suppression test (DST) 124, 200, 206
dextroamphetamine sulphate 68
diabetes mellitus 89, 95, 97, 106, 107, 109, 123
Diagnostic and Statistical Manual of Mental Disorders 10, 74
diffusion MRI tractography 67
disinhibition syndrome 86
divalproex 25, 26, 35
for bipolar depression treatment 39
for rapid cycling BD treatment 29
dopaminergic (D_2) hypothalamic neurons 95
double-blind placebo-controlled (DBPC) studies, of mood-stabilising pharmacotherapy 75
drug abuse 129, 130, 141, 142, 144, 146
drug–drug interactions 88, 114
DST *see* dexamethasone suppression test (DST)
dual mania 10
dysphoric mania 15

ECA *see* Epidemiologic Catchment Area (ECA)
ECT *see* electroconvulsive therapy (ECT)
EE *see* expressed emotion (EE)
egosyntonic hypomania 145
elderly patients
medications and other modalities in treatment of mania 16
suffering from bipolar disorder 117
electroconvulsive therapy (ECT) 15, 87
for antimanic treatment 20, 63
for bipolar depression treatment 40, 62–3
for catatonia treatment 64
continuation and maintenance in treatment-resistant BD 64
efficacy and safety in depressive illness 62
for mania treatment 63
in mixed affective states 63–4
for treatment of severe mania or depression during pregnancy 116
treatment-resistant BD 64
EMBLEM *see* European mania in bipolar longitudinal evaluation of medication (EMBLEM)
emotional instability 101, 141, 142, 148
endometrial carcinoma 94
end-organ dysfunction 199
Enhanced Relapse Prevention (ERP) programme 181
Epidemiologic Catchment Area (ECA) 129
epileptiform seizures 17
EPSs *see* extrapyramidal side effects (EPSs)
ethyl-eicosapentaenoic acid (EPA) 38
and BD type I depression treatment 27
euphoric mania 15, 16, 18
European mania in bipolar longitudinal evaluation of medication (EMBLEM) 10, 18
expressed emotion (EE) 46
extrapyramidal side effects (EPSs) 18, 19, 78, 110, 117
extrapyramidal symptoms 110

familial pure depressive disease (FPDD) 67
family and school supports, for bipolar disorder patients 80
family-focused therapy (FFT) 46–7, 79, 80, 156, 170, 186, 205
for adolescents 51
modules of 47
trials
review of 51
summary of 51–2
FAST *see* functioning assessment short test (FAST) scale
FBC *see* full blood count (FBC)
F-fluoro-deoxyglucose (FDG) 67
FFT *see* family-focused therapy (FFT)
fMRI *see* functional magnetic resonance imaging (fMRI)
FPDD *see* familial pure depressive disease (FPDD)
Framingham stroke risk score 86
fronto-temporal dementia (FTD) 85
full blood count (FBC) 100, 112, 113
functional magnetic resonance imaging (fMRI) 67, 123, 195, 201, 206
functioning assessment short test (FAST) scale 171, 190

gabapentin 41, 125, 146, 147
for late-life BD treatment 89
for treatment of acute mania 36
GAD *see* generalised anxiety disorder (GAD)
GAF *see* global assessment of functioning (GAF)
generalised anxiety disorder (GAD) 1, 120, 121, 122, 131, 188
generic community mental health teams 180
global assessment of functioning (GAF) scale 122, 132, 166, 189–90
globus pallidus 68
glucocorticoid receptors (GR) 27, 107, 200

HADS *see* hospital anxiety and depression scale (HADS)
haloperidol 13, 15, 17, 18, 35, 36, 63, 89, 165, 169, 170
and antimanic treatment 19–20
Hamilton depression rating (HAM-D) *see* Hamilton depression rating scale (HDRS)
Hamilton depression rating scale (HDRS) 27, 29, 62, 68, 78, 122, 187–8, 204
health care system 184
general and sexual 97

health issues, associated with bipolar disorder
anticonvulsants 109
antipsychotic drugs and extrapyramidal side effects 110
atypical antipsychotics 109–10
management of 111–13
medications 108
mood stabilisers 108–9
special populations 115–17
treatment-specific monitoring 113–15
weight gain 110
health-related quality of life (HRQOL) 163
hormones
affect on moods 199–200
replacement therapy 93, 97
hospital anxiety and depression scale (HADS) 189
HPA *see* hypothalamic-pituitary-adrenal (HPA) axis
HPT *see* hypothalamic-pituitary-thyroid (HPT)
HRQOL *see* health-related quality of life (HRQOL)
hypercortisolism 206
hyperlipidaemia 79, 89, 95, 97
hyperphagia 34
hyperprolactinaemia 19, 95, 96, 97, 109, 110, 114, 115, 117
hypersomnia 34, 139, 143, 156, 157
hypertension 95, 106, 107, 109, 111, 121, 123, 125
hypnotics 141
hypogonadotrophic hypogonadism 96
hypomania 9, 16, 63, 139, 184
antidepressant-induced 143
episode 140
symptoms of 131
hypothalamic-pituitary-adrenal (HPA) axis 106, 124, 199
hypothalamic-pituitary-thyroid (HPT) 106
hypothalamo-pituitary-gonadal axis dysfunction 94

ICM *see* intensive clinical management (ICM)
IDS *see* inventory of depressive symtomatology (IDS)
IFIT *see* integrated family and individual therapy (IFIT)
index depressive episode 41, 144
inferior thalamic peduncle, for patient with TRD 66, 68–9
inositol 25
insomnia 66, 156, 158
cognitive behavioural therapy for 157
integrated family and individual therapy (IFIT) 51, 57
integrated group therapy 135
intensive clinical management (ICM) 57, 155
vs. interpersonal and social rhythm therapy 52
International Classification of Diseases 10
International Society of Bipolar Disorders 184
interpersonal and social rhythm therapy (IPSRT) 40, 47, 51, 155, 156, 186, 205
vs. intensive clinical management (ICM) 52
phases of 48
for treatment of bipolar disorder 52
trials
review of 52
summary of 52–3
interpersonal therapy (IPT) 47, 79, 80
inventory of depressive symtomatology (IDS) 27
IPSRT *see* interpersonal and social rhythm therapy (IPSRT)
IPT *see* interpersonal therapy (IPT)

lamotrigine 26, 36, 76, 77, 98, 99, 100, 110, 113, 114, 116, 134
adverse effects in young people 78–9
as anticonvulsants 109
antidepressant properties 146
for antimanic treatment 17, 75
for BD type I depression treatment 24–5
breastfeeding, use during 117
for late-life BD treatment 89
monotherapy 24–5, 38, 40, 41
for rapid cycling BD treatment 29, 148
for treatment of adult bipolar depression 77, 109
late-life bipolar disorder
clinical features of
age of onset 84–5
illness course and outcome 85
neuroimaging studies and vascular disease 86–7
secondary mania and neurological co-morbidity 85–6
management of 87–90
levetiracetam 13, 17, 37
LFT *see* liver function tests (LFT)
LIFE *see* longitudinal interval follow-up evaluation (LIFE)
Life Goals Program 45, 49, 59, 178
lithium 13, 17, 18, 19, 26, 35, 36, 39, 44, 63, 76, 113, 129
adverse effects in young people 78–9
antidepressant properties 146
for antimanic treatment 16, 77
for bipolar depression treatment 24, 40, 77
clinics 179–80
for late-life BD treatment 88
management of physical health issues associated with 113
medications during breastfeeding 117
monotherapy 42
as mood stabilisers 108–9
for patients with mixed episodes 37
for pharmacotherapy of acute mania 75
for prophylaxis bipolar disorder treatment 41
for rapid cycling BD treatment 29
safe use in older people 87
teratogenic potential of 98
for treating acute bipolar depression 40
lithium–haloperidol combination pharmacotherapy 63
liver function tests (LFT) 112, 113, 185
longitudinal interval follow-up evaluation (LIFE) 49, 50, 53
lorazepam 13, 18, 20, 35, 88

MADRS *see* Montgomery–Asberg depression rating scale (MADRS)
magnetic resonance diffusion tensor imaging 194
magnetic resonance imaging (MRI) 194–5
major depression (MD) 10, 66, 84, 122, 147, 200, 206
core behavioural symptom of 68
major depressive disorder (MDD) 106, 121, 124
mania 9, 129 *see also individual entries*
Beck's cognitive model of 45
camphor-induced seizures in treatment of 62
clinical management of 36–7
diagnostic criteria for 11
dosing and duration of treatment 11–12
electroconvulsive therapy in treatment of 63
emergency management of acute 35

frequency of symptoms 12
medications and other modalities in treatment of
anticonvulsants 16–17
atypical antipsychotics 17–19
experimental medications 20
lithium 16
typical neuroleptics 19–20
monotherapy and skilful polypharmacy for treatment of 12–15
pharmacological and non-pharmacological treatments 13
pharmacological treatment of acute 35–6, 75
recommendations for 36
principles for treatment of 10–11
psychotic symptoms 37
secondary mania and neurological co-morbidity 85–6
subtype and severity of 15–16
symptoms of 131, 134
treatment algorithm 14, 37
types of 10
manic-depressive illness 139
MAOIs *see* monoamine oxidase inhibitors (MAOIs)
Mattis dementia rating scale 85
MBS *see* metabolic syndrome (MBS)
MCID *see* minimal clinically important difference (MCID)
MDD *see* major depressive disorder (MDD)
MDQ *see* mood disorders questionnaire (MDQ)
medications
antidiabetic 107
antihypertensive 95
antipsychotic 35, 36, 95, 114, 116, 117, 125, 206
combination 26, 35, 36, 146
evidence-based 186
health issues in bipolar disorders 108
prophylactic 63, 100, 115
psychotropic 17, 87, 94, 117, 124, 204
for treatment of mania 35
memory loss 63, 199
menstrual cycle dysfunction 94
mental health services 102, 136, 176, 181
metabolic disorders 106
metabolic syndrome (MBS) 106, 121, 123
NCEP ATP III criteria for 107
pathophysiology of 108
mifepristone 29
for BD type I depression treatment 27
Milkowitz's family treatment programme, for bipolar disorder 46
minimal clinically important difference (MCID) 170
mini-mental state examination (MMSE) 85
modafinil, for BD type I depression treatment 27
monoamine oxidase inhibitors (MAOIs) 147
Montgomery–Asberg depression rating scale (MADRS) 25, 26, 27, 28, 68, 122, 188
'Mood chart' self-monitoring 78
mood disorders, affect of hormone on 199–200
mood disorders questionnaire (MDQ) 33, 187
mood stabilisers 4, 36, 40, 77, 89, 117, 145, 148
agents as 20, 77
lithium 108–9
teratogenic potential of 98
therapy 125
MRI *see* magnetic resonance imaging (MRI)
MSIF *see* multidimensional scale for independent functioning (MSIF)
multidimensional scale for independent functioning (MSIF) 164

naltrexone 134, 135
National Cholesterol Education Program Adult Treatment Panel (NCEP ATP III) 106, 123
National Comorbidity Survey (NCS) 120, 129, 130
National Comorbidity Survey Replication (NCS-R) 129, 130
National Epidemiologic Survey on Alcohol and Related Conditions 122, 130
National Health Service 64
National Institute for Health and Clinical Excellence (NICE) 113, 115, 116
NCS *see* National Comorbidity Survey (NCS)
NCS-R *see* National Comorbidity Survey Replication (NCS-R)
neurocognitive impairment 85, 87, 200, 201, 202
neurodegenerative disorders 201
neurodevelopmental disorders 77
neuroleptic malignant syndrome 76, 78
neuroleptics 15
benzodiazepines 20
chlorpromazine 20
haloperidol 19–20
North American Association for the Study of Obesity 114

obsessive–compulsive disorder (OCD) 120, 121, 122, 124, 131, 147, 185
obsessive–compulsive symptoms 76, 142, 144, 147
OCD *see* obsessive–compulsive disorder (OCD)
OCPs *see* oral contraceptive pills (OCPs)
oestrogen therapy 93
olanzapine 110, 125, 169
for antimanic treatment 18
for BD type I depression treatment 26
olanzapine/fluoxetine combination (OFC) 26, 40
oral contraceptive pills (OCPs) 115
oral contraceptives (OCs) 78, 96, 99, 113, 115
oxcarbazepine, for antimanic treatment 17

paediatric bipolar disorder 74
panic attacks 142, 146
panic disorder (PD) 94, 120, 121, 122, 142, 146, 147, 185
Parkinsonism 89
paroxetine 25, 26, 27, 99, 100, 125, 147
pathological altruism 141
PD *see* panic disorder (PD)
personal medicine (PM)
categories, subtypes and unique features of 2
concept of 1–2
Deegan's cascading effect of 5
defined 2
discussion with patients to avoid non-adherence to 3
dual functionality of 2
interplay and treatment adherence 2–3
and removing barriers to treatment and medication adherence 3–4
self-perpetuating reinforcement phenomenon of alignment 4–5
types of 3
personality disorders (PDs) 48, 55, 94, 120, 131, 139, 141–5, 148, 184, 185
PET *see* positron emission tomography (PET)

pharmacological management, of
bipolar disorder 35–40
pharmacotherapy *see also specific medications*
of acute mania 75
adverse effects in young people 78–9
and choice of initial agent 75–7
for enhancing medication adherence 79
evidence and guiding symptoms for 76
evidence-based treatment 75
for maintenance mood stabilisers 77–8
maintenance therapy 77–8
for medical and mood monitoring 78
recommendations for bipolar disorder 41
for treatment of
BD type I depression 24–8, 77
BD type II depression 28–9, 77
rapid cycling BD 29
phenytoin 13, 37, 41
for antimanic treatment 17
phospholipid metabolism 135
Pittsburgh sleep quality index (PSQI) 157
pituitary cells 95
placebo adjunctive therapy 38, 41
placebo-controlled monotherapy 18, 25, 26
PM *see* personal medicine (PM)
polycystic ovary syndrome (PCOS) 17, 95–7, 109, 116
polydipsia 78, 79, 88
positron emission tomography (PET) 68, 195
post-traumatic stress disorder (PTSD) 121, 122, 124, 125, 131
pramipexole
for BD type I depression treatment 27
for BD type II depression treatment 27, 28
pregabalin 17
pregnancy in women, with bipolar disorder 115–16
prophylaxis bipolar disorder
index depressive episode 41
pharmacological treatment for 41
protein kinase C inhibition 20
PSQI *see* Pittsburgh sleep quality index (PSQI)
psychiatric disorders 6, 44, 62, 73, 97, 120, 132, 134, 157, 168, 190
psychiatric medications 3, 7
Deegan's cascading effect of 5
self-perpetuating reinforcement phenomenon of alignment 5
psychiatric syndromes 66
psychoeducation 44–5, 54
review of 47–9
Psychological Treatment of Bipolar Disorder (Johnson and Leahy) 165
psychosocial treatments, impact on cognitive outcome in bipolar patients 205
psychotherapy, evidence-based 79–80
psychotic mania 1, 10, 15, 18, 19, 76
PTSD *see* post-traumatic stress disorder (PTSD)
puerperal psychosis 94

QALY *see* quality-adjusted life year (QALY)
quality of life enjoyment and satisfaction questionnaire (Q-LES-Q) 169, 170, 190
quality of life (QoL) 187
assessment of 163
in bipolar disorder 166–8
abnormal mood states 169
degree of impairment 168–9
development of disorder-specific scale 171
treatment outcome 169–71
impact of treatment interventions on 189
in patients with schizophrenia 169
psychosocial functioning and 162
incorporating into clinical practice 171–2
QoL in bipolar disorder 166–71
rationale for assessing 164–5
psychosocial functioning for use in clinical practice 171–2
quality-adjusted life year (QALY) 196
quetiapine 125
for antimanic treatment 18
for BD type I depression treatment 25–6
for BD type II depression treatment 25–6, 28
for rapid cycling BD treatment 29

randomised controlled trials (RCTs) 15, 16, 18, 26, 33, 44, 46, 47, 49–53, 62, 63, 88, 109, 114, 129, 144, 146, 175, 183
rapid cycling BD 74, 77, 122, 125, 132, 143
pharmacotherapy for 29
in women 93
rapid mood fluctuations 142
RCTs *see* randomised controlled trials (RCTs)
resting energy expenditure (REE) 108
retigabine 13, 17
reversible inbibitors of monoamine oxidase (type A) (RIMAs) 147
risperidone 77
and antimanic treatment 18–19
as mood stabiliser 25

SADS-C *see* schedule for affective disorders and schizophrenia (SADS-C)
scales for assessing symptoms and severity of bipolar disorder
Hamilton depression rating scale 187–8
Montgomery–Asberg depression rating scale 188
Young mania rating scale 188
schedule for affective disorders and schizophrenia (SADS-C) 51
schizophrenia 15, 20, 26, 27, 46, 73, 144, 164, 180, 184
SCN *see* suprachiasmatic nucleus (SCN)
secondary mania and neurological co-morbidity 85–6
sedating antipsychotics 157
sedatives 141
selective serotonin reuptake inhibitor (SSRI) monotherapy 29, 38, 39, 40, 145
self-care strategies, for personal medicine 2
self-management strategies, in bipolar disorder 164
self-medication, for bipolar disorder 132
serotonin-norepinephrine reuptake inhibitors (SNRIs) 145
serum electrolytes 112
service delivery for bipolar disorder
assessment of key ingredients of 177
models for
assertive outreach teams 182
collaborative care model 177–9
determining implementation for service change 182–3
early intervention teams 182
extended role of lithium clinic 179–80
interface with primary care 181–2
specialist affective disorder 180
supporting generic community mental health teams 180–1

reasons for 175–6
universal needs of patients 176–7
severe mood and behavioural dysregulation (SMD) 74
sex steroids 93
sexual dysfunction 95, 96, 115, 145
sexually transmitted disease 97
SFBN *see* Stanley Foundation Bipolar Network (SFBN)
single photon emission computed tomography 195
sleep disturbances, assessment of 157
sleep quality, cognitive-behavioural strategies for improving 158–9
sleep/wake cycle 16, 152, 154
SMD *see* severe mood and behavioural dysregulation (SMD)
SNRIs *see* serotonin-norepinephrine reuptake inhibitors (SNRIs)
social anxiety 142, 147
social rhythm metric (SRM) 47, 48, 156
social rhythm therapy (SRT) 79, 80, 155, 170, 186
social rhythms 47
chronic instability in 156
social zeitgeber hypothesis and social rhythm management 154–6
sodium valproate 17, 98, 109, 125
somatic anxiety symptoms 188
specialist affective disorder services 180
specialist mood disorder services 180
SRM *see* social rhythm metric (SRM)
SRT *see* social rhythm therapy (SRT)
SSRI monotherapy *see* selective serotonin reuptake inhibitor (SSRI) monotherapy
Stanley Foundation Bipolar Network (SFBN) 122, 130, 131, 163
STEP-BD *see* Systematic Treatment Enhancement Program for Bipolar Disorder (STEP-BD)
Stevens–Johnson syndrome 40, 100, 109
structural brain imaging, in bipolar disorder 194
substance-abuse treatment programme 135
substance-use disorders (SUD) 130, 176
effect of bipolar disorder on 129, 133
substance-use treatment services 136
subsyndromal mania 11
SUD *see* substance-use disorders (SUD)
suprachiasmatic nucleus (SCN) 152
systematic care management program 49
Systematic Treatment Enhancement Program for Bipolar Disorder (STEP-BD) 25, 27, 53, 123, 130, 134, 169

tardive dyskinesia 19, 20, 76, 78, 87, 89, 110, 117
TAU *see* treatment as usual (TAU)
TCAs *see* tricyclic antidepressants (TCAs)
temperament and character inventory (TCI) 132
Texas Medication Algorithm Project (TMAP) 48, 49, 175
theoretical models, for bipolar disorder
cognitive behavioural therapy 45–6, 49–51
family-focused therapy 46–7, 51–2
interpersonal and social rhythm therapy 47
psychoeducation 44–5, 49
thyroid dysfunction 16, 76, 108, 199
thyroxine, for rapid cycling BD treatment 29
tiagabine 17, 36
TMAP *see* Texas Medication Algorithm Project (TMAP)
TMS *see* transcranial magnetic stimulation (TMS)
topiramate 17, 36, 38, 41, 75–9, 77, 99, 135
toxic epidermal necrolysis 109
training and assessment tools, for diagnosis of bipolar disorder
assessment of symptoms and severity of bipolar disorder
Hamilton depression rating scale 187
Montgomery–Asberg depression rating scale 188
assessment scales for clinical practice 186–7
Beck anxiety inventory 188–9
clinical global impression 188
common training/skill problems encountered by clinicians
clarification of 'personality' issues 185
diagnosis of bipolar disorder 184–5
identification of co-morbid anxiety disorders 185
identification of substance-abuse issues 185
side effects and adherence with treatment 185–6
functioning assessment short test 190
global assessment of functioning 189–90
hospital anxiety and depression scale 189
mood disorders questionnaire 187
psychosocial functioning and quality of life 189
quality of life enjoyment and satisfaction questionnaire 190
training in BD diagnosis, treatment and rating scales
basic clinical exposure 190
education for maintenance of competence in professional societies 191
key tools–rating scales and diagnostic instruments 190–1
Young mania rating scale 188
training in BD diagnosis, treatment and rating scales
basic clinical exposure 190
education for maintenance of competence in professional societies 191
key tools–rating scales and diagnostic instruments 190–1
transcranial magnetic stimulation (TMS) 64–5
in depressed patients with unipolar and bipolar course 65
transcranial stimulation, in bipolar disorder 9, 64–5
TRD *see* treatment-resistant depression (TRD)
treatment as usual (TAU) 49, 50, 56, 177
treatment for bipolar disorder
pharmacological 134–5
psychotherapy 135–6
young people 80
treatment-resistant depression (TRD) 66, 67, 125
tricyclic antidepressants (TCAs) 38, 99, 125, 145, 147
typical mania 10

unipolar depression 10, 28, 33, 47, 50, 62, 63, 84, 94, 106, 120, 142, 144, 153, 156, 166, 177, 181, 184, 194–6, 199

vagus nerve stimulation
antidepressant effect of 66
for bipolar disorder 65–6

valproate 113, 134
and antimanic treatment 17
and BD type I depression treatment 25
for late-life BD treatment 88
vascular depression 87
vascular mania 87

waist:hip ratio 114
weight gain 18, 19, 42, 77–9, 88, 89, 99, 106, 108–11, 114, 125, 146, 170
white matter hyperintensities (WMHs) 86, 87, 194
white matter lesions 86, 196
WMHs *see* white matter hyperintensities (WMHs)
World Health Organization (WHO) 9, 106, 123, 163, 176

years lost to disability (YLD) 199
Young mania rating scale (YMRS) 50, 78, 188

ziprasidone 13, 15, 41, 75, 79, 109, 110, 114, 115
for antimanic treatment 19, 35
zonisamide 13
for antimanic treatment 17

Printed in the United States
By Bookmasters